T0091842

Magnetoelectronics of Microwaves and Extremely High Frequencies in Ferrite Films

Alexander A. Ignatiev

Magnetoelectronics of Microwaves and Extremely High Frequencies in Ferrite Films

Professor Alexander A. Ignatiev
Saratov State University
Department of Physics
Astrakhanskaya 83
Saratov
Russia 410026

ISBN 978-0-387-85456-4 e-ISBN 978-0-387-85457-1
DOI:10.1007/978-0-387-85457-1

Library of Congress Control Number: 2008938181

Printed on acid-free paper

9 8 7 6 5 4 3 2 1

springer.com

Foreword

This book is devoted to physical bases of magnetoelectronic millimetric waves. Magnetoelectronic represents a direction on a joint of physics of the magnetic phenomena in magneto arranged environments, radiophysics of wave and oscillatory processes in layered screened bigirotropic (tensors dielectric $\overleftrightarrow{\varepsilon}$ and magnetic $\overleftrightarrow{\mu}$ penetrabilities) structures, semi-conductor microelectronics and circuitry. The millimetric range gives the certain specificity to researches. It, first of all: losses and their growth with frequency in spending screens and the metallized coverings, dielectrics, semi-conductor layers, ferrite, increase in effective internal magnetic fields and fields bias, reduction of the geometrical sizes of coverings, structures and toughening of admissions by manufacturing and the assembly, new methods of diagnostics of film structures of ferrite, including not destroying.

Promotion of researches was spent to a millimetric range in directions:

- Developments of methods of the theoretical analysis for studying properties of various types of converters;
- Development of various kinds and types of converters;
- Development of methods of researches of properties of waves at excitation, reception and distribution in layered structures of various kinds;
- Carrying out of experimental researches;
- Development of methods and means of not destroying control of parameters of film structures of ferrite;
- Development operated magnetoelectronic structures and devices of low and high levels of capacity.

In such order the theoretical and experimental material received is stated in the book. The serious contribution to separate researches have brought: the senior scientific employee, Dr. Lepestkin A.N. (experiment, physical modelling); the scientific employee, Dr. Mostovoj A.A. (the theory, physical modelling); the younger scientific employee, Dr. Beginin E.N. (programming, calculations, experiments, laboratory breadboard models).

The basic sections of the book were read to students of physical faculty of the Saratov State University by it. In a special course "Magnetoelectronics of microwave and extremely high frequencies" also were accompanied by N.G. Chernyshevskogo in corresponding laboratory installations, developed in course and degree works of students.

The second part of the book – "Heteromagnetic microelectronics (magnetoelectronics of the active devices)" which is in a stage of a writing and development of a problem, is devoted to a new direction on creation of multipurpose operated microdevices, the microsystems which are carrying out finished functions on formation of various kinds and spectra of signals in transistor-magnetic, ferrite-semi-conductor structures (strengthenings, generation, mixture, parametrical effects, multiplication, division, frequency modulation, magnetosensitive modes).

This part of works would be impossible without support from a management of JSC "Tantal", attraction of industrial technologies, material, hardware and the software of researches.

The important component in development of this direction is attraction of young and skilled science officers, students and post-graduate students.

Special gratitude to leading engineers of faculty of general physics SSU – for Danke O.G., Sirotinina T.N., Galanova O.N. for a computer set of the book and preliminary editor.

Annotation

Physical bases of magnetoelectronics in the millimetric range are considered. The results are given of theoretical and experimental research excitation processes in fast and slow electromagnetic waves of various types in bigirotropic and ferrite-dielectric structures with metallization and various kinds of magnetization, traditional and new types of converters, propagation of waves in structures and management of their dispersions, new methods and devices of nondestructive diagnostics of monocrystal film structures of ferrite, designing of passive magnetoelectronic controlled devices of low (milliwatt) and high (kilowatt) levels of capacity.

For scientists, developers of magnetoelectronic devices, lecturers, postgraduate student and student of high schools.

Books offered to the reader include materials of scientific investigations wave and oscillatory processes in layered magnetoarranged structures as passive type (volume 1 "Magnetoelectronics of microwave and extremely high frequencies in ferrite films", under such name it has been published the monograph in Russia in publishing house "Nauka" in the co-authorship from Alexander Lyashenko who has rendered the main financial support), and active type (volume 2 "Geteromagnetic microelectronics. Microsystems of the active type", under this name the monograph in Russia in same publishing house "Nauka" in the co-authorship with the same Alexander Lyashenko who again has rendered the main financial support already as in development of this direction in JSC "Tantal" at equipment of laboratories by the advanced techniques and the equipment so financial support at publishing charges) has been published.

Structures of the active type on the basis of ferrite films of various types were studied during with 1980–1992 at the Saratov State University (SSU) on faculty of the general physics under direction of the author. These investigations of a distance the detailed and profound understanding of physics of processes, have provided development of experimental samples of various devices with record-breaking high parameters in a millimetric range. At the same investigations directions of studying layered magneto semi-conductor structures of the active type (generators, amplifiers, amalgamators, converters, sensors) in centimetric and millimetric ranges of radiowaves have been predicted.

To century of Saratov State University and chair of general physics.

A number of new microfield interactions in magneto semi-conductor structures of various types in modes of generation has been found out in 1995, and detailed investigations are executed per 1996–2002 in SSU, during 2002–2006 in JSC "Tantal". The basic executors of these investigations – employees of faculty of the general physics with employees of Design office of critical technologies of JSC "Tantal" and JSC "Research Institute Tantal".

Following step of inestigations and development are heteromagnetic microelectronics and analog-digital microsystems on ferrite films, and also heteromagnetic nanoelectronics. These are the following (volume 3 and volume 4) above which work and a meeting with which is conducted expects readers.

Saratov
2008

Professor A. Ignatiev

Reviewers:
Full member of the Russian Academy of Natural Sciences N.I. Sinicyn
Doctor of Science S.G. Souchkov

Translator:
Dr. E.A. Ignatieva, Dr. S.L. Shmakov

Ignatiev A.A.
Magnetoelectronis of microwave and extremely high frequencies in ferrite films.
Moscow: Nauka, 2005. – 380 p. ISBN 5-02-033534-7.

Contents

Abbreviations

AFC	amplitude-frequency characteristics
BCM	beyond-cutoff mode
BHF	barium hexaferrite
CL	coplanar line
CLA	coplanar line antenna
CLC	coplanar line converter
DDL	dispersive delay line
DDDL	decreasing dispersive delay line
DFDL	dispersion-free delay line
DL	delay line (of a signal)
EHF	extremely high frequencies (30, 0–300, 0 GHz)
FDMT	ferrite-dielectric matching transformer
FDR	ferrite-dielectric resonator
FDS	ferrit-dielectric structure
FDT	ferrite-dielectric transformer
FFDS	film ferrit-dielectric structure
FIGF	Fourier image of Grin's function
FDLS	ferrite-dielectric layered structure
FMR	ferromagnetic resonance
FW	fast wave
GDT	group delay time
GGG	gallium–gadolinium garnet
HF	high frequency
HPL	high power level
HTSC	high-temperature superconductivity covering
IDDL	increasing dispersive delay line
IFB	infrared frequency band
IFM	instantaneous frequency measurement
LBS	layered bigyrotropic structure
LFDS	layered ferrit-dielectric structure
LMW	linear magnetic wave

LPL	low power level
MAW	medium-accuracy wavemeter
MED	magnetoelectronic device
MFS	magnetic field sensor
MMR	millimeter range (30, 0–300, 0 GHz)
MS	magnetic system
MSA	microstrip antenna
MSL	microstrip line
MSLC	microstrip line converter
MSW	magnetostatic (exchangeless) spin wave with $\kappa' \leq 10^3 - 10^4\,\mathrm{cm}^{-1}$
NMR	nuclear magnetic resonance
PA	phased array
PIWF	phase-inversion waveguide filter
RFDS	resonance-field sensor
RFS	resonance-frequency sensor
RFW	return fast wave
RSLMSW	return spatial magnetostatic wave
RSLW	return spatial wave
RSMSW	return surface magnetostatic wave
RSW	return surface wave
RSSW	return surface slow wave
SC	strip converter
SCW	spin (converted) wave (heavily decelerated wave with $\kappa' \geq 10^4 - 10^5\,cm^{-1}$)
SFG	sweep-frequency generator
SFSW	space fast surface wave
SFW	space fast wave
SGC	slotted-guide converter
SGCC	slotted-guide converter with conductor
SL	slotline
SLA	slotline antenna
SLC	slotline converter
SLMSW	spatial magnetostatic wave
SLSW	spatial slow wave
SMSMSW	semi-surface magnetostatic waves
SMSW	surface magnetostatic wave
SSLMSW	space spatial magnetostatic wave
SSLW	space spatial wave
SSMSW	space surface magnetostatic wave (Eshbah-Deimon's wave)
SSMSW*	semi-spatial magnetostatic wave
SSSLW	space slow spatial wave
SSSW	space slow surface wave
SSW	space slow wave
SSW*	space surface wave
SW	slow wave

SWR_e	standing wave-voltage ratio
TDC	three-dimensional chip
TF	transmitting filter
TFC	temperature frequency coefficient
TFDC	temperature field coefficient
TL	transmission line
TSL	transformer on SL
UHF	ultrahigh frequency $(0,3-3,0$ GHz)
WB	wobbulator
WBCC	waveguide beyond-cutoff converter
WC	waveguide converter
YIG	yttrium–iron garnet

Table of Symbols

M	magnetization
M_S	saturation magnetization
H_{0i}	internal magnetic field in ferrite
$\tilde{h}$	high-frequency (variable) magnetic field
$\tilde{m}$	high-frequency (variable) magnetization
χ	high-frequency magnetic susceptibility
χ'	real part of high-frequency susceptibility
χ''	imaginary part of high-frequency susceptibility
ΔH	linewidth (half-width) of ferromagnetic resonance
α	parameter of ferromagnetic losses
$\overleftrightarrow{\mu}$	tensor of degaussed factor
∇	nabla operator (∇_x – component to axis $0X$, ∇_y – component to axis $0Y$, ∇_z – component to axis $0Z$)
$\langle\rangle$	average to size, square
$\overleftrightarrow{N}_A$	anisotropy tensor
κ	wave number ($\kappa = \kappa' + j\kappa''$)
κ'	real part of wave number
κ''	imaginary part of wave number
m	meter
S	centi
W	width of microstrip line, high-frequency power
H_0	magnetostaticfield
E	electric field strength (E_x – component to axis $0X$, E_y – component to axis $0Y$, E_z – component to axis $0Z$)
H	magnetic intensity (H_x – component to axis $0X$, H_y – component to axis $0Y$, H_z – component to axis $0Z$)
ω	angular (cyclic) frequency
ε	dielectric penetrability ($\overleftrightarrow{\varepsilon}$ – tensor)
μ	magnetic penetrability ($\overleftrightarrow{\mu}$ – tensor)
X,Y,Z	space coordinates

ν	frequency
x, y, z	space coordinates
$\overset{\leftrightarrow}{\mu}_M$	vector of tangential electric and magnetic fields components
p	number
$\overset{\leftrightarrow}{I}_S$	vector from surface magnetic j_M and electric j_e currents
j_M	surface magnetic current
j_e	surface electric current
m	number
$\overset{\leftrightarrow}{G}$	tensor of Green's function
$\overset{\leftrightarrow}{\tilde{G}}$	Fourier image of Grin's function
$\overset{\leftrightarrow}{\tilde{I}}_S$	Fourier image of Grin's function on boundary of source
$\overset{\leftrightarrow}{F}$	tensor on source surface
$\overset{\leftrightarrow}{A}$	matrix, a – matrix component
$\overset{\leftrightarrow}{T}$	matrix of coupling
λ_κ	characteristic value
S	number
b	number
c	number
d	number
e	number
$\overset{\leftrightarrow}{B}$	half-space tensor of layered structure
$\overset{\leftrightarrow}{D}$	half-space tensor of layered structure
$\overset{\leftrightarrow}{F}$	tensor
$\overset{\leftrightarrow}{Z}$	tensor of surface impedance
$\overset{\leftrightarrow}{\beta}_E$	tensor for electric wall
$\overset{\leftrightarrow}{\beta}_H$	tensor for magnetic wall
Δ	detector
φ	angle
$\bar{R}$	radius-vector
ϑ	angle
δ	delta function
j	imaginary unit ($j = \sqrt{-1}$), $j_{M,E}$ – outside magnetic and electric currents
f	function
P	power flux density
Φ	phase
$\prod$	composition
$\prod_{LM}^{1,2,3}$	power flows of LM-wave in layer 1, 2, 3
a	size
h	size
F	function

$\overleftrightarrow{\eta}$	tensor
$\overleftrightarrow{\xi}$	tensor
E	function
T	tesla
Ψ	function
K	loss coefficient of power transfer
K	coefficient of power absorption
T	temperature
$K_{los.B}$	conversion coefficient of radiating power
$K_{los.R}$	conversion coefficient of active power
γ	gyromagnetic (magnetomechanic) attitude for electron
F	compactness of free energy
g	g–structure factor
m_e	rest electron mass
c	light speed
δ	delta
Ψ	high-frequency potential, function
τ	delay time
ν_H	ferromagnetic-resonance frequency
ν_M	characteristic frequency ($\nu_M = \frac{\omega_M}{2\pi}, \omega_M = 4\pi M_S$)
$\nu_\perp$	frequency of transverse resonance
κ_0	wave number
κ_{bor}	border frequency
ν	dependence of frequency for angular displacement φ in magnetic field
K	movable losses coefficient
L_a	converter effective length
L_{bcs}	length of beyond-cutoff section
$\Delta\nu_{3dB}$	passband in level -3dB
K	barrage level
P_f	accepted ferrite power
V	volume
P_{out}	output power level
sp (index)	spinel
st (index)	signal transmission
c (index)	conversion
im (index)	impulse
he (index)	heart exchange
sq (index)	squareness
cas (index)	cascade
cf (index)	carryover factor

Introduction

Recently, the wave and oscillatory processes in layered structures made of epitaxial ferrite films in the linear and nonlinear modes, aimed at the design of new controllable devices for signal processing in the UHF (0.3–3.0 GHz) and EHF (over 30 GHz) ranges are intensely studied.

High-quality films were created in the late 1970s – early 1980s after the appearance of solid ferrites, for which wave processes in various directing systems were analyzed in terms of gyromagnetic electrodynamics. Design principles for ferrite devices in the frequency range of 150–200 GHz, namely, irreversible phase shifters, gates, circulators, harmonic oscillators, power limiters, filters, tunable oscillators [1–26] have been developed. Most research and development of ferrite devices were carried out in subresonance (low in magnitude) magnetic fields [5, 6]. In the sphere of resonant magnetic fields, solid monocrystals (shanks, plates, spheres, disks, compound structures on their basis, including dielectric and semiconductor layers [5, 6]) were studied. By analogy with acoustoelectronics this lead was named microwave magnetoelectronics by the academician Y.V. Gulyaev at the early 1980s. Signal delay and filtration, phase rotation, multichannel analysis schemes were realized on the basis of **magnetostatic (slow) exchangeless spin waves** with their phase constants $\kappa' \leq 10^3-10^4\,\mathrm{cm}^{-1}$, loss compensation and signal multiplication [29–41] were achieved. The knowledge of **exchange spin (strongly delayed) waves** with $\kappa' \geq 10^4-10^5\,\mathrm{cm}^{-1}$ initiated the development of spin-wave microwave electronics [42–45]. These two terms are most common.

In going from solid to film ferrites:

- The layer-localized energy density increases.
- The wave group velocity decreases.
- The zone of homogeneous and internal magnetic fields extends.
- The boundary internal magnetic field gradient increases.
- The functionality and range of the variable parameters of the devices expand.
- The weight and dimensions of the devices decrease.
- Simple coupling with waveguide and strip lines, construction of planar and solid integrated circuits are provided.

A.A. Ignatiev, *Magnetoelectronics of Microwaves and Extremely High Frequencies in Ferrite Films*,
DOI: 10.1007/978-0-387-85457-1_1,

- The price decreases and the opportunity to make use of micron and submicron technologies and mass production of magnetoelectronic devices appears.

In comparison with acoustoelectronic devices providing signal processing in the UHF and long-wave parts of the microwave range, magnetoelectronic ones can be implemented in the microwave and EHF ranges with much lower losses introduced, they have working elements (converters, transmission lines, various kinds of loading) with their sizes not rigidly related to the frequency range, and possess a number of advantages, namely:

- High-speed tuning of frequency, signal phase, passband
- An expanded dynamic power range

Analysis of publications of this lead [46–103] has shown that the major part of the research was carried out over a frequency range up to 3–7 GHz. The published works in a frequency range up to 10–20 GHz are much fewer, and for a frequency range of 20–30 GHz and above the publications are fragmentary.

The stable tendency of vacuum and solid-state electronics to advance into the MMR of radiowaves [94–102] is due to the necessity:

- To expand the frequency range
- To increase the volume and speed of information transfer
- To increase the jamming protection and secrecy
- To improve the electromagnetic compatibility of many systems and channels
- To increase the spatial resolution of objects
- To reduce the beam aperture and antenna dimensions
- To provide wave propagation in bad conditions (dust, smoke, suspensions) with losses much lower than in the optical and IFB

Satellite and ground communication systems of various purposes, radiolocation and radio navigation, electronic warfare and reconnaissance equipment, measuring equipment, various economic and medico-biological applications are intensely developed [104–108].

When using ferrite films and layered structures on their basis in the MMR range:

- The "*magnetic rigidity*" increases and, hence, for weakly anisotropic materials the internal magnetic field H_{0i} essentially exceeds the demagnetizing fields $4\pi M_S(H_{0i}) >> 4\pi M_S$ for strongly anisotropic materials the crystallographic anisotropy fields $H_A >> 4\pi M_S$, and the HF fields $\widetilde{h}$ and magnetization $\widetilde{m}$ increase in a wide range of signal power $\widetilde{h} << H_{0i}$, $\widetilde{m} << 4\pi M_S$, which provides an expanded dynamic range in the linear mode.
- *Division of losses* connected with the domain mode and ferromagnetic resonance in the substance *intensifies*.
- The line width of ferromagnetic resonance $\Delta H(\nu)$ increases linearly and independently of the power level over a wide range, both in the continuous and pulse modes; at the resonant and tuned frequency the diagonal χ''_{res} and off-diagonal χ''_{ares} components of the HF susceptibility tensor $\overleftrightarrow{\chi}$ $\chi''_{res} = \chi''_{ares} = \frac{M_S}{2\Delta H(\nu)}$ decrease and increase, respectively.

- *The transverse gradient magnitude of the FMR line width* $\nabla_x \Delta H(\nu)$ *increases and its influence* intensifies, as well as the influence of the transverse gradient of the internal magnetic field $\nabla_x \overset{\leftrightarrow}{N} M_S$ and related gradients of the demagnetizing field ($\overset{\leftrightarrow}{N}$ being the tensor of demagnetizing factors) and anisotropy field $\nabla_x N_A$.
- *The vorticity of HF fields intensifies.*

Most interesting regularities and features of excitation and propagation of MMR waves of various types are observed in layered structures made of *high-quality ferrite films* and magnetic materials, they possess:

- *A low level of ferromagnetic losses* ($\alpha = \frac{\Delta H}{\langle H_{0i} \rangle} \leq 10^{-4}$) and a small *transverse gradient* $\nabla_x \alpha$ ($\frac{\Delta \alpha}{\langle \alpha \rangle} \leq 10^{-1}$, the symbol $\langle\,\rangle$ means averaging over the layer's thickness).
- *A small transverse gradient of the internal magnetic field* ($\nabla_x H_{0i} \leq 10^{-1}$) and related gradients – $\nabla_x \overset{\leftrightarrow}{N} M_S$, $\nabla_x N_A$($\frac{\Delta M_S}{\langle M_S \rangle} \leq 10^{-1}$, $\frac{\Delta H_A}{\langle H_A \rangle} \leq 10^{-1}$).

Wave processes in a wide class of modern materials (solid and film structures of YIG, spinels, barium hexaferrite) have been investigated. Nevertheless, the achieved success in the creation of new types of structures, including films of ferroelectrics, semiconductors, magnetic semiconductors, wide opportunities of dopation and implantation by means of various techniques, directed at the creation of structures with required laws of changing their parameters over the thickness and area of the structures, the started research of nonlinear wave processes, including magneto-optical interactions, require the development of generalized theoretical analysis [179, 182–186, 265, 266].

For a correct theoretical description of the processes of excitation and propagation of various types of waves in layered structures made of magnetico-ordered films in the short-wave part of the centimeter and millimeter ranges, self-consistent electrodynamic analysis of structures with a transverse change of their dissipative and magnetic parameters is necessary, which can be well described as a multilayer bigyrotropic structure in external electric and magnetic fields. These points are significant not far from the resonant frequencies. No problems of such a kind have been considered earlier. No features of excitation of various types of waves in magnetico-ordered film structures with a low but finite level of dissipation in the near and far zones of radiation, including the self-consistent approximation, have been investigated, no input–output signal transfer coefficients in transmission lines with losses have been calculated [346–366].

The electrodynamic approximate methods [107], with all the advantages of the analytical expressions derived in [108], allowed no analysis of wave excitation and propagation in waveguides with plates and, especially, ferrite films near their resonant frequencies.

At studying spin wave phenomena, approximate approaches based on the circuit theory and shortened (magnetostatic) Maxwell equations [109–178] have been widely applied till recently. An account of the back influence of the excited magnetostatic waves' fields on the distribution of extraneous electric current is presented in [128, 135, 137, 152, 168]. Analysis of MSW excitation by strip sources of various topologies in terms of Green's functions is developed in [143, 149, 157]. An electrodynamic approach to the self-consistent problem of excitation is undertaken in [180], where independent branches (magnetostatic and electrodynamic ones) in a real wave process are resolved.

Before our research no theoretical and experimental data on wave processes in layered FFDS in the millimeter range have been published. Excitation and propagation of waves of various types in compound structures made of solid ferrites were investigated first, then the study was extended to films.

The following basic leads of our theoretical and experimental research in the millimeter range have been formulated:

- Excitation and reception of electromagnetic waves, including magnetostatic waves in FFDS with various types of converters, constructed in sections of waveguides, slot, complanar and microstrip lines
- Propagation of waves of various types in partially and completely screened flat FFDS and selection of ways to control their dispersions
- Analysis of the properties of mono- and polycrystalline solid and film ferrites and layered structures on their basis by means of nondestructive methods in the microwave and EHF ranges
- Development of physical design principles, design of magnetoelectronic devices on FFDS and their prototyping

Before discussion of the state-of-the-art, we shall dwell on the terminology to be used. Inherently, any waves excited in LFDS are electromagnetic by nature. Now the widely applied term MSW reflects the essence of the *magnetostatic approximation* used. However, most of the domestic and foreign authors so have gotten used to this term that they treat these waves as physical reality. The real wave processes in LFDS at high frequencies may contain SCW branches which are exchangeless, strongly delayed electromagnetic waves with $\kappa' > 10^3 - 10^4\,\mathrm{cm}^{-1}$. More delayed waves with $\kappa' > 10^4 - 10^5\,\mathrm{cm}^{-1}$ are of the quantum nature and, as a matter of fact, SCW based on exchange interactions. The monograph considers waves of the electromagnetic nature with $\kappa' \leq 10^4\,\mathrm{cm}^{-1}$, including MSW. The theoretical approach used allows any types of waves excited and ducted in magnetico-ordered film structures to be investigated with due account of the short-wave range features. LE and LM waves have been chosen as basic ones.

Let us briefly discuss the status of the basic problems.

1. The problem of excitation and reception of electromagnetic waves of various types in layered structures made of magnetico-ordered films is a primary challenge at physical studies and development of magnetoelectronic devices [109–178]. A section of a waveguide or strip line, loaded with a layered LBS and magnetized by an external field, forms *a converter*, in which on the input of this magnetoelectronic device the power propagating in the supply line is directly transformed into the wave power in LBS, and on the output a reverse transformation to the wave power in the outlet line is done [462–464]. Depending on the orientation direction of the external field $\overline{H}_0$ with respect to the chosen plane

of a ferrite film/plate, tangent, normal and arbitrary (slanting) magnetization of the structure are distinguished. With reference to the direction of $\overline{H}_0$ with due account of the metal screen effect (ferrite loading), tangent – transverse magnetization with the wave to be able to propagate perpendicular to $\overline{H}_0$, and tangent – longitudinal magnetization with the wave to propagate along the field $\overline{H}_0$, are considered. Depending on the orientation of LBS to the exciting plane within the limits of which extraneous HF currents or fields are concentrated, we shall distinguish converters with a parallel and orthogonal orientation of the structures. Converters with the parallel (traditional) orientation of LBS are widely spread and have been under study in various approximations, mainly in a frequency range up to 10–15 GHz. No converters with the orthogonal orientation of LBS have been investigated so far.

The first converters, which are still widely spread, were made as thin conductors, microstrips connected to the central conductor with a coaxial or conducting strip (microstrip) line lying on the ferrite layer in contact with it. To provide the maximum concentration of HF magnetic fields in the modes close to ferromagnetic resonance, right-polarized electromagnetic waves were excited in the structure. Excitation of waves with their phase constants $\kappa_{\max} \leq \frac{2\pi}{W}$, W – being the width (diameter) of the exciting conductor, is possible in such converters. For $W \approx 6$ and $0.6\,\mu$m excitation of waves with $\kappa'_{\max} \approx 10^4\,$cm and $\approx 10^5\,$cm^{-1}, respectively, is possible, which provides exploration of quantum exchange interactions and wave processes as well as classical dipole–dipole interactions in a magnetico-ordered structure. A self-consistent theory of converters of various MSW types for LFDS in a MSW approximation is presented in [128, 135, 137, 152, 169]. The experimental studies have been mainly conducted in a frequency range up to 3–5 GHz.

Fewer works [145, 146, 155, 175] are devoted to analysis of MSW converters based on slot and coplanar lines. The first comprehensive theoretical and experimental investigations of such converters in the MMR are reported in [467–533].

The influence of the transverse gradient of saturation magnetization on the dispersion characteristics of an excited MSW, in view of exchange interactions, is discussed in [174].

The basic drawback of the known works is the absence of an adequate conformity between their theoretical approaches for MSW and SCW, including the electrodynamic approach [180], and their physical models of the LFDS under study, which entails essential differences at increased frequencies.

Research works in the design of magnetic semiconductors, the possibility to use of ferroelectric and semiconductor layers in magnetized structures in the MMR required a generalized problem of wave excitation in multilayer bigyrotropic structures to be formulated.

The variety of MMR converters including various types of strip and waveguide devices with the parallel and orthogonal orientations of their multilayer bigyrotropic structure has determined the necessity to solve the excitation problems for a system of arbitrary oriented point electric and magnetic currents.

2. The dispersion characteristics of MSW, SCW, and magnetoelastic waves in layered structures based on ferrite films have been thoroughly investigated in a frequency range up to 3–5 GHz. The theoretical approaches were mainly based on the MSW approximation and the account of exchange interactions in a spin system, they were directed to searching for effective ways to control wave dispersions and to provide some required characteristics of TL. Factors influencing MSW have been under intense study, namely, the presence of screens [184, 187], wave reflections and diffraction on heterogeneities [188–208], waves in structures with spatially-inhomogeneous internal [209–217] and external [218–222] magnetic fields, the impedance of loading [223], problems of thermostating [224–226], hybrid and exchange interactions [227–239], anisotropy [240–242]. Nonlinear [243–264] and parametric [267–272] phenomena in LFDS have been studied as well.

Recently, interest to studying MSW in waveguides with ferrites [273, 274] has increased.

The electrodynamic approaches to analysis of electromagnetic waves in tangent and normally magnetized ferrites and layered structures [270–284] consider no specific features of LFDS and LBS at increased frequencies, namely: the dispersion properties of ferromagnetic losses, their transverse gradients and gradients of the internal magnetic field. The approximate electrodynamic approaches in terms of a small perturbation of the waveguide by a ferrite sample give no information of wave properties near the resonant frequencies [281, 284]. At the same time, already in early works [285, 286] electrodynamic corrections to the resonant frequencies of ferromagnetic microwave spheres and ellipsoids were shown to possibly have values comparable with the influence of anisotropy fields H_A even in weakly anisotropic ferrites with $H_0 >> H_A$ and $H_0 >> 4\pi M_S$, $4\pi M_S >> H_A$. The electrodynamic approach proposed in [312] to explore single and bilayer ferrite structures in waveguides with tangent magnetization with due account of losses (high enough ones with a parameter $\alpha = \frac{\Delta H}{H_{0i}} \geq 10^{-2}$ were used) has allowed the amplitude and phase characteristics of the H_{10} type waves to be analyzed. For the first time the comprehensive investigation of magnetized LFDS with metallization in a broad frequency range was performed in A.N. Lepestkin's dissertation. In this work low ferromagnetic losses ($\alpha \leq 10^{-4}$) and small transverse gradients of saturation magnetization $\frac{\Delta M_S}{\langle M_S \rangle} \leq 0.05$–$0.10$ are shown to be of great importance in the millimeter range for wave processes near the transverse resonance frequency in the pre-limiting ($\nu >> \nu_{cr}$), limiting ($\nu \cong \nu_{cr}$), and post-limiting ($\nu << \nu_{cr}$) modes. Such structures, unlike those studied earlier, are called *high-quality ones*. The wave processes in high-quality ferrite films with $\alpha \leq 10^{-4}$ are essentially different from those in strongly dissipative ferrite films with $\alpha \leq 10^{-2}$ and have some features of their dispersion characteristics near the resonant frequencies. In particular, the processes of selective attenuation and transmission of signals in various modes are determined not by one (fast) but by two waves, namely, the fast and slow ones which can separately or simultaneously (together) determine the total effect.

The basic drawback of the experimental investigations of the dispersion MSW characteristics in the microwave range and most serious errors in the results' treatment are due to the absence of reliable techniques and devices for registration of reference points, i.e., the resonant frequencies of structures.

Miniaturized sensors for a frequency range up to 60 GHz (for YIG structures) and up to 140 GHz (for barium hexaferrite compound structures) with their accuracy by 1–3 orders of magnitude higher than that of the Hall sensors, having a spatial (by area) resolution $\sim 10^{-2}\,\mathrm{mm}^2$ and a linear one $\sim 10\,\mu\mathrm{m}$ have been designed. Studies performed in a frequency range up to 80–100 GHz show that the waves correctly described in the strict electrodynamic approach only constitute 60–80% of the total frequency band of the AFC of excited wave signals while MSW make up only 20–40% just in the long-wave part of the millimeter range.

At advance into the MMR the fraction of MSW decreases. These results have been approved in a broad frequency range (6–80 GHz) on TL made of various materials (YIG films, spinels, barium hexaferrite), including doped and multilayer (with five to seven layers) structures, at normal, tangent, and arbitrary magnetization, using various waveguide and strip converters, in both the narrow-band (near the resonant frequency) and broadband (highly above the resonant frequencies) modes.

Studying features of wave processes in structures containing magnetized layers of ferrites, ferroelectrics, semiconductors, magnetic semiconductors, plasma, aimed at the creation of highly effective controlled MMR devices, would require a substantiation of the adequacy of the structure's model used, specification of its most important attributes. It included:

- Development of a new theoretical approach to analysis of excitation, propagation, and reception of various types of waves in multilayer bigyrotropic structures
- Theoretical and experimental studies of the dispersion properties of TL based on a wide class of modern epitaxial films with the structures of YIG, spinel, barium hexaferrite, near the resonant frequencies and at tuning-out from them
- Giving recommendations on the application of waves of various types for the development of controlled devices in the microwave and EHF ranges

3. Promotion of physical research and creation of magnetoelectronic devices for new frequency ranges require:

- Development of wide-range nondestructive diagnostic techniques of the basic parameters of LFDS in the microwave and EHF ranges
- Development of techniques for experimental examination of the amplitude and phase constants for $\kappa' \leq 10^4\,\mathrm{cm}^{-1}$
- Design of resonant-frequency sensors for layered magnetico-ordered structures and the corresponding magnetic fields with an increased accuracy and spatial resolution

Our analysis of the publications shows that the most exact method for magnetic field measurement (NMR) does not meet the two following contradictory requirements:

- Small dimensions (for measurements in specific gaps thinner than 0.5–1.0 mm); an increased field heterogeneity ($\frac{\Delta H_0}{H_0} > 10^4$)

Hall sensors have a poor accuracy (>1–2%) at increased values of fields (above 10–20 kOe) and a sensor area of the order of $0.1\,\text{mm}^2$.

For correct binding to the resonant frequencies of arbitrary magnetized LFDS and LBS in the microwave and EHF ranges are required:

- An increase of the accuracy by not less than 1–2 orders of magnitude in comparison with Hall sensors
- An increase in the spatial (by area) resolution by not less than an order of magnitude, and that in the linear resolution by 1–2 orders of magnitude (up to $10\,\mu\text{m}$)

The known types of sensors do not meet the following requirements simultaneously:

- Small transverse dimensions
- An increased accuracy and spatial resolution
- Accuracy preservation in heterogeneous fields

This makes urgent the development of MMR sensors for the resonant frequencies and the corresponding magnetic fields for magnetoelectronics.

Now the known methods and devices for film ferrite diagnostics [287–315] are mainly for the microwave range, they provide determination of thickness-average dissipative (ΔH_{κ}, the ferromagnetic resonance line width of the κ number of spin-wave resonance with LFDS) and magnetic (the saturation magnetization M_s, the field of crystallographic $\overline{H}_A$ and axial $\overline{H}_s$ anisotropy) parameters.

For millimeter-range LFDS, nondestructive wide-range control of the ferromagnetic resonance line width ΔH and its transverse gradient $\nabla_x \Delta H$, transverse gradients of the internal magnetic field $\nabla_x \Delta H_{0i}$, magnetization – $\nabla_x M_S$ and fields of anisotropy $\nabla_x H_A$, $\nabla_x H_s$, the distribution of the boundary internal magnetic field over the sample $\nabla_y \Delta H_{0i}$ are required. Only the knowledge of these parameters, their admissible values with specific features for application in the selective and broadband modes of LFDS and MMR, can provide correct treatment of experimental results, point out most effective ways to achieve the required parameters (introduced losses, bandwidth, converter and transmission line, out-of-band screening, deviation of the dispersion characteristic and frequency range from the required ones, AFC shape, etc.).

4. There is a drastic gap between the intensity and range of research in microwave magnetoelectronics in our country and abroad. No industrial production of magnetoelectronic devices has been launched. This is due to the physical processes in this field being poorly studied, a low level of the theoretical approaches to the problems, a poor adequacy of the models in use and the treatment of experimental results. Most important and specific factors for various devices in both the microwave and EHF ranges should be revealed.

Our analysis of the domestic and foreign publications shows that the magnetoelectronic devices based on ferrite films very slowly leave the stage of basic research which covers the linear and nonlinear modes. Work is under way to design delay lines of various types [316–345], low-and-high-passing and low-and-high-blocking filters [368–383], resonators [384–414], directed tappers [415, 416], oscillators [420–439], amplifiers [440–445], convolvers [452–457], modulators [450], and squelches [451].

There are practically no data on the development of nonlinear MMR devices, devices of low and high power levels (LPL and HPL). There is no precise pattern of the priority application areas of these or those types of electromagnetic waves in LFDS, achievable parameters, requirements to the quality of ferrite films, possibilities of advance into the MMR [505, 506, 572–578].

The above circumstances have determined the necessity to develop the physical basis for designing of MMR magnetoelectronic devices. Our tasks were as follows:

- Design and optimization of various types of waveguide and strip converters
- Formation of TL with required dispersions
- Development of miniature magnetic systems with thermostabilization and rapid field reorganization
- Design of devices for coordination with supplying lines
- Search for ways to reduce the irregularity and to improve the AFC shape
- Search for ways to raise electric durability

Part I
Theory of Electromagnetic Wave Excitation in Bigyrotropic Structures

The known types of electromagnetic wave converters for layered structures on the basis of magnetico-ordered mediums are classified into converters with parallel and orthogonal orientations of their structures to the plane of exciting surface currents. The scientific problem is to study electromagnetic wave excitation on increased frequencies in screened structures on the basis of magnetico-ordered films, for which transverse gradients of losses, gradients of the electric and magnetic parameters are characteristic. These circumstances required a solution of the problems for multilayer structures. Expansion of the functionality of magnetoelectronic devices at the use of, besides ferrite films, those of ferroelectrics, semiconductors, magnetic semiconductors, has determined an approach to solution of the problems of excitation for multilayer bigyrotropic structures in an external magnetic field.

The aim of our theoretical analysis is in finding the tensor Green functions of waves for an arbitrary layered bigyrotropic structure limited by impedance surfaces, first of all, in finding of the self-consistent amplitudes of fields and power flows of waves excited by arbitrary surface electric and magnetic currents considering the basic factors of the millimeter range.

Chapter 1
Converters with Parallel Structure Orientation

The theory of electromagnetic wave excitation by surface electric and magnetic currents located on an internal interface of a multilayer bigyrotropic structure screened by impedance surfaces, is considered. Converters on the basis of microstrip, slot and coplanar lines, their various modifications and topologies fall into such a type.

1.1 Statement of the Problem: Boundary Conditions – Coupling Matrix

Let's examine a flat structure containing p layers with their electric and magnetic conductivity[1] defined by the tensors $\overleftrightarrow{\varepsilon}_n$ and $\overleftrightarrow{\mu}_n$, $n = 1, 2, \ldots, p$, magnetized by an external field $\overline{H}_0$ along an Cartesian axis (Fig. 1.1). The structure is homogeneous in the plane YOZ and screened by impedance surfaces at $x = 0$ and $x = x_p$.

The fields in the structure $\overline{E}$, $\overline{H}$ satisfy the equations

$$\begin{cases} [\nabla, \overline{E}_n] = -j\omega\mu_0 \overleftrightarrow{\mu}_n \overline{H}_n \\ [\nabla, H_n] = j\omega\varepsilon_0 \overleftrightarrow{\varepsilon}_n \overline{E}_n, \end{cases} \tag{1.1}$$

where $n = 1, 2, \ldots, p$ (p is the number of layers).

Let's introduce a column matrix $\overleftrightarrow{N}_n$ composed of the tangential constituents of the electric $E^n_{y,z}(x,y,z)$ and magnetic $H^n_{y,z}(x,y,z)$ fields

$$\overleftrightarrow{N}_n = \begin{Vmatrix} E^n_y(x,y,z) \\ E^n_z(x,y,z) \\ H^n_y(x,y,z) \\ H^n_z(x,y,z) \end{Vmatrix}. \tag{1.2}$$

[1] The type of the tensors for ferrites and conductive crystals considering dissipations at different directions of the field $\overline{H}_0$ is given in Appendix 1.

A.A. Ignatiev, *Magnetoelectronics of Microwaves and Extremely High Frequencies in Ferrite Films*.
DOI: 10.1007/978-0-387-85457-1_2,

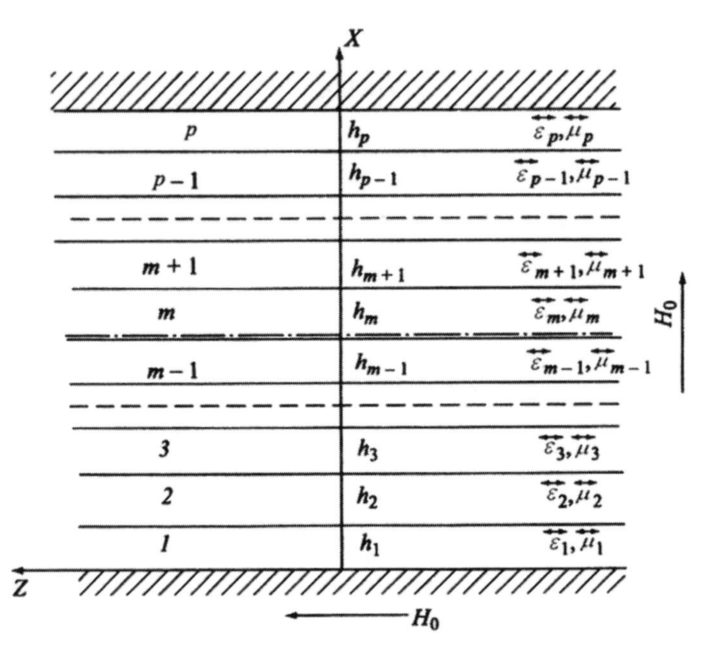

Fig. 1.1 Structure containing P layers with their electric and magnetic conductivity defined by the tensors $\overleftrightarrow{\varepsilon}_n$ and $\overleftrightarrow{\mu}_n, n = 1, 2, \ldots, p$, magnetized by an external field $\overline{H}_0$ along an Cartesian axis

Similarly we shall introduce a column matrix $\overleftrightarrow{I}_S$ of the constituents of the surface magnetic $j_M(j_{My}, j_{Mz})$ and electric $j_E(j_{Ey}, j_{Ez})$ currents located on the boundary of the structure at $x = x_m$

$$\overleftrightarrow{I}_s = \begin{Vmatrix} J_{Mz}(y,z) \\ J_{My}(y,z) \\ J_{Ez}(y,z) \\ J_{Ey}(y,z) \end{Vmatrix}. \tag{1.3a}$$

Using the Fourier transform by the coordinates y and z for $\overleftrightarrow{N}_n$ and $\overleftrightarrow{I}_S$ as

$$\begin{pmatrix} \overleftrightarrow{N}_n(x,y,z) \\ \overleftrightarrow{I}_s(x,y,z) \end{pmatrix} = \frac{1}{2\pi} \iint\limits_{-\infty}^{\infty} \begin{pmatrix} \widetilde{\overleftrightarrow{N}}_n(x, \kappa_y, \kappa_z) \\ \widetilde{\overleftrightarrow{I}}_s(\kappa_y, \kappa_z) \end{pmatrix} \cdot e^{-j(\kappa_y y + \kappa_z z)} \mathrm{d}\kappa_y \mathrm{d}\kappa_z, \tag{1.3b}$$

we have the following boundary conditions:

– On the screen at $x = 0$

$$\widetilde{\overleftrightarrow{N}}_n = \begin{Vmatrix} \widetilde{E}_y^1(0, \kappa_y, \kappa_z) \\ \widetilde{E}_z^1(0, \kappa_y, \kappa_z) \\ \widetilde{H}_y^1(0, \kappa_y, \kappa_z) \\ \widetilde{H}_z^1(0, \kappa_y, \kappa_z) \end{Vmatrix} \tag{1.4}$$

- On all the internal layer interfaces of the structure, except for the boundary with the sources $x = x_m$, the constituents of HF fields are continuous and at $x = x_m$, $n = 1, 2, \ldots, m-1, m+1, \ldots, p-1$

$$\overset{\leftrightarrow}{\widetilde{N}}_n(x_n) = \overset{\leftrightarrow}{\widetilde{N}}_{n+1}(x_n) \tag{1.5}$$

- On the boundary $x = x_m$ in the structure we have a discontinuity of the tangential constituents of HF fields, determined by extraneous surface currents

$$\overset{\leftrightarrow}{N}_{m+1}(x_m, \kappa_y, \kappa_z) - \overset{\leftrightarrow}{N}_m(x_m, \kappa_y, \kappa_z) = \overset{\leftrightarrow}{\widetilde{I}}_{s+1}(x_m, \kappa_y, \kappa_z) \tag{1.6}$$

- On the screen at $x = x_p$

$$\overset{\leftrightarrow}{\widetilde{N}}_p(x_p, \kappa_y, \kappa_z) = \left\| \begin{matrix} E_y^p(x_p, \kappa_y, \kappa_z) \\ E_z^p(x_p, \kappa_y, \kappa_z) \\ H_y^p(x_p, \kappa_y, \kappa_z) \\ H_z^p(x_p, \kappa_y, \kappa_z) \end{matrix} \right\|. \tag{1.7}$$

We shall introduce a column matrix of Green functions as

$$\overset{\leftrightarrow}{N}_n(x, y, z) = \iint\limits_S \overset{\leftrightarrow}{G}_n(x, y, z; y_s, z_s)\, \overset{\leftrightarrow}{I}\, \overset{\leftrightarrow}{I}_s(y_s, z_s)\mathrm{d}y_s\mathrm{d}z_s, \tag{1.8a}$$

where S is the surface on which the extraneous point currents are localized (for the electric current $\overline{j}_e$ we shall use the coordinates y_{se}, z_{se}, and for magnetic $\overline{j}_m - y_{sm}$, z_{sm}), $\overset{\leftrightarrow}{I}$ – an identity 4×4 matrix, and s is the source

$$\overset{\leftrightarrow}{G}_n(x, y, z; y_s, z_s) = \left\| \begin{matrix} G_1^n(x, y, z; y_s, z_s) \\ G_2^n(x, y, z; y_s, z_s) \\ G_3^n(x, y, z; y_s, z_s) \\ G_4^n(x, y, z; y_s, z_s) \end{matrix} \right\|. \tag{1.8b}$$

Passing to the Fourier transforms in Eq. (1.8a), we have

$$\overset{\leftrightarrow}{\widetilde{N}}(x, \kappa_y, \kappa_z) = \iint\limits_S \overset{\leftrightarrow}{\widetilde{G}}_n(x, \kappa_y, \kappa_z; y_s, z_s)\overset{\leftrightarrow}{I}\, \overset{\leftrightarrow}{I}_s(y_s, z_s)\mathrm{d}y_s\, \mathrm{d}z_s. \tag{1.9}$$

Let's formulate boundary conditions for FIGF.

On the impedance surface limiting the structure, considering Eqs. (1.4) and (1.9) at $x = 0$,

$$\overset{\leftrightarrow}{\widetilde{G}}_1(0, x, \kappa_y, \kappa_z; y_s, z_s) = \left\| \begin{matrix} \widetilde{G}_1^1(0) \\ \widetilde{G}_2^1(0) \\ \widetilde{G}_3^1(0) \\ \widetilde{G}_4^1(0) \end{matrix} \right\|. \tag{1.10}$$

On the internal boundaries of the structure, except for the boundary $x = x_m$, considering Eqs. (1.5) and (1.9), at $x = X_n$ we have

$$\overset{\leftrightarrow}{\widetilde{G}}_1(x_n, \kappa_y, \kappa_z; y_s, z_s) = \overset{\leftrightarrow}{\widetilde{G}}_{n+1}(x_{n+1}, \kappa_y, \kappa_z; y_s, z_s), \tag{1.11}$$

where $n = 1, 2, \ldots, m-1, m+1, \ldots, p$.

On the impedance surface limiting the structure, considering Eqs. (1.7) and (1.9) at $x = x_p$, we have

$$\overset{\leftrightarrow}{\widetilde{G}}_p(x_p, \kappa_y, \kappa_z; y_s, z_s) = \left\| \begin{matrix} \widetilde{G}_1^p(x_p) \\ \widetilde{G}_2^p(x_p) \\ \widetilde{G}_3^p(x_p) \\ \widetilde{G}_4^p(x_p) \end{matrix} \right\|. \tag{1.12}$$

Let's examine the condition for FIGF on the source's boundary $(x = x_m)$. From Eqs. (1.6) and (1.9) it follows that

$$\iint_S \left[\overset{\leftrightarrow}{\widetilde{G}}_{m+1}(x_m, \kappa_y, \kappa_z; y_s, z_s) - \overset{\leftrightarrow}{\widetilde{G}}_m(x_m, \kappa_y, \kappa_z; y_s, z_s) \right] \times \overset{\leftrightarrow\leftrightarrow}{I\,I}_s(y_s, z_s) \mathrm{d}y_s \, \mathrm{d}z_s = \overset{\leftrightarrow}{\widetilde{I}}_s. \tag{1.13}$$

If the right-hand side of Eq. (1.13) is represented by a Fourier integral, then

$$\iint_S \left[\overset{\leftrightarrow}{\widetilde{G}}_{m+1}(x_m, \kappa_y, \kappa_z; y_s, z_s) - \overset{\leftrightarrow}{\widetilde{G}}_m(x_m, \kappa_y, \kappa_z; y_s, z_s) \right] \times \overset{\leftrightarrow\leftrightarrow}{I\,I}_s(y_s, z_s) \mathrm{d}y_s \, \mathrm{d}z_s$$
$$= \frac{1}{2\pi} \iint_S \overset{\leftrightarrow}{I}_s(y_s, z_s) e^{j(\kappa_y y_s + \kappa_z z_s)} \mathrm{d}y_s \mathrm{d}z_s. \tag{1.14}$$

It should be mentioned that in Eq. (1.14) the right-hand side is nonzero only on the surface of the source S. Then from Eq. (1.14) it follows that

$$\left[\overset{\leftrightarrow}{\widetilde{G}}_{m+1}(x_m, \kappa_y, \kappa_z; y_s, z_s) - \overset{\leftrightarrow}{\widetilde{G}}_m(x_m, \kappa_y, \kappa_z; y_s, z_s) \right] \cdot \overset{\leftrightarrow\leftrightarrow}{I\,I}_s$$
$$= \overset{\leftrightarrow}{I}_s(y_s z_s) e^{j(\kappa_y y_s + \kappa_z z_s)}. \tag{1.15}$$

As for the source with its coordinates (y_{sM}, z_{sM}) and (y_{sE}, z_{sE}) the expression (1.3a) looks like

$$\overset{\leftrightarrow}{I}_s(y_s, z_s) = \left\| \begin{matrix} j_{My}(y_{sM}, z_{sM}) \\ j_{Mz}(y_{sM}, z_{sM}) \\ j_{Ey}(y_{sE}, z_{sE}) \\ j_{Ez}(y_{sE}, z_{sE}) \end{matrix} \right\|, \tag{1.16a}$$

then we shall reduce the right-hand side of Eq. (1.15) to a form, similar to its left side

$$\left[\overset{\leftrightarrow}{\widetilde{G}}_{m+1}(x_m, \kappa_y, \kappa_z; y_s, z_s) - \overset{\leftrightarrow}{\widetilde{G}}_m(x_m, \kappa_y, \kappa_z; y_s, z_s) \right] \cdot \overset{\leftrightarrow\leftrightarrow}{I\,I}_s(y_s, z_s)$$
$$= \overset{\leftrightarrow}{F}(y_s, z_s) \cdot \overset{\leftrightarrow\leftrightarrow}{I\,I}_s(y_s, z_s), \tag{1.16b}$$

where

$$\overleftrightarrow{F}(y_s, z_s) = \frac{1}{2\pi}\begin{Vmatrix} e^{j(\kappa_y y_{sM}+\kappa_z z_{sM})} \\ e^{j(\kappa_y y_{sM}+\kappa_z z_{sM})} \\ e^{j(\kappa_y y_{sE}+\kappa_z z_{sE})} \\ e^{j(\kappa_y y_{sE}+\kappa_z z_{sE})} \end{Vmatrix} = \frac{1}{2\pi}\begin{Vmatrix} F_1 \\ F_2 \\ F_3 \\ F_4 \end{Vmatrix}. \tag{1.17}$$

Equation (1.17) will be transformed to the form

$$\overleftrightarrow{\tilde{\tilde{G}}}_{m+1}(x_m, \kappa_y, \kappa_z; y_s, z_s) - \overleftrightarrow{\tilde{\tilde{G}}}_m(x_m, \kappa_y, \kappa_z; y_s, z_s) = \overleftrightarrow{F}(y_s, z_s), \tag{1.18}$$

which determines the boundary condition of FIGF $\overleftrightarrow{\tilde{\tilde{G}}}_{m+1}$ and $\overleftrightarrow{\tilde{\tilde{G}}}_m$ in the source's location $(x = x_m, y_s, z_s)$. We shall express the $\overleftrightarrow{\tilde{\tilde{G}}}_{m+1}$ and $\overleftrightarrow{\tilde{\tilde{G}}}_m$ functions through $\overleftrightarrow{\tilde{\tilde{G}}}_p(x_p)$ and $\overleftrightarrow{\tilde{\tilde{G}}}_1(0)$ by means of coupling matrixes. A similar method was used in [278] for the tangential constituents of HF fields in analysis of wave dispersions in multilayer bigyrotropic structures without losses. Recalculation of the tangential constituents of fields from layer to layer was carried out through so-called transmission matrixes [278]. As in the transverse direction (axis 0*X*) of the analyzed structure there is no power transfer (more precisely, no condition for wave propagation), it is reasonable to use the term "coupling matrix" to reflect the essence of the process more precisely. In addition, it is convenient to consider the Green functions defined on the limiting surfaces at $x = 0$ – Eq. (1.10) and $x = x_p$ – Eq. (1.12). Then, by means of the coupling matrixes for FIGF we shall express $\overleftrightarrow{\tilde{\tilde{G}}}_m(x_m)$ and $\overleftrightarrow{\tilde{\tilde{G}}}_{m+1}(x_m)$ through $\overleftrightarrow{\tilde{\tilde{G}}}_1(0)$ correspondingly. The compatibility condition for Eq. (1.18) gives equality to zero of the corresponding determinant made of the factors of the corresponding FIGF on the limiting surfaces, which allow own wave dispersions in screened multilayer bigyrotropic structures to be analyzed by numerical methods. On the next step of the specified extraneous current approximation, the FIGF on the limiting surfaces can be found. Then, by means of the structure coupling matrices the FIGF in all the layers of the structure $\overleftrightarrow{\tilde{\tilde{G}}}_n(x_n)$, $n = 1, 2, \ldots, p$ can be found through the values of $\overleftrightarrow{\tilde{\tilde{G}}}_1(0)$ and $\overleftrightarrow{\tilde{\tilde{G}}}_p(x_p)$. By Eqs. (1.9) and (1.4) the tangential constituents $E^n_{y,z}$ and $H^n_{y,z}$ of HF fields in the layers of the structure are determined. With the transverse constituents $E^n_x = f(E^n_{y,z}, H^n_{y,z})$ and $H^n_x = f(E^n_{y,z}, H^n_{y,z})$ of HF fields, by means of the found constituents $E^n_{x,y,z}$ and $H^n_{x,y,z}$ it is possible to determine the power flows of the waves excited in the structure, in the specified extraneous surface current approximation. Then, solving the integral equation for the exciting currents, we shall consider the back influence of the excited waves's fields.

We'll get the coupling matrix of the FIGF of a multilayer bigyrotropoic structure, when the tensors $\overleftrightarrow{\mu}_n$ and $\overleftrightarrow{\varepsilon}_n$ are defined in a general form

$$\overleftrightarrow{\mu}_n = \begin{Vmatrix} \mu^n_{11} & \mu^n_{12} & \mu^n_{13} \\ \mu^n_{21} & \mu^n_{22} & \mu^n_{23} \\ \mu^n_{31} & \mu^n_{32} & \mu^n_{33} \end{Vmatrix}, \tag{1.19}$$

$$\overleftrightarrow{\varepsilon}_n = \begin{Vmatrix} \varepsilon_{11}^n & \varepsilon_{12}^n & \varepsilon_{13}^n \\ \varepsilon_{21}^n & \varepsilon_{22}^n & \varepsilon_{23}^n \\ \varepsilon_{31}^n & \varepsilon_{32}^n & \varepsilon_{33}^n \end{Vmatrix}. \tag{1.20}$$

The particular form of $\overleftrightarrow{\mu}$ and $\overleftrightarrow{\varepsilon}$ for the models used is given in Appendix 1. Equation (1.1) in projections onto the coordinate axes after the use of Fourier transforms by y and z give

$$\begin{cases} j\kappa_y\widetilde{H}_z^n + j\kappa_z\widetilde{H}_y^n = j\omega\varepsilon_0(\varepsilon_{11}^n\widetilde{E}_x^n + \varepsilon_{12}^n\widetilde{E}_y^n + \varepsilon_{13}^n\widetilde{E}_z^n) & (1.21) \\ -j\kappa_z\widetilde{H}_x^n - \dfrac{\mathrm{d}\widetilde{H}_z^n}{\mathrm{d}x} = j\omega\varepsilon_0(\varepsilon_{21}^n\widetilde{E}_x^n + \varepsilon_{22}^n\widetilde{E}_y^n + \varepsilon_{23}^n\widetilde{E}_z^n), & (1.22) \\ \dfrac{\mathrm{d}\widetilde{H}_y^n}{\mathrm{d}x} + j\kappa_y\widetilde{H}_x^n = j\omega\varepsilon_0(\varepsilon_{31}^n\widetilde{E}_x^n + \varepsilon_{32}^n\widetilde{E}_y^n + \varepsilon_{33}^n\widetilde{E}_z^n) & (1.23) \end{cases}$$

$$\begin{cases} -j\kappa_y\widetilde{E}_z^n + j\kappa_z\widetilde{E}_y^n = -j\omega\mu_0(\mu_{11}^n\widetilde{H}_x^n + \mu_{12}^n\widetilde{H}_y^n + \mu_{13}^n\widetilde{H}_z^n) & (1.24) \\ -j\kappa_z\widetilde{E}_x^n - \dfrac{\mathrm{d}\widetilde{E}_z^n}{\mathrm{d}x} = -j\omega\mu_0(\mu_{21}^n\widetilde{H}_x^n + \mu_{22}^n\widetilde{H}_y^n + \mu_{23}^n\widetilde{H}_z^n). & (1.25) \\ \dfrac{\mathrm{d}\widetilde{E}_y^n}{\mathrm{d}x} + j\kappa_y\widetilde{E}_x^n = -j\omega\mu_0(\mu_{31}^n\widetilde{H}_x^n + \mu_{32}^n\widetilde{H}_y^n + \mu_{33}^n\widetilde{H}_z^n) & (1.26) \end{cases}$$

From Eqs. (1.21) and (1.24) for the transverse components of HF fields it follows that

$$\widetilde{E}_x^n = \frac{1}{\varepsilon_{11}^n}\left[\frac{1}{\omega\varepsilon_0}(\kappa_z\widetilde{H}_y^n - \kappa_y\widetilde{H}_z^n) - \varepsilon_{12}^n\widetilde{E}_y^n - \varepsilon_{13}^n\widetilde{E}_z^n\right], \tag{1.27}$$

$$\widetilde{H}_x^n = \frac{1}{\mu_{11}^n}\left[\frac{1}{\omega\mu_0}(\kappa_y\widetilde{E}_z^n - \kappa_z\widetilde{E}_y^n) - \mu_{12}^n\widetilde{H}_y^n - \mu_{13}^n\widetilde{H}_z^n\right]. \tag{1.28}$$

Eliminating E_x^n and H_x^n, by means of Eqs. (1.27) and (1.28) from Eqs. ((1.22) and (1.23)) and Eqs. ((1.25) and (1.26)), we derive for the transverse derivatives of the Fourier transforms of the tangential constituents of HF fields

$$\begin{aligned} \frac{\mathrm{d}\widetilde{E}_y^n}{\mathrm{d}x} = j\Bigg\{&\left(\frac{\varepsilon_{12}^n}{\varepsilon_{11}^n}\kappa_y + \frac{\mu_{31}^n}{\mu_{11}^n}\kappa_z\right)\widetilde{E}_y^n + \left(\frac{\varepsilon_{13}^n}{\varepsilon_{11}^n} - \frac{\mu_{31}^n}{\mu_{11}^n}\right)\kappa_y\widetilde{E}_z^n \\ &+ \left[\omega\mu_0\left(\frac{\mu_{12}^n\mu_{31}^n}{\mu_{11}^n} - \mu_{32}^n\right) - \frac{1}{\omega\varepsilon_0\varepsilon_{11}^n}\kappa_y\kappa_z\right]\widetilde{H}_y^n \\ &+ \left[\omega\mu_0\left(\frac{\mu_{31}^n\mu_{13}^n}{\mu_{11}^n} - \mu_{33}^n\right) + \frac{1}{\omega\varepsilon_0\varepsilon_{11}^n}\kappa_y^2\right]\widetilde{H}_y^n\Bigg\}, \end{aligned} \tag{1.29}$$

$$\frac{d\widetilde{E}_z^n}{dx} = j\left\{\left(\frac{\varepsilon_{12}^n}{\varepsilon_{11}^n} - \frac{\mu_{21}^n}{\mu_{11}^n}\right)\kappa_z\widetilde{E}_y^n + \left(\frac{\varepsilon_{13}^n}{\varepsilon_{11}^n}\kappa_z + \frac{\mu_{21}^n}{\mu_{11}^n}\kappa_y\right)\widetilde{E}_z^n \right.$$
$$+\left[\omega\mu_0\left(\mu_{22}^n - \frac{\mu_{21}^n\mu_{12}^n}{\mu_{11}^n}\right) - \frac{1}{\omega\varepsilon_0\varepsilon_{11}^n}\kappa_z^2\right]\widetilde{H}_y^n \tag{1.30}$$
$$\left.+\left[\omega\mu_0\left(\mu_{23}^n - \frac{\mu_{21}^n\mu_{13}^n}{\mu_{11}^n}\right) + \frac{1}{\omega\varepsilon_0\varepsilon_{11}^n}\kappa_y\kappa_z\right]\widetilde{H}_z^n\right\},$$

$$\frac{d\widetilde{H}_y^n}{dx} = j\left\{\left[\omega\varepsilon_0\left(\varepsilon_{32}^n - \frac{\varepsilon_{31}^n\varepsilon_{12}^n}{\varepsilon_{11}^n}\right) + \frac{1}{\omega\mu_0\mu_{11}^n}\kappa_y\kappa_z\right]\widetilde{E}_y^n\right.$$
$$+\left[\omega\varepsilon_0\left(\varepsilon_{33}^n - \frac{\varepsilon_{31}^n\varepsilon_{13}^n}{\varepsilon_{11}^n}\right) - \frac{1}{\omega\mu_0\mu_{11}^n}\kappa_y^2\right]\widetilde{E}_z^n \tag{1.31}$$
$$\left.+\left(\frac{\mu_{12}^n}{\mu_{11}^n}\kappa_y + \frac{\varepsilon_{31}^n}{\varepsilon_{11}^n}\kappa_z\right)\widetilde{H}_y^n + \left(\frac{\mu_{13}^n}{\mu_{11}^n} - \frac{\varepsilon_{31}^n}{\varepsilon_{11}^n}\right)\kappa_y\widetilde{H}_z^n\right\},$$

$$\frac{d\widetilde{H}_z^n}{dx} = j\left\{\left[\omega\varepsilon_0\left(\frac{\varepsilon_{21}^n\varepsilon_{12}^n}{\varepsilon_{11}^n} - \varepsilon_{22}^n\right) + \frac{1}{\omega\mu_0\mu_{11}^n}\kappa_z^2\right]\widetilde{E}_y^n\right.$$
$$+\left[\omega\varepsilon_0\left(\frac{\varepsilon_{21}^n\varepsilon_{13}^n}{\varepsilon_{11}^n} - \varepsilon_{23}^n\right) - \frac{1}{\omega\mu_0\mu_{11}^n}\kappa_y\kappa_z\right]\widetilde{E}_z^n \tag{1.32}$$
$$\left.+\left(\frac{\mu_{12}^n}{\mu_{11}^n} - \frac{\varepsilon_{21}^n}{\varepsilon_{11}^n}\right)\kappa_z\widetilde{H}_y^n + \left(\frac{\mu_{13}^n}{\mu_{11}^n}\kappa_z + \frac{\varepsilon_{21}^n}{\varepsilon_{11}^n}\kappa_y\right)\widetilde{H}_z^n\right\}.$$

From Eqs. (1.29) to (1.32) a matrix form of the equation follows

$$\frac{d\overset{\leftrightarrow}{\widetilde{N}}_n(x,\kappa_y,\kappa_z)}{dx} = j\overset{\leftrightarrow}{A}_n\overset{\leftrightarrow}{\widetilde{N}}_n(x,\kappa_y,\kappa_z), \tag{1.33}$$

where

$$\overset{\leftrightarrow}{N}_n(x,y,z) = \frac{1}{2\pi}\iint_{-\infty}^{\infty}\overset{\leftrightarrow}{\widetilde{N}}_n(x,\kappa_y,\kappa_z)\cdot e^{-j(\kappa_y y+\kappa_z z)}d\kappa_y d\kappa_z,$$

$\overset{\leftrightarrow}{A}_n$ is a 4×4 matrix, whose components (including losses) are complex and look like

$$a_{11}^n = \frac{\varepsilon_{12}^n}{\varepsilon_{11}^n}\kappa_y + \frac{\mu_{31}^n}{\mu_{11}^n}\kappa_z,$$
$$a_{12}^n = \left(\frac{\varepsilon_{13}^n}{\varepsilon_{11}^n} - \frac{\mu_{31}^n}{\mu_{11}^n}\right)\kappa_y,$$
$$a_{13}^n = \omega\mu_0\left(\frac{\mu_{12}^n\mu_{31}^n}{\mu_{11}^n} - \mu_{32}^n\right) - \frac{1}{\omega\varepsilon_0\varepsilon_{11}^n}\kappa_y\kappa_z,$$
$$a_{14}^n = \omega\mu_0\left(\frac{\mu_{31}^n\mu_{13}^n}{\mu_{11}^n} - \mu_{33}^n\right) + \frac{1}{\omega\varepsilon_0\varepsilon_{11}^n}\kappa_y^2,$$

$$
\begin{aligned}
a_{21}^n &= \left(\frac{\varepsilon_{12}^n}{\varepsilon_{11}^n} - \frac{\mu_{21}^n}{\mu_{11}^n}\right)\kappa_z,\\
a_{22}^n &= \frac{\varepsilon_{13}^n}{\varepsilon_{11}^n}\kappa_z + \frac{\mu_{21}^n}{\mu_{11}^n}\kappa_y,\\
a_{23}^n &= \omega\mu_0\left(\mu_{22}^n - \frac{\mu_{21}^n\mu_{12}^n}{\mu_{11}^n}\right) - \frac{1}{\omega\varepsilon_0\varepsilon_{11}^n}\kappa_z^2,\\
a_{24}^n &= \omega\mu_0\left(\mu_{23}^n - \frac{\mu_{21}^n\mu_{13}^n}{\mu_{11}^n}\right) + \frac{1}{\omega\varepsilon_0\varepsilon_{11}^n}\kappa_y\kappa_z,\\
a_{31}^n &= \omega\varepsilon_0\left(\varepsilon_{32}^n - \frac{\varepsilon_{31}^n\varepsilon_{12}^n}{\varepsilon_{11}^n}\right) + \frac{1}{\omega\mu_0\mu_{11}^n}\kappa_y\kappa_z,\\
a_{32}^n &= \omega\varepsilon_0\left(\varepsilon_{33}^n - \frac{\varepsilon_{31}^n\varepsilon_{13}^n}{\varepsilon_{11}^n}\right) - \frac{1}{\omega\mu_0\mu_{11}^n}\kappa_y^2,\\
a_{33}^n &= \frac{\mu_{12}^n}{\mu_{11}^n}\kappa_y + \frac{\varepsilon_{31}^n}{\varepsilon_{11}^n}\kappa_z,\\
a_{34}^n &= \left(\frac{\mu_{13}^n}{\mu_{11}^n} - \frac{\varepsilon_{31}^n}{\varepsilon_{11}^n}\right)\kappa_y,\\
a_{41}^n &= \omega\varepsilon_0\left(\frac{\varepsilon_{21}^n\varepsilon_{12}^n}{\varepsilon_{11}^n} - \varepsilon_{22}^n\right) + \frac{1}{\omega\mu_0\mu_{11}^n}\kappa_z^2,\\
a_{42}^n &= \omega\varepsilon_0\left(\frac{\varepsilon_{21}^n\varepsilon_{13}^n}{\varepsilon_{11}^n} - \varepsilon_{23}^n\right) - \frac{1}{\omega\mu_0\mu_{11}^n}\kappa_y\kappa_z.\\
a_{43}^n &= \left(\frac{\mu_{12}^n}{\mu_{11}^n} - \frac{\varepsilon_{21}^n}{\varepsilon_{11}^n}\right)\kappa_z,\\
a_{44}^n &= \frac{\varepsilon_{21}^n}{\varepsilon_{11}^n}\kappa_y + \frac{\mu_{13}^n}{\mu_{11}^n}\kappa_z.
\end{aligned}
\tag{1.34}
$$

Integrating Eq. (1.33) over the interval $[x,\ x_0]$, we get

$$
\overset{\leftrightarrow}{\widetilde{N}}_n(x, \kappa_y, \kappa_z) = \overset{\leftrightarrow}{T}_n(x - x_0)\widetilde{N}_n(x_0, \kappa_y, \kappa_z), \tag{1.35}
$$

where $\overset{\leftrightarrow}{T}_n(x - x_0) = e^{j\overset{\leftrightarrow}{A}_n(x-x_0)}$ is the 4×4 coupling matrix of the layer limited by the x_0 and x surfaces.

Following [466], we shall represent $\overset{\leftrightarrow}{T}_n$ as a spectral representation, i.e. expansion into a series about eigenvalues λ_κ or transverse wave numbers

$$
\overset{\leftrightarrow}{T}_n(x - x_0) = \sum_{\kappa=1}^{p} \frac{\prod\limits_{\substack{i=1\\ i\neq\kappa}}^{s}(\overset{\leftrightarrow}{A}_n - \lambda_i\overset{\leftrightarrow}{I})}{\prod\limits_{\substack{i=1\\ i\neq\kappa}}^{s}(\lambda_\kappa - \lambda_i)} e^{j\lambda_\kappa(x-x_0)}, \tag{1.36}
$$

where $i = 1, 2, 3, 4$; $n = 1, 2, \ldots, p$; λ_κ – is an eigenvalue of the matrix $\overleftrightarrow{A}_n$; S – the number of different eigenvalues (for a bigyrotropic medium $S = 4$, for a gyrotropic one $S = 2$).

The eigenvalues of the matrix $\overleftrightarrow{A}_n$ are determined from the condition

$$\det(\overleftrightarrow{A}_n - \lambda_\kappa \overleftrightarrow{I}) = 0. \tag{1.37}$$

For a bigyrotropic structure, from Eq. (1.37) in the general case we have a 4th-degree equation in eigenvalues λ_κ

$$\lambda_\kappa^4 + b_\kappa^n \lambda_\kappa^3 + c_\kappa^n \lambda_\kappa^2 + d_\kappa^n \lambda_\kappa + e_\kappa^n = 0, \tag{1.38}$$

where $b_\kappa^n = -\left(a_{11}^n + a_{22}^n + a_{33}^n + a_{44}^n\right)$,

$$\begin{aligned} c_\kappa^n = {} & a_{44}^n \left(a_{11}^n + a_{22}^n + a_{33}^n\right) + a_{33}^n \left(a_{11}^n + a_{22}^n\right) - a_{24}^n a_{42}^n - a_{34}^n a_{43}^n \\ & + a_{23}^n a_{32}^n - a_{12}^n a_{21}^n - a_{13}^n a_{31}^n - a_{14}^n a_{41}^n, \end{aligned}$$

$$\begin{aligned} d_\kappa^n = {} & -a_{44}^n \left[a_{11}^n \left(a_{22}^n + a_{33}^n\right) + a_{22}^n a_{33}^n\right] - a_{11}^n \left(a_{22}^n a_{33}^n - a_{24}^n a_{42}^n - a_{34}^n a_{43}^n - a_{23}^n a_{32}^n\right) \\ & - a_{34}^n \left(a_{23}^n a_{42}^n - a_{22}^n a_{43}^n\right) - a_{24}^n \left(a_{32}^n a_{43}^n - a_{33}^n a_{42}^n\right) \\ & + a_{12}^n \left[a_{21}^n \left(a_{33}^n + a_{44}^n\right) - a_{24}^n a_{41}^n - a_{23}^n a_{31}^n\right] \\ & - a_{13}^n \left[a_{31}^n \left(a_{22}^n + a_{44}^n\right) + a_{21}^n a_{32}^n + a_{34}^n a_{41}^n\right] \\ & + a_{14}^n \left[a_{41}^n \left(a_{22}^n + a_{33}^n\right) - a_{21}^n a_{42}^n - a_{31}^n a_{43}^n\right], \end{aligned}$$

$$\begin{aligned} e_\kappa^n = {} & a_{11}^n \left[a_{22}^n \left(a_{33}^n a_{44}^n - a_{34}^n a_{43}^n\right) + a_{42}^n \left(a_{23}^n a_{34}^n - a_{24}^n a_{33}^n\right) + a_{32}^n \left(a_{24}^n a_{43}^n - a_{23}^n a_{44}^n\right)\right] \\ & + a_{12}^n \left[a_{33}^n \left(a_{23}^n a_{41}^n - a_{21}^n a_{44}^n\right) + a_{34}^n \left(a_{21}^n a_{43}^n - a_{23}^n a_{41}^n\right) + a_{31}^n \left(a_{23}^n a_{44}^n - a_{24}^n a_{43}^n\right)\right] \\ & - a_{13}^n \left[a_{32}^n \left(a_{21}^n a_{44}^n - a_{24}^n a_{41}^n\right) + a_{34}^n \left(a_{22}^n a_{41}^n - a_{23}^n a_{42}^n\right) + a_{31}^n \left(a_{24}^n a_{42}^n + a_{22}^n a_{44}^n\right)\right] \\ & + a_{14}^n \left[a_{32}^n \left(a_{23}^n a_{41}^n - a_{21}^n a_{43}^n\right) + a_{33}^n \left(a_{21}^n a_{42}^n - a_{22}^n a_{41}^n\right) + a_{31}^n \left(a_{22}^n a_{43}^n - a_{23}^n a_{42}^n\right)\right]. \end{aligned}$$

From Eq. (1.38) familiar special cases for a boundless bigyrotropic structures follow. So, when the field $\bar{H}_0$ is oriented along the axis $0Z$ the components of tensors $\overleftrightarrow{\mu}_n$ and $\overleftrightarrow{\varepsilon}_n$ look like (the notation accepted in [8] is kept here)

$$\begin{aligned} & \mu_{11}^n = \mu_{22}^n = \mu^n,\ \mu_{33}^n = \mu_{11}^n, && \varepsilon_{11}^n = \varepsilon_{22}^n = \varepsilon^n,\ \varepsilon_{33}^n = \varepsilon_{11}^n, \\ & \mu_{12}^n = j\mu_a^n,\ \mu_{21}^n = -j\mu_a^n, && \varepsilon_{12}^n = j\varepsilon_a^n,\ \varepsilon_{21}^n = -j\varepsilon_a^n, \\ & \mu_{13}^n = \mu_{23}^n = \mu_{31}^n = \mu_{32}^n = 0; && \varepsilon_{13}^n = \varepsilon_{23}^n = \varepsilon_{31}^n = \varepsilon_{32}^n = 0. \end{aligned} \tag{1.39}$$

When a wave propagates along the axis $0Z$ ($k_y = 0$), from Eq. (1.34), considering Eq. (1.39), we have

$$\begin{aligned} & a_{11}^n = a_{12}^n = a_{13}^n = 0, && a_{31}^n = a_{33}^n = a_{34}^n = 0, \\ & a_{14}^n = -\omega\mu_0\mu_{11}^n, && a_{32}^n = \omega\varepsilon_0\varepsilon_{11}^n, \\ & a_{21}^n = j\kappa_z \left(\frac{\varepsilon_a^n}{\varepsilon^n} + \frac{\mu_a^n}{\mu^n}\right), && a_{41}^n = \frac{1}{\omega\mu_0\mu^n}\kappa_z^2 - \omega\varepsilon_0\varepsilon_\perp^n, \end{aligned}$$

$$\begin{aligned}
&a_{23}^n = a_{24}^n = 0, && a_{42}^n = a_{44}^n = 0,\\
&a_{43}^n = -\frac{1}{\omega\varepsilon_0\varepsilon^n}\kappa_z^2 + \omega\mu_0\mu_\perp^n, && a_{43}^n = j\kappa_z\left(\frac{\varepsilon_a^n}{\varepsilon^n} + \frac{\mu_a^n}{\mu^n}\right),\\
&\mu_\perp^n = \mu^n - \frac{(\mu_a^n)^2}{\mu^n}, && \varepsilon_\perp^n = \varepsilon^n - \frac{(\varepsilon_a^n)^2}{\varepsilon^n},
\end{aligned} \tag{1.40}$$

and when along the axis $0Y$ $(k_z = 0)$

$$\begin{aligned}
&a_{11}^n = j\frac{\varepsilon_a^n}{\varepsilon^n}\kappa_y, && a_{31}^n = a_{33}^n = a_{34}^n = 0,\\
&a_{12}^n = a_{13}^n = 0, && a_{32}^n = -\frac{1}{\omega\mu_0\mu^n}\kappa_y^2 + \omega\varepsilon_0\varepsilon^n,\\
&a_{14}^n = \frac{1}{\omega\varepsilon_0\varepsilon^n}\kappa_y^2 - \omega\mu_0\mu_{11}^n, && a_{41}^n = \omega\varepsilon_0\varepsilon_\perp^n,\\
&a_{21}^n = a_{24}^n = 0, && a_{42}^n = a_{43}^n = 0,\\
&a_{22}^n = -j\frac{\mu_a^n}{\mu^n}\kappa_y, && a_{44}^n = -j\frac{\varepsilon_a^n}{\varepsilon^n}\kappa_y.\\
&a_{23}^n = \omega\mu_0\mu_\perp^n,
\end{aligned} \tag{1.41}$$

For the case of $\kappa_y = 0$, from Eqs. (1.38) and (1.40) it follows that $b_\kappa^n = d_\kappa^n = 0$ and the equation for eigenvalues looks like

$$\lambda_\kappa^4 + c_{y\kappa}^n\lambda_\kappa^2 + e_{y\kappa} = 0, \tag{1.42}$$

where

$$c_{y\kappa}^n = \kappa_0^2\left(\varepsilon_{||}^n\mu_\perp^n + \mu_{||}^n\varepsilon_\perp^n\right) - \left(\frac{\varepsilon_{||}^n}{\varepsilon^n} + \frac{\mu_{||}^n}{\mu^n}\right)\kappa_z^2,$$

$$e_{y\kappa}^n = -\kappa_0^2\kappa_z^2\varepsilon_{||}^n\mu_{||}^n\left(\frac{\varepsilon_a^n}{\varepsilon^n} + \frac{\mu_a^n}{\mu^n}\right), \quad \kappa_0 = \frac{\omega}{c}.$$

The solution of Eq. (1.42) is

$$\begin{aligned}
(\lambda_{1,2})^2 = {} & \frac{1}{2}\left[\kappa_0^2\left(\varepsilon_{||}^n\mu_\perp^n + \mu_{||}^n\varepsilon_\perp^n\right) - \left(\frac{\varepsilon_{||}^n}{\varepsilon^n} + \frac{\mu_{||}^n}{\mu^n}\right)\kappa_z^2\right]\\
& \pm\left\{\frac{1}{4}\left[\kappa_0^2\left(\varepsilon_{||}^n\mu_\perp^n + \mu_{||}^n\varepsilon_\perp^n\right) - \left(\frac{\varepsilon_{||}^n}{\varepsilon^n} + \frac{\mu_{||}^n}{\mu^n}\right)\kappa_z^2\right]^2\right.\\
& \left.+\kappa_0^2\kappa_z^2\varepsilon_{||}^n\mu_{||}^n\left(\frac{\varepsilon_a^n}{\varepsilon^n} + \frac{\mu_a^n}{\mu^n}\right)^2\right\}^{\frac{1}{2}},
\end{aligned} \tag{1.43}$$

which coincide with the eigenvalues [8][2]. For a ferrite magnetized up to saturation $(\mu_{||}^n = 1)$, from Eq. (1.43) we have

[2] This work gives an expression with evident misprints.

$$(\lambda_{1,2})^2 = \frac{1}{2}\left[\kappa_0^2\left(\varepsilon_{||}^n\mu_\perp^n + \varepsilon_\perp^n\right) - \left(\frac{\varepsilon_{||}^n}{\varepsilon^n} + \frac{1}{\mu^n}\right)\kappa_z^2\right]$$
$$\pm\left\{\frac{1}{4}\left[\kappa_0^2\left(\varepsilon_{||}^n\mu_\perp^n + \varepsilon_\perp^n\right) - \left(\frac{\varepsilon_{||}^n}{\varepsilon^n} + \frac{1}{\mu^n}\right)\kappa_z^2\right]^2 + \kappa_0^2\kappa_z^2\varepsilon_{||}^n\left(\frac{\varepsilon_a^n}{\varepsilon^n} + \frac{\mu_a^n}{\mu^n}\right)^2\right\}^{\frac{1}{2}}. \tag{1.44}$$

For the case of $k_z = 0$ the coefficients in Eq. (1.38) are

$$b_\kappa^n = j\kappa_y\frac{\mu_a^n}{\mu^n},$$

$$c_\kappa^n = \frac{\varepsilon_a^n}{\varepsilon^n}\left(\frac{\varepsilon_a^n}{\varepsilon^n} - \frac{\mu_a^n}{\mu^n}\right)\kappa_y^2 + \omega\mu_0\mu_\perp^n\left[\left(\omega\varepsilon^n\varepsilon_0 - \frac{\kappa_y^2}{\omega\mu_0\mu_\perp^n}\right) - \left(\frac{\kappa_y^2}{\omega\varepsilon_0\varepsilon} - \omega\mu_0\mu_\perp^n\right)\right],$$

$$d_\kappa^n = j\kappa_y^3\frac{(\varepsilon_a^n)^2}{\varepsilon^n}\frac{\mu_a^n}{\mu^n} - \omega\mu_0\mu_\perp^n\left(\omega\varepsilon^n\varepsilon_0 - \frac{\kappa_y^2}{\omega\mu_0\mu_\perp^n}\right)$$
$$\times j\left(\frac{\kappa_y^2}{\omega\varepsilon_0\varepsilon^n} - \omega\mu_0\mu_{||}^n\right)\omega\varepsilon_0\varepsilon_\perp^n\kappa_y\frac{\mu_a^n}{\mu^n},$$

$$e_\kappa^n = -\kappa_y^2\frac{(\varepsilon_a^n)^2}{\varepsilon^n}\left(\omega\varepsilon_0\varepsilon^n - \frac{\kappa_y^2}{\omega\mu_0\mu^n}\right)\omega\mu_0\mu_\perp^n$$
$$+\varepsilon_0^2\mu_0\omega^3(\varepsilon_\perp^n)^2\mu_\perp\omega\varepsilon_0\varepsilon^n - \frac{\kappa_y^2}{\omega\mu_0\mu^n}, \tag{1.45}$$

and the full Eq. (1.38) can be solved, for example, by degree reduction with the use of a cubic resolvent [474].

To analyze the processes of wave excitation in layered structures we shall single out two classes of longitudinally inhomogeneous waves, namely, *LE*, for which $E_x = 0$, and *LM*, for which $H_x = 0$. For these waves the matrixes $\overleftrightarrow{A}_{LE}$ and $\overleftrightarrow{A}_{LM}$ can be determined from Eqs. (1.29) to (1.32).

The components of the matrix $\overleftrightarrow{A}_{LE}$ are

$$a_{11}^n = \frac{\mu_{31}^n}{\mu_{11}^n}\kappa_z, \qquad a_{21}^n = -\frac{\mu_{21}^n}{\mu_{11}^n}\kappa_z,$$
$$a_{12}^n = -\frac{\mu_{31}^n}{\mu_{11}^n}\kappa_y, \qquad a_{22}^n = \frac{\mu_{21}^n}{\mu_{11}^n}\kappa_y,$$
$$a_{13}^n = \omega\mu_0\left(\frac{\mu_{12}^n\mu_{31}^n}{\mu_{11}^n} - \mu_{32}^n\right), \quad a_{23}^n = \omega\mu_0\left(\mu_{22}^n - \frac{\mu_{12}^n\mu_{21}^n}{\mu_{11}^n}\right), \tag{1.46}$$
$$a_{14}^n = \omega\mu_0\left(\frac{\mu_{13}^n\mu_{31}^n}{\mu_{11}^n} - \mu_{33}^n\right), \quad a_{24}^n = \omega\mu_0\left(\mu_{23}^n - \frac{\mu_{13}^n\mu_{21}^n}{\mu_{11}^n}\right),$$

$$a_{31}^n = \omega\varepsilon_0\varepsilon_{32}^n + \frac{\kappa_y\kappa_z}{\omega\mu_0\mu_{11}^n}, \qquad a_{41}^n = -\omega\varepsilon_0\varepsilon_{22}^n + \frac{\kappa_z^2}{\omega\mu_0\mu_{11}^n},$$

$$a_{32}^n = \omega\varepsilon_0\varepsilon_{33}^n - \frac{\kappa_y^2}{\omega\mu_0\mu_{11}^n}, \qquad a_{42}^n = -\omega\varepsilon_0\varepsilon_{23}^n - \frac{\kappa_y\kappa_z}{\omega\mu_0\mu_{11}^n},$$

$$a_{33}^n = \frac{\mu_{12}^n}{\mu_{11}^n}\kappa_y, \qquad a_{43}^n = \frac{\mu_{12}^n}{\mu_{11}^n}\kappa_z,$$

$$a_{34}^n = \frac{\mu_{13}^n}{\mu_{11}^n}\kappa_y, \qquad a_{44}^n = \frac{\mu_{13}^n}{\mu_{11}^n}\kappa_z.$$

The components of the matrix $\overleftrightarrow{A}_{LM}$ are

$$\begin{aligned}
a_{11}^n &= \frac{\varepsilon_{12}^n}{\varepsilon_{11}^n}\kappa_y, & a_{21}^n &= \frac{\varepsilon_{12}^n}{\varepsilon_{11}^n}\kappa_z,\\
a_{12}^n &= \frac{\varepsilon_{13}^n}{\varepsilon_{11}^n}\kappa_y, & a_{22}^n &= \frac{\varepsilon_{13}^n}{\varepsilon_{11}^n}\kappa_z,\\
a_{13}^n &= -\omega\mu_0\mu_{32}^n - \frac{\kappa_y\kappa_z}{\omega\varepsilon_0\varepsilon_{11}^n}, & a_{23}^n &= \omega\mu_0\mu_{22}^n - \frac{\kappa_z^2}{\omega\varepsilon_0\varepsilon_{11}^n},\\
a_{14}^n &= -\omega\mu_0\mu_{33}^n + \frac{\kappa_y^2}{\omega\varepsilon_0\varepsilon_{11}^n}, & a_{24}^n &= \omega\mu_0\mu_{23}^n + \frac{\kappa_y\kappa_z}{\omega\varepsilon_0\varepsilon_{11}^n},\\
a_{31}^n &= \omega\varepsilon_0\varepsilon_{32}^n - \frac{\varepsilon_{31}^n\varepsilon_{12}^n}{\varepsilon_{11}^n}, & a_{41}^n &= -\omega\varepsilon_0\left(\varepsilon_{22}^n - \frac{\varepsilon_{21}^n\varepsilon_{12}^n}{\varepsilon_{11}^n}\right),\\
a_{32}^n &= \omega\varepsilon_0\varepsilon_{33}^n - \frac{\varepsilon_{31}^n\varepsilon_{13}^n}{\varepsilon_{11}^n}, & a_{42}^n &= -\omega\varepsilon_0\left(\varepsilon_{23}^n - \frac{\varepsilon_{21}^n\varepsilon_{13}^n}{\varepsilon_{11}^n}\right),\\
a_{33}^n &= \frac{\varepsilon_{31}^n}{\varepsilon_{11}^n}\kappa_z, & a_{43}^n &= -\frac{\varepsilon_{21}^n}{\varepsilon_{11}^n}\kappa_z,\\
a_{34}^n &= -\frac{\varepsilon_{31}^n}{\varepsilon_{11}^n}\kappa_y, & a_{44}^n &= \frac{\varepsilon_{21}^n}{\varepsilon_{11}^n}\kappa_y.
\end{aligned} \tag{1.47}$$

The formulated boundary conditions for FIGF and the coupling matrix $\overleftrightarrow{T}_n(x)$ found for the tangential constituents of HF fields in the multilayer bigyrotropic structure allow the Green functions in screened structures to be determined.

1.2 Inhomogeneous Matrix Equation for a Screened Multilayer Bigyrotropic Structure: Green Functions and Power Flux

Let's consider a bigyrotropic multilayer structure screened by surfaces with an arbitrary anisotropic surface impedance at $x = 0$ and $x = x_p$.

To use a boundary condition for FIGF in the sources' plane of Eq. (1.18) $\overleftrightarrow{\widetilde{G}}_m(x_m)$ and $\overleftrightarrow{\widetilde{G}}_{m+1}(x_m)$ should be expressed by means of the known coupling matrixes $\overleftrightarrow{T}_n(x)$

through FIGF on the screening coverings $\overset{\leftrightarrow}{\tilde{\tilde{G}}}_1(0)$ and $\overset{\leftrightarrow}{\tilde{\tilde{G}}}_p(x_p)$, with

$$\begin{aligned}\overset{\leftrightarrow}{\tilde{\tilde{G}}}_1(0) &= \overset{\leftrightarrow}{T}_1(-h_1)\cdot\ \dots\ \cdot\overset{\leftrightarrow}{T}_{m-1}(-h_{m-1})\overset{\leftrightarrow}{\tilde{\tilde{G}}}_m(x_m),\\ \overset{\leftrightarrow}{\tilde{\tilde{G}}}_p(x_p) &= \overset{\leftrightarrow}{T}_p(h_p)\cdot\ldots\cdot\overset{\leftrightarrow}{T}_{m+1}(h_{m+1})\overset{\leftrightarrow}{\tilde{\tilde{G}}}_{m+1}(x_m).\end{aligned} \tag{1.48}$$

From Eq. (1.48) it follows that

$$\begin{aligned}\overset{\leftrightarrow}{\tilde{\tilde{G}}}_m(x_m) &= \left[\overset{\leftrightarrow}{T}_1(-h_1)\cdot\ \dots\ \cdot\overset{\leftrightarrow}{T}_{m-1}(-h_{m-1})\right]^{-1}\overset{\leftrightarrow}{\tilde{\tilde{G}}}_1(0),\\ \overset{\leftrightarrow}{\tilde{\tilde{G}}}_{m+1}(x_m) &= \left[\overset{\leftrightarrow}{T}_p(h_p)\cdot\ \dots\ \cdot\overset{\leftrightarrow}{\tilde{\tilde{T}}}_{m+1}(h_{m+1})\right]^{-1}\overset{\leftrightarrow}{\tilde{\tilde{G}}}_p(x_p).\end{aligned} \tag{1.49}$$

Then, substituting Eq. (1.49) into Eq. (1.18), we get an inhomogeneous matrix equation

$$\begin{aligned}&\left[\overset{\leftrightarrow}{T}_p(h_p)\cdot\ \dots\ \cdot\overset{\leftrightarrow}{T}_{m+1}(h_{m+1})\right]^{-1}\overset{\leftrightarrow}{\tilde{\tilde{G}}}_p(x_p)\\ &-\left[\overset{\leftrightarrow}{T}_1(-h_1)\cdot\ \dots\ \cdot\overset{\leftrightarrow}{T}_{m-1}(-h_{m-1})\right]^{-1}\overset{\leftrightarrow}{\tilde{\tilde{G}}}_1(0) = \overset{\leftrightarrow}{F}(y_s,z_s).\end{aligned} \tag{1.50}$$

Using in Eq. (1.50) the following designation for the products of coupling matrices in the corresponding half-spaces of the layered structure:

– At $x_m \le x \le x_p$,

$$\overset{\leftrightarrow}{B} = b_{ij} = \left[\overset{\leftrightarrow}{T}_p(h_p)\cdot\ldots\cdot\overset{\leftrightarrow}{T}_{m+1}(h_{m+1})\right]^{-1} \tag{1.51a}$$

– At $0 \le x < x_p$,

$$\overset{\leftrightarrow}{D} = d_{ij} = \left[\overset{\leftrightarrow}{T}_1(-h_1)\cdot\ldots\cdot\overset{\leftrightarrow}{T}_{m-1}(-h_{m-1})\right]^{-1}, \tag{1.51b}$$

we obtain an inhomogeneous matrix equation as

$$\overset{\leftrightarrow}{B}\overset{\leftrightarrow}{\tilde{\tilde{G}}}_p(x_p) + \overset{\leftrightarrow}{D}\overset{\leftrightarrow}{\tilde{\tilde{G}}}_1(0) = \overset{\leftrightarrow}{F}. \tag{1.52}$$

Equation (1.52) is equivalent to a set of four equations in eight *multipliers* of FIGF, namely, four ones for $\overset{\leftrightarrow}{\tilde{\tilde{G}}}_1(0)$ and other four ones for $\overset{\leftrightarrow}{\tilde{\tilde{G}}}_p(x_p)$. To solve set (1.52), we introduce a surface impedance [11] on $x = 0$ and $x = x_p$, determining the relation between the tangential components of HF fields

$$\left\|\begin{matrix}E_y\\E_z\end{matrix}\right\| = \overset{\leftrightarrow}{Z}(x)\left\|\begin{matrix}H_y\\H_z\end{matrix}\right\|, \tag{1.53}$$

$$\left\|\begin{matrix}H_y\\H_z\end{matrix}\right\| = \overset{\leftrightarrow}{Z}^{-1}(x)\left\|\begin{matrix}E_y\\E_z\end{matrix}\right\|, \tag{1.54}$$

where $\overleftrightarrow{Z} = \begin{Vmatrix} Z_{11} & Z_{12} \\ Z_{21} & Z_{22} \end{Vmatrix}$ is the surface impedance of the screening covering at $x = 0$, $x = x_p$, $\overleftrightarrow{Z}^{-1}$ the surface conductivity of the screening covering at $x = 0$, $x = x_p$, which is inverse of the matrix $\overleftrightarrow{Z}$.

Which of Eqs. (1.53) or (1.54) should be used depends on the type of the limiting surface at $x = 0$, $x = x_p$. If the surface is an electric wall with $E_{y,\,z}\ (x = 0,\ x = x_p) = 0$ and $H_{y,\,z}\ (x = 0,\ x = x_p) \neq 0$, it is Eq. (1.53), if a magnetic one, then Eq. (1.54).

If matrices of the form

$$\overleftrightarrow{\beta}_E = \begin{Vmatrix} 1 & 0 \\ 0 & 1 \\ 0 & 0 \\ 0 & 0 \end{Vmatrix}, \quad \overleftrightarrow{\beta}_H = \begin{Vmatrix} 0 & 0 \\ 0 & 0 \\ 1 & 0 \\ 0 & 1 \end{Vmatrix}, \tag{1.55}$$

are introduced, then $\overleftrightarrow{N}_n(x)$ can be rewritten as

$$\overleftrightarrow{N}_n(x) = \left[\overleftrightarrow{\beta}_E \overleftrightarrow{Z}(x) + \overleftrightarrow{\beta}_H\right] \cdot \begin{Vmatrix} H_y^n(x) \\ H_z^n(x) \end{Vmatrix}, \tag{1.56}$$

$$\overleftrightarrow{N}_n(x) = \left[\overleftrightarrow{\beta}_E + \overleftrightarrow{\beta}_H \overleftrightarrow{Z}^{-1}(x)\right] \cdot \begin{Vmatrix} E_y^n(x) \\ E_z^n(x) \end{Vmatrix}, \tag{1.57}$$

considering Eqs. (1.53) and (1.54), respectively.

Using relation (1.9), with due account of Eqs. (1.56) and (1.57), we shall write FIGF on the screening coverings as:

– At $x = 0$

$$\overleftrightarrow{\tilde{G}}_1(0, \kappa_y, \kappa_z; y_s, z_s) = \left[\overleftrightarrow{\beta}_E + \overleftrightarrow{\beta}_H \overleftrightarrow{Z}^{-1}(0)\right] \cdot \begin{Vmatrix} \tilde{G}_1^1(0) \\ \tilde{G}_2^1(0) \end{Vmatrix}, \tag{1.58a}$$

$$\overleftrightarrow{\tilde{G}}_1(0, \kappa_y, \kappa_z; y_s, z_s) = \left[\overleftrightarrow{\beta}_E \overleftrightarrow{Z}(0) + \overleftrightarrow{\beta}_H\right] \cdot \begin{Vmatrix} \tilde{G}_3^1(0) \\ \tilde{G}_4^1(0) \end{Vmatrix} \tag{1.58b}$$

– At $x = x_p$

$$\overleftrightarrow{\tilde{G}}_p(x_p, \kappa_y, \kappa_z; y_s, z_s) = \left[\overleftrightarrow{\beta}_E + \overleftrightarrow{\beta}_H \overleftrightarrow{Z}^{-1}(x_p)\right] \cdot \begin{Vmatrix} \tilde{G}_1^p(x_p) \\ \tilde{G}_2^p(x_p) \end{Vmatrix}, \tag{1.59a}$$

$$\overleftrightarrow{\tilde{G}}_p(x_p, \kappa_y, \kappa_z; y_s, z_s) = \left[\overleftrightarrow{\beta}_E \overleftrightarrow{Z}(x_p) + \overleftrightarrow{\beta}_H\right] \cdot \begin{Vmatrix} \tilde{G}_3^p(x_p) \\ \tilde{G}_4^p(x_p) \end{Vmatrix} \tag{1.59b}$$

Substituting a pair of relationships, for example, Eqs. (1.58b) and (1.59b), into Eq. (1.52), we derive the following form of the inhomogeneous matrix equations

$$\overleftrightarrow{B}\left[\overleftrightarrow{\beta}_E\overleftrightarrow{Z}(0)+\overleftrightarrow{\beta}_H\right]\cdot\begin{Vmatrix}\widetilde{G}_3^p(x_p)\\ \widetilde{G}_4^p(x_p)\end{Vmatrix}-\overleftrightarrow{D}\left[\overleftrightarrow{\beta}_E\;\overleftrightarrow{Z}(x_p)+\overleftrightarrow{\beta}_H\right]\cdot\begin{Vmatrix}\widetilde{G}_3^1(0)\\ \widetilde{G}_4^1(0)\end{Vmatrix}=\overleftrightarrow{F}. \tag{1.60}$$

Or, for Eqs. (1.58a) and (1.59a) we have

$$\overleftrightarrow{B}\left[\overleftrightarrow{\beta}_E-\overleftrightarrow{\beta}_H\overleftrightarrow{Z}^{-1}(0)\right]\cdot\begin{Vmatrix}\widetilde{G}_1^p(x_p)\\ \widetilde{G}_2^p(x_p)\end{Vmatrix}-\overleftrightarrow{D}\left[\overleftrightarrow{\beta}_E-\overleftrightarrow{\beta}_H\overleftrightarrow{Z}^{-1}(x_p)\right]\cdot\begin{Vmatrix}\widetilde{G}_1^1(0)\\ \widetilde{G}_2^1(0)\end{Vmatrix}=\overleftrightarrow{F}. \tag{1.61}$$

Each of relations (1.60) and (1.61) is equivalent to a set of four inhomogeneous algebraic equations in four unknown FIGF (two ones on the screen at $x=0$ and two at $x=x_p$).

For $\overleftrightarrow{Z}=0$ from Eq. (1.58b) a particular case of screening coverings follows in the form of electric walls, for which FIGF are:

$$\text{At } x=0\ \overleftrightarrow{\widetilde{G}}{}_1^E(0)=\begin{Vmatrix}0\\0\\G_3^{1E}(0)\\G_4^{1E}(0)\end{Vmatrix},\ \text{and at } x=x_p\quad \overleftrightarrow{\widetilde{G}}{}_p^E(x_p)=\begin{Vmatrix}0\\0\\G_3^{pE}(x_p)\\G_4^{pE}(x_p)\end{Vmatrix}. \tag{1.62}$$

For $\overleftrightarrow{Z}^{-1}=0$ from Eqs. (1.58a) and (1.59a) a special case of screening coverings follows in the form of magnetic walls, for which FIGF are:

$$\text{At } x=0\ \overleftrightarrow{\widetilde{G}}{}_1^M(0)=\begin{Vmatrix}G_1^{1M}(0)\\G_2^{1M}(0)\\0\\0\end{Vmatrix},\ \text{and at } x=x_p\quad \overleftrightarrow{\widetilde{G}}{}_p^M(x_p)=\begin{Vmatrix}G_1^{PM}(x_p)\\G_2^{PM}(x_p)\\0\\0\end{Vmatrix}. \tag{1.63}$$

The developed approach allows any types of screening coverings to be analyzed, and the top $(x=x_p)$ and bottom $(x=0)$ coverings can be identical or different, they may possess the properties of an ideal covering in the form of an electric $(\sigma_E=\infty)$ or magnetic $(\sigma_M=\infty)$ screen, coverings with finite conductivities σ_1 and σ_2, and also may show anisotropic properties of $\overleftrightarrow{\sigma}_1$ and $\overleftrightarrow{\sigma}_2$.

In particular, for the structures screened by magnetic walls (Eq. (1.63)), from Eq. (1.61) we have

$$\begin{cases} b_{11}\widetilde{G}_1^M p(x_p)+b_{12}\widetilde{G}_2^M p(x_p)+d_{11}\widetilde{G}_1^M 1(0)+d_{12}\widetilde{G}_2^M 1(0)=F_1,\\ b_{21}\widetilde{G}_1^M p(x_p)+b_{22}\widetilde{G}_2^M p(x_p)+d_{21}\widetilde{G}_1^M 1(0)+d_{22}\widetilde{G}_2^M 1(0)=F_2,\\ b_{31}\widetilde{G}_1^M p(x_p)+b_{32}\widetilde{G}_2^M p(x_p)+d_{31}\widetilde{G}_1^M 1(0)+d_{32}\widetilde{G}_2^M 1(0)=F_3,\\ b_{41}\widetilde{G}_1^M p(x_p)+b_{42}\widetilde{G}_2^M p(x_p)+d_{41}\widetilde{G}_1^M 1(0)+d_{42}\widetilde{G}_2^M 1(0)=F_4, \end{cases} \tag{1.64}$$

and for the structures screened by electric walls (Eq. (1.62)), from Eq. (1.60) we derive

$$\begin{cases} b_{13}\widetilde{G}_3^{Ep}(x_p)+b_{14}\widetilde{G}_4^{Ep}(x_p)+d_{13}\widetilde{G}_3^{E1}(0)+d_{14}\widetilde{G}_4^{E1}(0)=F_1, \\ b_{23}\widetilde{G}_3^{Ep}(x_p)+b_{24}\widetilde{G}_4^{Ep}(x_p)+d_{23}\widetilde{G}_3^{E1}(0)+d_{24}\widetilde{G}_4^{E1}(0)=F_2, \\ b_{33}\widetilde{G}_3^{Ep}(x_p)+b_{34}\widetilde{G}_4^{Ep}(x_p)+d_{33}\widetilde{G}_3^{E1}(0)+d_{34}\widetilde{G}_4^{E1}(0)=F_3, \\ b_{43}\widetilde{G}_3^{Ep}(x_p)+b_{44}\widetilde{G}_4^{Ep}(x_p)+d_{43}\widetilde{G}_3^{E1}(0)+d_{44}\widetilde{G}_4^{E1}(0)=F_4. \end{cases} \tag{1.65}$$

Sets (1.64) and (1.65) have their determinants

$$\Delta^M = \begin{Vmatrix} b_{11} & b_{12} & d_{11} & d_{12} \\ b_{21} & b_{22} & d_{21} & d_{22} \\ b_{31} & b_{32} & d_{31} & d_{32} \\ b_{41} & b_{42} & d_{41} & d_{42} \end{Vmatrix}, \tag{1.66}$$

$$\Delta^E = \begin{Vmatrix} b_{13} & b_{14} & d_{13} & d_{14} \\ b_{23} & b_{24} & d_{23} & d_{24} \\ b_{33} & b_{34} & d_{33} & d_{34} \\ b_{43} & b_{44} & d_{43} & d_{44} \end{Vmatrix}. \tag{1.67}$$

For uniqueness of the solution of Eqs. (1.66) and (1.67) it is necessary that $\Delta^M = 0$ and $\Delta^E = 0$, which determines the dispersions of the characteristic waves in multilayer bigyrotropic structures screened by magnetic or electric walls.

Resolving sets (1.66) and (1.67) by Cramer's rule [458], we find components $\overset{\leftrightarrow}{\widetilde{G}}{}_1^{M,E}(0)$ and $\overset{\leftrightarrow}{\widetilde{G}}{}_1^{M,E}(x_p)$. For structures limited by magnetic or electric walls:

- At $x = 0$,

$$\overset{\leftrightarrow}{\widetilde{G}}{}_1^{M,E}(0) = \frac{1}{\Delta^{M,E}}\left(F_M \overset{\leftrightarrow}{\widetilde{q}}{}_1^{M,E} + F_E \overset{\leftrightarrow}{\widetilde{\eta}}{}_1^{M,E}\right) \tag{1.68}$$

- At $x = x_p$,

$$\overset{\leftrightarrow}{\widetilde{G}}{}_1^{M,E}(x_p) = \frac{1}{\Delta^{M,E}}\left(F_M \overset{\leftrightarrow}{\widetilde{q}}{}_p^{M,E} + F_E \overset{\leftrightarrow}{\widetilde{\eta}}{}_p^{M,E}\right), \tag{1.69}$$

$$F_M = \frac{1}{2\pi} e^{j(\kappa_y y_{sM} + \kappa_z z_{sM})},$$

$$F_E = \frac{1}{2\pi} e^{j(\kappa_y y_{sE} + \kappa_z z_{sE})},$$

for magnetic walls

$$\widetilde{q}_1^M = \begin{Vmatrix} \{\Delta^M\}_{11} + \{\Delta^M\}_{21} \\ \{\Delta^M\}_{12} + \{\Delta^M\}_{22} \\ 0 \\ 0 \end{Vmatrix}, \quad \widetilde{\eta}_1^M = \begin{Vmatrix} \{\Delta^M\}_{31} + \{\Delta^M\}_{41} \\ \{\Delta^M\}_{32} + \{\Delta^M\}_{42} \\ 0 \\ 0 \end{Vmatrix} \tag{1.70}$$

$$\tilde{q}_p^M = \begin{Vmatrix} \{\Delta^M\}_{13} + \{\Delta^M\}_{23} \\ \{\Delta^M\}_{14} + \{\Delta^M\}_{24} \\ 0 \\ 0 \end{Vmatrix}, \quad \tilde{\eta}_p^M = \begin{Vmatrix} \{\Delta^M\}_{33} + \{\Delta^M\}_{43} \\ \{\Delta^M\}_{34} + \{\Delta^M\}_{44} \\ 0 \\ 0 \end{Vmatrix}$$

$$\tilde{q}_1^E = \begin{Vmatrix} 0 \\ 0 \\ \{\Delta^E\}_{11} + \{\Delta^E\}_{12} \\ \{\Delta^E\}_{14} + \{\Delta^E\}_{22} \end{Vmatrix}, \quad \tilde{\eta}_1^E = \begin{Vmatrix} 0 \\ 0 \\ \{\Delta^E\}_{31} + \{\Delta^E\}_{41} \\ \{\Delta^E\}_{32} + \{\Delta^E\}_{42} \end{Vmatrix} \tag{1.71}$$

$$\tilde{q}_p^E = \begin{Vmatrix} 0 \\ 0 \\ \{\Delta^E\}_{13} + \{\Delta^E\}_{23} \\ \{\Delta^E\}_{14} + \{\Delta^E\}_{24} \end{Vmatrix}, \quad \tilde{\eta}_p^E = \begin{Vmatrix} 0 \\ 0 \\ \{\Delta^E\}_{33} + \{\Delta^E\}_{43} \\ \{\Delta^E\}_{34} + \{\Delta^E\}_{44} \end{Vmatrix}.$$

$\{\Delta^{M,E}\}_{i,j}$ are the cofactors derived from row i and column j of Eqs. (1.64) and (1.65), $i,j = 1,2,3,4$.

If Green's functions are determined on either electric or magnetic screen, then in an arbitrary layer $x = x_n$ of the FIGF structures:

– For $n \leq m$

$$\overset{\leftrightarrow}{\tilde{G}}{}_n^{M,E}(x_n) = \overset{\leftrightarrow}{T}_n(x - x_{n-1}) \cdot \ldots \cdot \overset{\leftrightarrow}{T}_1(h_1) \frac{F_M \overset{\leftrightarrow}{q}{}_1^{M,E} + F_E \overset{\leftrightarrow}{\eta}{}_1^{M,E}}{\Delta^{M,E}} \tag{1.72a}$$

– For $n > m$

$$\overset{\leftrightarrow}{\tilde{G}}{}_n^{M,E}(x_n) = \overset{\leftrightarrow}{T}_n(x - x_{n+1}) \cdot \ldots \cdot \overset{\leftrightarrow}{T}_p(-h_p) \frac{F_M \overset{\leftrightarrow}{q}{}_p^{M,E} + F_E \overset{\leftrightarrow}{\eta}{}_p^{M,E}}{\Delta^{M,E}} \tag{1.72b}$$

Using Eqs. (1.8a), (1.36), and (1.72a,), we find for the constituents of HF fields:

– In half-space I under the sources $(n \leq m)$

$$\overset{\leftrightarrow}{\tilde{N}}{}_n^{M,E}(x, \kappa_y, \kappa_z) = \overset{\leftrightarrow}{T}_n(x - x_{n-1}) \cdot \ldots \cdot \overset{\leftrightarrow}{T}_1(h_1) \frac{F_{0M} \overset{\leftrightarrow}{q}{}_1^{M,E} + F_{0E} \overset{\leftrightarrow}{h}{}_1^{M,E}}{\Delta^{M,E}} \tag{1.73a}$$

– In half-space II above the sources $(n > m)$

$$\overset{\leftrightarrow}{\tilde{N}}{}^{M,E}{}_n(x, \kappa_y, \kappa_z) = \overset{\leftrightarrow}{T}_n(x - x_{n-1}) \cdot \ldots \cdot \overset{\leftrightarrow}{T}_p(-h_p) \frac{F_{0M} \overset{\leftrightarrow}{q}{}_p^{M,E} + F_{0E} \overset{\leftrightarrow}{h}{}_p^{M,E}}{\Delta^{M,E}}, \tag{1.73b}$$

where

$$F_{0M} = \iint_S F_M(y_{sM}, z_{sM}) \mathrm{d}y_s \mathrm{d}z_s, \quad F_{0E} = \iint_S F_E(y_{sE}, z_{sE}) \mathrm{d}y_s \mathrm{d}z_s.$$

After taking the inverse Fourier transform of Eqs. (1.73a) and (1.73b) we have the column matrices of the tangential constituents of HF fields:

– In half-space I, $n \leq m$

$$\begin{aligned} \overset{\leftrightarrow}{N}{}_n^{M,E}(x,y,z) = &\frac{1}{2\pi} \iint \overset{\leftrightarrow}{T}_n(x - x_{n-1}) \cdot \ \ldots \ \cdot \overset{\leftrightarrow}{T}_1(h_1) \\ &\times \frac{F_{0M} \overset{\leftrightarrow}{q}{}_1^{M,E} - F_{0E} \overset{\leftrightarrow}{\eta}{}_p^{M,E}}{\Delta^{M,E}} \cdot e^{-j(\kappa_y y + \kappa_z z)} \mathrm{d}\kappa_y \mathrm{d}\kappa_z \end{aligned} \tag{1.74a}$$

– In half-space II, $n > m$

$$\begin{aligned} \overset{\leftrightarrow}{N}{}_n^{M,E}(x,y,z) = &\frac{1}{2\pi} \iint \overset{\leftrightarrow}{T}_n(x - x_{n+1}) \cdot \ldots \cdot \overset{\leftrightarrow}{T}_p(-h_p) \\ &\times \frac{F_{0M} \overset{\leftrightarrow}{q}{}_p^{M,E} - F_{0E} \overset{\leftrightarrow}{\eta}{}_p^{M,E}}{\Delta^{M,E}} \cdot e^{-j(\kappa_y y + \kappa_z z)} \mathrm{d}\kappa_y \mathrm{d}\kappa_z. \end{aligned} \tag{1.74b}$$

For application of asymptotic methods to calculate the HF fields of the excited waves we shall pass into cylindrical coordinates with the axis along $0X$. Introducing the replacements

$$\begin{aligned} y &= R\cos\varphi, \quad \kappa_y = \kappa\cos\theta, \\ z &= R\sin\varphi, \quad \kappa_z = \kappa\sin\theta, \end{aligned} \tag{1.75}$$

we get in the cylindrical coordinates:

– In half-space I, $n \leq m$

$$\begin{aligned} \overset{\leftrightarrow}{N}{}_n^{E,M}(R,\varphi,x) = &\frac{1}{2\pi} \int_0^{2\pi} \mathrm{d}\theta \int_0^{\infty} \kappa \mathrm{d}\kappa \cdot \overset{\leftrightarrow}{T}_n(x - x_n) \cdot \ldots \cdot \overset{\leftrightarrow}{T}_1(h_1) \\ &\times \frac{F_{0M} \overset{\leftrightarrow}{q}{}_1^{M,E} - F_{0E} \overset{\leftrightarrow}{\eta}{}_1^{M,E}}{\Delta^{M,E}} \cdot e^{-j\kappa R\cos(\theta - \varphi)} \end{aligned} \tag{1.76a}$$

– In half-space II, $n > m$

$$\begin{aligned} \overset{\leftrightarrow}{N}{}_n^{E,M}(R,\varphi,x) = &\frac{1}{2\pi} \int_0^{2\pi} \mathrm{d}\theta \int_0^{\infty} \kappa \mathrm{d}\kappa \cdot \overset{\leftrightarrow}{T}_n(x - x_n) \cdot \ldots \cdot \overset{\leftrightarrow}{T}_p(-h_p) \\ &\times \frac{F_{0M} \overset{\leftrightarrow}{q}{}_p^{M,E} - F_{0E} \overset{\leftrightarrow}{\eta}{}_p^{M,E}}{\Delta^{M,E}} \cdot e^{-j\kappa R\cos(\theta - \varphi)}, \end{aligned} \tag{1.76b}$$

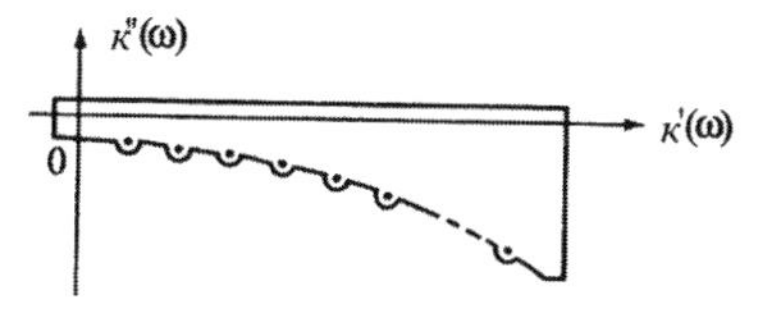

Fig. 1.2 The integration contour

where φ is the angle defining the direction $\bar{R}$ from the axis $0Y$; θ the angle defining the direction $\bar{\kappa}$ from the axis $0Y$.

Let's note that the column matrix $\overleftrightarrow{N}_n^{E,M}(R,\varphi,x)$ in the cylindrical coordinates looks as

$$\overleftrightarrow{N}_n^{E,M}(R,\varphi,x) = \begin{Vmatrix} E_y^n(R,\varphi,x) \\ E_z^n(R,\varphi,x) \\ H_y^n(R,\varphi,x) \\ H_z^n(R,\varphi,x) \end{Vmatrix}. \tag{1.77}$$

In the right-hand sides of Eqs. (1.76a) and (1.76b) we have l poles of first order by the wave number κ. The integration contour of this expression looks like Fig. 1.2, considering the dependence $\kappa''(\omega)$. Using the theorem of residues [458], we have:

– In half-space I, $n \le m$

$$\begin{aligned} \overleftrightarrow{N}_n^{E,M}(R,\varphi,x) &= j\sum_{j=1}^{l}\int_0^{2\pi} \mathrm{d}\theta \cdot \kappa_j \overleftrightarrow{T}_n(x-x_{n-1})\cdot\,\ldots \\ &\times \overleftrightarrow{T}_1(h_1)\frac{F_{0M}\overleftrightarrow{q}_1^{M,E} + F_{0E}\overleftrightarrow{\eta}_1^{M,E}}{\Delta_1^{M,E}} \cdot e^{-j\kappa_j\,R\,\cos(\theta-\varphi)} \end{aligned} \tag{1.78a}$$

– In half-space II, $n > m$

$$\begin{aligned} \overleftrightarrow{N}_n^{E,M}(R,\varphi,x) &= j\sum_{j=1}^{l}\int_0^{2\pi} \mathrm{d}\theta\,\kappa_j \overleftrightarrow{T}_n(x-x_{n-1})\cdot\ldots \\ &\times \overleftrightarrow{T}_p(-h_p)\frac{F_{0M}\overleftrightarrow{q}_p^{M,E} + F_{0E}\overleftrightarrow{\eta}_p^{M,E}}{\Delta_1^{M,E}} \cdot e^{-j\kappa_j R\cos(\theta-\varphi)}, \end{aligned} \tag{1.78b}$$

where $\Delta_1^{M,E} = \dfrac{\mathrm{d}\Delta^{M,E}}{\mathrm{d}\kappa_j}$, $j = 1,\ 2,\ \ldots,\ l$ – is the number of poles.

The expressions between square brackets of Eqs. (1.78a) and (1.78b) have no singularities by θ and are restricted within the limits of integration. This allows the asymptotic stationary-phase method [459] to be used for their calculation. The extremum of the phase $\Phi(\theta) = \kappa_j(\theta)cos(\theta-\varphi)$ is determined from the equation

$$\frac{\mathrm{d}\kappa_j(\theta)}{\mathrm{d}\theta}\cos(\theta-\varphi) - \kappa_j(\theta)\sin(\theta-\varphi) = 0. \tag{1.79}$$

If condition (1.79) is satisfied at $\theta = \theta_S$ and $\Phi''(\theta_s) \neq 0$, then Eqs. (1.78a) and (1.78b) have [459] their asymptotic value:

– In half-space I, $n \leq m$

$$\overleftrightarrow{N}_n^{M,E}(R,\varphi,x) = j\sum_{j=1}^{l} \kappa_j \overleftrightarrow{T}_n(x - x_{n-1}) \cdot \ldots$$
$$\times \overleftrightarrow{T}_1(h_1)\frac{F_{0M}\overleftrightarrow{q}_1^{M,E} - F_{0E}\overleftrightarrow{\eta}_1^{M,E}}{\Delta_1^{M,E}}\sqrt{\frac{2\pi}{jR\Phi''(\theta_s)}} \cdot e^{-j\,R\,\Phi(\theta_s(\varphi))} \tag{1.80a}$$

– In half-space II, $n > m$

$$\overleftrightarrow{N}_n^{M,E}(R,\varphi,x) = j\sum_{j=1}^{l} \kappa_j \overleftrightarrow{T}_n(x - x_{n+1}) \cdot \ldots$$
$$\times \overleftrightarrow{T}_p(-h_p)\frac{F_{0M}\overleftrightarrow{q}_p^{M,E} - F_{0E}\overleftrightarrow{\eta}_p^{M,E}}{\Delta_1^{M,E}}\sqrt{\frac{2\pi}{jR\Phi''(\theta_s)}} \cdot e^{-j\,R\,\Phi(\theta_s(\varphi))} \tag{1.80b}$$

The derived expressions (1.80a) and (1.80b) represent the components of HF fields of the excited wave in a multilayer structure in the cylindrical coordinates.

Let's switch over to the Cartesian coordinates for Eqs. (1.80a) and (1.80b) by means of the following relations

$$R = \sqrt{y^2 + z^2},$$
$$\varphi = arctg\frac{z}{y}.$$

Then we get:

– In half-space I, $n \leq m$

$$\overleftrightarrow{N}_n^{M,E}(R,\varphi,x) = j\sum_{j=1}^{l} \kappa_j \overleftrightarrow{T}_n(x - x_{n-1}) \cdot \ldots$$
$$\times \overleftrightarrow{T}_1(h_1)\frac{F_{0M}\overleftrightarrow{q}_1^{M,E} - F_{0E}\overleftrightarrow{\eta}_1^{M,E}}{\Delta_1^{M,E}}\sqrt{\frac{2\pi}{jR\Phi''(\theta_s)}} \cdot e^{-j\,R\,\Phi(\theta_s(\varphi))} \tag{1.81a}$$

– In half-space II, $n > m$

$$\overleftrightarrow{N}_n^{M,E}(R,\varphi,x) = j\sum_{j=1}^{l} \kappa_j \overleftrightarrow{T}_n(x - x_{n+1}) \cdot \ldots$$
$$\times \overleftrightarrow{T}_p(-h_p)\frac{F_{0M}\overleftrightarrow{q}_p^{M,E} - F_{0E}\overleftrightarrow{\eta}_p^{M,E}}{\Delta_1^{M,E}}\sqrt{\frac{2\pi}{jR\Phi''(\theta_s)}} \cdot e^{-j\,R\,\Phi(\theta_s(\varphi))}. \tag{1.81b}$$

These relations allow the tangential constituents of HF electric and magnetic fields to be determined in the corresponding half-space converters with a parallel orientation of the multilayer bigyrotropic structure screened by an arbitrary impedance surfaces, in the approximation of a specified surface current of excitation. Analysis of the properties of such converters in the near and far zones of radiation, considering most essential factors of the millimeter range – (losses in layers, arrangements of the external screens, transverse gradients of electric and magnetic parameters) is possible.

The reverse influence of HF fields of the waves excited in the structure on the extraneous sources of current and field will be considered at formulation of boundary conditions for the normal components of the electric and magnetic fields on the surfaces of the magnetic and electric currents, respectively. The electric current is localized within the limits of the area $S_E(j_{Ey},\ j_{Ez})$ on a perfectly conductive covering, and the magnetic current is within the limits of the metal-free area with $S_M(j_M Y,\ j_M Z)$ at $x = x_m$.

On the surface S_E the normal constituents of the resultant HF magnetic field are $H_{xm} = 0$, and on S_M the normal constituent of the resultant HF electric field is $E_{xm} = 0$.

The constituents H_{xm} and E_{xm} are determined by both the fields of the exciting current H_{xm}^T and E_{xm}^T, and those of the exciting wave H_{xm}^B and E_{xm}^B. On the surface of electric current

$$H_{xm}^T + H_{xm}^B = 0,$$

and on the surface of magnetic current

$$E_{xm}^T + E_{xm}^B = 0.$$

For the Fourier transforms these conditions are

$$\widetilde{H}_{xm}^B + \widetilde{H}_{xm}^T = 0, \tag{1.82}$$

$$\widetilde{E}_{xm}^B + \widetilde{E}_{xm}^T = 0. \tag{1.83}$$

If we introduce matrices of the form

$$S_E^m = \left\| -\frac{\varepsilon_{12}^m}{\varepsilon_{11}^m} \ -\frac{\varepsilon_{13}^m}{\varepsilon_{11}^m} \ -\frac{\kappa_z}{\omega\varepsilon_0\varepsilon_{11}^m} \ -\frac{\kappa_y}{\omega\varepsilon_0\varepsilon_{11}^m} \right\|,$$

$$S_H^m = \left\| \frac{\kappa_y}{\omega\mu_0\mu_{11}^m} \ -\frac{\kappa_z}{\omega\mu_0\mu_{11}^m} \ -\frac{\mu_{12}^m}{\mu_{11}^m} \ -\frac{\mu_{13}^m}{\mu_{11}^m} \right\|,$$

then the resultant constituents of the Fourier transforms of the fields are

$$\widetilde{H}_{xm} = \overleftrightarrow{S}_H^m \overleftrightarrow{\widetilde{N}}_m, \tag{1.84}$$

$$\widetilde{E}_{xm} = \overleftrightarrow{S}_E^m \overleftrightarrow{\widetilde{N}}_m. \tag{1.85}$$

Then on the surface of the source we have the conditions:

- For $\bar{j}_E(y,z) \in S_E$, $\overleftrightarrow{S}_H^m(\overleftrightarrow{N}_m^B + \overleftrightarrow{N}_m^T) = 0$
- For $\bar{j}_M(y,z) \in S_M$, $\overleftrightarrow{S}_E^m(\overleftrightarrow{N}_m^B + \overleftrightarrow{N}_m^T) = 0$

In the generalized form these conditions are

$$\overleftrightarrow{L}_s(\overleftrightarrow{N}_m^B + \overleftrightarrow{N}_m^T) = 0,$$

where $\overleftrightarrow{L}_s = \overleftrightarrow{S}_H^m$ for the electric current, $\overleftrightarrow{L}_s = \overleftrightarrow{S}_E^m$ for the magnetic current.

The Fourier transforms of the tangential constituents of HF fields excited in the structure by a specified extraneous current $\overleftrightarrow{J}$, on the surface of the source at $x = x_m$, according to Eqs. (1.73), looks as

$$\begin{aligned}\overleftrightarrow{\tilde{N}}_m^{M,E}(x,\kappa_y,\kappa_z) =& \overleftrightarrow{T}_m(-h_m)\cdot\ldots\cdot\overleftrightarrow{T}_p(-h_p)\frac{1}{\Delta^{M,E}}\frac{1}{2\pi}\iint e^{j(\kappa_y y_{sM}+\kappa_z y_{sM})} \\ &\times \overleftrightarrow{q}_p^{M,E} + e^{j(\kappa_y y_{sM}+\kappa_z y_{sM})}\cdot\overleftrightarrow{\eta}_p^{M,E}\overleftrightarrow{I}\overleftrightarrow{I}s(y_s,z_s)\mathrm{d}y_s\mathrm{d}z_s.\end{aligned} \tag{1.86}$$

From the inhomogeneous set of Maxwell's equations

$$\begin{cases}[\nabla E_n] = -j\omega\mu_0\overleftrightarrow{\mu}_n H_n - j_M^T \\ [\nabla H_n] = j\omega\varepsilon_0\overleftrightarrow{\varepsilon}_n E_n + j_E^T\end{cases},$$

where $j_{M,E}^T$ is the extraneous magnetic and electric currents, we obtain a matrix equation for the Fourier transforms of the tangential constituents of the fields of exciting current

$$\overleftrightarrow{\tilde{N}}_m^T(x_m,\kappa_y,\kappa_z) = \left[f(\kappa_y,\kappa_z)\overleftrightarrow{I} - j\overleftrightarrow{A}_m(\kappa_y,\kappa_z)\right]^{-1}\overleftrightarrow{\tilde{I}}s(\kappa_y,\kappa_z)e^{f(\kappa_y,\kappa_z)x_m},$$

where $f(\kappa_y,\kappa_z)$ is – the function determining the transverse wave number in the structure.

Let's introduce a function of the form

$$\overleftrightarrow{\tilde{J}}(\kappa_y,\kappa_z) = \left[f(\kappa_y,\kappa_z)\overleftrightarrow{I} - j\overleftrightarrow{A}_m(\kappa_y,\kappa_z)\right]^{-1}\overleftrightarrow{\tilde{I}}s(\kappa_y,\kappa_z). \tag{1.87}$$

Then the matrix equation is

$$\overleftrightarrow{\tilde{N}}_m^T(x_m,\kappa_y,\kappa_z) = \overleftrightarrow{\tilde{J}}(\kappa_y,\kappa_z),$$

from which

$$\overleftrightarrow{N}_m(x_m,y,z) = \overleftrightarrow{J}(y,z)$$

follows.

We introduce Green's function for the fields of exciting current

$$\overleftrightarrow{N}_m^T(x_m,y,z) = \iint\limits_S \overleftrightarrow{G}^T(x_m,y,z;y^T,z^T)\overleftrightarrow{J}(y^T,z^T)\mathrm{d}y^T\mathrm{d}z^T. \tag{1.88}$$

From the last relation with the use of the δ function we derive

$$\overleftrightarrow{G}^T(x_m,y,z;\; y^T,z^T) = \delta(y-y^T)\delta(z-z^T).$$

Taking the Fourier transforms, we get

$$\begin{aligned}&\frac{1}{2\pi}\iint\limits_{-\infty}^{\infty} \overleftrightarrow{\tilde{G}}^T(x_m,\kappa_y,\kappa_z,y^T,z^T)e^{-j(\kappa_y y+\kappa_z z)}\mathrm{d}\kappa_y\mathrm{d}\kappa_z\\ &= \frac{1}{2\pi}\iint\limits_{-\infty}^{\infty} e^{-j[\kappa_y(y-y^T)+\kappa_z(z-z^T)]}\mathrm{d}\kappa_y\mathrm{d}\kappa_z.\end{aligned} \tag{1.89}$$

From the last relation it follows that

$$\overleftrightarrow{\tilde{G}}^T(x_m,\kappa_y,\kappa_z;\; y^T,z^T) = \frac{1}{2\pi}e^{j(\kappa_y y^T+\kappa_z z^T)}. \tag{1.90}$$

Then Eq. (1.88), considering Eq. (1.87), can be rewritten as

$$\begin{aligned}\overleftrightarrow{\tilde{N}}_m^T(x_m,\kappa_y,\kappa_z) = \iint \frac{1}{2\pi}e^{j(\kappa_y y^T+\,\kappa_z z^T)}e^{f(\kappa_y,\kappa_z)x_m}\\ \times\left[f(\kappa_y,\kappa_z)\overleftrightarrow{I} - j\overleftrightarrow{A}_m(\kappa_y,\kappa_z)\right]^{-1}\overleftrightarrow{I}s(y_s,z_s)\mathrm{d}y_s\mathrm{d}z_s.\end{aligned}$$

For extraneous electric and magnetic currents the matrices are

$$\overleftrightarrow{I}_s = \begin{Vmatrix} 0\\ 0\\ j_{Ez}\\ j_{Ey}\end{Vmatrix},\quad \overleftrightarrow{I}_s = \begin{Vmatrix} j_{Mz}\\ j_{My}\\ 0\\ 0\end{Vmatrix}.$$

Then the integral equation for the converter with electric current (for example, MSL of excitation) is

$$\begin{aligned}&\overleftrightarrow{S}_H^m\left\{\iint\limits_{s_e}\frac{1}{2\pi}e^{j(\kappa_y y_{sM}+\kappa_z z_{sM})}\left[\overleftrightarrow{f}(\kappa_y,\kappa_z)\overleftrightarrow{I} - j\overleftrightarrow{A}_m(\kappa_y,\kappa_z)\right]^{-1}\begin{Vmatrix} 0\\ 0\\ 0\\ j_{Ez}\\ j_{Ey}\end{Vmatrix}\right.\\ &\times e^{f(\kappa_y,\kappa_z)x_m}\mathrm{d}y_{sM}\mathrm{d}z_{sM} + \overleftrightarrow{T}_m(-h_m)\cdot\;\ldots\;\cdot\overleftrightarrow{T}_p(-h_p)\cdot\frac{1}{\Delta^{M,E}}\frac{1}{2\pi}\\ &\times\left.\iint\limits_{s_E} e^{j(\kappa_y y_{sE}+\kappa_z z_{sE})}q_p^{M,E}\overleftrightarrow{I}\begin{Vmatrix} 0\\ 0\\ j_{Ez}\\ j_{Ey}\end{Vmatrix}\mathrm{d}y_{sM}\mathrm{d}z_{sM}\right\} = 0.\end{aligned} \tag{1.91a}$$

Similarly, the integral equation for the converter with magnetic current (for example, TSL of excitation) is

$$\overleftrightarrow{S}_E^m \left\{ \iint\limits_{s_E} \frac{1}{2\pi} e^{j(\kappa_y y_{sM} + \kappa_z z_{sM})} \left[\overleftrightarrow{f}(\kappa_y, \kappa_z) \overleftrightarrow{I} - j \overleftrightarrow{A}_m(\kappa_y, \kappa_z) \right]^{-1} \left\| \begin{matrix} j_{Mz} \\ j_{My} \\ 0 \\ 0 \end{matrix} \right\| \right.$$
$$\times\, e^{f(\kappa_y,\kappa_z)x_m} \mathrm{d}y_{sM} \mathrm{d}z_{sM} + \overleftrightarrow{T}_m(-h_m) \cdot \ \ldots \ \cdot \overleftrightarrow{T}_p(-h_p) \cdot \frac{1}{\Delta^{M,E}} \frac{1}{2\pi} \tag{1.91b}$$
$$\left. \times \iint\limits_{s_M} e^{j(\kappa_y y_{sM} + \kappa_z z_{sM})} q_p^{M,E} \overleftrightarrow{I} \left\| \begin{matrix} j_{Mz} \\ j_{My} \\ 0 \\ 0 \end{matrix} \right\| \mathrm{d}y_{sM} \mathrm{d}z_{sM} \right\} = 0.$$

Possible procedures to solve such an equation will be examined in Section 1.3.

Let's examine the power fluxs of the excited waves transferred in the structure. The power flux density along the axes $0Y$ and $0Z$ are

$$\begin{aligned} p_{yn} &= E_{xn} H_{zn}^* - E_{zn} H_{xn}^*, \\ p_{zn} &= E_{yn} H_{xn}^* - E_{xn} H_{yn}^* \,. \end{aligned} \tag{1.92}$$

To make use of the matrix form we introduce

$$\overleftrightarrow{S}_H^n = \left\| \begin{matrix} 0 & \dfrac{\kappa_0}{\omega \mu_0 \mu_{11}^n} & -\dfrac{\mu_{12}^n}{\mu_{11}^n} & -\dfrac{\mu_{13}^n}{\mu_{11}^n} \end{matrix} \right\|, \tag{1.93}$$

$$\overleftrightarrow{S}_E^n = \left\| \begin{matrix} -\dfrac{\varepsilon_{12}^n}{\varepsilon_{11}^n} & -\dfrac{\varepsilon_{13}^n}{\varepsilon_{11}^n} & 0 & -\dfrac{\kappa_0}{\omega \varepsilon_0 \varepsilon_{11}^n} \end{matrix} \right\|, \tag{1.94}$$

$$\overleftrightarrow{\beta}_{E_y} = \| 1 \quad 0 \quad 0 \quad 0 \|, \tag{1.95}$$

$$\overleftrightarrow{\beta}_{E_z} = \| 0 \quad 1 \quad 0 \quad 0 \|, \tag{1.96}$$

$$\overleftrightarrow{\beta}_{H_y} = \| 0 \quad 0 \quad 1 \quad 0 \|, \tag{1.97}$$

$$\overleftrightarrow{\beta}_{H_z} = \| 0 \quad 0 \quad 0 \quad 1 \|. \tag{1.98}$$

Then, using Eqs. (1.93) and (1.98), we rewrite Eq. (1.92) in the matrix form

$$P_{yn} = \overleftrightarrow{S}_E^n \overleftrightarrow{N}_n^{M,E} \overleftrightarrow{\beta}_{H_z} \left(\overleftrightarrow{N}_n^{M,E} \right)^* - \overleftrightarrow{\beta}_{E_z} \overleftrightarrow{N}_n^{M,E} \left(\overleftrightarrow{S}_H^n \overleftrightarrow{N}_n^{M,E} \right)^*, \tag{1.99a}$$

$$P_{zn} = \overleftrightarrow{\beta}_{E_y} \overleftrightarrow{N}_n^{M,E} \left(\overleftrightarrow{S}_H^n \overleftrightarrow{N}_n^{M,E} \right)^* - \overleftrightarrow{S}_E^n \overleftrightarrow{N}_n^{M,E} \left(\overleftrightarrow{\beta}_{H_y} \overleftrightarrow{N}_n^{M,E} \right)^*. \tag{1.99b}$$

To determine power fluxes it is necessary to integrate Eqs. (1.99a) and (1.99b) over the cross-section of the structure in the corresponding direction.

Let's admit the area of sources to be a rectangular of the size $a_y \times a_z$ (a_y is the size along the axis $0Y$, a_z is along the axis $0Z$). Then the power fluxes are:

- Along axis 0Y

$$\Pi_{yn} = \int_{a_{z1}}^{a_{z2}} \mathrm{d}z \int_{x_{n-1}}^{x_n} \mathrm{d}x P_{yn}, \quad a_{z2} - a_{z1} = a_z \tag{1.100}$$

- Along axis 0Z

$$\Pi_{zn} = \int_{a_{y1}}^{a_{y2}} \mathrm{d}z \int_{x_{n-1}}^{x_n} \mathrm{d}x P_{zn}, \quad a_{y2} - a_{y1} = a_y. \tag{1.101}$$

The power flux of the waves excited in the structure is

$$\Pi_y = \sum_{n=1}^{p} \Pi_{yn}, \tag{1.102}$$

$$\Pi_z = \sum_{n=1}^{p} \Pi_{zn}, \tag{1.103}$$

where $n = 1, 2, \ldots, p$ is the number of layers.

The derived expressions allow the following things to be investigated in multilayer bigyrotropic structures screened by impedance surfaces in the specified current approximation and self-consistent approximation:

- Dispersions of characteristic waves
- Amplitudes of the excited HF-constituents of the fields and the transferred partial power fluxes
- Total power fluxes

for various topologies of extraneous surface electric and magnetic currents.

1.3 Electromagnetic Wave Excitation in a Three-Layer Bigyrotropic Structure by Strip Converters

Let's examine a three-layer structure containing bigyrotrpic layers loaded with conductive coverings with $\sigma_E = \infty$ (Fig. 1.3). In a particular case, most interesting for practice, the average layer with $n = 2$ can be bigyrotropic (a magnetic semiconductor in external electric E_0 and magnetic H_0 fields, a medium with tensors $\overleftrightarrow{\mu}_2, \overleftrightarrow{\varepsilon}_2$), can have only magnetic (a ferrite with tensors $\overleftrightarrow{\mu}_2$ and $\overleftrightarrow{\varepsilon}_2$ in a field $\bar{H}_0$) or only electric (a conductive crystal with tensors $\overleftrightarrow{\varepsilon}_2$ in a field $\bar{H}_0$) gyrotropy (Fig. 1.4). The external

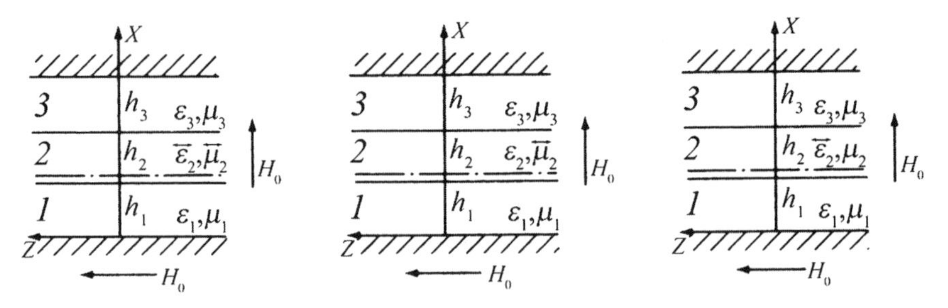

Fig. 1.3 A three-layer structure containing bigyrotrpic layers loaded with conductive coverings with $\sigma_э = \infty$

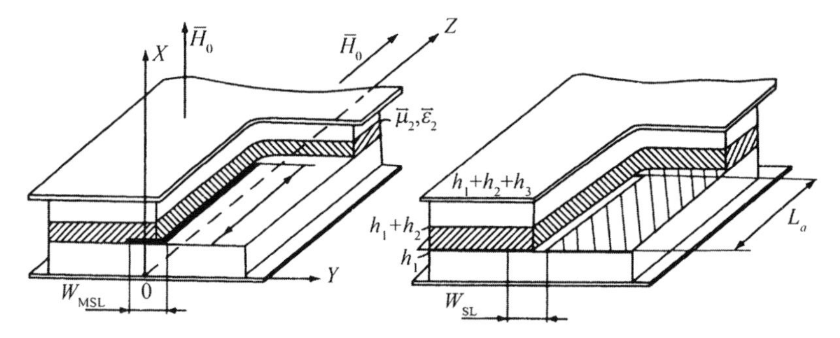

Fig. 1.4 Structures with magnetic (a ferrite with tensors $\overleftrightarrow{\mu}_2$ and $\overleftrightarrow{\varepsilon}_2$ in a field $\bar{H}_0$) or electric (a conductive crystal with tensors $\overleftrightarrow{\varepsilon}_2$ in a field $\bar{H}_0$) gyrotropy

field $\bar{H}_0$ is applied along the axis 0Z. On the interface at $x = h_1$ a microstrip or slot line of a width W_{MSL}, is placed along the axis 0Z, its length being L_0 (Fig. 1.4).

For the MSL at normalization on $F(y_{0E}, z_{0E})$ from Eq. (1.65) cofactors of FIGF on the screens follow:

– At $x = 0$

$$\Delta^E_{\tilde{G}^1_3(0,\bar{j}_E)} = \begin{Vmatrix} b_{13} & b_{14} & 0 & d_{14} \\ b_{23} & b_{24} & 0 & d_{24} \\ b_{33} & b_{34} & 1 & d_{34} \\ b_{43} & b_{44} & 1 & d_{44} \end{Vmatrix} \quad \Delta^E_{\tilde{G}^1_4(0,\bar{j}_E)} = \begin{Vmatrix} b_{13} & b_{14} & d_{13} & 0 \\ b_{23} & b_{24} & d_{23} & 0 \\ b_{33} & b_{34} & d_{33} & 1 \\ b_{43} & b_{44} & d_{44} & 1 \end{Vmatrix} \tag{1.104}$$

– At $x = x_3 = h_1 + h_2 + h_3$

$$\Delta^E_{\tilde{G}^3_3(x_3\bar{j}_E)} = \begin{Vmatrix} 0 & b_{14} & d_{13} & d_{14} \\ 0 & b_{24} & d_{23} & d_{24} \\ 1 & b_{34} & d_{33} & d_{34} \\ 1 & b_{44} & d_{43} & d_{44} \end{Vmatrix} \qquad \Delta^E_{\tilde{G}^3_4(x_3,\bar{j}_E)} = \begin{Vmatrix} b_{13} & 0 & d_{13} & d_{14} \\ b_{23} & 0 & d_{23} & d_{24} \\ b_{33} & 1 & d_{33} & d_{34} \\ b_{43} & 1 & d_{43} & d_{44} \end{Vmatrix}. \tag{1.105}$$

For TSL at normalization on $F(y_{0M}, Z_{0M})$ from Eq. (1.64) cofactors of FIGF on the screens follow:

- At $x = 0$

$$\Delta^E_{\widetilde{G}^1_3(0,\bar{j}_M)} = \begin{Vmatrix} b_{13} & b_{14} & 1 & d_{14} \\ b_{23} & b_{24} & 1 & d_{24} \\ b_{33} & b_{34} & 0 & d_{34} \\ b_{43} & b_{44} & 0 & d_{44} \end{Vmatrix} \quad \Delta^E_{\widetilde{G}^1_4(0,\bar{j}_M)} = \begin{Vmatrix} b_{13} & b_{14} & d_{13} & 1 \\ b_{23} & b_{24} & d_{23} & 1 \\ b_{33} & b_{34} & d_{33} & 0 \\ b_{43} & b_{44} & d_{44} & 0 \end{Vmatrix} \tag{1.106}$$

- At $x = x_3$

$$\Delta^E_{\widetilde{G}^3_3(x_3,\bar{j}_M)} = \begin{Vmatrix} 1 & b_{14} & d_{13} & d_{14} \\ 1 & b_{24} & d_{23} & d_{24} \\ 0 & b_{34} & d_{33} & d_{34} \\ 0 & ,b_{44} & d_{43} & d_{44} \end{Vmatrix} \quad \Delta^E_{\widetilde{G}^3_4(x_3,\bar{j}_M)} = \begin{Vmatrix} b_{13} & 1 & d_{13} & d_{14} \\ b_{23} & 1 & d_{23} & d_{24} \\ b_{33} & 0 & d_{33} & d_{34} \\ b_{43} & 0 & d_{43} & d_{44} \end{Vmatrix}. \tag{1.107}$$

Using Eqs. (1.104) and (1.105), we can determine FIGF for the MSL on the screens:

- At $x = 0$

$$\widetilde{G}^E_1(0, j_E) = \frac{F(y_E, z_E)}{\Delta^E_1} \cdot \begin{Vmatrix} 0 \\ 0 \\ \Delta^E_{\widetilde{G}^1_3}(0, \bar{j}_E) \\ \Delta^E_{\widetilde{G}^1_4}(0, \bar{j}_E) \end{Vmatrix} = F(y_E, z_E)\, \overleftrightarrow{W}_1, \tag{1.108}$$

where

$$\overleftrightarrow{W}_1 = \frac{1}{\Delta^E_1} \begin{Vmatrix} 0 \\ 0 \\ \Delta^E_{\widetilde{G}^1_3}(0, \bar{j}_E) \\ \Delta^E_{\widetilde{G}^1_4}(0, \bar{j}_E) \end{Vmatrix} \tag{1.108a}$$

- At $x = x_3$

$$\widetilde{G}^E_3(x_3, j_E) = \frac{F(y_E, z_E)}{\Delta^E_1} \cdot \begin{Vmatrix} 0 \\ 0 \\ \Delta^E_{\widetilde{G}^3_3}(x_3, \bar{j}_E) \\ \Delta^E_{\widetilde{G}^3_4}(x_3, \bar{j}_E) \end{Vmatrix} = F(y_E, z_E)\, \overleftrightarrow{W}_2, \tag{1.109}$$

where

$$\overleftrightarrow{W}_2 = \frac{1}{\Delta^E_1} \begin{Vmatrix} 0 \\ 0 \\ \Delta^E_{\widetilde{G}^3_3}(x_3, \bar{j}_E) \\ \Delta^E_{\widetilde{G}^3_4}(x_3, \bar{j}_E) \end{Vmatrix} \tag{1.109a}$$

For a SL it is necessary to find FIGF for the surface at $x = x_3$ only

$$\widetilde{G}_3^E(x_3, j_M) = \frac{F(y_M, z_M)}{\Delta_1^E} \cdot \begin{Vmatrix} \widetilde{G}_3^3(x_3, \bar{j}_M) \\ \widetilde{G}_4^3(x_3, \bar{j}_M) \\ 0 \\ 0 \end{Vmatrix} = F(y_M, z_M)\, \overleftrightarrow{W}_3, \tag{1.110}$$

where

$$\overleftrightarrow{W}_3 = \frac{1}{\Delta_1^E} \begin{Vmatrix} \widetilde{G}_3^3(x_3, \bar{j}_M) \\ \widetilde{G}_4^3(x_3, \bar{j}_M) \\ 0 \\ 0 \end{Vmatrix}. \tag{1.110a}$$

Equations (1.108) and (1.110) allow finding FIGF of the constituents of HF fields in the layers.

For a MSL we have

$$\begin{aligned} \overleftrightarrow{N}_1^E(x, \bar{j}_E) &= \overleftrightarrow{T}_1(x)\overleftrightarrow{W}_1 F_{0E}, \\ \overleftrightarrow{N}_3^E(x, \bar{j}_E) &= \overleftrightarrow{T}_3(x - x_3)\overleftrightarrow{W}_2 F_{0E}, \\ \widetilde{\overleftrightarrow{N}}_2^E(x, \bar{j}_E) &= \overleftrightarrow{T}_2(x - x_2)\overleftrightarrow{T}_3(-h_3)\overleftrightarrow{W}_2 F_{0E}. \end{aligned} \tag{1.111}$$

For a SL we have

$$\begin{aligned} \widetilde{\overleftrightarrow{N}}_3^E(x, \bar{j}_M) &= \overleftrightarrow{T}_3(x - x_3)\overleftrightarrow{W}_3 F_{0M}, \\ \widetilde{\overleftrightarrow{N}}_2^E(x, \bar{j}_M) &= \overleftrightarrow{T}_2(x - x_2)\overleftrightarrow{T}_3(-h_3)\overleftrightarrow{W}_3 F_{0M}. \end{aligned} \tag{1.112}$$

After taking the inverse Fourier transform, from Eqs. (1.111) and (1.112) by Eq. (1.83) we derive:

- For a MSL

$$\begin{aligned} \widetilde{\overleftrightarrow{N}}_1^E(x, \bar{j}_E) &= \overleftrightarrow{T}_1(x) \cdot 2\pi \cdot j\overleftrightarrow{W}_1 F_{0E}, \\ \widetilde{\overleftrightarrow{N}}_3^E(x, \bar{j}_E) &= \overleftrightarrow{T}_3(x - x_3) \cdot 2\pi \cdot j\overleftrightarrow{W}_2 F_{0E}, \\ \overleftrightarrow{N}_2^E(x, \bar{j}_E) &= \overleftrightarrow{T}_2(x - x_2)\overleftrightarrow{T}_3(-h_3) \cdot 2\pi \cdot j\overleftrightarrow{W}_2 F_{0E} \end{aligned} \tag{1.113}$$

- For a SL

$$\begin{aligned} \widetilde{\overleftrightarrow{N}}_3^E(x, \bar{j}_M) &= \overleftrightarrow{T}_3(x - x_3) 2\pi j\overleftrightarrow{W}_3 F_{0M}, \\ \widetilde{\overleftrightarrow{N}}_2^E(x, \bar{j}_M) &= \overleftrightarrow{T}_2(x - x_2)\overleftrightarrow{T}_3(-h_3) 2\pi j\overleftrightarrow{W}_3 F_{0M} \end{aligned} \tag{1.114}$$

For our bigyrotropic structure the coupling matrices are

$$\begin{aligned} \overleftrightarrow{T}^1(x) &= \overleftrightarrow{T}_1^1 e^{j\kappa_{x1}^1 x} + \overleftrightarrow{T}_2^1 e^{j\kappa_{x2}^1 x} + \overleftrightarrow{T}_3^1 e^{j\kappa_{x3}^1 x} + \overleftrightarrow{T}_4^1 e^{j\kappa_{x4}^1 x}, \\ \overleftrightarrow{T}^2(x) &= \overleftrightarrow{T}_1^2 e^{j\kappa_{x1}^2 x} + \overleftrightarrow{T}_2^2 e^{j\kappa_{x2}^2 x} + \overleftrightarrow{T}_3^2 e^{j\kappa_{x3}^2 x} + \overleftrightarrow{T}_4^2 e^{j\kappa_{x4}^2 x}, \\ \overleftrightarrow{T}^3(x) &= \overleftrightarrow{T}_1^3 e^{j\kappa_{x1}^3 x} + \overleftrightarrow{T}_2^3 e^{j\kappa_{x2}^3 x} + \overleftrightarrow{T}_3^3 e^{j\kappa_{x3}^3 x} + \overleftrightarrow{T}_4^3 e^{j\kappa_{x4}^3 x}. \end{aligned} \tag{1.115}$$

For the gyrotropic structure with layer 2 – being a ferrite with $\overleftrightarrow{\mu}_2$

$$\overleftrightarrow{T}^1(x) = \overleftrightarrow{T}_1^1 e^{j\kappa_{x1}^1 x} + \overleftrightarrow{T}_2^1 e^{j\kappa_{x2}^1 x},$$
$$\overleftrightarrow{T}^2(x) = \overleftrightarrow{T}_1^2 e^{j\kappa_{x1}^2 x} + \overleftrightarrow{T}_2^2 e^{j\kappa_{x2}^2 x}, \qquad (1.116)$$
$$\overleftrightarrow{T}^3(x) = \overleftrightarrow{T}_1^3 e^{j\kappa_{x1}^3 x} + \overleftrightarrow{T}_2^3 e^{j\kappa_{x2}^3 x}.$$

$$\overleftrightarrow{T}_1^n = \frac{\overleftrightarrow{A}_n - \kappa_{x2}^n \overleftrightarrow{I}}{\kappa_{x_1}^n - \kappa_{x_2}^n}, \qquad \overleftrightarrow{T}_2^n = \frac{\overleftrightarrow{A}_n - \kappa_{x1}^n \overleftrightarrow{I}}{\kappa_{x_2}^n - \kappa_{x_1}^n}, \qquad n = 1,2,3.$$

For the gyrotropic structure with layer 2 being a conductive crystal with $\overleftrightarrow{\varepsilon}_2$

$$\overleftrightarrow{T}^1(x) = \overleftrightarrow{T}_3^1 e^{j\kappa_{x3}^1 x} + \overleftrightarrow{T}_4^1 e^{j\kappa_{x4}^1 x},$$
$$\overleftrightarrow{T}^2(x) = \overleftrightarrow{T}_3^2 e^{j\kappa_{x3}^2 x} + \overleftrightarrow{T}_4^2 e^{j\kappa_{x4}^2 x}, \qquad (1.117)$$
$$\overleftrightarrow{T}^3(x) = \overleftrightarrow{T}_3^3 e^{j\kappa_{x3}^3 x} + \overleftrightarrow{T}_4^3 e^{j\kappa_{x4}^3 x}.$$

$$\overleftrightarrow{T}_3^n = \frac{\overleftrightarrow{A}_n - \kappa_{x4}^n \overleftrightarrow{I}}{\kappa_{x3}^n - \kappa_{x4}^n}, \quad \overleftrightarrow{T}_4^n = \frac{\overleftrightarrow{A}_n - \kappa_{x3}^n \overleftrightarrow{I}}{\kappa_{x4}^n - \kappa_{x3}^n}, \quad n = 1,\ 2,\ 3.$$

From Eq. (1.92) the power flux density transferred in the y direction in the structure is

$$P_{yn} = -\overleftrightarrow{S}_E^n \overleftrightarrow{N}_n^n \overleftrightarrow{\eta} \overleftrightarrow{N}_n^* - \overleftrightarrow{S}_H^n \overleftrightarrow{N}_n^* \overleftrightarrow{\xi} \overleftrightarrow{N}_n, \qquad (1.118)$$

where $\overleftrightarrow{\eta} = \|0\ 0\ 0\ 1\|$, $\quad \overleftrightarrow{\xi} = \|0\ 1\ 0\ 0\|$.

The first term in Eq. (1.112) is determined by LM waves ($H_x = 0$), and the second one is by LE waves ($E_x = 0$). The power transferred by waves in the layers of the structure $n = 1,\ 2,\ 3$ is

$$\Pi_{LM}^1 = -\overleftrightarrow{S}_E^1 \Bigg[\overleftrightarrow{T}_1^1 \overleftrightarrow{\eta} \overleftrightarrow{T}_1^{1*} \frac{e^{j(\kappa_{x_1}^1 - \kappa_{x_1}^{1*})h_1} - 1}{j(\kappa_{x_1}^1 - \kappa_{x_1}^{1*})} + \overleftrightarrow{T}_1^1 \overleftrightarrow{\eta} \overleftrightarrow{T}_2^1 \frac{e^{j(\kappa_{x_1}^1 - \kappa_{x_2}^{1*})h_1} - 1}{j(\kappa_{x_1}^1 - \kappa_{x_2}^{1*})}$$
$$+ \overleftrightarrow{T}_2^1 \overleftrightarrow{\eta} \overleftrightarrow{T}_1^{1*} \frac{e^{j(\kappa_{x_2}^1 - \kappa_{x_1}^{1*})h_1} - 1}{j(\kappa_{x_2}^1 - \kappa_{x_1}^{1*})} + \overleftrightarrow{T}_2^1 \overleftrightarrow{\eta} \overleftrightarrow{T}_1^{1*} \frac{e^{j(\kappa_{x_2}^1 - \kappa_{x_2}^{1*})h_1} - 1}{j(\kappa_{x_2}^1 - \kappa_{x_2}^{1*})} \Bigg] \overleftrightarrow{W}_1 2\pi \cdot jF_{0E},$$

$$\Pi_{LM}^2 = -\overleftrightarrow{S}_E^2 \Bigg[\overleftrightarrow{T}_1^2 \overleftrightarrow{\eta} \overleftrightarrow{T}_1^{3*} \cdot \frac{e^{j(\kappa_{x_3}^2 - \kappa_{x_3}^{2*})h_3} - e^{j(\kappa_{x_1}^2 - \kappa_{x_1}^{2*})h_2}}{j(\kappa_{x_1}^2 - \kappa_{x_1}^{2*})} + \overleftrightarrow{T}_1^2 \overleftrightarrow{\eta} \overleftrightarrow{T}_2^{3*}$$
$$\times \frac{e^{j(\kappa_{x_3}^2 - \kappa_{x_2}^{2*})h_1} - e^{j(\kappa_{x_3}^2 - \kappa_{x_2}^{2*})h_3}}{j(\kappa_{x_1}^2 - \kappa_{x_2}^{2*})} + \overleftrightarrow{T}_2^2 \overleftrightarrow{\eta} \overleftrightarrow{T}_1^{3*} \cdot \frac{e^{j(\kappa_{x_2}^2 - \kappa_{x_1}^{2*})h_3} - e^{j(\kappa_{x_2}^2 - \kappa_{x_1}^{2*})h_2}}{j(\kappa_{x_2}^2 - \kappa_{x_1}^{2*})}$$
$$+ \overleftrightarrow{T}_2^2 \overleftrightarrow{\eta} \overleftrightarrow{T}_2^{3*} \cdot \frac{e^{j(\kappa_{x_2}^2 - \kappa_{x_2}^{2*})h_3} - e^{j(\kappa_{x_2}^2 - \kappa_{x_2}^{2*})h_2}}{j(\kappa_{x_2}^2 - \kappa_{x_2}^{2*})} \Bigg] \times \overleftrightarrow{W}_2 2\pi \cdot jF_{0E},$$

$$\Pi_{LM}^3 = -\overleftrightarrow{S}_E^3 \left[\overleftrightarrow{T}_1^3 \overleftrightarrow{\eta} \overleftrightarrow{T}_1^{3*} \cdot \frac{e^{j(\kappa_{x_3}^3 - \kappa_{x_3}^{3*})h_2} - e^{j(\kappa_{x_1}^3 - \kappa_{x_1}^{3*})h_3}}{j(\kappa_{x_1}^2 - \kappa_{x_1}^{2*})} + \overleftrightarrow{T}_1^3 \overleftrightarrow{\eta} \overleftrightarrow{T}_2^{3*} \right.$$
$$\times \frac{e^{j(\kappa_{x_1}^3 - \kappa_{x_2}^{3*})h_2} - e^{j(\kappa_{x_1}^3 - \kappa_{x_2}^{3*})h_3}}{j(\kappa_{x_1}^3 - \kappa_{x_2}^{3*})} + \overleftrightarrow{T}_2^3 \overleftrightarrow{\eta} \overleftrightarrow{T}_1^{3*} \cdot \frac{e^{j(\kappa_{x_2}^3 - \kappa_{x_1}^{3*})h_2} - e^{j(\kappa_{x_2}^3 - \kappa_{x_1}^{3*})h_3}}{j(\kappa_{x_2}^3 - \kappa_{x_2}^{3*})}$$
$$\left. + \overleftrightarrow{T}_2^3 \overleftrightarrow{\eta} \overleftrightarrow{T}_2^{3*} \cdot \frac{e^{j(\kappa_{x_2}^3 - \kappa_{x_2}^{3*})h_2} - e^{j(\kappa_{x_2}^3 - \kappa_{x_2}^{3*})h_3}}{j(\kappa_{x_2}^3 - \kappa_{x_2}^{3*})} \right] \times \overleftrightarrow{W}_3 2\pi \cdot jF_{0E}. \tag{1.119}$$

On replacement in Eq. (1.119), considering Eqs. (1.93) and (1.94),

$$\begin{aligned} \overleftrightarrow{S}_E^n &\to -\overleftrightarrow{S}_H^n \\ \overleftrightarrow{\eta} &\to \overleftrightarrow{\xi} \end{aligned},$$

we get the power fluxes Π_{LE}^n, $n = 1, 2, 3$ in the layers of the structure. The total power flux in the three-layer bigyrotropic structure is

$$\Pi = \sum_{i=1}^{3} (\Pi_{LE}^i + \Pi_{LM}^i). \tag{1.120}$$

Let's examine, for example, self-consistent excitation of waves in the three-layer bigyrotropic structure (Fig. 1.4) by a surface electric current $j_{Ez}(y^T)$, along the axis 0Z.

The integral equation, according to Eq. (1.91a), looks as

$$\frac{j}{2\pi} \int_{-w/2}^{w/2} \frac{j_{E2}(y^T)\mathrm{d}y^T}{y - y^T} j\overleftrightarrow{S_H^2} \left[\overleftrightarrow{T}_2\ (\kappa_0, -h_2) \cdot \overleftrightarrow{T}_3(\kappa_0, -h_3) e^{-j\kappa_0 y \frac{1}{\Delta_1^E(\kappa_0)}} \right.$$
$$\left. \times \begin{Vmatrix} 0 \\ 0 \\ \{\Delta_E\}_{33} \\ \{\Delta_E\}_{34} \end{Vmatrix} \cdot \int_{-w/2}^{w/2} j_2(y^T) e^{j\kappa_0 y^T} \mathrm{d}z \right] = 0. \tag{1.121}$$

Transforming this equation to a form convenient for calculations, by means of the replacements $\xi = \frac{2y}{w}$ and $\xi^T = \frac{2y^T}{w}$, we come to

$$\int_{-1}^{1} \frac{j_{Ez}(\xi^T)}{\xi - \xi^T} \mathrm{d}\xi^T + E \cdot e^{-j\frac{\kappa_0 w}{2}\xi^T} \int_{-1}^{1} j_{EZ}(\xi^T)\, e^{j\frac{\kappa_0 w}{2}\xi^T} \mathrm{d}\xi^T, \tag{1.122}$$

where

$$E = 2\pi \overleftrightarrow{S_H^2} \left[\overleftrightarrow{T}_2(\kappa_0, -h_2) \cdot \overleftrightarrow{T}_3(\kappa_0, -h_3) \frac{1}{\Delta^E(\kappa_0)} \begin{Vmatrix} 0 \\ 0 \\ \{\Delta_E\}_{33} \\ \{\Delta_E\}_{34} \end{Vmatrix} \right].$$

Carrying out the procedure of Cauchy integral inversion and getting rid of singularity, we get a new form of Eq. (1.122)

$$j_{Ez}(\xi) = \frac{1}{\sqrt{1-\xi^2}} \left[M + K(\xi) \int_{-1}^{1} j_{Ez}(\xi^T) e^{j\frac{\kappa_0 w}{2}\xi^T} \mathrm{d}\xi^T \right], \tag{1.123}$$

where

$$K(\xi) = \int_{-1}^{1} E \frac{e^{-j\frac{\kappa_0 w}{2}\xi^T} \sqrt{1-(\xi^T)^2}}{\xi^T - \xi} \mathrm{d}\xi^T, \tag{1.124}$$

M is a constant related to the surface current.

Introducing the replacement $j_{Ez}(\xi) = \frac{M}{\sqrt{1-\xi^2}} I(\xi)$, from Eq. (1.123), we derive an expression for the total current:

$$I(\xi) = 1 + K(\xi) \int_{-1}^{1} \frac{I(\xi)}{\sqrt{1-(\xi^T)^2}} \, e^{j\frac{\kappa_0 w \xi^T}{2}} \mathrm{d}\xi^T, \tag{1.125}$$

which we reduce to a standard form by another replacement $\xi^T = \sin t$

$$I(t) = 1 + K(t) \int_{-\pi/2}^{\pi/2} I(t^T) \, e^{j\frac{\kappa_0 w \sin t^T}{2}} \mathrm{d}t^T. \tag{1.126}$$

Equation (1.126) is a second-kind inhomogeneous integral Fredholm equation [461].

To calculate $K(t)$ in Eq. (1.126), we shall use Helder's inequality [465], which gives an upper bound of the Cauchy-type integral

$$K(\xi) = E \left[\int_{-1}^{1} \frac{f(\xi^T) - f(\xi)}{\xi^T - \xi} \mathrm{d}\xi^T - f(\xi) \cdot \ln \frac{1-\xi}{1-\xi} \right], \tag{1.127}$$

where

$$f(\xi) = e^{-j\frac{\kappa_0 w}{2}\xi} \sqrt{1-\xi^2},$$
$$f(\xi^T) = e^{-j\frac{\kappa_0 w}{2}\xi^T} \sqrt{1-(\xi^T)^2},$$
$$\xi^T = \xi + \varepsilon.$$

To satisfy Helder's condition, it is necessary to estimate the difference between $f(\xi^T)$ and $f(\xi)$ from above, i.e. to get an inequality of the form

$$\left| f(\xi^T) - f(\xi) \right| < A \left| \xi^T - \xi \right|^{\lambda}, \tag{1.128}$$

where A is Helder's constant, λ Helder's parameter. It is possible to show that

$$K(\xi) \leq E\left(e^{-j\frac{\kappa_0 w}{2}\xi}\sqrt{1-\xi^2}\ln\frac{1-\xi}{1-\xi} - \xi\right). \tag{1.129}$$

Considering Eq. (1.129), Eq. (1.126) will look as

$$I(t) = 1 + E\left(e^{-j\frac{\kappa_0 w}{2}\sin t}\cos t \cdot \ln\frac{1-\sin t}{1+\sin t} - \sin t\right) \cdot \int_{-\pi/2}^{\pi/2} I(t^T)\, e^{j\frac{\kappa_0 w \sin t^T}{2}}\, dt^T, \tag{1.130}$$

which is solved by one of the standard methods [461].

The solution of Eq. (1.130) gives the distribution of the surface exciting current including the back influence of the waves excited in the structure. Using the found solution, by Eq. (1.81) we find $\overleftrightarrow{N}_n$ in the corresponding half-spaces of the converter, which allow the power fluxes of the waves excited in the structure to be determined by Eq. (1.99).

The number of iterations required was determined by convergence between $I(t)$ and $\overleftrightarrow{N}(I(t))$. To calculate the power flux Π transferred in the structure, a basic software package for scientific estimations was used. The value Δ' (see Eq. 1.59a) determining the group speed of the power flux in the structure, was calculated by global spline approximation.

Let's examine the results of calculation of the surface exciting current distribution in a MSL in the self-consistent approach in the centimeter ($\nu_H = 3\,\text{GHz}$) and millimeter ($\nu_H = 30\,\text{GHz}$) ranges of radio waves.

Figure 1.5 depicts the dependences of the distribution of the current magnitude $|j_{Ez}| = \frac{|j|}{I_0}$ by the width of the converter (*Fig. 1.5a*) and its phase $\beta = arctg\frac{\text{Im}\, j_{EZ}}{\text{Re}\, j_{EZ}}$ (*Fig. 1.5b*) *on* W for different values of the wave phase constant κ' in a tangent magnetized structure.

$1 - w = 1 \cdot 10^{-5} m,\quad \kappa' = 8.68 \cdot 10^3 m^{-1};$
$2 - w = 1 \cdot 10^{-5} m,\quad \kappa' = 9.95 \cdot 10^4 m^{-1};$
$3 - w = 5 \cdot 10^{-5} m,\quad \kappa' = 8.68 \cdot 10^3 m^{-1};$
$4 - w = 5 \cdot 10^{-5} m;\quad \kappa' = 5.43 \cdot 10^4 m^{-1};$
$5 - w = 1 \cdot 10^{-4} m,\quad \kappa' = 8.68 \cdot 10^3 m^{-1};$
$6 - w = 1 \cdot 10^{-4} m,\quad \kappa' = 9.95 \cdot 10^4 m^{-1};$
$\nu_H = 3\text{GHz},\quad \alpha_{11} = 10^{-5},\quad h_1 = h_2 = 0.35 \cdot 10^{-3} m,$
$h_2 = 25 \cdot 10^{-6} m,\quad \varepsilon_{1,2} = 14,\quad \varepsilon_3 = 1,$
$4\pi M_s = 0.176\ \text{T}.$

We note in passing that the uniform current distribution by the MSL width was the initial, originally specified one. In the self-consistent approach the following regularities were observed.

At excitation of weakly delayed waves with $\kappa' \leq 10^4 m^{-1}$ the distribution $|j_{Ez}(w)|$ slightly changes at variation of the width w of the converter (cf. curves 1–5,

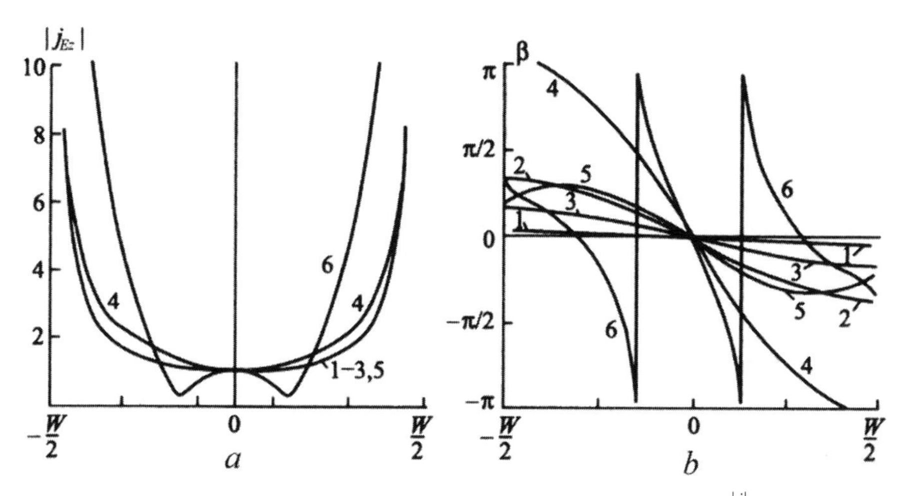

Fig. 1.5 The dependences of the distribution of the current magnitude $|j_{Ez}| = \frac{|j|}{I_0}$ by the width of the converter (*a*) and its phase $\beta = arctg\frac{\mathrm{Im}\, j_{Ez}}{\mathrm{Re}\, j_{Ez}}$ (*b*) on W for different values of the wave phase constant κ' in a tangent magnetized structure

Fig. 1.5*a*). The most essential change is experienced by the dependencies $\beta(w)$, and with increasing κ' their disturbance (cf. curves 1–3 and 4, Fig. 1.5*a, b*) becomes more intense. The wider the MSL used in the converters and the higher the wave delay, the stronger the disturbance $|j_{Ez}|$ and β (cf. curves 1–5 and 6 on Fig. 1.5*a, b*). In the range of strongly delayed electromagnetic – (magnetostatic) waves ($\kappa' \geq 10^5\, m^{-1}$) the *variations of* $|j_{Ez}(w)|$ agree with the data by B.A. Kalinikos et al. The sign-alternating character of the dependencies $\beta(w)$ means that the reverse influence of the fields of the excited waves on the exciting current leads to transformation of the potential character of the *original* current into the vortical one. The stronger the wave is delayed (higher κ'), the more pronounced this feature.

At transition into the millimeter range a decrease of that value of the parameter $(\kappa' w)_0$ is noted, from which the back influence on the exciting current is observed. So, for a structure with $\alpha = 10^{-5}$ at $\nu_{\mathrm{H}} = 3\,\mathrm{GHz} - (\kappa' w)_0 \approx 2.7$, and at $\nu_{\mathrm{H}} = 30\,\mathrm{GHz} - (\kappa' w)_0 \approx 1.4$. On Fig. 1.6*a, b* the analyzed dependencies for $w = 5 \cdot 10^{-4}\,\mathrm{m}$, $\kappa' = 10.8 \cdot 10^2\,\mathrm{m}^{-1}$, $\kappa'' - 0.259\,\mathrm{m}^{-1}$, $\alpha = 10^{-5}$, $\nu_{\mathrm{H}} = 3\,\mathrm{GHz}$, $4\pi M_s = 0.176\,\mathrm{T}$ are given.

At decreasing w the disturbance of $|j_{Ez}|$ and β reduces. On Fig. 1.7*a, b* these dependencies for $w = 10^{-4}\,\mathrm{m}$ are shown. On going into the range of more delayed waves some increase in the disturbance of $|j_{Ez}|$ and β was observed. Figure 1.8*a, b* presents the dependence for a converter with the parameters $w = 10^{-4}\,\mathrm{m}$, $\kappa' = 1.4 \times 10^5\,\mathrm{m}^{-1}$, $\kappa'' = -257\,\mathrm{m}^{-1}$, $\alpha = 10^{-5}$, $4\pi M_s = 0.176\,\mathrm{T}$.

Unlike the centimeter range on transition into the range of magnetostatic waves in the millimeter range a weak disturbance of $|j_{Ez}|$ and β is noted, and this dependence has a weakly-expressed sign-alternating character. At expansion of the MSL for magnetostatic waves the disturbance of the dependences $|j_{Ez}|$ and $\beta(w)$ was

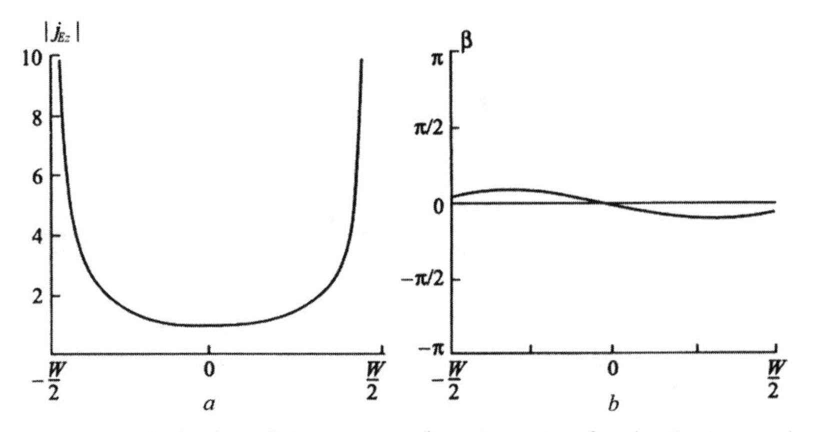

Fig. 1.6 Dependencies $|j_{Ez}|$ and β for $w = 5 \cdot 10^{-4}$ m, $\kappa' = 10.8 \cdot 10^2\,\mathrm{m}^{-1}$, $\kappa'' - 0.259\mathrm{m}^{-1}$, $\alpha = 10^{-5}$, $\nu_H = 3\,\mathrm{GHz}$, $4\pi M_s = 0.176\,\mathrm{T}$

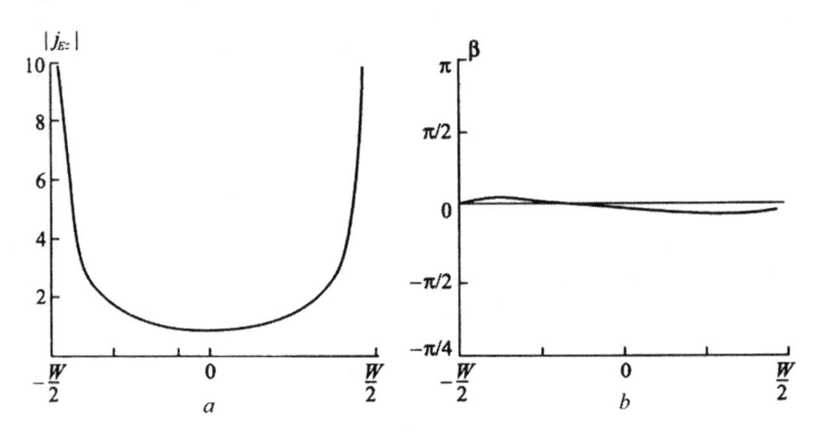

Fig. 1.7 Dependencies $|j_{Ez}|$ and β for $w = 10^{-4}$ m

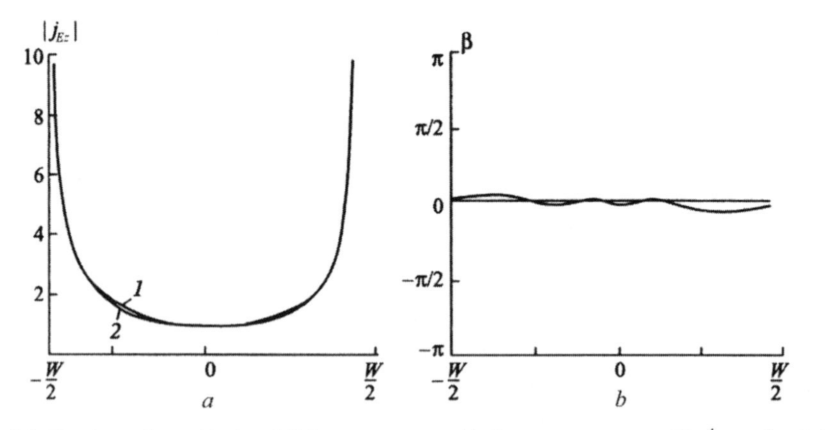

Fig. 1.8 The dependence $|j_{Ez}|$ and β for a converter with the parameters $w = 10^{-4}$ m, $\kappa' = 1.4 \times 10^5\,\mathrm{m}^{-1}$, $\kappa'' = -257\,\mathrm{m}^{-1}$, $\alpha = 10^{-5}$, $4\pi M_s = 0.176\,\mathrm{T}$

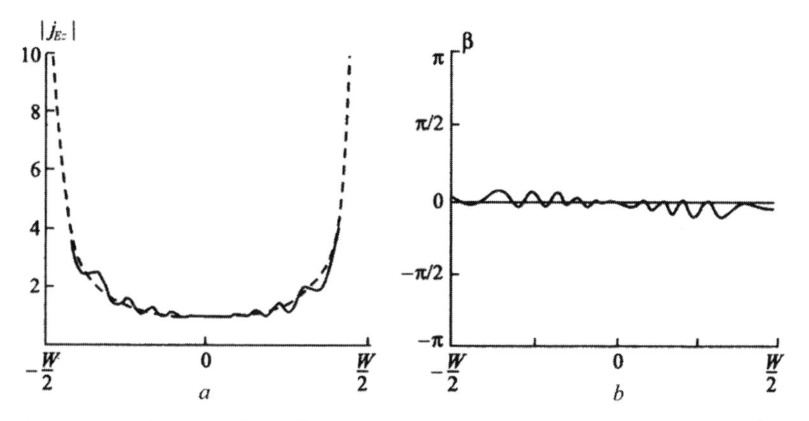

Fig. 1.9 The dependence $|j_{Ez}|$ and β for a converter with the parameters $w = 5\cdot 10^{-4}\,\mathrm{m}$, $\kappa' = 1.4\cdot 10^5\,\mathrm{m}^{-1}$, $\kappa'' = -257\,\mathrm{m}^{-1}$, $\alpha = 10^{-5}$, $4\pi M_s = 0.176\,\mathrm{T}$ (the dependencies for weakly delayed waves are shown as the dotted lines)

somewhat intensified. Figure 1.9a, b shows data of the dependence for a converter with the parameters $w = 5\cdot 10^{-4}\,\mathrm{m}$, $\kappa' = 1.4\cdot 10^5\,\mathrm{m}^{-1}$, $\kappa'' = -257\,\mathrm{m}^{-1}$, $\alpha = 10^{-5}$, $4\pi M_s = 0.176\,\mathrm{T}$ (the dependencies for weakly delayed waves are shown as the dotted lines). In comparison with the centimeter range, the account of the back influence of the fields of the excited waves gives less essential corrections and manifests itself as a finer structure of $|j_{Ez}|$ and $\beta(w)$.

1.4 Conclusions

1. A new theoretical approach to analysis of electromagnetic wave converters in layered structures on the basis of film multilayer bigyrotropic structures with their parallel orientation to the exciting plane has been developed.
2. The method is based on tensor Green functions, for which boundary conditions were formulated on screens with arbitrary impedances on the boundaries of the layers and in the plane of extraneous electric and magnetic surface currents, for which, by means of the coupling matrices of the tangential constituents of HF fields, an inhomogeneous matrix equation has been derived to determine wave dispersions and Green functions in the layers; the integral equation considers the back influence of HF fields of the waves excited in the structure on extraneous surface currents.
3. Expressions for the amplitudes of HF fields and power fluxes excited in the structure of *LE* and *LM* waves both in the near and far zones of radiation, considering basic factors of the short-wave range for ferrite film structures, namely, transverse gradients of dissipation, electric and magnetic parameters, screens conductance.
4. The solution of the problem for multilayer bigyrotropic structures allows analysis of the wave properties of structures containing ferrite, ferroelectric, conductive

crystal, magnetic semiconductors with due account of some specific factors appearing near the resonant frequencies and at tuning-out from them, including the range of existence of strongly delayed (magnetostatic) spin waves.

5. The developed approach provides analysis of strip converters of various topologies with the parallel orientation of their structures screened by surfaces with an arbitrary impedance, e.g. the converters on the basis of microstrip, slot, coplanar lines, including their various modifications in the form of mono-and multi-element topologies.

Chapter 2
Converters with Orthogonal Structure Orientation

Excitation of electromagnetic waves in a thin-plate multilayered bigyrotropic structure by extraneous point sources located on its end surface is considered. The structure is screened by impedance surfaces. Strip-line and wave devices with their layered structures orthogonally oriented to the plane of the sources fall into such a type of converters. Expressions were derived. They permit investigating key parameters of such converters with due account of the features of the millimeter range. As well as for the parallel orientation, the problem is solved by the method of tensor Green functions with using of the coupling matrices of multilayered bigyrotropic structures. These matrices allow determination of Green's functions in any layer of the structure in the approach of given sources. A matrix equation for a given extraneous current has been derived in the first approximation. The integral equation considers the back influence of the field of excited waves on the sources. The expressions gotten allow analysis of power fluxes of the basic types of waves in such converters in the millimeter range.

2.1 Statement of the Problem: Non-uniform Matrix Equation

Consider a layered bigyrotropic structure with its tensors $\overleftrightarrow{\varepsilon}_n$, $\overleftrightarrow{\mu}_n, n = 1, 2, \ldots, p$, on the end surface of which $(y = 0)$ extraneous point electric and magnetic currents are located with their surface densities $\bar{j}_E(j_{Ex}, j_{Ez})$ and $\bar{j}_M(j_{Mx}, j_{Mz})$, respectively (Fig. 2.1). The structure is screened by impedance walls at $x = 0$ and $x = x_p$. Sources $\bar{j}_E$, $\bar{j}_M$ can be located:

- Within the limits of one layer $x_{m-1} < x \leq x_{m+1}$, including its border $x = x_{m-1}$
- Within the limits of several layers

The fields in the structure satisfy Maxwell's equation (1.1). Unlike converters with parallel orientation (Chapter 1), in this case, boundary conditions must be three-dimensional and be satisfied on the internal borders of the section in the plane $Y0Z$ and in the plane of the sources $y = 0$, simultaneously; moreover,

A.A. Ignatiev, *Magnetoelectronics of Microwaves and Extremely High Frequencies in Ferrite Films.*
DOI: 10.1007/978-0-387-85457-1_3,

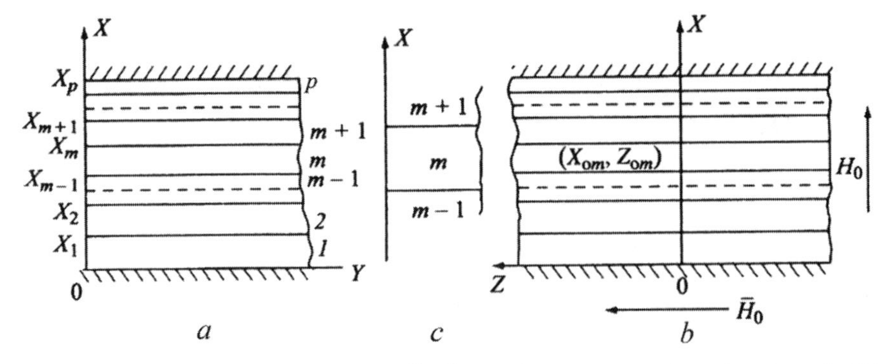

Fig. 2.1 Bigyrotropic structure with tensors $\overleftrightarrow{\varepsilon}_n, \overleftrightarrow{\mu}_n, n = 1, 2, \ldots, p$, on the end surface of which $(y = 0)$ extraneous point electric and magnetic currents are located with their surface densities $\bar{j}_E(j_{Ex}, j_{Ez})$ and $\bar{j}_M(j_{Mx}, j_{Mz})$

$$D_y^n \Bigg|_{y=0} = \frac{\nabla j_E}{j\omega}, \quad B_y^n \Bigg|_{y=0} = \frac{\nabla j_M}{j\omega}, \; n = 1, 2, \ldots, p. \tag{2.1}$$

Let's note that, generally, each of the p areas of the structure can have its own distribution of extraneous surface currents.

Considering

$$\begin{aligned} D_y^n &= \varepsilon_0(\varepsilon_{21}^n E_x^n + \varepsilon_{22}^n E_y^n + \varepsilon_{23}^n E_z^n), \\ B_y^n &= \mu_0(\mu_{21}^n H_x^n + \mu_{22}^n H_y^n + \mu_{23}^n H_z^n), \end{aligned} \tag{2.2}$$

after taking the Fourier-transforms of Eq. (2.1) by all the three coordinates, we obtain

$$(\varepsilon_{21}^n \widetilde{E}_x^n + \varepsilon_{22}^n \widetilde{E}_y^n + \varepsilon_{23}^n \widetilde{E}_z^n) \Bigg|_{y=0} = \frac{\kappa_x^n \tilde{j}_{Ex} + \kappa_z^n \tilde{j}_{Ez}}{j\omega}, \tag{2.3a}$$

$$(\mu_{21}^n \widetilde{H}_x^n + \mu_{22}^n \widetilde{H}_y^n + \mu_{23}^n \widetilde{H}_z^n) \Bigg|_{y=0} = \frac{\kappa_x^n \tilde{j}_{Mx} + \kappa_z^n \tilde{j}_{Mz}}{j\omega}. \tag{2.3b}$$

Using the relation between $\widetilde{E}_x^n(\widetilde{H}_y^n, \widetilde{H}_z^n, \widetilde{E}_y^n, \widetilde{E}_z^n)$ and $\widetilde{H}_x^n(\widetilde{E}_y^n, \widetilde{E}_z^n, \widetilde{H}_y^n, \widetilde{H}_z^n)$, by substitution of Eqs. (1.27) and (1.28) into Eqs. (2.3a) and (2.3b), respectively, we get

$$\left\{ \frac{\varepsilon_{21}^n}{\varepsilon_{11}^n} \left[\frac{1}{\varepsilon_0 \omega} (\kappa_z \widetilde{H}_y^n - \kappa_y \widetilde{H}_z^n) - \varepsilon_{12}^n \widetilde{E}_y^n - \varepsilon_{13}^n \widetilde{E}_z^n \right] + \varepsilon_{22}^n \widetilde{E}_y^n + \varepsilon_{23}^n \widetilde{E}_z^n \right\} \Bigg|_{y=0}$$

$$= \frac{\kappa_x^n \tilde{j}_{Ex} + \kappa_z^n \tilde{j}_{Ez}}{j\omega}, \tag{2.4a}$$

$$\left\{\frac{\mu_{21}^n}{\mu_{11}^n}\left[\frac{1}{\mu_0\omega}\left(\kappa_y\widetilde{E}_z^n-\kappa_z\widetilde{E}_y^n\right)-\mu_{12}^n\widetilde{H}_y^n-\mu_{13}^n\widetilde{H}_z^n\right]+\mu_{22}^n\widetilde{H}_y^n+\mu_{23}^n\widetilde{H}_z^n\right\}\Bigg|_{y=0}$$

$$=\frac{\kappa_x^n\tilde{j}_{Mx}+\kappa_z^n\tilde{j}_{Mz}}{j\omega}. \tag{2.4b}$$

After a similar transformation of Eqs. (2.4a), (2.4b) we have

$$\left\{\left(\varepsilon_{22}^n-\frac{\varepsilon_{21}^n\varepsilon_{12}^n}{\varepsilon_{11}^n}\right)\widetilde{E}_y^n+\left(\varepsilon_{23}^n-\frac{\varepsilon_{21}^n\varepsilon_{13}^n}{\varepsilon_{11}^n}\right)\widetilde{E}_z^n+\frac{\varepsilon_{21}^n\kappa_z}{\omega\varepsilon_0\varepsilon_{11}^n}\widetilde{H}_y^n-\frac{\varepsilon_{21}^n\kappa_y}{\omega\varepsilon_0\varepsilon_{11}^n}\widetilde{H}_z^n\right\}\Bigg|_{y=0}$$

$$=\frac{\kappa_x^n\tilde{j}_{Ex}+\kappa_z^n\tilde{j}_{Ez}}{j\omega}, \tag{2.5a}$$

$$\left\{\frac{\mu_{21}^n\kappa_y}{\omega\mu_0\mu_{11}^n}\tilde{E}_z^n-\frac{\mu_{21}^n\kappa_z}{\omega\mu_0\mu_{11}^n}\widetilde{E}_y^n+\left(\mu_{22}^n-\frac{\mu_{21}^n\mu_{12}^n}{\omega\mu_0\mu_{11}^n}\right)\widetilde{H}_y^n-\left(\mu_{23}^n-\frac{\mu_{21}^n\mu_{13}}{\omega\mu_0\mu_{11}^n}\right)\widetilde{H}_z^n\right\}\Bigg|_{y=0}$$

$$=\frac{\kappa_x^n\tilde{j}_{Mx}+\kappa_z^n\tilde{j}_{Mz}}{j\omega}. \tag{2.5b}$$

Let's introduce a 4×2 matrix $\overleftrightarrow{S}_n$

$$\overleftrightarrow{S}_n=\left\|\begin{matrix}\varepsilon_{22}^n-\frac{\varepsilon_{21}^n\varepsilon_{12}^n}{\varepsilon_{11}^n}; & \varepsilon_{23}^n-\frac{\varepsilon_{21}^n\varepsilon_{13}^n}{\varepsilon_{11}^n}; & \frac{\varepsilon_{21}^n\kappa_z}{\omega\varepsilon_0\varepsilon_{11}^n}; & -\frac{\varepsilon_{21}^n\kappa_y}{\omega\varepsilon_0\varepsilon_{11}^n}\\ \frac{\mu_{21}^n\kappa_y}{\omega\mu_0\mu_{11}^n}; & -\frac{\mu_{21}^n\kappa_z}{\omega\mu_0\mu_{11}^n}; & \mu_{22}^n-\frac{\mu_{21}^n\mu_{12}^n}{\omega\mu_0\mu_{11}^n}; & \mu_{23}^n-\frac{\mu_{21}^n\mu_{13}}{\omega\mu_0\mu_{11}^n}\end{matrix}\right\|, \tag{2.6}$$

a 1×2 matrix $\overleftrightarrow{\mathrm{K}}_n$ and a 2×2 matrix $\overleftrightarrow{\tilde{J}}$

$$\overleftrightarrow{\mathrm{K}}_n=\left\|\begin{matrix}\kappa_x^n\\ \kappa_z\end{matrix}\right\|,\quad \overleftrightarrow{\tilde{J}}=\left\|\begin{matrix}\tilde{j}_{Ex}\tilde{j}_{Ez}\\ \tilde{j}_{Mx}\tilde{j}_{Mz}\end{matrix}\right\|. \tag{2.7}$$

Then boundary conditions Eqs. (2.5a) and (2.5b) in view of Eqs. (2.6) and (2.7) in the matrix form are

$$\overleftrightarrow{\tilde{N}}_n\big|_{y=0}\overleftrightarrow{S}_n=\overleftrightarrow{\mathrm{K}}_n\overleftrightarrow{J}_n, \tag{2.8}$$

giving 1×2 matrices in both the left-hand and right-hand sides.

For change-over of Eq. (2.8) to 1×4 column matrices let's use the following transformation. Take a square 2×4 matrix $\overleftrightarrow{X}_n$, for which

$$\overleftrightarrow{S}_n\overleftrightarrow{X}_n=\overleftrightarrow{I}, \tag{2.9}$$

where $\overleftrightarrow{I}$ is a 4×4 identity matrix.

Then, after multiplication of Eq. (2.8) by $\overleftrightarrow{X}_n$ including Eq. (2.9), we have

$$\overleftrightarrow{\tilde{N}}_n\big|_{y=0}\overleftrightarrow{I} = \overleftrightarrow{K}_n \overleftrightarrow{\tilde{J}}_n \overleftrightarrow{X}_n. \tag{2.10}$$

The last relation is the matrix form of the boundary condition on the sources' surface of the converter, and

$$\overleftrightarrow{\tilde{N}}_n(\kappa_x^n, \kappa_y, \kappa_z) = \overleftrightarrow{\tilde{J}}_1(\kappa_x^n, \kappa_y, \kappa_z), \tag{2.11}$$

where $\overleftrightarrow{\tilde{J}}_1 = \overleftrightarrow{K}_n \overleftrightarrow{\tilde{J}} \overleftrightarrow{X}_n$.

As both functions in Eq. (2.11) are integrable, we can take the inverse Fourier transform. Then

$$\overleftrightarrow{N}_n(x,y,z) = \overleftrightarrow{J}_1(x,z). \tag{2.12}$$

Allowing for Eqs. (2.7) and (2.11) we represent Eq. (2.12) as

$$\overleftrightarrow{N}_n(x,y,z)\Bigg|_{y=0} = \overleftrightarrow{J}_{1E}(x,z) + \overleftrightarrow{J}_{1M}(x,z), \tag{2.13}$$

where $\overleftrightarrow{J}_{1E}(x,z) = \overleftrightarrow{f}_x(j_{Ex}) + \overleftrightarrow{f}_z(j_{Ez})$, $\overleftrightarrow{J}_{1M}(x,z) = \overleftrightarrow{g}_x(j_{Mx}) + \overleftrightarrow{g}_z(j_{Mz})$.

Introduce Green's function in the form of

$$\begin{aligned}\overleftrightarrow{N}_{nE}(x,y,z) = &\iint\limits_{S_E} \overleftrightarrow{G}_{nE}(x,y,z;x_E,z_E)\,\overleftrightarrow{J}_{1E}(x_E,z_E)\mathrm{d}x_E\mathrm{d}z_E \\ &+ \iint\limits_{S_M} \overleftrightarrow{G}_{nM}(x,y,z;x_M,z_M)\,\overleftrightarrow{J}_{1M}(x_M,z_M)\mathrm{d}x_M\mathrm{d}z_M,\end{aligned} \tag{2.14a}$$

where

$$\overleftrightarrow{G}_{nE} = \begin{Vmatrix} G_{1E}^n \\ G_{2E}^n \\ G_{3E}^n \\ G_{4E}^n \end{Vmatrix} \quad \mathit{and} \quad \overleftrightarrow{G}_{nM} = \begin{Vmatrix} G_{1M}^n \\ G_{2M}^n \\ G_{3M}^n \\ G_{4M}^n \end{Vmatrix}.$$

Considering the similarity of both Green's functions and integrals in Eq. (2.14a), below we shall consider only one of the terms, omitting the "E" and "M" subscripts (electric and magnetic currents of excitation) at Green's function and using an "s" subscript (source).

Then we define Green's function as

$$\overleftrightarrow{N}_n(x,y,z) = \iint\limits_{S_s} \overleftrightarrow{G}_n(x,y,z;x_s,z_s)\,\overleftrightarrow{J}_{1s}(x_s,z_s)\mathrm{d}x_s\mathrm{d}z_s. \tag{2.14b}$$

Similarly, use boundary condition (2.13) in the form of

$$\overleftrightarrow{N}_n(x,y,z)\Bigg|_{y=0} = \overleftrightarrow{J}_{1s}(x,z). \tag{2.15}$$

Transforming the right-hand side of Eq. (2.15) and using the properties of δ-functions, in view of Eq. (2.14b), we have

$$\iint\limits_{S_s} \overleftrightarrow{G}_n(x,y,z;x_s,z_s)\Bigg|_{y=0} \overleftrightarrow{J}_{1s}(x_s,z_s)\mathrm{d}x_s\mathrm{d}z_s = \iint\limits_{S_s} \delta(x-x_s),\delta(z-z_s)\,\overleftrightarrow{J}_{1s}(x_s,z_s)\mathrm{d}x_s\mathrm{d}z_s,$$

from which the boundary condition for Green's function follows:

$$\overleftrightarrow{G}_n(x,y,z;x_s,z_s)\Bigg|_{y=0} = \delta(x-x_s)\delta(z-z_s). \tag{2.16}$$

Let's transform Eq. (2.16) by taking the Fourier transform by the coordinates x and z, then we have

$$\begin{aligned}&\frac{1}{2\pi}\iint \overleftrightarrow{\tilde{G}}_n(\kappa_x^n,y,\kappa_z;x_s,z_s)\Big|e^{-j(\kappa_x^n x+\kappa_z z)}\,\mathrm{d}\kappa_x^n\mathrm{d}\kappa_z\\ &= \frac{\overleftrightarrow{F}}{2\pi}\iint e^{-j[\kappa_x^n(x-x_s)+\kappa_z(z-z_s)]}\mathrm{d}\kappa_x^n\mathrm{d}\kappa_z,\end{aligned} \tag{2.17}$$

with the column matrix $\overleftrightarrow{F} = \begin{Vmatrix}1\\1\\1\\1\end{Vmatrix}$.

Equation (2.17) is followed by

$$\overleftrightarrow{\tilde{G}}_n(\kappa_x^n,\kappa_y,\kappa_z;x_s,z_s)\Bigg|_{y=0} = \frac{\overleftrightarrow{F}}{2\pi}e^{-j(\kappa_x^n x_s+\kappa_z z_s)}. \tag{2.18}$$

After taking the inverse Fourier transform of Eq. (2.18) we have Green's functions on the plane of the sources $y=0$.

$$\overleftrightarrow{\tilde{G}}_n(x,y,z;x_s,z_s) = \frac{\overleftrightarrow{F}}{2\pi}\int\limits_{-\infty}^{\infty}\int e^{-j[\kappa_x^n(x-x_s)+\kappa_z(z-z_s)]}\mathrm{d}\kappa_x^n\mathrm{d}\kappa_z. \tag{2.19}$$

Assume that the required Green function can be represented as

$$\overleftrightarrow{\tilde{\tilde{G}}}_n(x,y,z;x_s,z_s) = \overleftrightarrow{\tilde{G}}_n(x,y,z;x_s,z_s)\Big|_{y=0} \times \psi(y), \tag{2.20}$$

where $\psi(y)$ is the function satisfying the condition of radiation behavior at infinity $(\lim_{y\to\infty} \psi(y) = 0)$.

Multiplying Eq. (2.19) by $\psi(y)$ and considering Eq. (2.20), after taking the Fourier transform by "y" in the right-hand side we have

$$\overleftrightarrow{\tilde{G}}_n(x,y,z;x_s,z_s) = \frac{\overleftrightarrow{F}}{(2\pi)^{5/2}} \int\int_{-\infty}^{\infty}\int e^{-j[\kappa_x^n(x-x_s)+\kappa_y y+\kappa_z(z-z_s)]}\psi(\kappa_y)\mathrm{d}\kappa_z^n \mathrm{d}\kappa_y \mathrm{d}\kappa_z. \tag{2.21}$$

As the transverse wave number $\kappa_x = \kappa_x(\kappa_y, \kappa_z)$, that in Eq. (2.21) after integration with respect to κ_x^n we have

$$\overleftrightarrow{G}_n(x,y,z;x_s,z_s) = \frac{\overleftrightarrow{F}}{(2\pi)^{5/2}} \int_{-\infty}^{\infty}\int e^{q_n(x-x_s)} e^{-j[\kappa_y y+\kappa_z(z-z_s)]}\tilde{\psi}(\kappa_y)\mathrm{d}\kappa_y \mathrm{d}\kappa_z, \tag{2.22}$$

where q_n corresponds to the transverse wave number (four and two for bigyrotropic and gyrotropic structures, respectively), the sign before which, in view of the sign before $(x-x_s)$, should satisfy the condition of energy attenuation of a wave excited in the structure.

We shall note that Eq. (2.22) defines Green's function, which is distinct from zero only in a region with an extraneous source and satisfies the boundary conditions on the surface of this source. On the other hand, in each area there are also own waves satisfying the boundary conditions by the tangential components on the interfaces of the layers of the bigyrotropic structure. Consequently, for linear wave processes (note that in the millimeter range the dynamic range in ferrites of the linear and near-linear modes is expanded to continuous and average power levels of few Watts in the pulse mode) Green's $\overleftrightarrow{G}_n$ function in each layer of the structure is

$$\overleftrightarrow{G}_n = \overleftrightarrow{G_n^i} + \overleftrightarrow{G_n^s} \; (n = 1,2,\ldots,p), \tag{2.23}$$

where $\overleftrightarrow{G_n^s}$ is a function, related to the own waves of the structure, should satisfy standard boundary conditions on interfaces, it is defined in shielded bigyrotropic structures in any section X through coupling matrices $\overleftrightarrow{T}_n(x-x_0)$.

$\overleftrightarrow{G_n^i}$ is a function, related to the sources, should satisfy the boundary conditions on the surface $y = 0$ of localization of extraneous current sources; if there are no sources within some area, then $\overleftrightarrow{G_\kappa^i} = 0$.

Extraneous current sources can be in any area of the layered structure on the surface $y = 0$; their number can be any and, in particular, equal to the number of layers p. This means that in this case it is necessary to consider P independent subtasks, including extraneous surface electric and magnetic currents in each task. According to the principle of superposition, a resultant Green's function appears:

$$\overleftrightarrow{G}_{rE} = \sum_{n=1}^{p} \overleftrightarrow{G}_n.$$

Assume the source to e on a plane within layer m, and for FIGF

– At $x_{m-1} < x \leq x_m$

$$\overleftrightarrow{\tilde{G}}_m = \overleftrightarrow{\tilde{G}}^i_m + \overleftrightarrow{\tilde{G}}^s_m \tag{2.24a}$$

– At $x > x_m$ and $x < x_{m-1}$

$$\overleftrightarrow{\tilde{G}}_n = \overleftrightarrow{\tilde{G}}^s_n \tag{2.24b}$$

Further we shall apply a procedure to get a non-uniform matrix equation, similar to that in Section 1.2.

On the surfaces of layer m the boundary conditions for FIGF will be as follows:

– At $x = x_m$

$$\overleftrightarrow{\tilde{G}}_{m+1}(x_m) = \overleftrightarrow{\tilde{G}}^s_m(x_m) + \overleftrightarrow{\tilde{G}}^i_m(x_m) \tag{2.25a}$$

– At $x = x_{m-1}$

$$\overleftrightarrow{\tilde{G}}_{m-1}(x_{m-1}) = \overleftrightarrow{\tilde{G}}^s_m(x_{m-1}) + \overleftrightarrow{\tilde{G}}^i_m(x_{m-1}) \tag{2.25b}$$

Let's express the quantities $\overleftrightarrow{\tilde{G}}_p(x_p)$ and $\overleftrightarrow{\tilde{G}}_1(0)$ on the limiting surfaces at $x = x_p$ and $x = 0$ through coupling matrixes in the corresponding half-spaces of the structure and the value of FIGF $\overleftrightarrow{\tilde{G}}_{m-1}(x_{m-1})$ and $\overleftrightarrow{\tilde{G}}_{m+1}(x_m)$ on the surfaces of layer m with sources

$$\overleftrightarrow{\tilde{G}}_p(x_p) = \overleftrightarrow{T}_p(h_p) \cdot \ldots \cdot \overleftrightarrow{T}_{m+1}(h_{m+1})\overleftrightarrow{\tilde{G}}_{m+1}(x_m), \tag{2.26a}$$

$$\overleftrightarrow{\tilde{G}}_1(0) = \overleftrightarrow{T}_1(-h_1) \cdot \ldots \cdot \overleftrightarrow{T}_{m-1}(-h_{m-1})\overleftrightarrow{\tilde{G}}_{m-1}(x_{m-1}). \tag{2.26b}$$

Let's express $\overleftrightarrow{\tilde{G}}_{m+1}(x_m)$ and $\overleftrightarrow{\tilde{G}}_{m-1}(x_{m-1})$ from Eq. (2.26a) as

$$\overleftrightarrow{\tilde{G}}_{m+1}(x_m) = [\overleftrightarrow{T}_p(h_p) \cdot \ldots \cdot \overleftrightarrow{T}_{m+1}(h_{m+1})]^{-1}\overleftrightarrow{\tilde{G}}_p(x_p), \tag{2.27a}$$

$$\overleftrightarrow{\tilde{G}}_{m-1}(x_{m-1}) = [\overleftrightarrow{T}_1(-h_1) \cdot \ldots \cdot \overleftrightarrow{T}_{m-1}(-h_{m-1})]^{-1}\overleftrightarrow{\tilde{G}}_1(0). \tag{2.27b}$$

Then, in view of Eqs. (2.27a), (2.27b), we get boundary conditions from Eqs. (2.26a), (2.26b) in the form of:

– At $x = x_m$

$$[\overleftrightarrow{T}_p(h_p)\cdot\ldots\cdot\overleftrightarrow{T}_{m+1}(h_{m+1})]^{-1}\overleftrightarrow{\tilde{G}}_p(x_p) - \overleftrightarrow{\tilde{G}}^i_m(x_m) = \overleftrightarrow{\tilde{G}}^s_m(x_m) \tag{2.28a}$$

– At $x = x_{m-1}$

$$[\overleftrightarrow{T}_1(-h_1)\cdot\ldots\cdot\overleftrightarrow{T}_{m-1}(-h_{m-1})]^{-1}\overleftrightarrow{\tilde{G}}_1(0) - \overleftrightarrow{\tilde{G}}^i_m(x_{m-1}) = \overleftrightarrow{\tilde{G}}^s_m(x_{m-1}). \tag{2.28b}$$

The relation between $\overleftrightarrow{\tilde{G}}^s_m(x_m)$ and $\overleftrightarrow{\tilde{G}}^s_m(x_{m-1})$ on the borders of the layer is defined through the coupling matrix $\overleftrightarrow{T}_m(x_m)$, and

$$\overleftrightarrow{\tilde{G}}^s_m(x_m) = \overleftrightarrow{T}_m(x_m)\overleftrightarrow{\tilde{G}}^s_m(x_{m-1}). \tag{2.29}$$

Subject to Eq. (2.29), the boundary condition at $x = x_m$ Eq. (2.28a) will be

$$\overleftrightarrow{T}^{-1}_m(x_m)[\overleftrightarrow{T}_p(h_p)\cdot\ldots\cdot\overleftrightarrow{T}_{m+1}(h_{m+1})]^{-1}\overleftrightarrow{\tilde{G}}_p(x_p) - \overleftrightarrow{T}^{-1}(x_m)\overleftrightarrow{\tilde{G}}^i_m(x_m) = \overleftrightarrow{\tilde{G}}^s_m(x_{m-1}), \tag{2.30}$$

from which, in view of Eq. (2.28b), a matrix equation follows

$$\begin{aligned}&\overleftrightarrow{T}^{-1}_m(x_m)[\overleftrightarrow{T}_p(h_p)\cdot\ldots\cdot\overleftrightarrow{T}_{m+1}(h_{m+1})]^{-1}\overleftrightarrow{\tilde{G}}_p(x_p) - \overleftrightarrow{T}^{-1}(x_m)\overleftrightarrow{\tilde{G}}^i_m(x_m)\\ &= [\overleftrightarrow{T}_1(-h_1)\cdot\ldots\cdot\overleftrightarrow{T}_{m-1}(-h_{m-1})]\cdot\overleftrightarrow{\tilde{G}}_1(0) - \overleftrightarrow{\tilde{G}}^i_m(x_{m-1}).\end{aligned} \tag{2.31}$$

Equation (2.31) can be rewritten as

$$\overleftrightarrow{B}\cdot\overleftrightarrow{\tilde{G}}_p(x_p) + \overleftrightarrow{C}\cdot\overleftrightarrow{\tilde{G}}_1(0) = \overleftrightarrow{T}^{-1}_m(h_m)\overleftrightarrow{\tilde{G}}^i_m(x_m) - \overleftrightarrow{\tilde{G}}^i_m(x_{m-1}), \tag{2.32}$$

where

$$\begin{aligned}\overleftrightarrow{B} &= \overleftrightarrow{T}^{-1}_m(x_m)[\overleftrightarrow{T}_p(h_p)\cdot\ldots\cdot\overleftrightarrow{T}_{m+1}(h_{m+1})]^{-1},\\ \overleftrightarrow{C} &= -[\overleftrightarrow{T}_1(-h_1)\cdot\ldots\cdot\overleftrightarrow{T}_{m-1}(-h_{m-1})]^{-1}.\end{aligned}$$

According to the specified FIGF $\overleftrightarrow{\tilde{G}}^i_m$, the right-hand side of Eq. (2.32) is distinct from zero. We shall transform it to a form similar to Eq. (1.50). Using Eq. (2.22), we have

$$\begin{aligned}&\overleftrightarrow{T}^{-1}_m(h_m)\overleftrightarrow{\tilde{G}}^i_m(x_m) - \overleftrightarrow{\tilde{G}}^i_m(x_{m-1}) = \overleftrightarrow{T}_m(h_m)\frac{\overleftrightarrow{F}}{(2\pi)^{5/2}}e^{q_m(x_m-x_s)}e^{-j[\kappa_y y+\kappa_z(z-z_s)]}\\ &\times\tilde{\psi}(\kappa_y) - \frac{\overleftrightarrow{F}}{(2\pi)^{5/2}}e^{q_m x_{m-1}}e^{-j[\kappa_y y-\kappa_z(z-z_s)]}\tilde{\psi}(\kappa_y)\\ &= \frac{e^{-j[\kappa_y y+\kappa_z(z-z_s)]}}{(2\pi)^{3/2}}\times\psi(\kappa_y)\left[\overleftrightarrow{T}^{-1}_m(h_m)\overleftrightarrow{F}\cdot e^{q_m(x_m x_s)} - \overleftrightarrow{F}\cdot e^{q_m(x_{m-1}x_s)}\right].\end{aligned} \tag{2.33}$$

Then, subject to the last transformations, the non-uniform matrix Eq. (2.33) will become

$$\overleftrightarrow{B}\overleftrightarrow{\tilde{G}}_p(x_p) - \overleftrightarrow{C}\overleftrightarrow{\tilde{G}}_1(0) = \overleftrightarrow{F}_1, \quad (2.34)$$

where

$$\overleftrightarrow{F}_1 = \frac{e^{j\kappa_z z_s}}{(2\pi)^{3/2}}\left[\overleftrightarrow{\tilde{T}}_m^{-1}(h_m)\overleftrightarrow{F}e^{q_m(x_m - x_s)} - \overleftrightarrow{F}e^{q_m(x_{m-1} - x_s)}\right].$$

The derived equation allows FIGF $\overleftrightarrow{\tilde{G}}_p(x_p)$ and $\overleftrightarrow{\tilde{G}}_1(0)$ on the structure-limiting screens to be determined in the approximation of given current, and, by means of coupling matrices, allows finding FIGF in the layers of the structure and, finally, determination of the amplitudes of HF fields of the excited waves and power fluxes.

2.2 Green's Functions for a Structure Shielded by Impedance Surfaces: Self-consistent Excitation

Generally, screens limiting a layered bigyrotropic structure at $x = 0$ and $x = x_p$ can have an arbitrary surface impedance Z.

If the screens are electric walls $(E_{y,z}\ (x = 0, x = x_p) = 0)$, FIGF is:

- At $x = 0$

$$\overleftrightarrow{\tilde{G}}{}_1^E(0) = \begin{Vmatrix} 0 \\ 0 \\ \tilde{G}_{13}^E(0) \\ \tilde{G}_{14}^E(0) \end{Vmatrix} \quad (2.35)$$

- At $x = x_p$

$$\overleftrightarrow{\tilde{G}}{}_p^E(x_p) = \begin{Vmatrix} 0 \\ 0 \\ \tilde{G}_{p3}^E(0) \\ \tilde{G}_{p4}^E(0) \end{Vmatrix}$$

and the determinant, corresponding to this case of the homogeneous equation set (2.34), looks as

$$\Delta_E^m = \det\begin{Vmatrix} b_{13}^m & b_{14}^m & c_{13}^m & c_{14}^m \\ b_{23}^m & b_{24}^m & c_{23}^m & c_{24}^m \\ b_{33}^m & b_{34}^m & c_{33}^m & c_{34}^m \\ b_{43}^m & b_{44}^m & c_{43}^m & c_{44}^m \end{Vmatrix} \quad (2.36)$$

If the screens are magnetic walls $(H_{y,z}(x=0,\ x=x_p=0))$, FIGF will be:

– At $x=0$

$$\overleftrightarrow{\widetilde{G}}{}_1^M(0) = \begin{Vmatrix} \widetilde{G}_{11}^M(0) \\ \widetilde{G}_{12}^M(0) \\ 0 \\ 0 \end{Vmatrix} \tag{2.37}$$

– At $x=x_p$

$$\overleftrightarrow{\widetilde{G}}{}_p^M(x_p) = \begin{Vmatrix} \widetilde{G}_{11}^M(x_p) \\ \widetilde{G}_{12}^M(x_p) \\ 0 \\ 0 \end{Vmatrix},$$

and the determinant of the corresponding homogeneous equation set (2.34) looks as

$$\Delta_E^m = \det \begin{Vmatrix} b_{11}^m & b_{12}^m & c_{11}^m & c_{12}^m \\ b_{21}^m & b_{22}^m & c_{21}^m & b_{22}^m \\ b_{31}^m & b_{32}^m & c_{31}^m & c_{32}^m \\ b_{41}^m & b_{42}^m & c_{41}^m & c_{42}^m \end{Vmatrix} \tag{2.38}$$

Let's note that the conditions $\Delta_E^m = 0$ and $\Delta_M^m = 0$ define wave dispersions in structures shielded by electric and magnetic walls, and

$$\begin{aligned} \Delta_E^m &= \Delta_E^p = \Delta_E \\ \Delta_M^m &= \Delta_M^p = \Delta_M \end{aligned} \tag{2.39}$$

for all p.

Consider two cases of the shielding surfaces being electric and magnetic walls.

2.2.1 Electrical Walls

In this case the corresponding determinants of the unknown constituents of Green's function tensor in Eq. (2.34) are

$$\Delta_{\widetilde{G}_{P3}^E}^m = \det \begin{Vmatrix} F_{11}^m & b_{14}^m & c_{13}^m & c_{14}^m \\ F_{12}^m & b_{24}^m & c_{23}^m & c_{24}^m \\ F_{13}^m & b_{34}^m & c_{33}^m & c_{34}^m \\ F_{14}^m & b_{44}^m & c_{43}^m & c_{44}^m \end{Vmatrix} \quad \Delta_{\widetilde{G}_{P4}^E}^m = \det \begin{Vmatrix} lb_{13}^m & F_{11}^m & c_{13}^m & c_{14}^m \\ b_{23}^m & F_{12}^m & c_{23}^m & b_{24}^m \\ b_{33}^m & F_{13}^m & c_{33}^m & c_{34}^m \\ b_{43}^m & F_{14}^m & c_{43}^m & c_{44}^m \end{Vmatrix},$$

$$\Delta_{\widetilde{G}_{13}^E}^m = \det \begin{Vmatrix} b_{13}^m & b_{14}^m & F_{11}^m & c_{14}^m \\ b_{23}^m & b_{24}^m & F_{12}^m & c_{24}^m \\ b_{33}^m & b_{34}^m & F_{13}^m & c_{34}^m \\ b_{43}^m & b_{44}^m & F_{14}^m & c_{44}^m \end{Vmatrix} \quad \Delta_{\widetilde{G}_{14}^E}^m = \det \begin{Vmatrix} b_{13}^m & b_{14}^m & c_{13}^m & F_{11}^m \\ b_{23}^m & b_{24}^m & c_{23}^m & F_{12}^m \\ b_{33}^m & b_{34}^m & c_{33}^m & F_{13}^m \\ b_{43}^m & b_{44}^m & c_{43}^m & F_{14}^m \end{Vmatrix}, \tag{2.40}$$

and, being represented through cofactors,

$$\Delta^m_{\widetilde{G}^E_{p3}} = \sum_{n=1}^{4} F^m_{1n}\{\Delta^m_E\}_{1n} \quad \Delta^m_{\widetilde{G}^E_{p4}} = \sum_{n=1}^{4} F^m_{1n}\{\Delta^m_E\}_{2n},$$
$$\Delta^m_{\widetilde{G}^E_{13}} = \sum_{n=1}^{4} F^m_{1n}\{\Delta^m_E\}_{3n} \quad \Delta^m_{\widetilde{G}^E_{14}} = \sum_{n=1}^{4} F^m_{1n}\{\Delta^m_E\}_{4n}. \tag{2.41}$$

Then from Eqs. (2.36) and (2.41) the FIGF components follow:

$$\widetilde{G}^E_{p_3}(x_p) = \frac{\Delta^m_{\widetilde{G}^E_{p3}}}{\Delta_E} = \frac{\sum\limits_{n=1}^{4} F^m_{1n}\{\Delta^m_E\}_{1n}}{\Delta_E},$$
$$\widetilde{G}^E_{p_4}(x_p) = \frac{\Delta^m_{\widetilde{G}^E_{p4}}}{\Delta_E} = \frac{\sum\limits_{n=1}^{4} F^m_{1n}\{\Delta^m_E\}_{2n}}{\Delta_E}, \tag{2.42}$$
$$\widetilde{G}^E_{13}(0) = \frac{\Delta^m_{\widetilde{G}^E_{13}}}{\Delta_E} = \frac{\sum\limits_{n=1}^{4} F^m_{1n}\{\Delta^m_E\}_{3n}}{\Delta_E}, \quad \widetilde{G}^E_{14}(0) = \frac{\Delta^m_{\widetilde{G}^E_{14}}}{\Delta_E} = \frac{\sum\limits_{n=1}^{4} F^m_{1n}\{\Delta_E m\}_{4n}}{\Delta_E}.$$

From Eq. (2.34), subject to Eq. (2.42), we have FIGF tensors on the electric walls:

– At $x=0$

$$\overset{\leftrightarrow}{\widetilde{G}}_1(0) = \frac{1}{\Delta_E}\sum_{n=1}^{4} F^m_{1n}\left\|\begin{matrix} 0 \\ 0 \\ \{\Delta^m_E\}_{3n} \\ \{\Delta^m_E\}_{4n} \end{matrix}\right\| \tag{2.42a}$$

– At $x=x_p$

$$\overset{\leftrightarrow}{\widetilde{G}}_p(x_p) = \frac{1}{\Delta_E}\sum_{n=1}^{4} F^m_{1n}\left\|\begin{matrix} 0 \\ 0 \\ \{\Delta^m_E\}_{1n} \\ \{\Delta^m_E\}_{2n} \end{matrix}\right\| \tag{2.42b}$$

Then from Eq. (2.14b) the spectral densities of the tangential components are

$$\overset{\leftrightarrow}{\widetilde{N}}{}^E(x,\kappa_y,\kappa_z) = \overset{\leftrightarrow}{T}_q(x_q)\cdot\overset{\leftrightarrow}{T}_{q-1}(h_{q-1})\cdot\ldots\cdot\overset{\leftrightarrow}{T}_1(h_1)\overset{\leftrightarrow}{\widetilde{V}}_E, \tag{2.43}$$

where

$$\overset{\leftrightarrow}{\widetilde{V}}_E = \sum_{m=1}^{p}\int\!\!\int_{S_s}\sum_{n=1}^{4} F^m_{1n}\left\|\begin{matrix} 0 \\ 0 \\ \{\Delta^E\}_{3n} \\ \{\Delta^E\}_{4n} \end{matrix}\right\| \overset{\leftrightarrow}{\widetilde{J}}_{1\perp}\mathrm{d}x_s\mathrm{d}z_s,$$

where the subscript "⊥" means the case of a converter with orthogonal orientation of its structure to the exciting plane, or, in another form, $\overset{\leftrightarrow}{\tilde{V}}_E = \sum_{m=1}^{p} \sum_{n=1}^{L} \left\| \begin{matrix} 0 \\ 0 \\ \{\Delta^E\}_{3n} \\ \{\Delta^E\}_{4n} \end{matrix} \right\| \overset{\leftrightarrow}{\tilde{F}}{}^m_{\kappa}$,

where $\overset{\leftrightarrow}{\tilde{F}}{}^m_{\kappa} = \int\int_{S_s} F^m_{1n} \overset{\leftrightarrow}{\tilde{J}}_{1\perp} \mathrm{d}x_s \mathrm{d}z_s$.

Equation (2.43) is similar to Eq. (1.73). After taking the inverse Fourier transform, by analogy with Eqs. (1.74) to (1.77), from Eq. (2.43) we get

$$\overset{\leftrightarrow}{\tilde{N}}{}^E_q(x,y,z) = j \sum_{j=1}^{l} \overset{\leftrightarrow}{T}_q(x) \cdot \ldots \cdot \overset{\leftrightarrow}{T}_1(h_1) \cdot \frac{e^{-j\left\{\sqrt{y^2+z^2}\Phi\left[arctg\left(\frac{z}{y}\right)\right]\right\}}}{\Delta^m_E} \times \sqrt{\frac{2\pi}{j\sqrt{y^2+z^2}\Phi''(\theta_s)}} \overset{\leftrightarrow}{V}_E. \tag{2.44}$$

Equation (2.44) allows the tangential components of HF fields of the excited wave in a structure limited by electric walls to be found.

2.2.2 Magnetic Walls

In this case the corresponding determinants in Eq. (2.34) look as

$$\Delta^m_{\tilde{G}^M_{p1}} = \det \left\| \begin{matrix} F^m_{11} & b^m_{12} & c^m_{11} & c^m_{22} \\ F^m_{12} & b^m_{22} & c^m_{21} & c^m_{22} \\ F^m_{13} & b^m_{32} & c^m_{31} & c^m_{32} \\ F^m_{14} & b^m_{42} & c^m_{41} & c^m_{42} \end{matrix} \right\|, \quad \Delta^m_{\tilde{G}^M_{p2}} = \det \left\| \begin{matrix} b^m_{11} & F^m_{11} & c^m_{11} & c^m_{12} \\ b^m_{21} & F^m_{12} & c^m_{21} & b^m_{22} \\ b^m_{31} & F^m_{13} & c^m_{31} & c^m_{32} \\ b^m_{41} & F^m_{14} & c^m_{41} & c^m_{42} \end{matrix} \right\|,$$

$$\Delta^m_{\tilde{G}^M_{11}} = \det \left\| \begin{matrix} b^m_{11} & b^m_{12} & F^m_{11} & c^m_{12} \\ b^m_{21} & b^m_{22} & F^m_{12} & c^m_{22} \\ b^m_{31} & b^m_{32} & F^m_{13} & c^m_{32} \\ b^m_{41} & b^m_{42} & F^m_{14} & c^m_{42} \end{matrix} \right\|, \quad \Delta^m_{\tilde{G}^M_{12}} = \det \left\| \begin{matrix} b^m_{11} & b^m_{12} & c^m_{11} & F^m_{11} \\ b^m_{21} & b^m_{22} & c^m_{21} & F^m_{12} \\ b^m_{31} & b^m_{32} & c^m_{31} & F^m_{13} \\ b^m_{41} & b^m_{42} & c^m_{41} & F^m_{14} \end{matrix} \right\|, \tag{2.45}$$

or, through cofactors,

$$\Delta^m_{\tilde{G}^M_{p1}} = \sum_{n=1}^{4} F^m_{1n} \{\Delta^m_M\}_{1n}, \quad \Delta^m_{\tilde{G}^M_{p2}} = \sum_{n=1}^{4} F^m_{1n} \{\Delta^m_M\}_{2n},$$
$$\Delta^m_{\tilde{G}^M_{11}} = \sum_{n=1}^{4} F^m_{1n} \{\Delta^m_M\}_{3n}, \quad \Delta^m_{\tilde{G}^M_{12}} = \sum_{n=1}^{4} F^m_{1n} \{\Delta^m_M\}_{4n}. \tag{2.46}$$

Then from Eqs. (2.38) and (2.44) the FIGF components follow

$$\tilde{G}^M_{p_1} = \frac{\Delta^m_{\tilde{G}^M_{p1}}}{\Delta_M} = \frac{\sum\limits_{n=1}^{4} F^m_{1n}\{\Delta^m_M\}_{1n}}{\Delta_M}, \quad \tilde{G}^M_{p_2} = \frac{\Delta^m_{\tilde{G}^M_{p2}}}{\Delta_M} = \frac{\sum\limits_{n=1}^{4} F^m_{1n}\{\Delta^m_M\}_{2n}}{\Delta_M},$$

$$\tilde{G}^M_{p_{11}} = \frac{\Delta^m_{\tilde{G}^M_{p11}}}{\Delta_M} = \frac{\sum\limits_{n=1}^{4} F^m_{1n}\{\Delta^m_M\}_{3n}}{\Delta_M}, \quad \tilde{G}^E_{p_{12}} = \frac{\Delta^m_{\tilde{G}^M_{p12}}}{\Delta_M} = \frac{\sum\limits_{n=1}^{4} F^m_{1n}\{\Delta^m_M\}_{4n}}{\Delta_M}. \quad (2.47)$$

From Eqs. (1.1) and (1.2), subject to Eq. (2.45), we have FIGF tensors of on the magnetic walls:

- At $x = 0$

$$\overset{\leftrightarrow}{\tilde{G}}_1(0) = \frac{1}{\Delta_M}\sum_{n=1}^{4} F^m_{1n} \left\| \begin{matrix} \{\Delta^m_M\}_{3n} \\ \{\Delta^m_M\}_{4n} \\ 0 \\ 0 \end{matrix} \right\| \quad (2.48)$$

- At $x = x_p$

$$\overset{\leftrightarrow}{\tilde{G}}_p(x_p) = \frac{1}{\Delta_M}\sum_{n=1}^{4} F^m_{1n} \left\| \begin{matrix} \{\Delta^m_M\}_{3n} \\ \{\Delta^m_M\}_{4n} \\ 0 \\ 0 \end{matrix} \right\| \quad (2.48a)$$

From Eq. (2.34) the spectral densities of the tangential components are

$$\overset{\leftrightarrow}{\tilde{N}}{}^M_q(x, \kappa_y, \kappa_z) = \overset{\leftrightarrow}{T}_q(x_q) \cdot \overset{\leftrightarrow}{T}_{q-1}(h_{q-1}) \cdot \ldots \cdot \overset{\leftrightarrow}{T}_1(h_1) \overset{\leftrightarrow}{\tilde{V}}_M, \quad (2.49)$$

where

$$\overset{\leftrightarrow}{\tilde{V}}_M = \sum_{m=1}^{p} \int\!\!\int_{S_m} \sum_{n=1}^{4} F^m_{1n} \left\| \begin{matrix} \{\Delta^M\}_{1n} \\ \{\Delta^M\}_{2n} \\ 0 \\ 0 \end{matrix} \right\| \cdot \overset{\leftrightarrow}{\tilde{J}}_{1\perp} dx_s dz_s. \quad (2.50)$$

After taking the inverse Fourier transform of Eq. (2.49) we have a column matrix of the tangential constituents of the fields of the waves excited by the converter with orthogonal structure orientation at the presence of screens as magnetic walls

$$\overset{\leftrightarrow}{\tilde{N}}{}^M_q(x,y,z) = j\sum_{0=1}^{l} \overset{\leftrightarrow}{T}_q(x) \cdot \ldots \cdot \overset{\leftrightarrow}{T}_1(h_1) \cdot \frac{e^{-j\left\{\sqrt{y^2+z^2}\Phi\left[arctg\left(\frac{z}{y}\right)\right]\right\}}}{\Delta^m_M} \times \sqrt{\frac{2\pi}{j\sqrt{y^2+z^2}\Phi''(\theta_s)}} \overset{\leftrightarrow}{V}_M. \quad (2.51)$$

Equation (2.50), like Eq. (2.43), is similar to Eq. (1.73). Taking the inverse Fourier transform, like Eqs. (1.74) to (1.77), we have

$$\begin{aligned}\overset{\leftrightarrow}{\tilde{N}}{}_p^M(x,y,z) &= j\sum_{j=1}^{l}\overset{\leftrightarrow}{T}_p(x_p)\cdot\ldots\cdot\overset{\leftrightarrow}{T}_1(h_1)\cdot\frac{1}{\Delta_1^M}\sum_{m=1}^{p}\sum_{n=1}^{L}F_n^m(j_E)\\ &\times\sqrt{\frac{2\pi}{j\sqrt{y^2+z^2}\Phi''(\theta_s)}}e^{-j\left\{\sqrt{y^2+z^2}\Phi(\theta)arctg\left(\frac{z}{y}\right)\right\}}\begin{Vmatrix}\{\Delta^M\}_{34}^m\\ \{\Delta^M\}_{44}^m\\ 0\\ 0\end{Vmatrix}.\end{aligned} \tag{2.52}$$

Equations (2.44) and (2.52) allow the power fluxes of a wave excited in the layered structure by orthogonal surface currents to be found.

The back influence of HF fields of the excited waves on the fields and currents of extraneous sources will be considered by analogy with converters with parallel structure orientation, but the normal components of the resulting fields are zero $S_E - H_y = 0$:

- On the surface of electric current $S_E - H_y = 0$
- On the surface of magnetic current $S_M - E_y = 0$

For example, consider a converter with the orthogonal orientation of a three-layer structure ($n = 3$). Adopt the extraneous currents $\bar{j}_E$ or $\bar{j}_M$ in each layer occupying the areas S_{En} and S_{Mn}, $n = 1,2,3$.

Then $H_{yn} = 0$ at $(x,z) \in S_{EM}$; $E_{yn} = 0$ at $(x,z) \in S_{Mn}$, and $n = 1,2,3$.

We can similarly write the matrix formula of the Fourier images of HF fields in the given-current approximation

$$\begin{aligned}\overset{\leftrightarrow}{\tilde{N}}{}_n^T(x,0,z) = \frac{1}{2\pi}\int\int e^{q_m(x-x_s)}\cdot\left[q_n(\kappa_y,\kappa_z)\overset{\leftrightarrow}{I} - j\overset{\leftrightarrow}{A}_n(\kappa_y,\kappa_z)\right]^{-1}\\ \times e^{j\kappa_z z_s}I_s(x_s,z_s)\mathrm{d}x_s\mathrm{d}z_s.\end{aligned} \tag{2.53}$$

The form of function $\overset{\leftrightarrow}{\tilde{N}}{}_n^B(x,0,z)$ for the source is similar to Eq. (1.81).

To derive the components $\tilde{E}_y^{B,T}$ and $\tilde{H}_y^{B,T}$ from $\overset{\leftrightarrow}{\tilde{N}}{}_n^B$ and $\overset{\leftrightarrow}{\tilde{N}}{}_n^T$, use matrices like

$$S_{H_y}^{\perp} = \|0\ \ 0\ \ 1\ \ 0\|, \tag{2.54}$$

$$S_{E_y}^{\perp} = \|1\ \ 0\ \ 0\ \ 0\|. \tag{2.55}$$

The integral equation for a converter with an electric current is

$$\begin{aligned}&\sum_{n=1}^{p}\overset{\leftrightarrow}{\tilde{S}}{}_{H_y}^{\perp}\left\{\overset{\leftrightarrow}{T}_n(x_n)\cdot\ldots\cdot\overset{\leftrightarrow}{T}_1(h_1)\tilde{\psi}(\kappa_y)\int\int_{S_E}\sum_{l=1}^{4}F_{1l}^n\cdot\begin{Vmatrix}0\\0\\ \{\Delta^E\}_{3l}\\ \{\Delta^E\}_{4l}\end{Vmatrix}\cdot\overset{\leftrightarrow}{\tilde{J}}(\bar{j}_E)\mathrm{d}x_s\mathrm{d}z_s\right.\\ &\left.+\frac{1}{2\pi}\int\int_{S_E}E^{q_n(x-x_s)}[q_n(\kappa_y,\kappa_z)\bar{I} - j\overset{\leftrightarrow}{A}_n(\kappa_y,\kappa_z)]^{-1}e^{j\kappa_z z_s}I_s(\bar{j}_\varepsilon)\,\mathrm{d}x_s\mathrm{d}z_s\right\} = 0.\end{aligned} \tag{2.56}$$

The integral equation for a converter with a magnetic current is

$$\sum_{n=1}^{p} \bar{S}_{E_y}^{\perp} \left\{ \overleftrightarrow{T}_n(x_n) \cdot \ldots \cdot \overleftrightarrow{T}_1(h_1) \tilde{\psi}(\kappa_y) \int\limits_{S_M}\int \sum_{l=1}^{4} F_{1l}^{n} \cdot \left\| \begin{matrix} 0 \\ 0 \\ \{\Delta^M\}_{3l} \\ \{\Delta^M\}_{4l} \end{matrix} \right\| \cdot \overleftrightarrow{\tilde{J}}(\bar{j}_M) \mathrm{d}x_s \mathrm{d}z_s \right. \tag{2.57}$$

$$\left. + \frac{1}{2\pi} \int\limits_{S_M}\int e^{q_n(x-x_s)} [q_n(\kappa_y, \kappa_z)\bar{I} - j\overleftrightarrow{A}_n(\kappa_y, \kappa_z)]^{-1} e^{j\kappa_z z_s} I_s(\bar{j}_M)\, \mathrm{d}x_s \mathrm{d}z_s \right\} = 0.$$

2.3 Three-Layered Structures

Consider the case, most widespread in practice, of a converter based on a three-layer ferrite-dielectric structure with shielding by electric walls. Assume the exciting current to be localized in the region of $y = 0$ of the layer with $n = 2$. Then

$$\overleftrightarrow{\tilde{V}}_E = \sum_{n=1}^{4} \left\| \begin{matrix} 0 \\ 0 \\ \left\{\Delta^E\right\}_{3n} \\ \left\{\Delta^E\right\}_{4n} \end{matrix} \right\| \cdot \overleftrightarrow{\tilde{F}}{}_n^2 \tag{2.58}$$

where $\overleftrightarrow{\tilde{F}}{}_n^2 = \int\limits_{S_2}\int \overleftrightarrow{\tilde{F}}{}_{1n}^2 \overleftrightarrow{\tilde{J}}{}_n^2 \mathrm{d}x_s \mathrm{d}z_s$.

The exciting electric current $\bar{j}_E$ or magnetic current $\bar{j}_M$ in the simplest case has only one of its components:

$$\overleftrightarrow{J}_{21} = \left\| \begin{matrix} 0 & j_{Ez} \\ 0 & 0 \end{matrix} \right\|, \quad \overleftrightarrow{J}_{22} = \left\| \begin{matrix} j_{Ex} & 0 \\ 0 & 0 \end{matrix} \right\|,$$
$$\overleftrightarrow{J}_{23} = \left\| \begin{matrix} 0 & 0 \\ 0 & j_{Mz} \end{matrix} \right\|, \quad \overleftrightarrow{J}_{24} = \left\| \begin{matrix} 0 & 0 \\ j_{Mx} & 0 \end{matrix} \right\|. \tag{2.59}$$

Let's note that the form of Eq. (2.59) will not change, be the exciting current in areas 1 or 2, 2 or 3 of the converter.

For our case we transform the right-hand side of Eq. (2.11), related with the source, to the form

$$\overleftrightarrow{\tilde{J}}{}_1^2 = j_\beta \overleftrightarrow{K}_2 \overleftrightarrow{J}_\beta \overleftrightarrow{x}_2, \tag{2.60}$$

where β specifies one of possible particular cases by Eq. (2.59)

$$\overleftrightarrow{J}_1 = \left\| \begin{matrix} 0 & 1 \\ 0 & 0 \end{matrix} \right\|, \quad \overleftrightarrow{J}_2 = \left\| \begin{matrix} 1 & 0 \\ 0 & 0 \end{matrix} \right\|,$$
$$\overleftrightarrow{J}_3 = \left\| \begin{matrix} 0 & 0 \\ 0 & 1 \end{matrix} \right\|, \quad \overleftrightarrow{J}_4 = \left\| \begin{matrix} 0 & 0 \\ 1 & 0 \end{matrix} \right\|. \tag{2.61}$$

Let the exciting current on the surface $y = 0$ be uniformly distributed within the limits of an area $W \times L$ (Fig. 2.2a, b), located at the center of the layer $n = 2$, and

$$J_\beta(x_s, z_s) = \begin{cases} j_{\beta 0} & \text{at } \frac{h_1+h_2+h_3}{2} - \frac{L}{2} \le x_s \le \frac{h_1+h_2+h_3}{2} + \frac{L}{2}, -\frac{w}{2} \le z_s \le \frac{2w}{2}, \\ 0 & \text{at other senses} \end{cases}$$

Subject to Eqs. (2.60) and (2.62), Eq. (2.43) looks as

$$\overleftrightarrow{\widetilde{F}}_s^2 = \overleftrightarrow{K}_2 \overleftrightarrow{\widetilde{I}}_1^2 \overleftrightarrow{x}_2 \int_{-w/2}^{w/2} dz_s \int_{-\frac{h_1+h_2+h_3}{2} - \frac{L}{2}}^{\frac{h_1+h_2+h_3}{2} + \frac{L}{2}} (F_{in})^2 \mathrm{d}x_s = \overleftrightarrow{K}_2 \overleftrightarrow{\widetilde{I}}_1^2 \overleftrightarrow{x}_2 I_2.$$

The function of sources in Eq. (2.34) in this case has the form

$$F_{1n}^2 = \frac{e^{j\kappa_z z_s}}{(2\pi)^{3/2}} \overleftrightarrow{T}_2^{-1}(h_2) F \cdot e^{q_2(h_1+h_1-x_s)} - F e^{q_2 h_1}.$$

Then integral I_2 in the previous expression is

$$I_2 = j_{\beta_0} \int_{-w/2}^{w/2} \mathrm{d}z_s \int_{-\frac{h_1+h_2+h_3}{2} - \frac{L}{2}}^{\frac{h_1+h_2+h_3}{2} + \frac{L}{2}} \frac{e^{j\kappa_z z}}{(2\pi)^{3/2}} \left[-\overleftrightarrow{T}_2^{-1}(h_2) F e^{-q_2(h_1+h_2-x_s)} - F e^{q_2(x_s-h_1)} \right] \mathrm{d}x_s.$$

The choice of sign before the cross-section wave numbers q_2 should provide proper behavior of radiation at infinity.

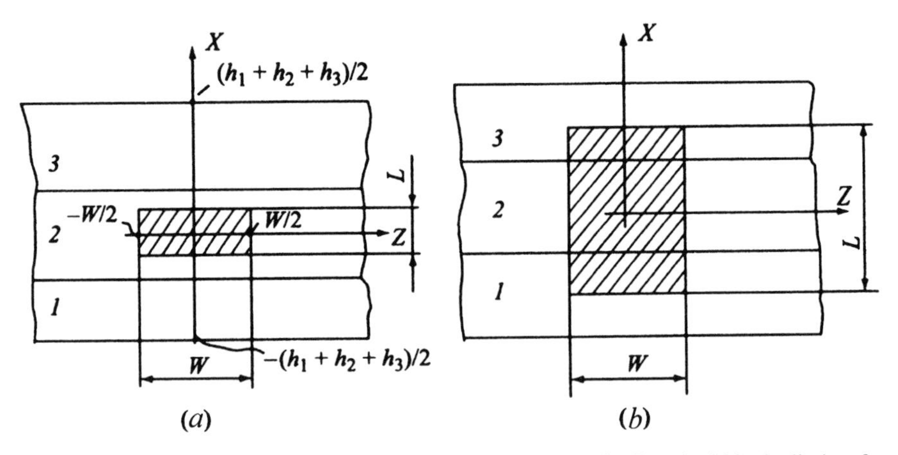

Fig. 2.2 The exciting current on the surface $y = 0$ be uniformly distributed within the limits of an area $W \times L$

Taking the integral gives

$$I_2 = \frac{j\beta_0}{(2\pi)^{3/2}} \frac{\sin\frac{\kappa_z w}{2}}{\frac{\kappa_z w}{2}} - \overset{\leftrightarrow}{\tilde{T}}{}_2^{-1}(h_2)\frac{F}{q_2} e^{-q_2\left(h_1+h_2-\frac{h_1+h_2+h_3}{2}-\frac{L}{2}\right)}$$
$$-e^{-q_2\left(h_1+h_2-\frac{h_1+h_2+h_3}{2}-\frac{L}{2}\right)} + \frac{F}{q_2}\left[e^{-q_2\left(\frac{h_1+h_2+h_3}{2}+\frac{L}{2}-h_1\right)} - e^{-q_2\left(-\frac{h_1+h_2+h_3}{2}-\frac{L}{2}-h_1\right)}\right]$$
$$= -\frac{j\beta_0 w}{(2\pi)^{3/2}} \frac{\sin\frac{\kappa_z w}{2}}{\frac{\kappa_z w}{2}} Sh\frac{q_2 L}{2}\cdot\frac{2}{q_2}e^{-q_2\frac{h_2+h_3}{2}}\cdot\left(\overset{\leftrightarrow}{\tilde{T}}{}_2^{-1}(h_2)Fe^{-\frac{q_2 h_1}{2}} + Fe^{\frac{q_2 h_1}{2}}\right).$$

Further we shall consider a case, when the area with the source falls outside the thickness of the layer $n = 2$, that is the source can be on the face surface ($y = 0$) within the limits of layers $n = 1$ and $n = 3$.

We keep the designations I_1 and I_3 of the integrals of a source in Eq. (2.58) for the corresponding areas. Then for area $n = 1$

$$I_1 = -\frac{j\beta_0 w}{q_1(2\pi)^{3/2}} \frac{\sin\frac{\kappa_z w}{2}}{\frac{\kappa_z w}{2}} F_{an}^1\left[1 - e^{-q_1\frac{h_1-h_2-h_3+L}{2}}\right]$$
$$-\left[e^{-q_1 h_1} - e^{-q_1\frac{h_1+h_2+h_3-L}{2}}\right] \tag{2.62}$$

and for area $n = 3$

$$I_3 = -\frac{j\beta_0 w}{q_3(2\pi)^{3/2}} \frac{\sin\frac{\kappa_z w}{2}}{\frac{\kappa_z w}{2}} F_{an}^3\left[e^{-q_3\frac{h_1+h_2+h_3-L}{2}} - e^{-q_3 h_3}\right]$$
$$+\left[e^{-q_3\left(-\frac{h_1+h_2+h_3+L}{2}\right)} - 1\right]. \tag{2.63}$$

Then Eq. (2.58) in different regions of the converter will look as:

– In layer $n = 1$

$$\overset{\leftrightarrow}{V}_1 = \sum_{n=1}^{L} \left\| \begin{matrix} 0 \\ 0 \\ \{\Delta^E\}_{3n} \\ \{\Delta^E\}_{4n} \end{matrix} \right\| \frac{j\beta_0 w}{q_1(2\pi)^{3/2}} \frac{\sin\frac{\kappa_z w}{2}}{\frac{\kappa_z w}{2}} \overset{\leftrightarrow}{\tilde{\vartheta}}{}_n^1, \tag{2.64}$$

$$\overset{\leftrightarrow}{\tilde{\vartheta}}{}_n^1 = \overset{\leftrightarrow}{K}_1 \overset{\leftrightarrow}{j}_{0\beta} \overset{\leftrightarrow}{x}_1 \left\{F_{an}^1\left[1 - e^{-q_1\frac{h_1-h_2-h_3+L}{2}}\right] - \left[e^{-q_1 h_1} - e^{-q_1\frac{h_1+h_2+h_3-L}{2}}\right]\right\}$$

– In layer $n = 2$

$$\overset{\leftrightarrow}{V}_2 = \sum_{n=1}^{4} \left\| \begin{matrix} 0 \\ 0 \\ \{\Delta^E\}_{3n} \\ \{\Delta^E\}_{4n} \end{matrix} \right\| - \frac{2 j\beta_0 w}{q_2(2\pi)^{3/2}} \frac{\sin\frac{\kappa_z w}{2}}{\frac{\kappa_z w}{2}} \overset{\leftrightarrow}{\tilde{\vartheta}}{}_n^2,$$

$$\overset{\leftrightarrow}{\tilde{\vartheta}}{}_n^2 = \overset{\leftrightarrow}{K}_2 \overset{\leftrightarrow}{j}_{0\beta} \overset{\leftrightarrow}{x}_2 Sh\frac{q_2 h_2}{2}\left(F_{an}^2 e^{-\frac{q_2 h_1}{2}} + e^{\frac{q_2 h_1}{2}}\right) e^{-q_2\frac{h_2+h_3}{2}} \tag{2.65}$$

– In layer $n = 3$

$$\overset{\leftrightarrow}{V}_3 = \sum_{n=1}^{4} \begin{Vmatrix} 0 \\ 0 \\ \{\Delta^E\}_{3n} \\ \{\Delta^E\}_{4n} \end{Vmatrix} \frac{j\beta_0 w}{q_3(2\pi)^{3/2}} \frac{\sin\frac{\kappa_z w}{2}}{\frac{\kappa_z w}{2}} \overset{\leftrightarrow}{\tilde{\vartheta}}{}^3_n, \tag{2.66}$$

$$\overset{\leftrightarrow}{\tilde{\vartheta}}{}^3_n = \overset{\leftrightarrow}{K}_3 \overset{\leftrightarrow}{j}_{0\beta} \overset{\leftrightarrow}{x}_3 \left\{ F^3_{an} \left[e^{-q_3 \frac{h_1+h_2+h_3-L}{2}} - e^{-q_3 h_3} \right] - \left[e^{-q_3 \frac{h_1-h_2-h_3+L}{2}} - e^{-q_3 h_3} \right] \right\}.$$

Using Eqs. (2.64) to (2.66), we derive from Eq. (2.44) an expression for the tangential constituents of HF fields of the basic mode of excited wave

$$\begin{aligned} \overset{\leftrightarrow}{\tilde{N}}{}^E_q(x,y,z) &= \overset{\leftrightarrow}{T}_q(x) \cdot \ldots \cdot \overset{\leftrightarrow}{T}_1(h_1) e^{-j\sqrt{y^2+z^2}\,\Phi(\theta_s) arctg \frac{z}{y}} \cdot \psi(y) \\ &\times \sqrt{\frac{2\pi}{j\sqrt{y^2+z^2}\Phi^*(\theta_s)}} \cdot \sum_{n=1}^{4} \begin{Vmatrix} 0 \\ 0 \\ \{\Delta^E\}_{3n} \\ \{\Delta^E\}_{4n} \end{Vmatrix} \frac{j\beta_0 w}{(2\pi)^{3/2}} \frac{\sin\frac{\kappa_z w}{2}}{\frac{\kappa_z w}{2}} \sum_{m=1}^{s} \overset{\leftrightarrow}{\tilde{\vartheta}}{}^m_n. \end{aligned} \tag{2.67}$$

Note that, depending on how the source is located in this or that layer of the converter, $\sum_{m=1}^{3} \overset{\leftrightarrow}{\tilde{\vartheta}}{}^m_n$ in the last expression will have different forms. For example, if the source is located only in field 2, this sum will be replaced by $\overset{\leftrightarrow}{V}_2$ from Eq. (2.65).

Consider cases of propagation of the waves excited by such a converter along the axes $0Y$ and $0Z$.

2.3.1 Wave Excitation Along 0Y

In this case $\kappa_z = 0$, and Eq. (2.67) looks as

$$\overset{\leftrightarrow}{\tilde{N}}{}^E_q = \overset{\leftrightarrow}{T}_n(x_n) \cdot \ldots \cdot \overset{\leftrightarrow}{T}_1(h_1)(w_q)_{oy},$$

$$(\overset{\leftrightarrow}{w}_q)_{oy} = 2\pi j \frac{e^{-j\kappa_y y}}{\Delta^E_1} \psi(y) \begin{Vmatrix} 0 \\ 0 \\ \{\Delta^E\}_{3n} \\ \{\Delta^E\}_{4n} \end{Vmatrix} \frac{j\beta_0 w}{(2\pi)^{3/2}} \sum_{m=1}^{3} \overset{\leftrightarrow}{\tilde{\vartheta}}{}^m_n. \tag{2.68}$$

The power flux will be determined from Eq. (1.118). Then, using Eqs. (1.116) and (1.117), we have for *LM* waves in the corresponding layers of the converter

$$\begin{aligned} \Pi^1_{LM} = -S^1_E \Bigg[& T^1_1 w_1 \eta T^{1^*}_1 w^*_1 \frac{e^{j\left(\kappa^1_{x_1} - \kappa^{1^*}_{x_1}\right)h_1} - 1}{j(\kappa^1_{x_1} - \kappa^{1^*}_{x_1})} \\ & + T^1_1 w_1 \eta T^{1^*}_2 w^*_1 \frac{e^{j\left(\kappa^1_{x_1} - \kappa^{1^*}_{x_2}\right)h_1} - 1}{j(\kappa^1_{x_1} - \kappa^{1^*}_{x_2})} + T^1_2 w_1 \eta T^{1^*}_1 w^*_1 \frac{e^{j\left(\kappa^1_{x_2} - \kappa^{1^*}_{x_1}\right)h_1} - 1}{j(\kappa^1_{x_2} - \kappa^{1^*}_{x_1})} \end{aligned}$$

$$+ T_2^1 w_1 \eta T_2^{1^*} w_1^* \frac{e^{j\left(\kappa_{x_2}^1 - \kappa_{x_2}^{1^*}\right)h_1} - 1}{j(\kappa_{x_2}^1 - \kappa_{x_2}^{1^*})}\Bigg], \tag{2.69a}$$

$$\Pi_{LM}^2 = -S_E^2 \left[T_1^2 w_2 \eta_2 T_1^{2^*} w_2^* \frac{e^{j\left(\kappa_{x_1}^2 - \kappa_{x_1}^{2^*}\right)h_2} - 1}{j\left(\kappa_{x_1}^2 - \kappa_{x_1}^{2^*}\right)} \right.$$

$$+ T_1^2 w_2 \eta_2 T_2^{2^*} w_2^* \frac{e^{j\left(\kappa_{x_1}^2 - \kappa_{x_2}^{2^*}\right)h_2} - 1}{j\left(\kappa_{x_1}^2 - \kappa_{x_2}^{2^*}\right)} + T_2^2 w_2 \eta\ T_1^{2^*}\ w_2^* \frac{e^{j\left(\kappa_{x_2}^2 - \kappa_{x_1}^{2^*}\right)h_2} - 1}{j\left(\kappa_{x_2}^2 - \kappa_{x_1}^{2^*}\right)}$$

$$\left. + T_2^2 w_2 \eta\ T_2^{2^*}\ w_2^* \frac{e^{j\left(\kappa_{x_2}^2 - \kappa_{x_2}^{2^*}\right)h_2} - 1}{j\left(\kappa_{x_2}^2 - \kappa_{x_2}^{2^*}\right)} \right], \tag{2.69b}$$

$$\Pi_{LM}^3 = -\overset{\leftrightarrow}{\tilde{S}}{}_E^3 \left[\overset{\leftrightarrow}{\tilde{T}}{}_1^3 \overset{\leftrightarrow}{w}_3 \overset{\leftrightarrow}{\eta}\ \overset{\leftrightarrow}{T}{}_1^3 \overset{\leftrightarrow}{\tilde{w}}{}_3^* \frac{e^{j\left(\kappa_{x_1}^3 - \kappa_{x_1}^{3^*}\right)h_3} - 1}{j\left(\kappa_{x_1}^3 - \kappa_{x_1}^{3^*}\right)} \right.$$

$$+ \overset{\leftrightarrow}{\tilde{T}}{}_1^3 \overset{\leftrightarrow}{w}_3 \overset{\leftrightarrow}{\eta}\ \overset{\leftrightarrow}{\tilde{T}}{}_2^{3*} w_3^* \frac{e^{j\left(\kappa_{x_1}^3 - \kappa_{x_2}^{3^*}\right)h_3} - 1}{j\left(\kappa_{x_1}^3 - \kappa_{x_2}^{3^*}\right)} + \overset{\leftrightarrow}{\tilde{T}}{}_2^3 \overset{\leftrightarrow}{w}_3 \overset{\leftrightarrow}{\eta} \overset{\leftrightarrow}{\tilde{T}}{}_1^{3*} \overset{\leftrightarrow}{\tilde{w}}{}_3^* \frac{e^{j\left(\kappa_{x_2}^3 - \kappa_{x_1}^{3^*}\right)h_3} - 1}{j\left(\kappa_{x_2}^3 - \kappa_{x_1}^{3^*}\right)}$$

$$\left. + \overset{\leftrightarrow}{\tilde{T}}{}_2^3 \overset{\leftrightarrow}{w}_3 \overset{\leftrightarrow}{\eta} \overset{\leftrightarrow}{\tilde{T}}{}_2^{3*} \overset{\leftrightarrow}{\tilde{w}}{}_3^* \frac{e^{j\left(\kappa_{x_2}^3 - \kappa_{x_2}^{3^*}\right)h_3} - 1}{j\left(\kappa_{x_2}^3 - \kappa_{x_2}^{3^*}\right)} \right]. \tag{2.69c}$$

To get the power fluxes of the *LE* wave Eq. (2.69) should be subjected to replacements $\overset{\leftrightarrow}{\tilde{S}}{}_E^n \to -\overset{\leftrightarrow}{\tilde{S}}{}_H^n$ and $\overset{\leftrightarrow}{\eta} \to \overset{\leftrightarrow}{\xi}$.

The total power flux in the converter with orthogonal orientation is

$$\Pi_{\perp 0y} = \sum_{i=1}^{3} (\Pi_{LE}^i + \Pi_{LM}^i). \tag{2.70}$$

2.3.2 Wave Excitation Along 0Z

In this case $\kappa_y = 0$, and Eq. (2.76) looks like

$$\overset{\leftrightarrow}{\tilde{N}}{}_q^E(x,y,z) = \overset{\leftrightarrow}{T}_q(x) \cdot \ldots \cdot \overset{\leftrightarrow}{T}_1(h_1)(w_q)_{0z},$$

$$(\overset{\leftrightarrow}{w}_q)_{0z} = 2\pi j \frac{\psi(y)}{\Delta_1^E} \sum_{n=1}^{4} \begin{Vmatrix} 0 \\ 0 \\ \left\{\Delta^E\right\}_{3n} \\ \left\{\Delta^E\right\}_{4n} \end{Vmatrix} \frac{j\beta_0 w}{(2\pi)^{3/2}} \frac{\sin\frac{\kappa_z w}{2}}{\frac{\kappa_z w}{2}} \sum_{m=1}^{3} \overset{\leftrightarrow}{\tilde{\vartheta}}{}_n^m. \tag{2.71}$$

Then, using Eq. (2.71) in Eqs. (2.69) and (2.70) in the corresponding layers of the converter, we find

$$\Pi_{\perp 0z} = \sum_{i=1}^{3} (\Pi_{LE}^{i} + \Pi_{LM}^{i})_{0z}.$$

2.4 Conclusions

1. A new theoretical approach to analysis of electromagnetic wave converters on layered structures on the basis of film multi-layered bigyrotropic structures with orthogonal orientation to the exciting plane has been developed.
2. The problem has been solved by the method of tensor Green functions, for which three-dimensional boundary conditions are formulated to consider the arrangement of extraneous sources at the end face of the layered structure and on its internal interfaces. Then by means of coupling matrices, a non-uniform matrix equation was obtained for determination of wave dispersion, Green's functions in the layers and the amplitudes of HF fields; the integral equation considers the influence of HF fields of the excited waves on the surface excitation current.
3. Several cases with excitation current to be located within the limits of one or several layers of the structure are considered.
4. Expressions have been derived to allow investigation of the properties of most widespread three-layer ferrite-dielectric converters with orthogonal orientation, shielded with metal coverings.
5. Our solution of the excitation problem for converters with orthogonal orientation of their multi-layered bigyrotropic structures allows analyzing the properties of structures containing ferrite, a ferroelectric crystal, a magnetic semiconductor in view of most specific factors for the millimeter range (losses in the layers, the influences of screens, cross-section gradients of electric and magnetic parameters of the structures).
6. The electrodynamic approach used allows features of excitation and propagation of waves in various structures near to the resonant frequencies at weak delay as well as in the field of strongly delayed electromagnetic (magnetostatic) waves to be analyzed.
7. The proposed approach is useful at studying wave processes in heterostructures in an external magnetic field, in development of selective devices with preassigned characteristics, sensors of resonant frequencies and fields, promising for the millimeter range.

Part II
Studies of Electromagnetic Wave Excitation in Film Ferrite Structures

Consider features of excitation and propagation of electromagnetic waves in layered structures containing ferrite films magnetized by an external field (tensor $\overleftrightarrow{\mu}$), conductive crystal films (tensor $\overleftrightarrow{\varepsilon}$), bigyrotropic films (tensors $\overleftrightarrow{\mu}$ and $\overleftrightarrow{\varepsilon}$) near the resonant frequencies and at tuning-out from them in the millimeter range. Dispersive characteristics of *LE* – and *LM* – will be treated in view of the factors most essential to the millimeter range (electric and magnetic losses in layers, the cross-section sizes of screened structures, the influence of screens, charge carrier concentration and effective weight, the direction of magnetization of an external field $\bar{H}_0$). We shall examine the properties of high-quality ferrites made of YIG with the cubic spinel structure $(Li\text{–}Zn)$, and those made of hexagonal ferrites in a frequency range up to 160 GHz. Characteristics of excited waves will be analyzed for various types of strip and waveguide converters. Experimental research was to confirm or correct the validity of the used theoretical models, to establish the applicability limits of MSW approach. Under study were the shape of AFC envelope, the frequency passbands and dispersions of excited waves, the position of resonant frequencies for tangentially and normally magnetized structures in narrow-band and broadband modes in UHF and EHF ranges.

The basic result of our research was substantiation of the applicability field of various types of weakly and strongly delayed waves with respect to the phase constant of wave propagation $\kappa_0' = \frac{\omega}{c}$ in free space within the millimeter range.

Waves with $\kappa' < \kappa_0'$ are promising for construction of selective devices, sensors of resonant frequencies, diagnostics of structures on frequencies close to the resonant ones. Waves with $\kappa' > \kappa_0' > (10^4\text{–}10^6)\text{m}^{-1}$ and, in particular, surface waves in tangentially or normally magnetized structures are promising for nondestructive control of film structures with a raised spatial resolution, including distribution of parameters over the thickness of a film. Waves with $\kappa' < 10^6\text{m}^{-1}$ can be used for devices with an expanded frequency band (delay lines, broadband signal receivers and analyzers of a, transversal filters).

Chapter 3
Waves in Planar Waveguides with Ferrite Films

Our analysis of the status of the problem has shown that in both centimeter and millimeter ranges of radiowaves:

- No dispersions of *LE* – and *LM* – waves have been investigated in tangentially and normally magnetized magneto-ordered structures with losses.
- No features of the dispersions of *LE* – and *LM* – waves near the resonant frequencies have been analyzed.
- No applicability limits of the MSW approach at both low $(\kappa' < 10^3\mathrm{m}^{-1})$ and high $(\kappa' > 10^3\text{–}10^5\mathrm{m}^{-1})$ wave numbers have been determined.
- No layered structures with conductive crystals, semiconductors, bigyirotropic media in external magnetic fields close to the resonant and cyclotron frequencies have been investigated in an electrodynamic approximation.

3.1 General Remarks

Before passing to consideration of the results of theoretical and experimental researches, we shall make general remarks.

The research subject are *LE* and *LM* waves in tangentially and normally magnetized layered structures with due account of major factors of the millimeter range (losses in layers, the sizes of a structure, the influence of screens, etc.). Dispersive characteristics near the resonant frequencies and at tuning-out from them are analyzed. The applicability limits of the MSW approximation in the UHF and EHF ranges are defined. Under study are structures with losses, containing ferrite (tebsor $\overleftrightarrow{\mu}$), a conductive crystal (tensor $\overleftrightarrow{\varepsilon}$), and a bigyrotropic medium (tensors $\overleftrightarrow{\mu}$ and $\overleftrightarrow{\varepsilon}$) in an external magnetic field.

The dispersions of electromagnetic waves in bigyrotropic layered structures were determined from the compatibility condition of matrix equation (1.52); for structures screened by magnetic walls, from Eq. (1.64) the condition $\Delta^M = 0$, follows, and for structures screened by electric walls, from Eq. (1.65) – $\Delta^E = 0$. The

A.A. Ignatiev, *Magnetoelectronics of Microwaves and Extremely High Frequencies in Ferrite Films*.
DOI: 10.1007/978-0-387-85457-1_4,

components of these determinants are the elements of products of Eqs. (1.51a) and (1.51b) of coupling matrices $\overleftrightarrow{B}^{E,M}$ and $\overleftrightarrow{D}^{E,M}$ in the corresponding half-spaces of the layered shielded structures for $n \geq m$ and $n < m$, defined by spectral decomposition on own values to cross-section wave numbers of matrices $(\overleftrightarrow{A}_n)^{E,M}$ depending on the direction of an external magnetic field $\overline{H}_0$ through $\overleftrightarrow{\varepsilon}_n$ and $\overleftrightarrow{\mu}_n$. For tangentially magnetized structures (the field $\overline{H}_0$ is directed along axis 0Z) a wave extends in plane $Z0Y$ and the matrices will be $(\overleftrightarrow{A}_t^n)^{E,M}$. For normally magnetized structures (the field $\overline{H}_0$ is along $0X$) a wave extends perpendicularly to the field. In matrices $(\overleftrightarrow{A}_\kappa^n)^{E,M}$ tensors $\overleftrightarrow{\mu}_{nz}$ and $\overleftrightarrow{\varepsilon}_{nz}$ from Appendix 1 (Ap.1a) are used, and in matrices $(\overleftrightarrow{A}_\mathrm{n}^n)^{E,M}$ tensors $\overleftrightarrow{\mu}_{nx}$ and $\overleftrightarrow{\varepsilon}_{nx}$ from (Ap.1b) are. For *LE* and *LM* waves the matrices $(\overleftrightarrow{A}_{LE_{\mathrm{t,n}}}^n)^{E,M}$ and $(\overleftrightarrow{A}_{LM_{\mathrm{t,n}}}^n)^{E,M}$ will contain their components defined by (1.46) and (1.47) in view of (Ap.1a) and (Ap.1b), respectively. (Subscripts "t", "n" for *LE* and *LM* mean tangential and normal magnetization). For calculation of wave dispersions in the complex plane $\kappa(\nu) = \kappa'(\nu) - j\kappa''(\nu)$ FORTRAN programs are used. The program contained the following modules:

- Initial data input
- Calculation of the matrices of dielectric $\overleftrightarrow{\varepsilon}_n$ and $\overleftrightarrow{\mu}_n$ magnetic penetrabilities of each layer of the structure
- Calculation of the coupling matrices of the tangential components of HF fields of the layers
- Formation of the dispersion matrix and calculation of its determinant
- Subroutines to calculate the eigenvalues of the matrix of each layer of cross-section wave numbers κ_x^n
- Subroutines to find the domain of the dispersive equation
- Output of results

The subroutine of finding complex eigenvalues of a matrix is realized by the Jacoby method with norm reduction for complex matrices [458].

The subroutine of solving the dispersive equation in a complex plane is realized by the modified secant method on a preset interval of longitudinal wave numbers. For real variables the method is described in [459].

The matrix subroutines are used from *SSPEST* library for CM computers.

Below layered structures used in physical research of wave processes and in various magnetoelectronics UHF and EHF devices will be considered. Of interest are transmission lines as planar waveguides limited by ideally conductive metal screens between which a three-layer structure consisting of an isotropic dielectric layer with its permeability ε_1 and thickness h_1, a layer with its tensor $\overleftrightarrow{\varepsilon}_2$ and scalar μ_2, tensor $\overleftrightarrow{\mu}_2$ and scalar ε_2 or tensors $\overleftrightarrow{\mu}_2$, $\overleftrightarrow{\varepsilon}_2$ of thickness h_2, and a layer with its permeability ε_3 and thickness h_3, all the components magnetized by a field $\overline{H}_0$, is placed.

Let's consider the calculation results of the dispersive dependences $\kappa(\nu) = \kappa'(\nu) - j\kappa''(\nu)$, where $\kappa'(\nu)$ and $\kappa''(\nu)$ are the dispersion of the phase and amplitude wave constants, respectively, in the structure along axis $0Y$. The dispersions of

$LE_{t,n}$ and $LM_{t,n}$ waves in structures with unilateral and bilaterial metallization in a frequency range of $\nu = 3 - 120$ GHz were investigated.

For various directions of an external field $\overline{H}_0$ and kinds of structure magnetization, such types of longitudinal-nonuniform waves will be excited for which the medium with magnetic properties influences the characteristics of waves in the structure stronger. The basic criterion is the requirement of orthogonality of the HF components of magnetic fields to the field $\overline{H}_0$ [8]. For example, at tangential magnetization the most essential change of dispersive characteristics will be observed for LE_t waves, for which $\overline{H}_{x,y} \perp \overline{H}_0$, and for LE_n waves, for which $\overline{H}_{y,z} \perp \overline{H}_0$, while at normal magnetization for LM_n waves only, for which $\overline{H}_{y,z} \perp \overline{H}_0$.

3.2 Waves in Tangentially Magnetized Structures

For the chosen type of structures at tangentially magnetization ($\overline{H}_0$ along the axis 0Z) the dispersions of LE_t waves should have the most essential change. Consider the results of our research of the dispersive properties of these waves in layered structures on the basis of YIG films. To reveal basic tendencies, calculations were made in a frequency range of $\nu = 3 \div 120$ GHz.

At the first step, the dispersive characteristics of LE_κ waves for hypothetical flat dielectric-ferrite-dielectric structures were investigated at one- and bilaterial metallization, when $\varepsilon_{1,2,3} = 1$ and the ferrite had losses $\alpha_{||} = \frac{\Delta H_{||}}{H_{0i}} = 10^{-6}$, where the subscript "$||$" denotes tangential magnetization of the structures and is common. The choice of these parameters is caused by the necessity of comparison with the MSW approximation and definition of criteria of its validity.

Figure 3.1 shows the dispersive characteristics of phase $(\kappa'_y)_{LE_t}$ – 1, 3, 5 and amplitude $(\kappa''_n)_{LE_t}$ – 2,4 constants of propagation of LE_t waves in structures with symmetrically located screens – 1, 2 – $h_1 = h_3 = 0.5$ m and unilateral metallization –3, 4 – $h_3 = 0.5$ m, $h_1 = 5 \cdot 10^{-4}$ m at parameters $\nu_H = 3 \cdot 10^9$ Hz, $\alpha_{||} = 10^{-6}$, $H_0 = 85.261$ kA/m, $4\pi M_S = 0.176$ T, $h_2 = 1 \cdot 10^{-5}$ m, $\varepsilon_{1,2,3} = 1$. Dotted line 5 represents the curve of SSMSW, i.e. – Eshbach-Damon's wave in the MSW approximation [8], existing in a frequency band

$$\nu_\perp = \sqrt{\nu_H(\nu_H + \nu_M)} \le \nu \le \nu_H + \frac{1}{2}\nu_M \quad \nu_M = \frac{\omega_M}{2\pi} = \frac{4\pi \cdot \gamma \cdot M_S}{2\pi}.$$

The course of the dependence $(\kappa'_y)_{LE_t}$ for a ferrite film with rather distant screens $h_1 = h_3 = 0.5$ m (curve 1) is close to the curve $(\kappa'_y)_{SSMSW}$. At unilateral loading of the ferrite film through the dielectric layer the following features in the frequency band $\nu > \nu_\perp$ are observed. At tending from the high-frequency range to $\nu_\perp$ for the dependence $(\kappa'_y)_{LE_t}$ (curve 3) in a frequency band of the order of magnitude 45–50 MHz in the beginning the reduction of a steepness corresponding reduction of group speed of a wave, and then, after passage of a point of an excess, its increase are observed. Near the point of inflection of the dependence $(\kappa'_y)_{LE_t}$ a selective increase

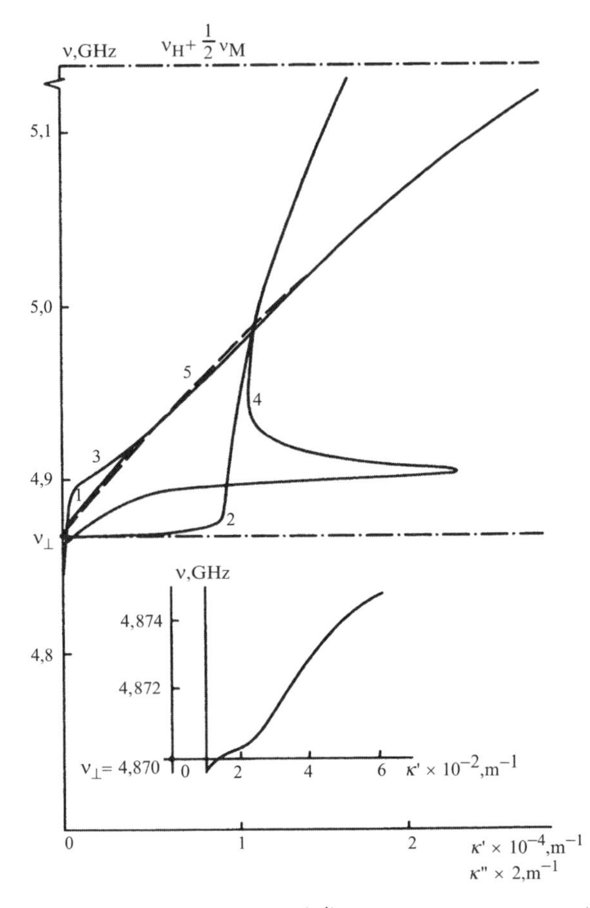

Fig. 3.1 The dispersive characteristics of phase $(\kappa_y')_{LE_t}$ – 1,3,5 and amplitude $(\kappa_H'')_{LE_t}$ – 2,4 constants of propagation of LE_t waves in structures with symmetrically located screens – $1,2 - h_1 = h_3 = 0.5\,\text{m}$ and unilateral metallization $-3,4 - h_3 = 0.5\,\text{m},\ h_1 = 5\cdot10^{-4}\,\text{m}$ at parameters $\nu_H = 3\cdot10^9\,\text{Hz},\ \alpha_{||} = 10^{-6},\ H_0 = 85.261\,\text{kA/m},\ 4\pi M_S = 0.176\,\text{T},\ h_2 = 1\cdot10^{-5}\,\text{m},\ \varepsilon_{1,2,3} = 1$. Dotted line 5 represents the curve of SSMSW, i.e. – Eshbach-Damon's wave in the MSW approximation [8]

in the peak constant $(\kappa_y'')_{LE_t}$ curve 4 with a central frequency $\nu_1 > \nu_\perp$, for which $\left|\kappa_y''(\nu_1)\right|_{\max}$ is observed. The dependence $\kappa''(\nu)$ is of asymmetric character, and the steepness of frequencies $\nu > \nu_1$ by ~3–5 times higher than that for frequencies $\nu < \nu_1$. Selective attenuation at $\nu \cong \nu_1$ is caused by two factors:

- Ferromagnetic losses in the film $(\alpha = 10^{-6})$
- Influence of the metal screen (cf. dependences: two for a free ferrite film, four for a ferrite film unilaterally loaded with a metal screen)

Let's note that near the frequency $\nu_\perp$, $(\kappa_y'(\nu \to \nu_\perp))_{LE_t} \to \kappa_0'$, where κ_0' – is the phase wave constant in a structure with an isotropic dielectrics, and $(\kappa_y'(\nu \to \nu_\perp))_{SSMSW} \to 0$ and practically independent of the arrangement of metal screens.

Analysis of the calculated dependences $(\kappa_y')_{LE_t}$ and $(\kappa_y'')_{LE_t}$ and for reasonable values of penetrabilities $\varepsilon_{1,2} = 14$, $\varepsilon_3 = 1$ has revealed no divergences with a hypothetical case of $\varepsilon_{1,2,3} = 1$.

From the dependences shown in Fig. 3.1 (3 and 5) one can see that the MSW approximation well enough coincides with the course of $(\kappa_y')_{LE_t}$ for electrodynamic calculation at $\kappa_y' > 4 \cdot 10^3\mathrm{m}^{-1}$ which value we shall accept for the lowerlower bound of the MSW approximation of κ_b'. For $\kappa_y' < 4 \cdot 10^3\mathrm{m}^{-1}$ an essential divergence with the MSW approximation by the dependences κ_y' and κ_y'' is observed, which, in particular, will lead to significant errors at estimation of the value and dispersions of group speeds and GDT of a signal, to attenuation a signal attenuation near the frequency $\nu_\perp$.

The influence of the second metal screen located at a distance $h_1 \cong h_3$ from the ferrite film on the dispersive characteristics of LE_t waves is illustrated in Fig. 3.2, where $(\kappa_y')_{LE_t} - 1$, $(\kappa_y'')_{LE_t} - 2$ at $\alpha_{||} = 10^{-6}$, $h_2 = 1 \cdot 10^{-5}\,\mathrm{m}$, $h_1 = h_3 = 5 \cdot 10^{-4}\,\mathrm{m}$, $\varepsilon_{1,2,3} = 1, 4\pi M_S = 0.176\mathrm{T}$. The dependence $\kappa_{SSMSW}' - 3$ corresponds to Eshbach-Damon's wave. Unlike the structure with unilateral metallization (Fig. 3.1), the dispersive characteristics in the frequency ranges $\nu < \nu_1$ shows a characteristic break which corresponds to the lowerlower cut-off frequency border ν_1 by

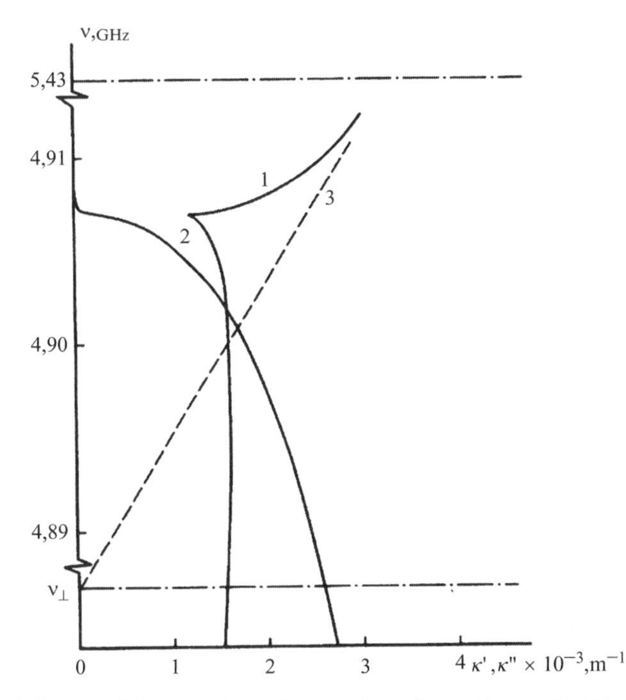

Fig. 3.2 The influence of the second metal screen located at a distance $h_1 \cong h_3$ from the ferrite film on the dispersive characteristics of LE_t waves, where $(\kappa_y')_{LE_t} - 1, (\kappa_y'')_{LE_t} - 2$ at $\alpha_{||} = 10^{-6}$, $h_2 = 1 \cdot 10^{-5}\,\mathrm{m}$, $h_1 = h_3 = 5 \cdot 10^{-4}\,\mathrm{m}$, $\varepsilon_{1,2,3} = 1$, $4\pi M_S = 0.176\mathrm{T}$

$(\kappa_y'')_{LE_t}$, and in the frequency range $\nu < \nu_1$ the course of the dependence $(\kappa_y')_{LE_t}$ gives $U_p \to \infty$, that is characteristic for an postlimit mode of a bilaterally-metallized ferrite-dielectric structure with $\nu \ll \nu_{cr}$.

Figure 3.3 presents dependences $(\kappa_y')_{LE_t}$ – 1 and $(\kappa_y'')_{LE_t}$ – 2 for a unilateral metallized structure with a ferrite film of $h_2 = 5 \cdot 10^{-5}$ m, the parameters being $\nu_{\mathrm{H}} = 3 \cdot 10^9 E$, $\alpha_{||} = 10^{-6}$, $H_0 = 85.261\,\mathrm{kA/m}$, $4\pi M_S = 0.176\mathrm{T}, h_3 = 0.5\,\mathrm{m}$, $h_1 = 5 \cdot 10^{-4}\,\mathrm{m}$, $\varepsilon_{1,2} = 14$, $\varepsilon_3 = 1$. The dotted line depicts Eshbach-Damon's dependence – 3. An increase in the thickness of the ferrite layer h_2 leads to an increase of the group speed of wave, a displacement of the frequency ν_1 towards higher frequencies of tuning-out from $\nu_\perp$ and to reduction of $\left|\kappa_y''\right|_{\max}$. The lowerlower border of the MSW approximation was also displaced towards higher values, and $\kappa_b' \cong 7 \cdot 10^3\,\mathrm{m}^{-1}$. The frequency range has also extended, it makes already 300 MHz in the field of $\nu > \nu_1$, where the electrodynamic approach is required for a correct description of the dispersive characteristics of LE_t waves.

In Fig. 3.4 dependences of the tuning-out $\Delta\nu$ (curve 1) of the frequencies of selective attenuation signal attenuation $\nu_\perp$ on the frequency of cross-section resonance of an idealized structures $\nu_\perp (\Delta\nu = \nu_1 - \nu_\perp)$ and value $\left|(\kappa_y''(\nu_1))_{LE_t}\right|_{\max}$ (curve 2) on the thickness of the ferrite layer h_2 are shown at $\alpha = const$ for structures with unilateral metallization of the ferrite film.

Figure 3.5 presents the dependence of $\left|(\kappa_y''(\nu_1))_{LE_t}\right|_{\max}$ – 1 on the parameter of ferromagnetic losses α for such structures with $h_2 = 1 \cdot 10^{-5}$ m. The value of selective attenuation signal attenuation on frequency ν_1 decreases with incrattenuation the ferrite layer thickness h_2 (Fig. 3.4) and growth of the parameter of losses α (Fig. 3.4), and its tuning-out $\Delta\nu$ increases with incrattenuation of the film thickness h_2 (Fig. 3.4).

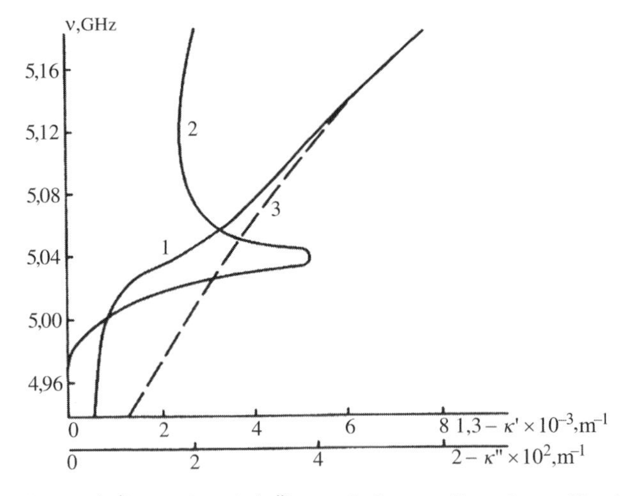

Fig. 3.3 Dependences $(\kappa_y')_{LE_t}$ – 1 and $(\kappa_y'')_{LE_t}$ – 2 for a unilateral metallized structure with a ferrite film of $h_2 = 5 \cdot 10^{-5}$ m, the parameters being $\nu_{\mathrm{H}} = 3 \cdot 10^9$ Hz, $\alpha_{||} = 10^{-6}$, $H_0 = 85.261\,\mathrm{kA/m}$, $4\pi M_S = 0.176\mathrm{T}, h_3 = 0.5\,\mathrm{m}$, $h_1 = 5 \cdot 10^{-4}$ m, $\varepsilon_{1,2} = 14$, $\varepsilon_3 = 1$. The dotted line depicts Eshbach-Damon's dependence – 3

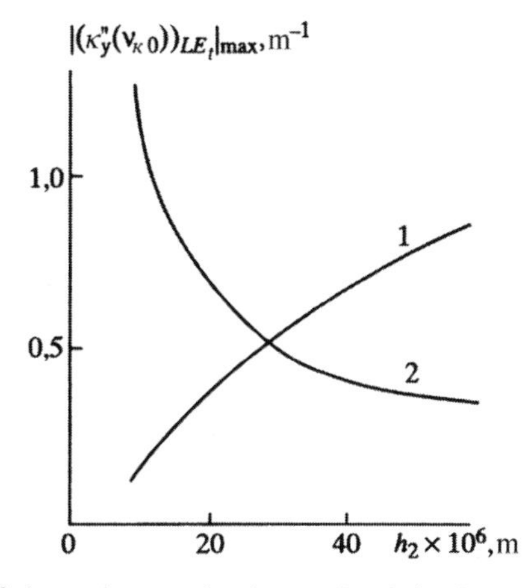

Fig. 3.4 Dependences of the tuning-out $\Delta\nu$ (curve 1) of the frequencies of selective attenuation signal attenuation $\nu_{\perp}$ on the frequency of cross-section resonance of an idealized structures $\nu_{\perp}(\Delta\nu = \nu_1 - \nu_{\perp})$ and value $\left|(\kappa_y''(\nu_1))_{LE_t}\right|_{max}$ (curve 2) on the thickness of the ferrite layer h_2 are shown at $\alpha = const$ for structures with unilateral metallization of the ferrite film

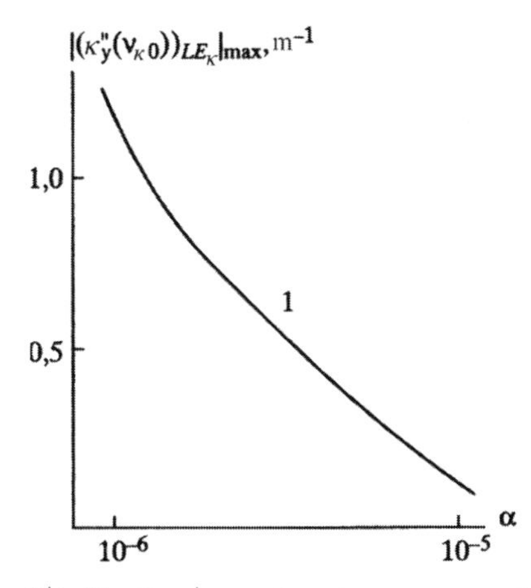

Fig. 3.5 The dependence of $\left|(\kappa_y''(\nu_1))_{LE_t}\right|_{max} - 1$ on the parameter of ferromagnetic losses α for such structures with $h_2 = 1 \cdot 10^{-5}$ m

The ferromagnetic losses characterized by the parameter $\alpha_{||} = \frac{\Delta H_{||}}{H_{0i}}$, are a most essential factor determining the level of introduced losses, due to wave propagation in a dissipative medium.

On Fig. 3.6 dependences $(\kappa_y')_{LE_t} - 1$ and $(\kappa_y'')_{LE_t} - 2$ for a unilaterally metallized structure in which the ferrite layer has $\alpha_{||} = 10^{-4}(\Delta H_{||} \approx 8.5\,\text{A/m})$ that corresponds to a high quality of the material at parameters $\nu_{\text{H}} = 3 \cdot 10^9\,\text{Hz}$, $h_3 = 0.5\,\text{m}$, $h_2 = 10^{-5}\,\text{m}$, $h_1 = 5 \cdot 10^{-4}\,\text{m}$, $\varepsilon_{1,2} = 14$, $\varepsilon_3 = 1$, $H_0 = 85.261\,\text{kA/m}$, $4\pi M_S = 0.176\,\text{T}$ are shown. Dotted curve 3 corresponds to Eshbach-Damon's wave. You see that an increase of the level of ferromagnetic losses up to real values (the data considered

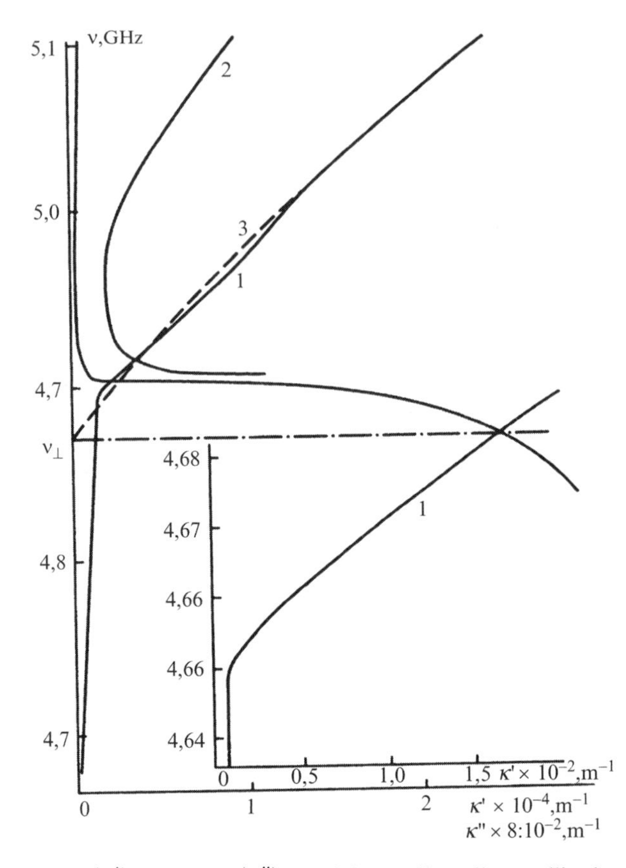

Fig. 3.6 Dependences $(\kappa_y')_{LE_t} - 1$ and $(\kappa_y'')_{LE_t} - 2$ for a unilaterally metallized structure in which the ferrite layer has $\alpha_{||} = 10^{-4}$ $(\Delta H_{||} \approx 8.5\,\mathrm{A/m})$ that corresponds to a high quality of the material at parameters $\nu_{\mathrm{H}} = 3 \cdot 10^9\,\mathrm{GHz}$, $h_3 = 0.5\,\mathrm{m}$, $h_2 = 10^{-5}\,\mathrm{m}$, $h_1 = 5 \cdot 10^{-4}\,\mathrm{m}$, $\varepsilon_{1,2} = 14$, $\varepsilon_3 = 1$, $H_0 = 85.261\,\mathrm{kA/m}$, $4\pi M_S = 0.176\,\mathrm{T}$ are shown. Dotted curve 3 corresponds to Eshbach-Damon's wave

above relate to a hypothetical structure as to the level of losses in ferrite) makes essential changes in the course of the dispersive dependences, and $(\kappa_y'(\nu))_{LE_t}$ at $\kappa_y' \to 0$ asymptotically tends to lower frequencies $\nu < \nu_\perp$. The dependences $\kappa_0' - a$ (the dot-and-dash straight line) and $(\kappa_y'(\nu))_{LE_t}$ are crossed at a frequency $\nu_1^0 < \nu_\perp$. Therefore the part of the dispersive characteristics with $(\kappa_y'(\nu))_{LE_t} < (\kappa_y'(\nu_1^0))_{LE_\kappa}$ corresponds to fast waves, and that with $(\kappa_y'(\nu))_{LE_t} \gg (\kappa_y'(\nu_1^0))_{LE_t}$ – to slow waves. For extremely low losses ($\alpha = 10^{-6}$, see Fig. 3.1) there was a precise enough lowerlower cut-off frequency, coinciding with $\nu_\perp$. In our case of low but reasonable losses $\alpha_{||} = 10^{-4}$ the course of the dependence $(\kappa_y'')_{LE_t}$ changes most essentially. In the frequency range $\nu > \nu_\perp$ corresponding to a bend of the dispersive dependence $(\kappa_y')_{LE_t}$ a sharp increase of the values $(\kappa_y'')_{LE_t}$ is observed. The analyzed structure represents, as a matter of fact, a filter of the lowerlower frequencies of LE_t waves, the lowerlower cut-off frequency of which ν_1^1 lays above $\nu_\perp$ of an idealized structures with $\alpha_{||} < 10^{-6}$ on 30–35 MHz.

Let's consider the dispersive characteristics of LE_t waves near $\nu_H = 10\,\text{GHz}$. Figure 3.7 presents the dependences $(\kappa_y')_{LE_t} - 1, 3, 5$ and $(\kappa_y'')_{LE_t} - 2, 4, 6$, reflecting the influence of the dielectric permeability of layers (usually the permeability of the materials used in the structures $\varepsilon_{1,2} \cong 14$) and the thickness of the ferrite layer h_2 $1, 2 - \varepsilon_{1,2,3} = 1$ and $3, 4$ $\varepsilon_{1,2} = 14$, $\varepsilon_3 = 1$ at $h_2 = 5 \cdot 10^{-5}\,\text{m}$ for $h_3 = 0.5\,\text{m}$, $h_1 = 5 \cdot 10^{-4}\,\text{m}$, $\alpha_{||} = 10^{-6}$, $H_0 = 284\,\text{kA/m}$, $4\pi M_S = 0.176\,\text{T}$, the dotted lines to denote Eshbach-Damon's waves for the corresponding cases. One can see that at transition into the middle wave part of the centimeter range the course of dependence $(\kappa_y'')_{LE_t}$ in the idealized structure $(\alpha_{||} = 10^{-6})$ is similar to the influence of real ferromagnetic losses ($\alpha_{||} = 10^{-4}$, see Fig. 3.6) or two metal screens close to the ferrite (Fig. 3.2). The lower cut-off frequency ν_1^1 is displaced towards higher frequencies with respect to $\nu_\perp$ by ca. 25 MHz. For really used materials with $\varepsilon_{1,2} = 14$ the cut-off frequency $(\nu_1^1 \rightarrow \nu_\perp)$ for thin films with $h_2 = 10^{-5}\text{m}$. An increase of the ferrite layer thickness (curves 5, 6) in structures with $\varepsilon_{1,2} = 14$ results

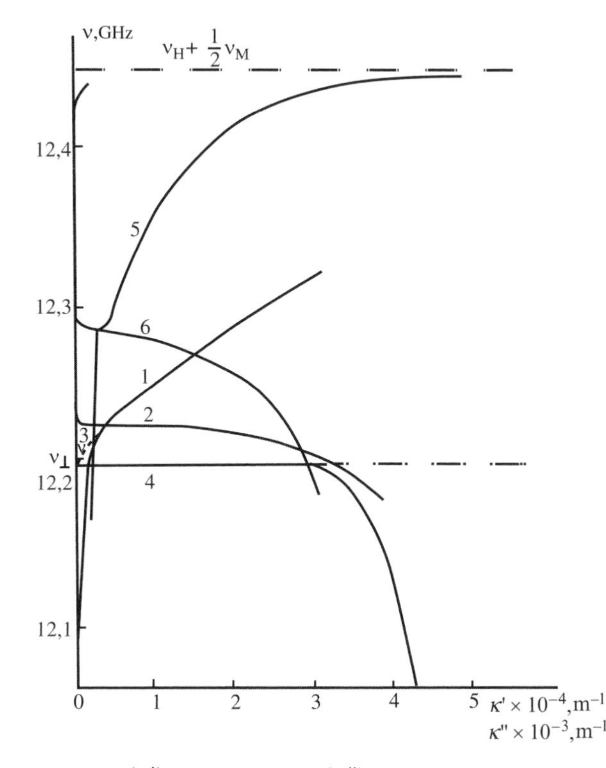

Fig. 3.7 The dependences $(\kappa_y')_{LE_t} - 1, 3, 5$ and $(\kappa_y'')_{LE_t} - 2, 4, 6$, reflecting the influence of the dielectric permeability of layers (usually the permeability of the materials used in the structures $\varepsilon_{1,2} \cong 14$) and the thickness of the ferrite layer h_2 $1, 2 - \varepsilon_{1,2,3} = 1$ and $3, 4$ $\varepsilon_{1,2} = 14$, $\varepsilon_3 = 1$ at $h_2 = 5 \cdot 10^{-5}\,\text{m}$ for $h_3 = 0.5\,\text{m}$, $h_1 = 5 \cdot 10^{-4}\,\text{m}$, $\alpha_{||} = 10^{-6}$, $H_0 = 284\,\text{kA/m}$, $4\pi M_S = 0.176\,\text{T}$, the dotted lines to denote Eshbach-Damon's waves for the corresponding cases

in enhancement of the displacement of frequency ν_1^1 towards higher frequencies (the displacement of ν_1^1 makes ca. +80 MHz).

For thick ferrite films ($h_2 = 5 \cdot 10^{-5}$ m) near ν_1^1 a sharp change of the steepness of the dependence $(\kappa_y')_{LE_t}$ (curve 5) is observed.

Let's analyse the properties of ferrite-dielectric structures in the millimeter range.

In Figs. 3.8 and 3.9 dispersive dependences $(\kappa_y')_{LE_t}$ (1, 2) and $(\kappa_y'')_{LE_t}$ (3, 4) near frequencies $\nu_H = 20$ and 30 GHz are shown: 1, 3 – $\varepsilon_{1,2,3} = 1$, 2, 4 – $\varepsilon_{1,2} = 14$, $\varepsilon_3 = 1$ at parameters $\alpha_{\|} = 10^{-6}$, $h_2 = 10^{-5}$ m, $h_3 = 0.5$ m, $h_1 = 5 \cdot 10^{-4}$ m, $4\pi M_S = 0.176$ T. As is obvious, for $\varepsilon_{1,2,3} = 1$ the dispersive characteristics $(\kappa_y'')_{LE_t}$ in the field of frequencies $(\nu_1^1 \to \nu_\perp)$ have a sharp increase, and ν_1^1 is tuned-off from $\nu_\perp$ by +10 MHz. The real permeability $\varepsilon_{1,2} = 14$, $\varepsilon_3 = 1$ lowers ν_1^1, and with growth of frequency ($\nu_H > 30$ GHz, $\nu_H = \gamma H_{0i}$) the value $\nu_1^1 < \nu_\perp$ and the tuning-out makes already −10 MHz.

In the field of frequency $\nu_\perp$ the dispersive characteristics $(\kappa_y')_{LE_t}$ experience an insignificant perturbation with a corresponding high-Q selective attenuation signal attenuation and $\left|(\kappa_y'(\nu_\perp))_{LE_t}\right|_{\max}$ (4 in Fig. 3.9). At increase of the ferrite layer thickness h_2 the perturbation of the dispersive characteristic $(\kappa_y')_{LE_t}$ near the frequency $\nu < \nu_\perp$ amplifies (Fig. 3.10). The size of the site $(\kappa_y')_{LE_t}$ with an abnormal dispersion is ca. $4 \cdot 10^3\,\text{m}^{-1}$, and it is located at $\nu < \nu_\perp$. To the middle part of this site with the central frequency $\nu_1^0 < \nu_\perp$ there corresponds a selective increase in dispersion $(\kappa_y'')_{LE_t}$. Unlike the case of a thin ferrite layer (Fig. 3.9) the frequency band of selective attenuation signal attenuation $(\kappa_y''(\nu_1^0))_{LE_t}$ has extended by ν_1^0 and the value of $\left|(\kappa_y''(\nu_1^0))_{LE_t}\right|_{\max}$ has decreased by five to six times. At tuning out from $\nu_\perp$ by −50 MHz in the frequency range $\nu < \nu_\perp < \nu_1^0$ the phase

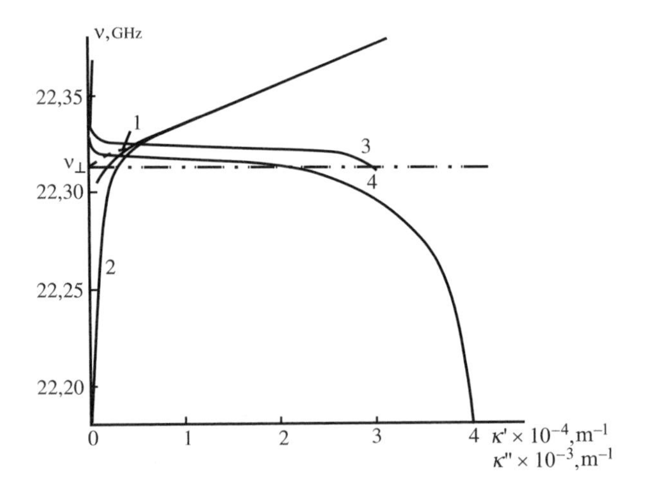

Fig. 3.8 Dependences $(\kappa_y')_{LE_t}$ (1, 2) and $(\kappa_y'')_{LE_t}$ (3, 4) near frequencies $\nu_H = 20$ GHz are shown: 1, 3 – $\varepsilon_{1,2,3} = 1$, 2, 4 – $\varepsilon_{1,2} = 14$, $\varepsilon_3 = 1$ at parameters $\alpha_{\|} = 10^{-6}$, $h_2 = 10^{-5}$ m, $h_3 = 0.5$ m, $h_1 = 5 \cdot 10^{-4}$ m, $4\pi M_S = 0.176$ T

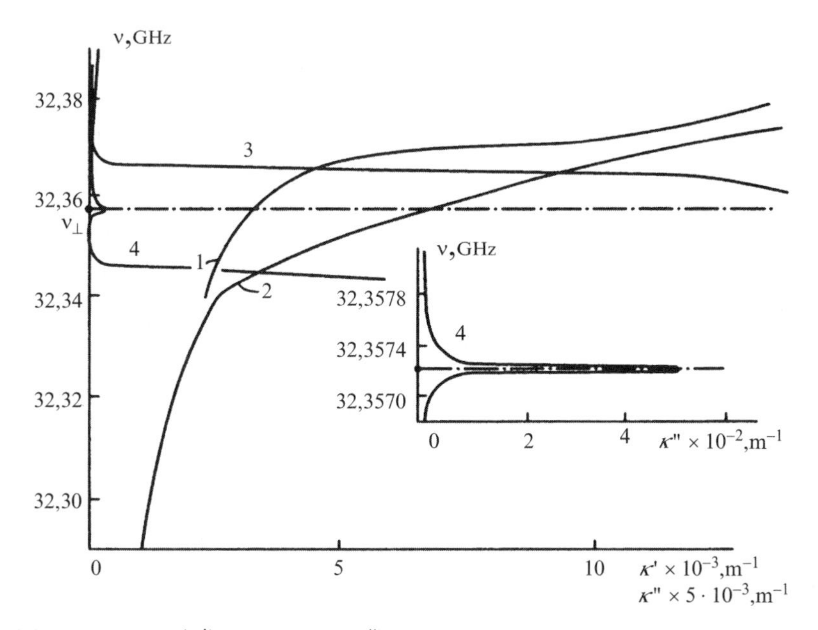

Fig. 3.9 Dependences $(\kappa_y')_{LE_t}$ (1, 2) and $(\kappa_y'')_{LE_t}$ (3, 4) near frequencies $\nu_{\mathrm{H}} = 30\,\mathrm{GHz}$ are shown: 1, 3 – $\varepsilon_{1,2,3} = 1$, 2, 4 – $\varepsilon_{1,2} = 14$, $\varepsilon_3 = 1$ at parameters $\alpha_{\|} = 10^{-6}$, $h_2 = 10^{-5}\,\mathrm{m}$, $h_3 = 0.5\,\mathrm{m}$, $h_1 = 5 \cdot 10^{-4}\,\mathrm{m}$, $4\pi M_S = 0.176\,\mathrm{T}$

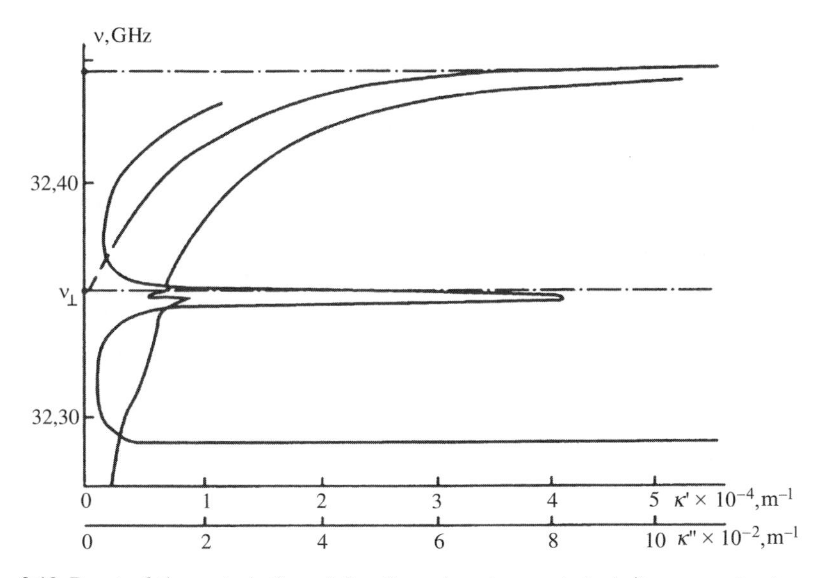

Fig. 3.10 Boost of the perturbation of the dispersive characteristic $(\kappa_y')_{LE_t}$ near the frequency $\nu < \nu_\perp$ at increase of the ferrite layer thickness h_2

constant $(\Delta\kappa_y')_{LE_t} \cong (10\text{–}20)10^2\,\mathrm{m}^{-1}$ is perturbed with a sharp increase in the value of $\left|(\kappa_y'')_{LE_t}\right|_{\max}$ corresponding to the point of inflection ν_1.

At increase of ferromagnetic losses a reduction of the abnormal part of the phase constant $(\kappa_y')_{LE_t}$ was observed and this region was displaced to a frequency range $\nu > \nu_\perp$ with a selective increase in the value of $\left|(\kappa_y''(\nu_1^0))_{LE_t}\right|$ corresponding to the point of inflection – Fig. 3.11 ($\alpha_\| = 10^{-4}$). The frequency $\nu_{1\mathrm{H}} < \nu_\perp$ corresponding to the sharp increase of $(\kappa_y'')_{LE_t}$ was simultaneously displaced towards lower values. The dot-and-dash line in Fig. 3.11 depicts the dispersion of κ_0'. One can see that in the band of SSW* existence for frequencies $\nu_{1\mathrm{H}} < \nu < \nu_\mathrm{H} + \frac{1}{2}\nu_\mathrm{M}$ we have a SSSW with $(\kappa_y')_{LE_t} > \kappa_0'$. In the frequency band $(\nu < \nu_{1\mathrm{H}})$ we have a SFSW, for which $(\kappa_y')_{LE_t} < \kappa_0'$.

Thus, ferrite-dielectric structures with unilateral metallization are characterized by:

- A frequency band of LE_t wave existence within $\nu_{1\mathrm{H}} < \nu < \nu_\mathrm{H} + \frac{1}{2}\nu_\mathrm{M}$, and in the long-wave part of the centimeter range $\nu_{1\mathrm{H}} \geq \nu_\perp$ while and in its short-wave part and in the millimeter range $\nu_{1\mathrm{H}} < \nu_\perp$.

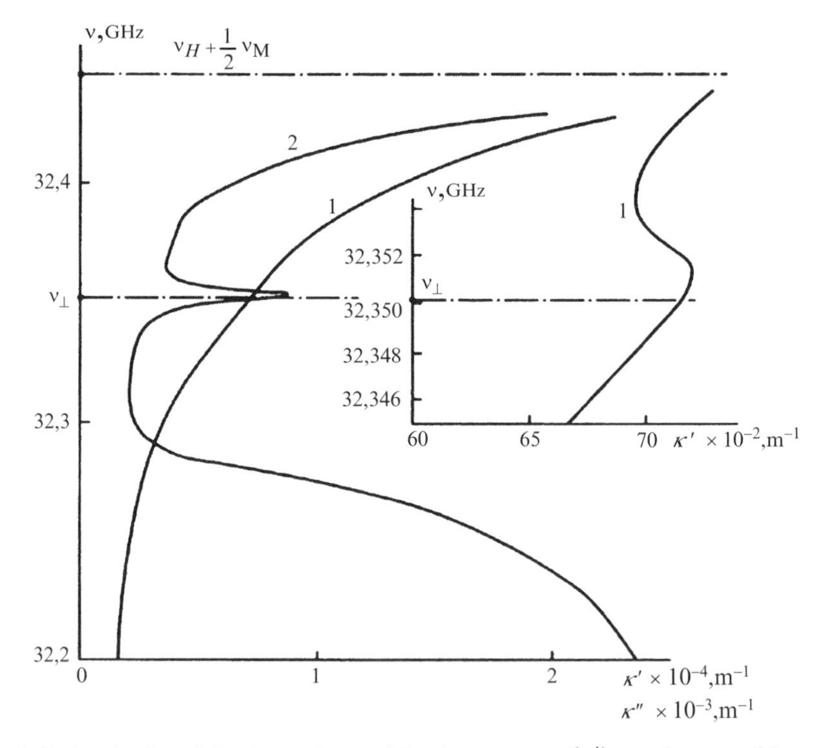

Fig. 3.11 A reduction of the abnormal part of the phase constant $(\kappa_y')_{LE_t}$ at increase of ferromagnetic losses was observed and its was displacement to a frequency range $\nu > \nu_\perp$

- displacement of the lower bound frequency $\nu_{1\mathrm{H}}$ towards lower frequencies with respect to $\nu_\perp$ and the more, the higher the working frequency and losses in the structure $\alpha_{||}$.
- A selective attenuation signal attenuation on frequencies $\nu_1^0 \cong \nu_\perp$, the stronger, the higher are the parameter of losses $\alpha_{||}$ and the thickness of the ferrite film h_2, and for thin films with $h_2 \cong 1 \cdot 10^{-5}$m, $\nu_1^0 < \nu_\perp$, while for thick ones with $h_2 \cong (2.5\text{–}5.0) \cdot 10^{-4}$m frequency $\nu_1^0 \geq \nu_\perp$.
- In the range $\nu_\perp < \nu_{SSW^*} < \nu_\mathrm{H} + \frac{1}{2}\nu_\mathrm{M}$ the value $\left|(\kappa_y'')_{LE_t}\right|_{\min}$ for surface field distribution in the structure approx. two times as greater as a similar one for the solid field distribution in the band $\nu_{1\mathrm{H}} \leq \nu_{SSLW} < \nu_\perp$.

Let's pass to consideration of the dispersive properties of planar waveguides with ferrite-dielectric filling, which are bilaterally-metallized layered structures.

Consider the properties of such waveguides in the prelimiting $(\nu \gg \nu_{cr})$ and beyond-cutoff $(\nu \ll \nu_{cr})$ modes in the presence of weakly dissipative $(\alpha_{||} < 10^{-4}\text{–}10^{-6})$ ferrite films of various thickness.

In structures on the basis of thick ferrite films, besides the above LE_κ waves passing at high κ_y' $(10^5\text{–}10^6\,\mathrm{m}^{-1})$ into SSMSW or, at least, tending to their upper bound frequencies in the mm range (we can name these slow ones with a corresponding superscript "S"–LE_t^S), in the range of small wave numbers $(\kappa_y')_{LE_t} <$ $(20\text{–}30) \cdot 10^2\,\mathrm{m}^{-1}$, there are waves with $\kappa_y' < \kappa_0'$, – to be called fast one with a corresponding superscript "F"–LE_t^F. For LE_t^F waves a weak dispersion of the phase constant in the ranges of frequencies $\nu > \nu_\perp$ and $\nu < \nu_\perp$ is characteristic.

Figure 3.12 compares the dispersions of phase and amplitude constants of a flat ferrite-dielectric waveguide with bilaterial metallization for thick and thin ferrite films $-(\kappa_y')_{LE_t^F}$ $-$ 1, $(\kappa_y'')_{LE_t^F}$ $-$ 2, 3 $-$ $(\kappa_y')_{LE_t^S}$, 4 $-$ $(\kappa_y'')_{LE_t^S}$ for $h_2 = 10^{-4}$m, and $(\kappa_y')_{LE_t^S}$ $-$ 5 and $(\kappa_y'')_{LE_t^S}$ $-$ 6 for $h_2 = 10^{-5}$m in structures with parameters $\nu_\mathrm{H} = 3 \cdot 10^{10}$Hz, $\alpha_{||} = 10^{-4}$, $h_1 = h_3 = 3.6 \cdot 10^{-3}$m, $\varepsilon_{1,2,3} = 1$, $H_0 = 2.387\,\mathrm{MA/m}$, $4\pi M_S = 0.176$T. One can see that in weakly-dissipative structures with thin ferrite films $(h_2 \cong 10^{-5}\,\mathrm{m})$ the steepness of the dispersive characteristics decreases, the group speed of a wave in the structure decreases, and only one single-wave mode with LE_t^S waves is observed, these waves being analogous to the above considered LE_t waves in structures with unilateral metallization. In weakly-dissipative structures with thick ferrite films $h_2 \cong 10^{-4}\,$m a two-wave mode with fast LE_t^F and slow LE_t^S waves is observed. In the field of wave numbers $\kappa_y' > 5 \cdot 10^3\,\mathrm{m}^{-1}$ the dispersions of the phase constants for thick and thin ferrite films practically coincide.

Near the frequency of cross-section resonance $(\nu_\perp \approx 32.357\,\mathrm{GHz})$ the dispersion of phase constant $(\kappa_y')_{LE_t^S}$ exhibits a region with an abnormal dispersion.

From LE_t waves a special case of H_{n0} waves in such structures follows. The influence of metal screens on the dispersive characteristics of flat ferrite-dielectric waveguides for H_{n0} waves was investigated in the UHF and EHF ranges for the prelimiting $(\nu \gg \nu_{cr})$, limiting $(\nu \cong \nu_{cr})$ and beyond-cutoff $\nu \ll \nu_{cr}$ modes of the structure. The conclusions made for tangentially magnetized structures on the basis of weakly-dissipative ferrites will be discussed in Chapter 4.

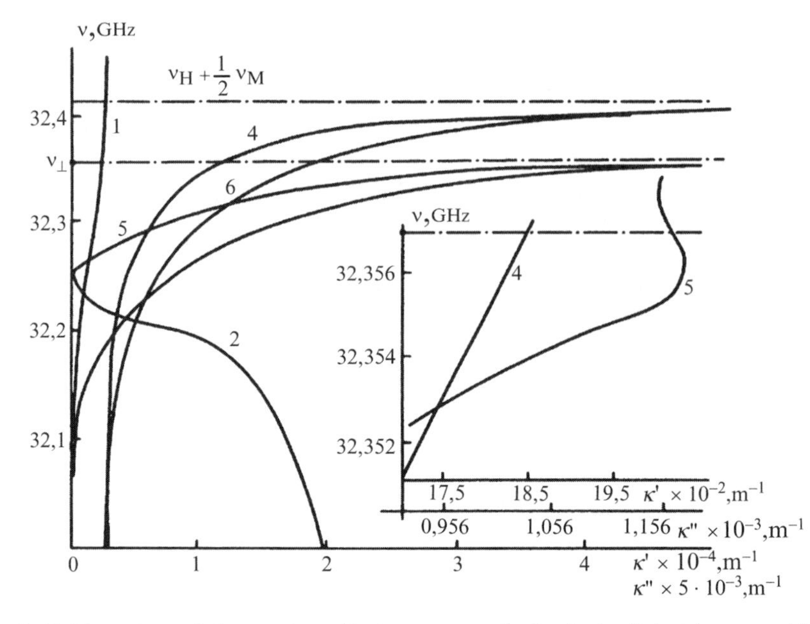

Fig. 3.12 Dispersions of phase and amplitude constants of a flat ferrite-dielectric waveguide with bilaterial metallization for thick and thin ferrite films $-(\kappa_y')_{LE_t^F} - 1$, $(\kappa_y'')_{LE_t^F} - 2$, $3 - (\kappa_y')_{LE_t^S}$, $4 - (\kappa_y'')_{LE_t^S}$ for $h_2 = 10^{-4}$ m, and $(\kappa_y')_{LE_t^S} - 5$ and $(\kappa_y'')_{LE_t^S} - 6$ for $h_2 = 10^{-5}$ m in structures with parameters $\nu_H = 3 \cdot 10^{10}$ Hz, $\alpha_\parallel = 10^{-4}$, $h_1 = h_3 = 3.6 \cdot 10^{-3}$ m, $\varepsilon_{1,2,3} = 1$, $H_0 = 2.387$ mA/m, $4\pi M_S = 0.176$ T

In the prelimiting mode $(\nu \gg \nu_{cr})$ the main effect in such structures is due to the interference mechanism of selective attenuation signal attenuation on frequency $\nu_{03} < \nu_\perp$ at interaction between LE_t^F and LE_t^S waves, and for frequency ν_{03} equality of the peak constants of waves $(\kappa_y''(\nu))_{LE_t^F} = (\kappa_y''(\nu))_{LE_t^S}$ and the validity of phase conditions $\Delta\kappa_y' L = \left|(\kappa_y'(\nu))_{LE_t^S} - (\kappa_H(\nu))_{LE_t^F}\right| L = n\pi$, $n = 1, 2, \ldots, L$ being the length of the ferrite-dielectric structure, are characteristic.

In the beyond-cutoff mode $(\nu \ll \nu_{cr})$ the main effect, i.e. selective passage of a signal on frequency $\nu > \nu_\perp$, is caused by competition of selective reduction of attenuation of signal attenuation:

- In the long-wave region of the UHF range $\nu \leq 3$ GHz on LE_t^S waves, for which $(\kappa_y''(\nu))_{LE_t^S} \ll (\kappa_y''(\nu))_{LE_t^F}$
- In the middle-wave region of the UHF range $(\nu \leq 15\text{–}20\,\text{GHz})$ on LE and LE_t waves equally, for which $(\kappa_y''(\nu))_{LE_t^S} \cong (\kappa_y''(\nu))_{LE_t^F}$
- In the short-wave region of the UHF range and in the EHF range – only on LE_t^F waves, for which $(\kappa_y''(\nu))_{LE_t^F} \ll (\kappa_y''(\nu))_{LE_t^S}$

Besides, an essential factor for frequencies close to $\nu_\perp$ is localization of HF fields with a semi-solid distribution of LE_t^F – and LE_t^S waves on opposite surfaces of the ferrite film, that is especially important in structures with an asymmetrical arrangement of the film between the metal screens ($h_1 \neq h_3$) and it is related, naturally, to the direction of an external field $\overline{H}_0(\overline{H}_0 \uparrow\uparrow 0z, \overline{H}_0 \uparrow\downarrow 0z)$.

In the limiting modes ($\nu \cong \nu_{cr}$, being the most complicated ones from the point of view of analysis of the dispersive characteristics of LE_t waves in such structures, on frequencies $\nu_{03} < \nu_\perp$ and $\nu_{0n} > \nu_\perp$ the effects of selective attenuation and passage of signals are combined.

In Fig. 3.13 the dispersive characteristics of LE_t^F waves in the beyond-cutoff mode ($\nu \ll \nu_{cr}$) of a ferrite-dielectric structure with closely located metal screens $(\kappa_y')_{LE_t^F} - 1$, $(\kappa_y'')_{LE_t^F} - 2$ are depicted for parameters $\nu_{\mathrm{H}} = 3 \cdot 10^{10}\,\mathrm{Hz}$, $\alpha_{\|} = 10^{-4}$, $h_1 = h_3 = 3.5 \cdot 10^{-4}\,\mathrm{m}$, $h_2 = 10^{-5}\,\mathrm{m}$, $\varepsilon_{1,2} = 14$, $\varepsilon_3 = 1$ $H_0 = 2.387\,\mathrm{MA/m}$, $4\pi M_S = 0.176\,\mathrm{T}$. The dotted line shows the dispersion κ_y' of Eshbach-Damon's wave. The mode of selective signal passage on a frequency $\nu > \nu_\perp$ is related to the surface distribution of HF fields of LE_t^F waves on frequency $\nu \cong \nu_n$. Reduction of the thickness of the ferrite layer leads to an increase of attenuation on the central frequency ν_n and to reduction of the passband, that is connected with reduction of the group of speed LE_t^F waves.

At analysis of the dispersions of LE_t waves in a frequency range $\nu_{\mathrm{H}} = 60$–115 GHz the following regularities have been found. In Fig. 3.14 the dispersive characteristics $(\kappa_y')_{LE_t^{F1}} - 1$, $(\kappa_y'')_{LE_t^{F1}} - 2$, $(\kappa_y')_{LE_t^{F2}} - 3$, $(\kappa_y'')_{LE_t^{F2}} - 4$ in a structure with parameters $\nu_{\mathrm{H}} = 1.15 \cdot 10^{11}\,\mathrm{GHz}$, $\alpha_{\|} = 5 \cdot 10^{-4}$, $h_3 = 0.5\,\mathrm{m}$, $h_2 = 25 \cdot 10^{-6}\,\mathrm{m}$, $h_1 = 5 \cdot 10^{-4}\,\mathrm{m}$, $\varepsilon_{1,2} = 14$, $\varepsilon_3 = 1$, $H_0 = 3.268\,\mathrm{MA/m}$, $4\pi M_S = 0.5\,\mathrm{T}$ are presented.

The dot-and-dash line depicts the dependence κ_o' while the dotted line does Eshbach-Damon's wave. For the LE_t^{F1} wave (1) having the dispersion of phase

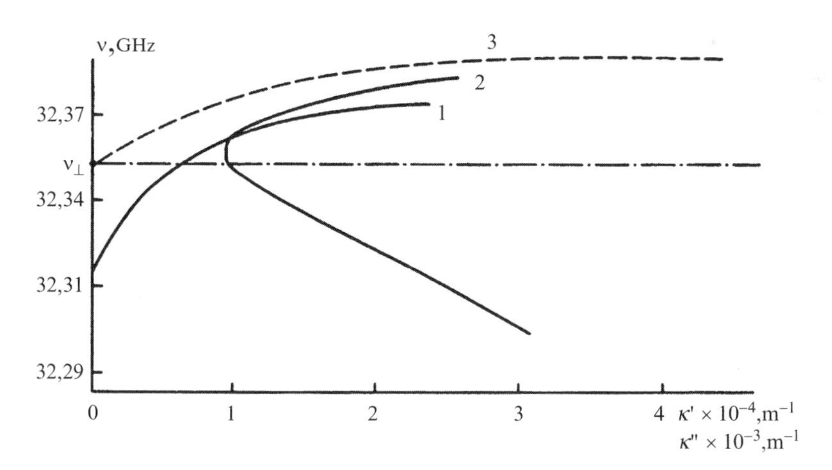

Fig. 3.13 The dispersive characteristics of LE_t^F waves in the beyond-cutoff mode ($\nu \ll \nu_{cr}$) of a ferrite-dielectric structure with closely located metal screens $(\kappa_y')_{LE_t^F} - 1$, $(\kappa_y'')_{LE_t^F} - 2$ are depicted for parameters $\nu_{\mathrm{H}} = 3 \cdot 10^{10}\,\mathrm{Hz}$, $\alpha_{\|} = 10^{-4}$, $h_1 = h_3 = 3.5 \cdot 10^{-4}\,\mathrm{m}$, $h_2 = 10^{-5}\,\mathrm{m}$, $\varepsilon_{1,2} = 14$, $\varepsilon_3 = 1$ $H_0 = 2.387\,\mathrm{MA/m}$, $4\pi M_S = 0.176\,\mathrm{T}$. The dotted line shows the dispersion κ_y' of Eshbach-Damon's wave

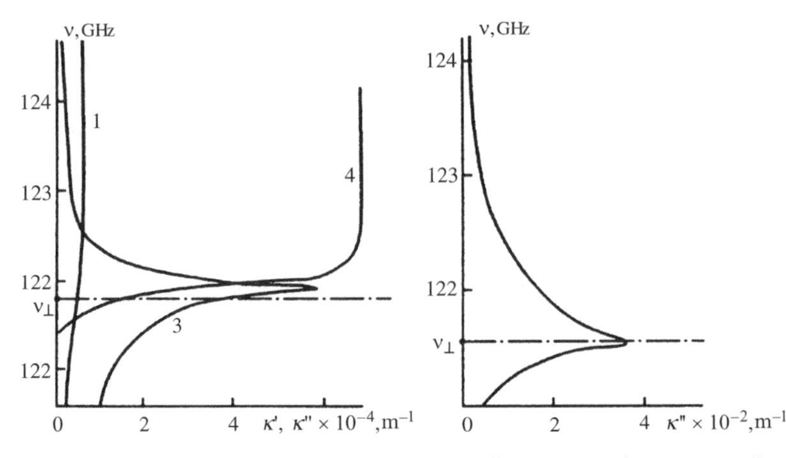

Fig. 3.14 The dispersive characteristics $(\kappa_y')_{LE_t^{F1}} - 1$, $(\kappa_y'')_{LE_t^{F1}} - 2$, $(\kappa_y')_{LE_t^{F2}} - 3$, $(\kappa_y'')_{LE_t^{F2}} - 4$ in a structure with parameters $\nu_H = 1.15 \cdot 10^{11}$ GHz, $\alpha_\| = 5 \cdot 10^{-4}$, $h_3 = 0.5$ m, $h_2 = 25 \cdot 10^{-6}$ m, $h_1 = 5 \cdot 10^{-4}$ m, $\varepsilon_{1,2} = 14$, $\varepsilon_3 = 1$, $H_0 = 3.268$ MA/m, $4\pi M_S = 0.5$ T

constant, close to that of κ_0', we have selective attenuation (2) on the frequency $\nu_{11} < \nu_\perp$ and $\left|(\kappa_y''(\nu_{11}))_{LE_t^{F1}}\right|_{\max} = 3.9 \cdot 10^2\,\mathrm{m}^{-1}$, and at tuning out from $\nu_\perp$ by $\pm 1\%$ the value of $(\kappa_y'')_{LE_t^{F1}}$ decreases more than by 20 times. For the LE_t^{F2} wave (3) having a dispersion similar to that of SSMSW in the field of $(\kappa_y')_{LE_t^{F2}} < 6 \cdot 10^4\,\mathrm{m}^{-1}$ for frequencies $\nu < \nu_{12}, \nu_{12} > \nu_\perp$, in the field of frequencies $\nu > \nu_{12}$ the dispersion becomes abnormal. The peak constant $\left|(\kappa_y''(\nu))_{LE_t^{F2}}\right|$ sharply increases in the field of frequency $\nu_\perp$, and $\left|(\kappa_y''(\nu))_{LE_t^{F2}}\right|_{\max} \gg \left|(\kappa_y''(\nu))_{LE_t^{F1}}\right|_{\max}$. Figure 3.15 compares the dispersive characteristics $(\kappa_y')_{LE_t^{F1}} - 1$, $(\kappa_y'')_{LE_t^{F1}} - 2$, $(\kappa_y')_{SSMSW} - 3$ in the analyzed structure. It is obvious that even within $\nu_{11} < \nu < \nu_{12}$, making no more than 0.3% of $\nu_\perp$ in a range of wave numbers $\Delta\kappa_y' < 2 \cdot 10^4\ \mathrm{m}^{-1}$ the divergence between the electrodynamic and MSW approximations for κ_y' is essential, and for group speeds the divergence in a band of frequencies $\nu_\perp < \nu < \nu_{12}$ reaches 160%, and in a band $\nu_{11} < \nu < \nu_\perp$ it does 560%.

Our theoretical and experimental (Chapters 4 and 5) researches have shown that at advance into the mm range:

- The frequency range of the existence of propagating LE_t waves in weakly-dissipative $\alpha < 10^{-4}$ ferrite films and structures on their basis essentially falls outside the bounds established by the MSW approximation.
- An essential divergence of the dispersions of phase and peak wave constants experimentally measured and theoretically calculated in the MSW – approximation at wave numbers $\kappa' < 10^4\ \mathrm{m}^{-1}$ is observed.

One of the tasks of our theoretical analysis was in delimitation of the applicability of the MSW approximation in the UHF and EHF ranges.

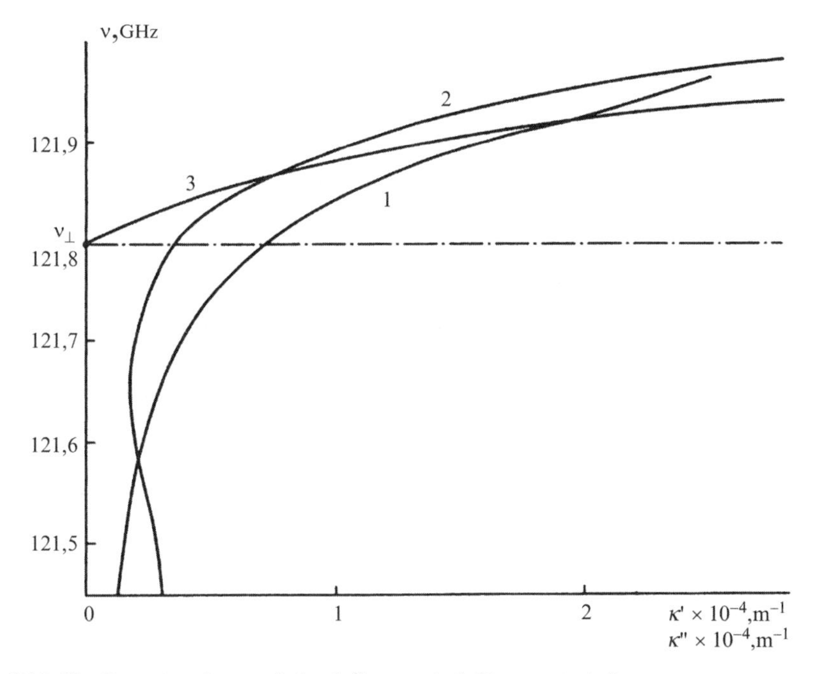

Fig. 3.15 The dispersive characteristics $(\kappa_y')_{LE_t^{F1}} - 1$, $(\kappa_y'')_{LE_t^{F1}} - 2$, $(\kappa_y')_{SSMSW} - 3$ in the analyzed structure

The criteria were:

- The value of the relative divergence of phase constants $(\kappa_y')_{LE_t}$ and $(\kappa_y')_{SSMSW}$ $\frac{\Delta\kappa_t'}{\kappa_t'} = \frac{\kappa_{LE_t}' - \kappa_{SSMSW}'}{\kappa_{LE_t}'}$ determining the quantitative divergence of the dispersions
- The lower border of phase constants $\kappa_{b.t}'$, from which the dispersions $(\kappa_y')_{LE_t}$ and $(\kappa_y')_{SSMSW}$ coincide, determining the qualitative divergence of the dispersions

Figure 3.16 presents the dependences $(\kappa_y')_{LE_t}$ – solid line and $(\kappa_y')_{SSMSW}$ – dashed line, illustrating the chosen criteria. On Fig. 3.17a, b the frequency dependences $\kappa_{cr.t}'$ ν are shown. The area in which the MSW approximation is valid is shaded. From Fig. 3.17b it is obvious that the traditionally accepted value of the lower border of the MSW approximation $(\kappa_{b.t}') \cong 10^3\,\mathrm{m}^{-1}$ (shown by the dotted line for tangentially magnetized structures) is valid only in the range of frequencies $\nu \leq 3 \cdot 10^9\,\mathrm{Hz}$.

In Fig. 3.18 the dependences $\frac{\Delta\kappa_t'}{\kappa_t'}$ on relative frequency $\frac{\Delta\nu}{\nu_\perp}$ in two frequency ranges $1 - \nu_H = 3 \cdot 10^{10}\,\mathrm{Hz}$, $\varepsilon_{1,2} = 14$, $\varepsilon_3 = 1$, $2 - \nu_H = 2 \cdot 10^{10}\,\mathrm{Hz}$, $\varepsilon_{1,2} = 14$, $\varepsilon_3 = 1$, $3 - \nu_H = 3 \cdot 10^{10}\,\mathrm{Hz}$, $\varepsilon_{1,2,3} = 1$ are shown for parameters $\alpha_{\|} = 10^{-6}$, $h_3 = 0.5\,\mathrm{m}$, $h_2 = 10^{-5}\,\mathrm{m}$, $h_1 = 5 \cdot 10^{-4}\,\mathrm{m}$, $4\pi M_S = 0.176\,\mathrm{T}$. One can see that for the case of $\varepsilon_{1,2,3} = 1$ (curve 3) the dispersive curves $(\kappa_y')_{LE_t}$ and $(\kappa_y')_{SSMSW}$ intersect twice. For real penetrabilities (curve 1) in the same range of frequencies an essentially greater

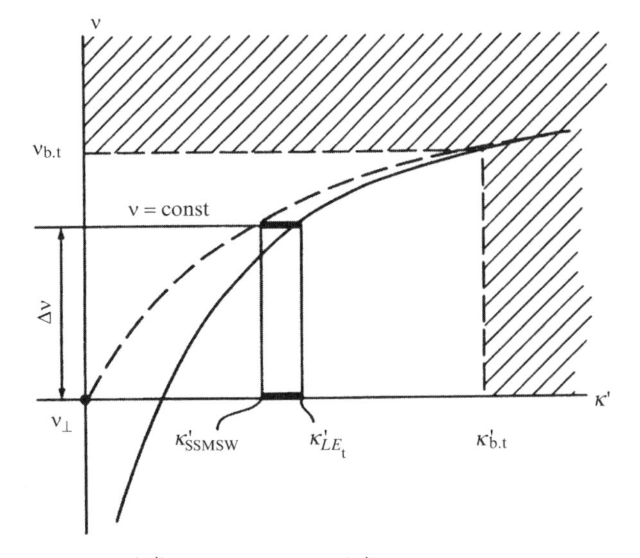

Fig. 3.16 The dependences $(\kappa_y')_{LE_t}$ – solid line and $(\kappa_y')_{SSMSW}$ – dashed line, illustrating the chosen criteria

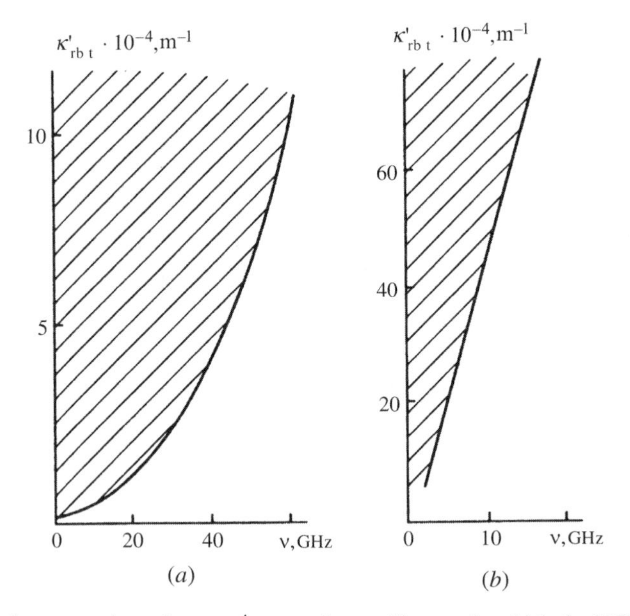

Fig. 3.17 The frequency dependences $\kappa_{b.t}'\ \nu$ are shown. The area in which the MSW approximationach is valid is shaded

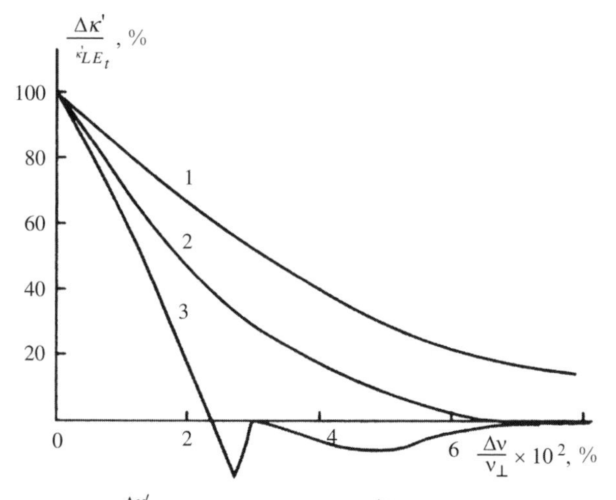

Fig. 3.18 The dependences $\frac{\Delta\kappa_t'}{\kappa_t'}$ on relative frequency $\frac{\Delta\nu}{\nu_\perp}$ in two frequency ranges $1-\nu_H = 3\cdot 10^{10}\,\text{Hz}$, $\varepsilon_{1,2} = 14$, $\varepsilon_3 = 1$, $2-\nu_H = 2\cdot 10^{10}\,\text{Hz}$, $\varepsilon_{1,2} = 14$, $\varepsilon_3 = 1$, $3-\nu_H = 3\cdot 10^{10}\,\text{Hz}$, $\varepsilon_{1,2,3} = 1$ are shown for parameters $\alpha_{||} = 10^{-6}$, $h_3 = 0.5\,\text{m}$, $h_2 = 10^{-5}\,\text{m}$, $h_1 = 5\cdot 10^{-4}\,\text{m}$, $4\pi M_S = 0.176\,\text{T}$

frequency tuning-out from $\nu_\perp$ is required, and for $\frac{\Delta\nu}{\nu_\perp} \cong 0.1\%$ ($\Delta\nu \approx 30\,\text{MHz}$) the value $\frac{\Delta\kappa_t'}{\kappa_t'} \cong 10\%$.

In Fig. 3.19 are shown, for comparison, the dependence $\frac{\Delta\kappa_t'}{\kappa_t'}$ of $\frac{\Delta\nu}{\nu_\perp}$ for films of YIG-1 and spinel -2 in a range of frequencies $\nu_H = 3\cdot 10^9\,\text{Hz}$, where $1-\alpha_{||} = 10^{-6}$, $h_1 = 5\cdot 10^{-4}\,\text{m}$, $h_2 = 5\cdot 10^{-4}\,\text{m}$, $h_3 = 0.5\,\text{m}$, $\varepsilon_{1,2} = 14$, $\varepsilon_3 = 1$, $4\pi M_S = 0.176\,\text{T}$; $2-\alpha_{||} = 10^{-6}$, $h_1 = 5\cdot 10^{-4}\,\text{m}$, $h_2 = 10^{-5}\text{m}$, $h_3 = 0.5\,\text{m}$, $\varepsilon_{1,2} = 14$, $\varepsilon_3 = 1$, $4\pi M_S = 0.5\,\text{T}$.

You see that for thin films of spinels the error is lower though only a frequency band not closer by 100 MHz to $\nu_\perp$ can be considered correct.

The error value of the dependence $\frac{\Delta\kappa_t'}{\kappa_t'}$ on frequency ν for various tunings-out $\frac{\Delta\nu}{\nu_\perp}$ is illustrated by Fig. 3.20, where $1-0.005\%$, $2-0.01\%$, $3-0.02\%$, $4-0.04\%$, $5-0.06\%$, $6-0.08\%$, $7-0.10\%$, $8-0.11\%$, $9-0.12\%$, $10-0.13\%$ for parameters $\alpha_{||} = 10^{-6}$, $h_1 = 5\cdot 10^{-4}\,\text{m}$, $h_2 = 10^{-5}\,\text{m}$, $h_3 = 0.5\,\text{m}$, $4\pi M_S = 0.176\,\text{T}$. It is obvious that for the given geometry of the structure in a frequency range 50–60 GHz the minimum value takes place at tunings-out from $\nu_\perp$ by $\frac{\Delta\nu}{\nu_\perp} > 0.1\%(\Delta\nu > 50\text{–}60\,\text{MHz})$.

Thus, our calculations have shown that both in the UHF and in EHF ranges the selective effects of attenuation and passage of signals near the frequency of cross-section resonance can be correctly described only with the electrodynamic approach. Any treatment of these effects in the MSW approximationach would be incorrect.

LM_t waves in tangentially-magnetized ferrite-dielectric structures. These waves are, as a matter of fact, longitudinal-non-uniform magnetic or *E* waves with

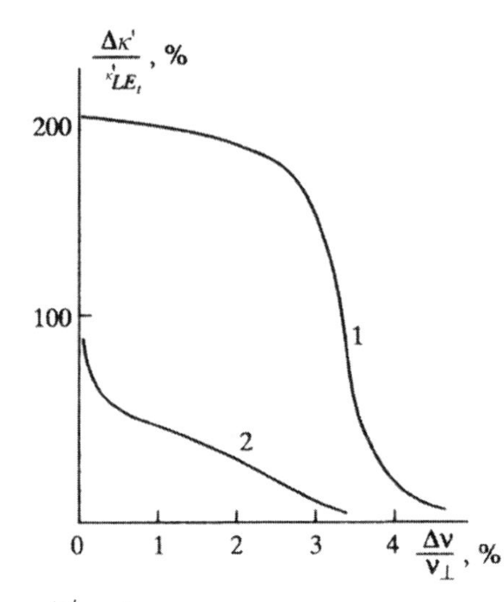

Fig. 3.19 The dependence $\frac{\Delta\kappa_t'}{\kappa_t'}$ of $\frac{\Delta\nu}{\nu_\perp}$ for films of YIG-1 and spinel – 2 in a range of frequencies $\nu_\mathrm{H} = 3\cdot10^9$ Hz, where $1 - \alpha_\| = 10^{-6}$, $h_1 = 5\cdot10^{-4}$ m, $h_2 = 5\cdot10^{-4}$ m, $h_3 = 0.5$ m, $\varepsilon_{1,2} = 14$, $\varepsilon_3 = 1$, $4\pi M_S = 0.176$ T; $2 - \alpha_\| = 10^{-6}$, $h_1 = 5\cdot10^{-4}$ m, $h_2 = 10^{-5}$ m, $h_3 = 0.5$ m, $\varepsilon_{1,2} = 14$, $\varepsilon_3 = 1$, $4\pi M_S = 0.5$ T

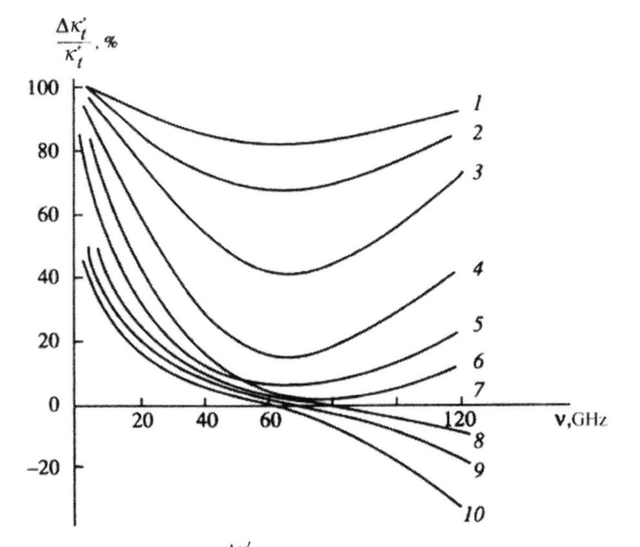

Fig. 3.20 Value of the dependence $\frac{\Delta\kappa_t'}{\kappa_t'}$ on frequency ν for various tunings-out $\frac{\Delta\nu}{\nu_\perp}$, where 1 – 0.005%, 2 – 0.01%, 3 – 0.02%, 4 – 0.04%, 5 – 0.06%, 6 – 0.08%, 7 – 0.10%, 8 – 0.11%, 9 – 0.12%, 10 – 0.13% for parameters $\alpha_\| = 10^{-6}$, $h_1 = 5\cdot10^{-4}$ m, $h_2 = 10^{-5}$ m, $h_3 = 0.5$ m, $4\pi M_S = 0.176$ T

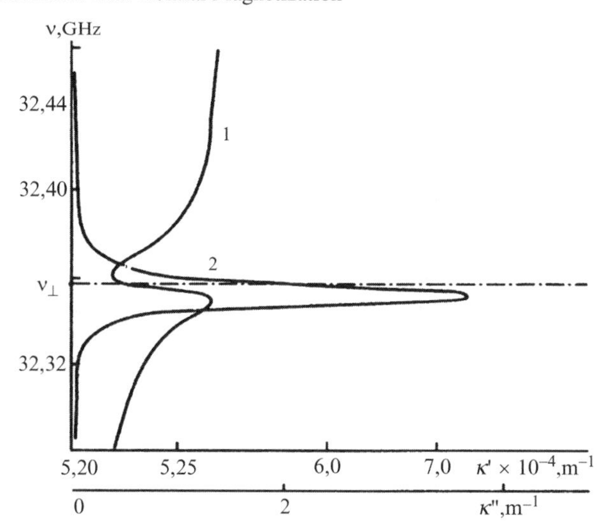

Fig. 3.21 The dispersive characteristics $(\kappa_y')_{LE_t} - 1$ and $(\kappa_y'')_{LEt_t} - 2$ in the structure are presented for parameters $\nu_H = 3 \cdot 10^{10}$ Hz, $\alpha_{\|} = 10^{-4}$, $h_1 = h_3 = 3.6 \cdot 10^{-3}$ m, $h_2 = 10^{-5}$ m, $\varepsilon_{1,2,3} = 1$, $H_0 = 2.387\,\mathrm{MA/m}$, $4\pi M_S = 0.176\,\mathrm{T}$

characteristic properties of the dielectric modes having perturbation in the field of frequency near $\nu_\perp$.

In Fig. 3.21 the dispersive characteristics $(\kappa_y')_{LE_t} - 1$ and $(\kappa_y'')_{LE_t} - 2$ in the structure are presented for parameters $\nu_H = 3 \cdot 10^{10}$ Hz, $\alpha_{\|} = 10^{-4}$, $h_1 = h_3 = 3.6 \cdot 10^{-3}$ m, $h_2 = 10^{-5}$ m, $\varepsilon_{1,2,3} = 1$, $H_0 = 2.387\,\mathrm{MA/m}$, $4\pi M_S = 0.176\,\mathrm{T}$. The dependence $(\kappa_y')_{LM_t}$ near frequency $\nu_\perp$ has a section with an abnormal dispersion whose middle part on frequency $\nu_1^0 < \nu_\perp$ corresponds to $\left|(k_y'')_{LM_t}\right|_{\max}$. Perturbation of the dispersive characteristics $(\kappa_y')_{LM_t}$ near $\nu_\perp$ is associated with the change of sign of $\mu_\perp = \mu - \frac{\mu_0^2}{\mu}$. Inherently, LM_t^F waves are fast waves with $(\kappa_y')_{LM_t^F} \ll \kappa_0'$.

In Fig. 3.22 the dispersive characteristics $(\kappa_y')_{LM_t^F} - 1$ and $(\kappa_y'')_{LM_t^F} - 2$ in a similar structure with an YIG film are shown with $h_2 = 10^{-4}$ m. One can note that the character of perturbation $(\kappa_y')_{LM_t^F}$ near $\nu_\perp$ has changed, and for frequency $\nu < \nu_\perp$ selective attenuation takes place with the value of $(\kappa_y''(\nu))_{LM_t^F}$ being maximum.

3.3 Waves in Structures with Normal Magnetization

For planar waveguides containing layered structures on the basis of ferrite films with normal magnetization (a field $\overline{H}_0$ along $0X$), the most essential change should have LM_n waves, where the subscript "n" means normal magnetization. As well as paragraph in the previous case (paragraph 3.1), we shall distinguish:

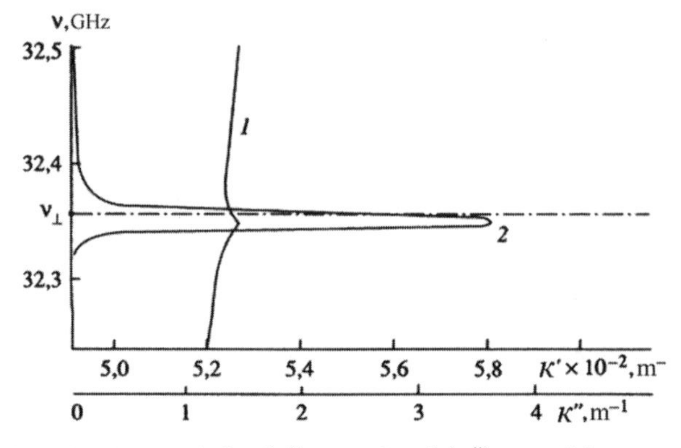

Fig. 3.22 The dispersive characteristics $(\kappa_y')_{LM_t^F} - 1$ and $(\kappa_y'')_{LM_t^F} - 2$ in structure with an YIG film are shown with $h_2 = 10^{-4}$ m

- LM_n waves tending at great κ_y' to the limiting, boundary frequencies of SSLMSW to be called *slow* with an appropriate superscript "*S*" – LM_n^S waves.
- LM_n waves which frequency beyond all bounds increases with increasing κ_y' with phase constants $(\kappa_y')_{LM_n} \ll \kappa_0'$. These waves will be named *fast* with an appropriate superscript "*F*" – LM_n^F waves.

In Fig. 3.23a–c the dispersive characteristics of LM_n^S – and LM_n^F waves in layered structures with parameters $\nu_n = 3 \cdot 10^9$ Hz, $\alpha_\perp = 10^{-4}$, $h_1 = h_3 = 5 \cdot 10^{-4}$ m, $h_2 = 10^{-5}$ m, $\varepsilon_{1,2} = 14$, $\varepsilon_3 = 1$, $H_0 = 0.225$ MA/m, $4\pi M_S = 0.176$ T are presented. In Fig. 3.23a, b the following dependences are shown: $1 - (\kappa_y')_{LM_n^S}$, $2 - (\kappa_y'')_{LM_n^S}$, $3 - (\kappa_y')_{SSLMSW}$. Figure 3.23c depicts $1 - (\kappa_y')_{LM_n^F}$, $2 - (\kappa_y'')_{LM_n^F}$. From Fig. 3.23a, b it is obvious that in the beyond-cutoff mode $(\nu \ll \nu_{cr})$ in a planar waveguide with normal magnetization of the ferrite film the MSW approximation is well enough valid with $\kappa_{b.n}' \geq 30 \cdot 10^2$ m^{-1}. The influence of metal screens leads to perturbation of the dispersion of phase constant $(\kappa_y')_{LM_n^S}$, which is much higher than that of SSLMSW, and a selective increase in the attenuation of the peak constant and $\left|(\kappa_y''(\nu))_{LM_n^S}\right|_{\max}$ corresponds to its point of inflection $\nu_{||} > \nu_n$. On frequency $\nu_{12} > \nu_{11}$ we observe $\left|(\kappa_y''(\nu))_{LM_n^S}\right|_{\max}$, corresponding to a broadband transmission mode LM_n^S waves passing at $\kappa_y' > 3 \cdot 10^3$ m^{-1} in SSLMSW.

From Fig. 3.23c it is obvious that the dispersive characteristics of phase constant has a complex character in the field of frequencies $\nu > \nu_n$. Our calculation has shown that the phase constant of wave in the planar waveguide of the above geometry with partial filling with a homogeneous isotropic dielectric ($h_1 = 5 \cdot 10^{-4}$ m, $h_2 = 10^{-5}$ m, $\varepsilon_{1,2} = 14$, $h_3 = 5 \cdot 10^{-4}$ m, $\varepsilon_3 = 1$) in the analyzed range of frequencies $\kappa_0' \approx 0.63 \cdot 10^2$ m^{-1}, and $(\kappa_y')_{LM_n} \ll \kappa_0'$, that allows to detect this wave as a fast one, LM_n^F wave.

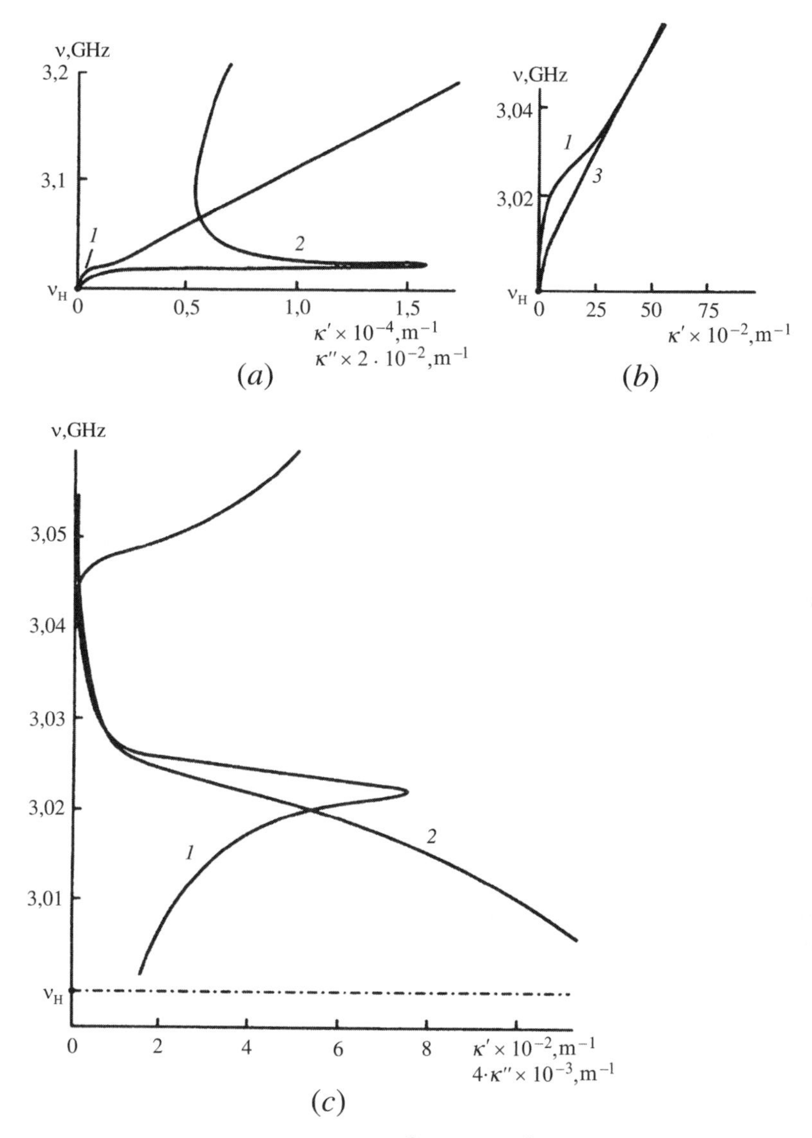

Fig. 3.23 The dispersive characteristics of LM_n^S – and LM_n^F waves in layered structures with parameters $\nu_{\mathrm{H}} = 3 \cdot 10^9\,\mathrm{Hz}$, $\alpha_{\perp} = 10^{-4}$, $h_1 = h_3 = 5 \cdot 10^{-4}\,\mathrm{m}$, $h_2 = 10^{-5}\,\mathrm{m}$, $\varepsilon_{1,2} = 14$, $\varepsilon_3 = 1$, $H_0 = 0.225\,\mathrm{MA/m}$, $4\pi M_S = 0.176\,\mathrm{T}$

For LM_n^F waves in the field of frequencies $\nu \geq \nu_{\mathrm{H}}$ characteristic are:

- The existence of parts with normal and abnormal dispersions $(\kappa_y')_{LM_n^F}$
- The presence of points of inflection on the dispersive characteristics $(\kappa_y')_{LM_n}$ with an extremely high steepness, which are marked by vertical lines A and B
- A "*beak*" on the dependence $(\kappa_y')_{LM_n^F}$ near the point of inflection on frequency ν of the dispersion $(\kappa_y')_{LM_n^F}$ that is due to the action of the metal screens

- The peak constants in the field of frequencies $\left|(\kappa_y'')_{LM_n^F}\right|_{\min} < \left|(\kappa_y'')_{LM_n^S}\right|_{\min}$
  Earlier such features of the dispersive characteristics of LM_n waves in planar waveguides containing a weakly-dissipative ferrite film were not described.

Let's note that the existence of fast LM_n^F and slow LM_n^S waves with close dispersion laws in the same frequency band at $\nu > \nu_{\mathrm{H}}$ can provide interference selective attenuation of a signal for frequencies, on which the condition of antiphase synchronism $\Delta\kappa' L = \left|(\kappa_y')_{LM_n^S} - (\kappa_y')_{LM_n^F}\right| L = n\pi$ will be satisfied.

At transition into a range of frequencies $\nu_{\mathrm{H}} = 10^{10}\,\mathrm{Hz}$ the considered features for LM_n^S and LM_n^F waves were kept. In Fig. 3.24 the following dependences are presented: $1-(\kappa_y')_{LM_n^S}$, $2-(\kappa_y')_{SSLMSW}$, $3-(\kappa_y'')_{LM_n^F}$, $4-(\kappa_y')_{LM_n^F}$, $5-(\kappa_y'')_{SSLMSW}$. One can note widening of the frequency band in which selective attenuation a signal on frequency ν_{11} and an increase, in comparison with $\nu_{\mathrm{H}} = 3\cdot 10^9\,\mathrm{Hz}$, of the value $\left|(\kappa_y''(\nu_{11}))_{LM_n}\right|_{\max}$ is observed, that is related to the growth of the line width of ferromagnetic resonance $\Delta H_\perp$ ($\alpha_\perp = const$). The range of wave numbers of the region with an abnormal dispersion of LM_n^F waves extends also, and the band of frequencies comprising this region, has not changed ($\sim 2\cdot 10^7\,\mathrm{Hz}$). Naturally, the arrangement of the points of inflection A and B with an extremely high steepness of the dispersive dependence of $\kappa'(\nu)$ has not changed too.

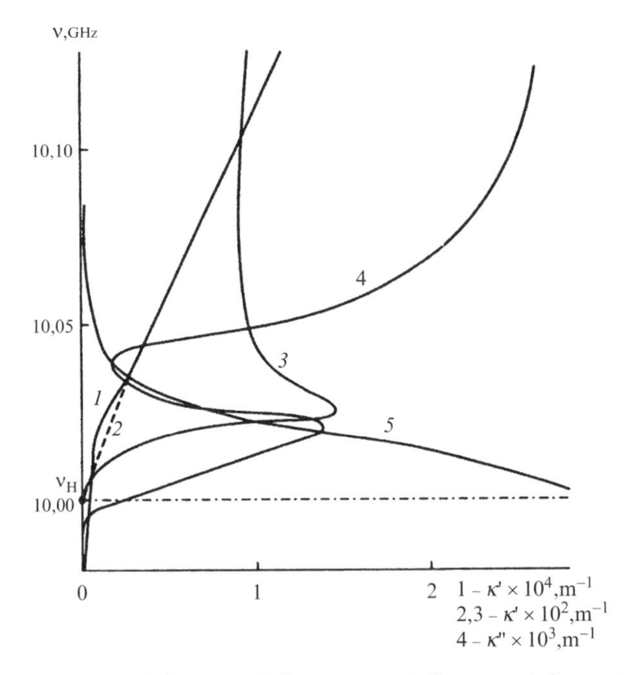

Fig. 3.24 Dependences: $1-(\kappa_y')_{LM_n^S}$, $2-(\kappa_y')_{SSLMSW}$, $3-(\kappa_y'')_{LM_n^F}$, $4-(\kappa_y')_{LM_n^F}$, $5-(\kappa_y'')_{SSLMSW}$

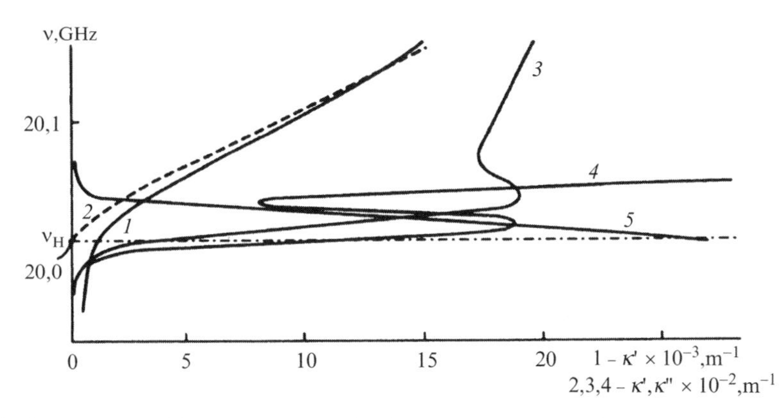

Fig. 3.25 In a range of frequencies $\nu_H = 20\,\text{GHz}$ the most essential change of waves dispersion was observed for $(\kappa''_y)_{LM_n^F}$ on frequency ν_{11} and at $\nu > \nu_{11}$ – reduction of the depth of selective attenuation and expansion of the band of frequencies were observed

In a range of frequencies $\nu_H = 20\,\text{GHz}$ (Fig. 3.25) the most essential change for the given structure was observed for $(\kappa''_y)_{LM_n^F}$ on frequency ν_{11} and at $\nu > \nu_{11}$ – reduction of the depth of selective attenuation and expansion of the band of frequencies were observed. Besides, for frequencies $\nu > \nu_{11}$ an increase in the steepness of the course of the dispersion $(\kappa''_y)_{LM_n^F}$ is noted.

In a range of frequencies (near?) $\nu_H = 3 \cdot 10^{10}\,\text{Hz}$ the role of the value of dielectric permeability of the layered structure for the dispersions of LM_n^F and LM_n^S waves was investigated. In Fig. 3.26a the dependences $1 - (\kappa'_y)_{LM_n^S}$ and $2 - (\kappa''_y)_{LM_n^S}$ for $\varepsilon_{1,2,3} = 1$, $3 - (\kappa'_y)_{LM_n^F}$ and $4 - (\kappa''_y)_{LM_n^F}$ for $\varepsilon_{1,2} = 14$, $\varepsilon_3 = 1$, and $5 - (\kappa'_y)_{SSLMSW}$ of a structure with parameters $\nu_H = 3 \cdot 10^{10}\,\text{Hz}$, $\alpha_\perp = 10^{-4}$, $h_2 = 10^{-5}\,\text{m}$, $h_1 = h_3 = 5 \cdot 10^{-4}\,\text{m}$, $4\pi M_S = 0.176\,\text{T}$ are shown.

From the presented dependences it is obvious that:

- The permeability value of the ferrite-dielectric structure $\varepsilon_{1,2} = 1 \div 14$ practically does not influence the course of the dispersion of phase constant $(\kappa'_y)_{LM_n^S}$.
- With increasing $\varepsilon_{1,2}$ the value of selective attenuation on frequency $\nu_{11} > \nu_H$ decreases.
- At tuning-out more than by $\pm 1\%$ from ν_H there is practically no influence of the value of dielectric permeability of layers $\varepsilon_{1,2,3}$.
- On frequency ν_H we have a spasmodic increase (approximately by 20 dB/cm of the attenuation of signal, related to the action of losses in the bilaterally-metallized ferrite-dielectric waveguide, and at frequencies $\nu > \nu_H$ attenuation builds up as 20 dB/cm.
- On frequencies $\nu > \nu_H$ – a peak constant $(\kappa''_y)_{LM_n}$.

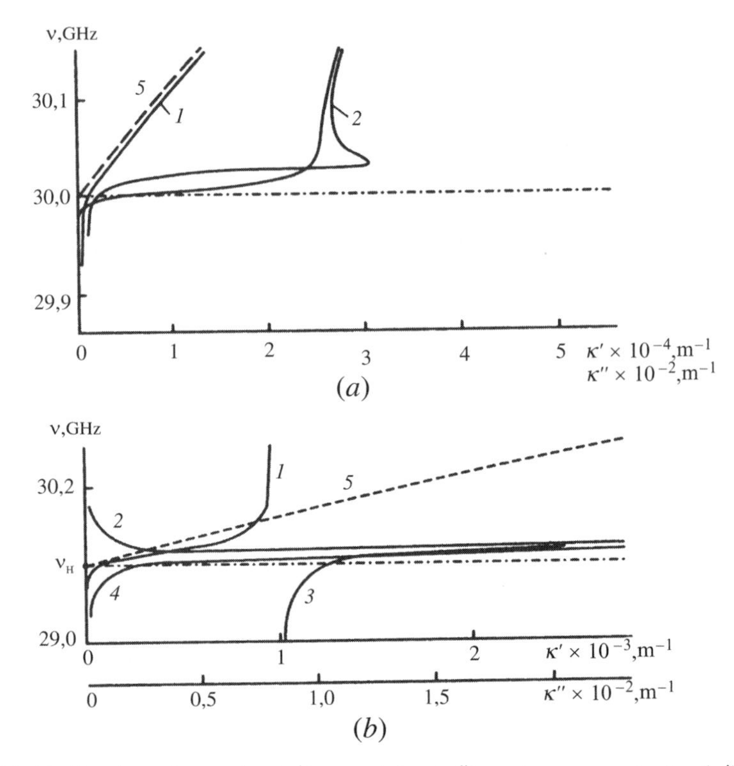

Fig. 3.26 (*a*) The dependences $1-(\kappa_y')_{LM_n^S}$ and $2-(\kappa_y'')_{LM_n^S}$ for $\varepsilon_{1,2,3}=1$, $3-(\kappa_y')_{LM_n^F}$ and $4-(\kappa_y'')_{LM_n^F}$ for $\varepsilon_{1,2}=14, \varepsilon_3=1$, and $5-(\kappa_y')_{SSLMSW}$ of a structure with parameters $\nu_H = 3\cdot 10^{10}\,\text{Hz}$, $\alpha_\perp = 10^{-4}$, $h_2 = 10^{-5}\,\text{m}$, $h_1 = h_3 = 5\cdot 10^{-4}\,\text{m}$, $4\pi M_S = 0.176\,\text{T}$; (*b*) The dispersive characteristics of LM_n^F waves in the analyzed structures

The dotted line shows the course of SSLMSW dispersion, from which it follows that at $\nu'_{b.H} > 20\cdot 10^2\,\text{m}^{-1}$ agreement with the electrodynamic calculation for phase constants $(\kappa_y')_{LM_n^S} = (\kappa_y')_{SSLMSW}$ takes place, but the calculated $(\kappa_y'')_{LE_n^S}$ essentially diverges with $(\kappa_y'')_{SSLMSW}$.

In Fig. 3.26b the dispersive characteristics of LM_n^F waves in the analyzed structures are shown. Splitting of the dispersive characteristics of phase constant LM_n^F waves into two one, one of which, fast (LM_n^{F1} – a wave), is a wave in one-sidedly metallized ferrite-dielectric structure, and another, slow (LM_n^{F2} – a wave) in a limit passes at $\kappa_y' \to \infty$ in magnetostatic. This process of splitting of the dispersion $(\kappa_y')_{LM_n^F}$ is related to hybridization of the ferrite-dielectric modes in a frequency range close to $\nu_\perp$, where their phase speeds are closest to each other. In Fig. 3.26b are shown: $1-(\kappa_y')_{LM_n^{F1}}$ and $2-(\kappa_y'')_{LM_n^{F1}}$, $3-(\kappa_y')_{LM_n^{F2}}$, $4-(\kappa_y'')_{LM_n^{F2}}$. Near the frequency $\nu_\perp$ in the field of $\kappa_0' \approx 10^3\text{m}^{-1}$ the dispersive characteristics $(\kappa_y')_{LM_n^{F1}}$ and $(\kappa_y')_{LM_n^{F2}}$ converge to the utmost.

For a LM_n^{F2} wave in the field of frequencies $\nu < \nu_H$ there is a weakly-dispersive region $(\kappa_y')_{LM_n^{F2}}$ with an extremely high steepness corresponding to the group speed value $(\nu_b)_{LM_n^{F2}} \to \infty$. At passing over frequency ν_H the steepness of the dispersive characteristics of phase constant of this wave decreases, and in the field of wave numbers κ_0' its steepness sharply increases again. The peak constant of the LM_n^{F2} wave is small in the field of frequencies $\nu > \nu_H$, and on frequency $\nu \approx \nu_H$ its value $\left|(\kappa_y''(\nu_H))_{LM_n^{F2}}\right|_{max}$.

For a LM_n^{F1} wave in the field of frequencies $\nu < \nu_H$ the dispersion of phase constant $(\kappa_y')_{LM_n^{F1}}$ has the maximum steepness and $(\nu_b)_{LM_n^{F1}} \to \infty$. At frequencies $\nu \to \nu_H$ the steepness of $(\kappa_y')_{LM_n^{F1}}$ decreases and for $\kappa_y' \gg \kappa_0'$ or $\kappa_y' \to \infty$ tends to the dispersion of SSLMSW. The peak constant is small in the field of frequencies $\nu < \nu_H$, and at $\nu \approx \nu_H$ the value $\left|(\kappa_y''(\nu_H))_{LM_n^{F1}}\right|_{max}$.

Thus, the effect of splitting of the phase constant of a LM_n^F wave in the structures on the basis of weakly-dissipative ferrite films in the field of $\kappa_y' \approx \kappa_0'$ provides selective attenuation of a signal in a planar waveguide on frequency $\nu \approx \nu_H$, and with reduction of the parameter of ferromagnetic losses $\alpha_\perp$ the dispersions of phase constants in the field of κ_0' converge, their steepnesses in the field of frequencies $\nu > \nu_H$ and $\nu < \nu_H$ increases, the band of frequencies decreases, and attenuation of signal rejection on frequency $\nu \approx \nu_H$ increases.

Our study of the properties of LM_n^F waves in normally-magnetized structures in a band of frequencies $\nu < \sqrt{\nu_H(\nu_H + \nu_M)} = \nu_\perp$ has shown that for rather thick films (tens microns) near the frequency of cross-section resonance $\nu_\perp$ the dispersive characteristics of LM_n^F waves pass into LM_t^F waves in tangentially-magnetized concerning other side defined by thickness of a film.

In Fig. 3.27 the dispersive characteristics of a LM_t^F wave with a "$\perp$" subscript assigned near the frequency $\nu_\perp(\nu < \nu_\perp)1 - (\kappa_y')_{LM_\perp}$, $2 - (\kappa_y'')_{LM_\perp}$, $3 - \kappa_0'$ are presented for parameters $\nu_H = 3 \cdot 10^{10}\,\text{Hz}$, $\alpha_{||} = 10^{-4}$, $h_2 = 10^{-4}\,\text{m}$, $h_1 = h_3 = 3.6 \cdot 10^{-3}\,\text{m}$, $\varepsilon_{1,2} = 14, \varepsilon_3 = 1$, $4\pi M_S = 0.176\text{T}$. The dispersive characteristics of the analyzed LM_t^F wave at $\nu \to \nu_H$ pass into the characteristics of a LM_t^F wave (Fig. 3.26b). For a LM_t^F wave at $\nu < \nu_\perp$ characteristic are:

- The occurrence of a region with an abnormal dispersion (κ_y') at tuning-out by -0.13% from $\nu_\perp(-47\,\text{MHz})$, which belongs to a slow LM_t wave $((\kappa_y')_{LM_t} > \kappa_0')$.
- A high steepness of the dispersion of phase constant $(\kappa_y')_{LM_t}$ passing into the dispersion $(\kappa_y')_{LM_n} \approx \kappa_0'$ in the band $\nu_H < \nu < \nu_\perp$.
- A practically linear dispersive characteristics $(\kappa_y')_{LM_t}$ at approaching $\nu_\perp$ from lower frequencies.
- The presence of selective attenuation of signal at tuning-out from $\nu_\perp$ by $-55\,\text{MHz}$, due to ferromagnetic losses $\alpha_{||}$ and a sharp increase of the value (κ_y'') near the upper boundary frequency of SSLMSW $\nu_\perp$.

Thus, in unilaterally-metallized normally-magnetized ferrite-dielectric structures with a low level of ferromagnetic losses $(\alpha < 10^{-4})$ in the range of frequencies

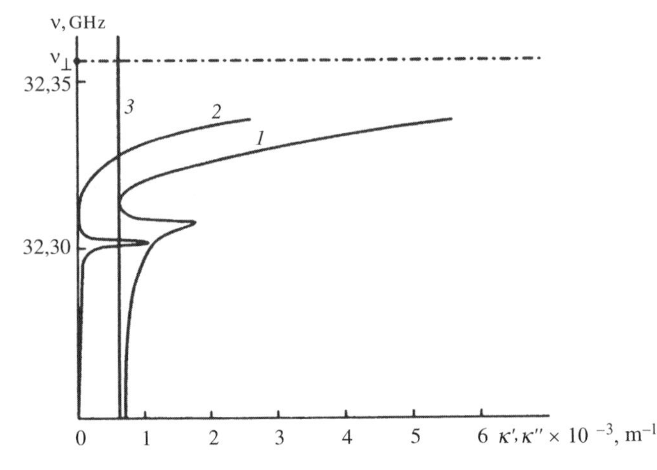

Fig. 3.27 The dispersive characteristics of a LM_t^F wave with a "⊥" subscript assigned near the frequency $\nu_\perp(\nu < \nu_\perp)$ $1-(\kappa_y')_{LM_\perp}$, $2-(\kappa_y'')_{LM_\perp}$, $3-\kappa_0'$ are presented for parameters $\nu_{\mathrm{H}} = 3\cdot 10^{10}\,\mathrm{Hz}$, $\alpha_{\|} = 10^{-4}$, $h_2 = 10^{-4}\,\mathrm{m}$, $h_1 = h_3 = 3.6\cdot 10^{-3}\,\mathrm{m}$, $\varepsilon_{1,2} = 14, \varepsilon_3 = 1$, $4\pi M_S = 0.176\,\mathrm{T}$

$\nu < \nu_{\mathrm{H}}$ there are dispersions of LM_n waves, for which $(\kappa_y')_{LM_n} < \kappa_0'$ and $\left|(\kappa_y'')_{LM_{\mathrm{n}}}\right|_{\min}$.
In bilaterally-metallized structures:

- In the prelimit mode $(\nu \gg \nu_{cr})$ at tangential magnetization of the structures on the basis of thick films, with respect to the side defining the thickness of the ferrite layer, the dispersive characteristics of LM_t waves go into the range of frequencies $\nu < \nu_\perp$, which in the band of frequencies $\nu_{\mathrm{H}} < \nu < \nu_\perp$ pass into LM_n waves in normal magnetization concerning a surface of a film for which $\left|(\kappa_y'')_{LM_n}\right|_{\min}$
- In the beyond-cutoff mode $(\nu \ll \nu_{cr})$ at normal magnetization of the layered structure on frequency $\sim\nu_{\mathrm{H}}$ there is a selective reduction of the peak constants of LE_{n} and LM_{n} waves.

Let's discuss the borders of applicability of the MSW approximation for normally-magnetized bilaterally-metallized layered structures on the basis of ferrite films in the UHF and EHF ranges.

Figure 3.28 depicts the dependences $\frac{\Delta\kappa_{\mathrm{n}}'}{\kappa_{\mathrm{n}}'} = \frac{\kappa_{KM_{\mathrm{n}}}' - \kappa_{SSLMSW}'}{\kappa_{LM_{\mathrm{n}}}'}$ on the value of frequency tuning-out $\frac{\Delta\nu_{\mathrm{H}}}{\nu_{\mathrm{H}}} = \frac{\nu-\nu_{\mathrm{H}}}{\nu_{\mathrm{H}}}$ in the following ranges of frequencies $1-\nu_{\mathrm{H}} = 3\cdot 10^9\,\mathrm{Hz}$, $H_0 = 85.26\,\mathrm{kA/m}$, $2-\nu_{\mathrm{H}} = 10^{10}\,\mathrm{Hz}$, $H_0 = 284.2\,\mathrm{kA/m}$, $3-\nu_{\mathrm{H}} = 2\cdot 10^{10}\,\mathrm{Hz}$, $H_0 = 852.6\,\mathrm{kA/m}$, for structures with parameters $\alpha_\perp = 10^{-4}$, $h_1 = h_3 = 5\cdot 10^{-4}\,\mathrm{m}$, $h_2 = 10^{-4}\,\mathrm{m}$, $\varepsilon_{1,2} = 14$, $\varepsilon_3 = 1$, $4\pi M_S = 0.176\mathrm{T}$.

It is obvious that at advance into the short-wave part of the centimeter range the band of frequencies close to ν_{H} in which the electrodynamic approach is only required for calculation of the dispersive characteristics extends.

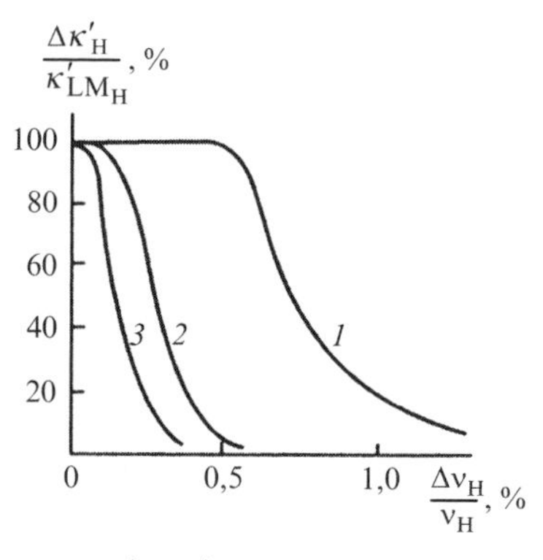

Fig. 3.28 The dependences $\frac{\Delta\kappa'_n}{\kappa'_n} = \frac{\kappa'_{LMn} - \kappa'_{SSLMSW}}{\kappa'_{LMn}}$ on the value of frequency tuning-out $\frac{\Delta\nu_H}{\nu_H} = \frac{\nu - \nu_H}{\nu_H}$ in the following ranges of frequencies $1 - \nu_H = 3 \cdot 10^9\,\text{Hz}$, $H_0 = 85.26\,\text{kA/m}$, $2 - \nu_H = 10^{10}\,\text{Hz}$, $H_0 = 284.2\,\text{kA/m}$, $3 - \nu_H = 2 \cdot 10^{10}\,\text{Hz}$, $H_0 = 852.6\,\text{kA/m}$, for structures with parameters $\alpha_\perp = 10^{-4}$, $h_1 = h_3 = 5 \cdot 10^{-4}\,\text{m}$, $h_2 = 10^{-4}\,\text{m}$, $\varepsilon_{1,2} = 14$, $\varepsilon_3 = 1$, $4\pi M_S = 0.176\,\text{T}$

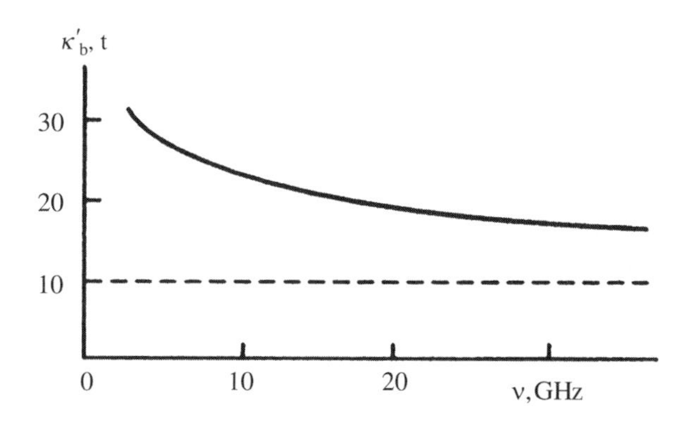

Fig. 3.29 The dependence of the lower boundary frequency of the MSW approximation for a normally-magnetized bilaterally-metallized layered structure on frequency ν_H is shown for $\alpha_\perp = 10^{-4}$, $h_2 = 50$, $\varepsilon_{1,2} = 14$, $\varepsilon_3 = 1$, $4\pi M_S = 0.176\,\text{T}$. The dotted line shows $\kappa'_{b.\text{H}} \approx 10^3\,\text{m}^{-1}$, which is accepted in [8]

In Fig. 3.29 the dependence of the lower boundary frequency of the MSW approximation for a normally-magnetized bilaterally-metallized layered structure on frequency ν_H is shown for $\alpha_\perp = 10^{-4}$, $h_2 = 50$, $\varepsilon_{1,2} = 14$, $\varepsilon_3 = 1$, $4\pi M_S = 0.176\text{T}$. The dotted line shows $\kappa'_{b.\text{H}} \approx 10^3\text{m}^{-1}$, which is accepted in [8]. One can see that at advance into the millimeter range, $\kappa'_{b.\text{H}}$ decreases a little and makes $\kappa'_{b.\text{H}} \approx 2 \cdot 10^3\,\text{m}^{-1}$, however, for peak constants in this range the electrodynamic approach gives results principally distinct from the MSW-approximation.

3.4 Waves in Structures on the Basis of Ferrite-Spinel, Magnetized Conducting Crystals, and Bigyrotropic Films

3.4.1 Structures on the Basis of Spinel Films

For advance into the short-wave range ferrites with a raised magnetization $4\pi M_S$ and fields of anisotropy $\overline{H}_A$ may be promising. Consider the dispersions of LE_t waves in structures on the basis of spinel films with $4\pi M_S = 0.5$T.

In Figs. 3.30 and 3.31 are given the dispersive characteristics $(\kappa_y')_{LE_t^S}$ and $(\kappa_y'')_{LE_t^S}$ for structures with parameters $\alpha_{\|} = 10^{-4}$, $h_2 = 10^{-5}$ m, $h_3 = 0.5$ m, $h_1 = 5 \cdot 10^{-4}$ m, $\varepsilon_{1,2} = 14$, $\varepsilon_3 = 1$, $4\pi M_S = 0.5$T in two frequency ranges $\nu_{\text{H}} = 3$ GHz and $\nu_{\text{H}} = 30$ GHz at $H_0 = 85.3$ kA/m, and $H_0 = 8.53$ MA/m, respectively. The increase of saturation magnetization of the ferrite layer results in, in comparison with YIG ($4\pi M_S = 0.176$T), to expansion of the band of frequencies and an increase in the average steepness $(\kappa_y')_{LE_t^S}$ that is related to the increase of the group speed of a LE_t^S wave in the structure. Perturbation on $(\kappa_y'')_{LE_t^S}$ close to $\nu_\perp$ in a range of frequencies $\nu_{\text{H}} = 3$ GHz is more considerable than that in a range $\nu_{\text{H}} = 30$ GHz, that is related to the increase in the line width of FMR, as $\alpha = \frac{\Delta H_{\|}}{H_{0i}} = 10^{-4}$. Thus, in a range of frequencies $\nu_{\text{H}} = 30$ GHz the band of frequencies within which there are LE_t^S waves is much wider.

In Fig. 3.32 the dispersions of phase constants of LE_t^S–wave −1 and SSMSW −2 for a structure with parameters $\nu_{\text{H}} = 3 \cdot 10^9$ Hz, $\alpha_{\|} = 10^{-4}$, $h_1 = 5 \cdot 10^{-4}$m, $h_2 = 10^{-5}$m, $h_3 = 0.5$m, $\varepsilon_{1,2} = 14$, $\varepsilon_3 = 1$, $4\pi M_S = 0.5$T are shown. It is obvious that for ferrite-spinels $\kappa_{b.t}' \approx 5 \cdot 10^4$ m^{-1} from which the MSW approximationach can be

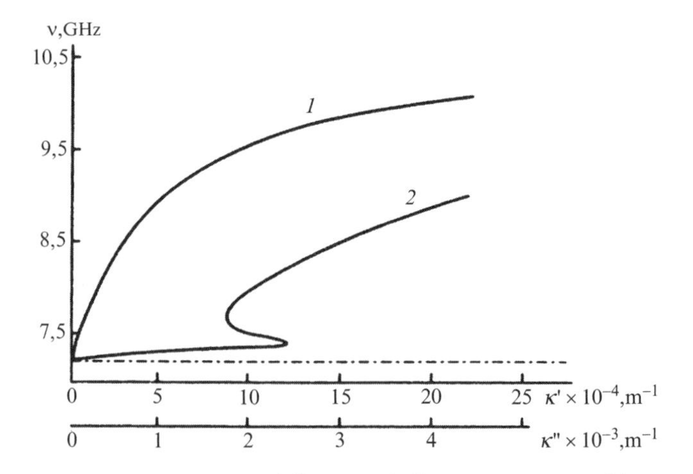

Fig. 3.30 The dispersive characteristics $(\kappa_y')_{LE_t^S}$ and $(\kappa_y'')_{LE_t^S}$ for structures with parameters $\alpha_{\|} = 10^{-4}$, $h_2 = 10^{-5}$ m, $h_3 = 0.5$ m, $h_1 = 5 \cdot 10^{-4}$ m, $\varepsilon_{1,2} = 14$, $\varepsilon_3 = 1$, $4\pi M_S = 0.5$T in frequency range $\nu_{\text{H}} = 3$ GHz

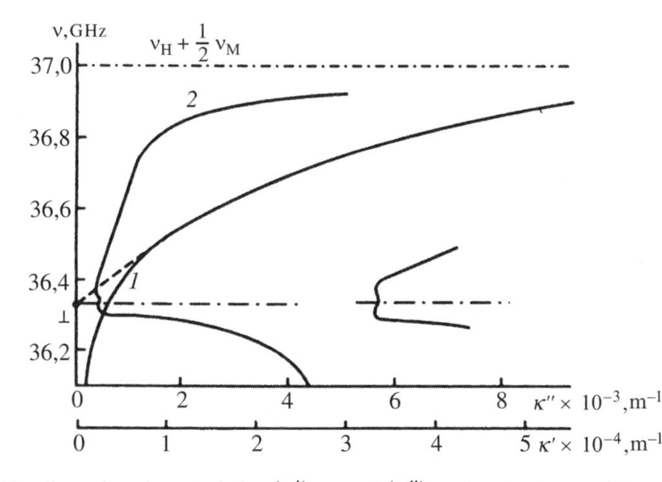

Fig. 3.31 The dispersive characteristics $(\kappa_y')_{LE_t^S}$ and $(\kappa_y'')_{LE_t^S}$ for structures with parameters $\alpha_{\|} = 10^{-4}$, $h_2 = 10^{-5}\,\text{m}$, $h_3 = 0.5\,\text{m}$, $h_1 = 5 \cdot 10^{-4}\,\text{m}$, $\varepsilon_{1,2} = 14$, $\varepsilon_3 = 1$, $4\pi M_S = 0.5\,\text{T}$ in frequency range $\nu_H = 30\,\text{GHz}$

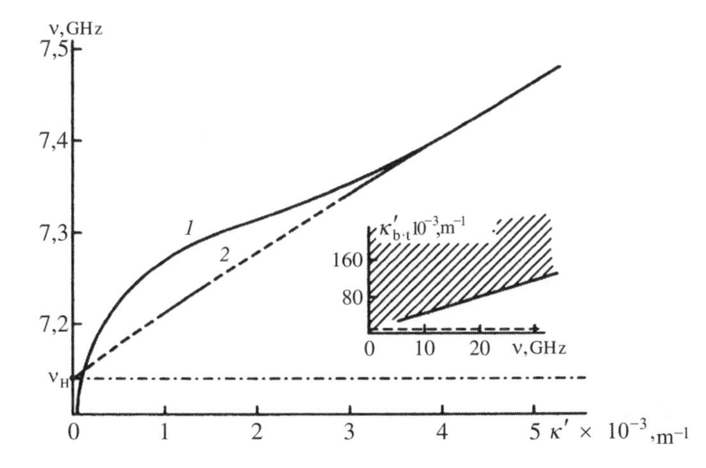

Fig. 3.32 The dispersions of phase constants of LE_t^S – wave – 1 and SSMSW – 2 for a structure with parameters $\nu_H = 3 \cdot 10^9\,\text{Hz}$, $\alpha_{\|} = 10^{-4}$, $h_1 = 5 \cdot 10^{-4}\,\text{m}$, $h_2 = 10^{-5}\,\text{m}$, $h_3 = 0.5\,\text{m}$, $\varepsilon_{1,2} = 14$, $\varepsilon_3 = 1$, $4\pi M_S = 0.5\,\text{T}$

used. In the field of $\kappa' < \kappa_{b.t}'$ an essential difference between $(\kappa_y')_{LE_t^S}$ and $(\kappa_y')_{SSMSW}$ takes place. In Fig. 3.32 (as a fragment) the dependence $\kappa_{b.t}'$ on frequency ν for structures-spinels is shown. The dotted line, as above, marks the level $\kappa' \approx 10^3\,\text{m}^{-1}$ traditionally accepted as the lower boundary frequency of the MSW approximation.

3.4.2 A Conducting Crystal in an External Magnetic Field

Analysis of HF-fields in such a waveguide shows that the layered structure containing a material (film) with tensor $\overleftrightarrow{\varepsilon}$ in an external field $\overline{H}_0$ most essentially changes the dispersive properties of a LM_t wave in the field of the resonant frequency [460], depending on the magnetic field H_0; $\nu_{\perp\overleftrightarrow{\varepsilon}} = \sqrt{(\nu_c^*)^2 + (\nu_p^+)^2}$, $\nu_c^* = \frac{e}{m^*}H_0$ is the cyclotron frequency, m^* is the effective weight of an electron, e is the charge of an electron, $\nu_p^+ = \frac{\nu_p}{\sqrt{\varepsilon_L}}$, $\nu_p^2 = \frac{e^2 n_e}{\varepsilon_0 m^*}$ – the plasma frequency, n_e is the concentration of electrons, ε_0 – the permeability of vacuum, ε_L the dielectric constant of the lattice of a polar crystal. We shall note that the resonant frequency $\nu_{\perp\overleftrightarrow{\varepsilon}}$ has a singularity close to $\varepsilon_\perp = \varepsilon - \frac{\varepsilon_a^2}{\varepsilon}$ and we shall remind that the frequency of cross-section resonance for a ferrite with its tensor $\overleftrightarrow{\mu}$ $-\nu_{\perp\overleftrightarrow{\mu}} = \sqrt{\nu_{\mathrm{H}}(\nu_{\mathrm{H}} + \nu_{\mathrm{M}})}$ has a singularity close to $\mu_\perp = \mu - \frac{\mu_a^2}{\mu}$. At small charge concentrations n_e or at greater fields H_0 the resonant frequency is $\nu_{\perp\overleftrightarrow{\varepsilon}} \approx \nu_c$.

In Fig. 3.33 the dispersive characteristics of phase constants $(\kappa_y')_{LM_t}$ for conducting film crystals $1 - h_2 = 10^{-4}\,\mathrm{m}$, $2 - h_2 = 5 \cdot 10^{-4}\,\mathrm{m}$, with parameters $h_1 = h_3 = 5 \cdot 10^{-3}\,\mathrm{m}, \varepsilon_{1,2} = 16, \varepsilon_3 = 1$, the frequency of relaxation $\nu_\tau = 0$, the concentration of electrons $n_e = 10^{16}\,\mathrm{m}^{-3}$, $H_0 = 852\,\mathrm{kA/m}$, the cyclotron frequency $\nu_c = 3 \cdot 10^{10}\mathrm{Hz}$ and $\kappa_\varepsilon' \approx 25 \cdot 10^2\,\mathrm{m}^{-1}$ are presented. Similarly to the case of waveguides with ferrite-dielectric normally-magnetized layered structures for which the phase constants of a LM_n wave in the field of frequency $\nu_{\perp\overleftrightarrow{\varepsilon}}$ split, we have splitting of the phase constants of LM_t waves in the field of frequency $\nu_{\perp\overleftrightarrow{\varepsilon}}$. Below we shall consider a case of close frequencies $\nu_{\perp\overleftrightarrow{\varepsilon}}$ and $\nu_{\perp\overleftrightarrow{\mu}} \approx \nu_c^*$ for a bigyrotropic layer on LM_t^1 and LM_t^2 waves. The thicker the film made of a conducting crystal, the stronger splitting into

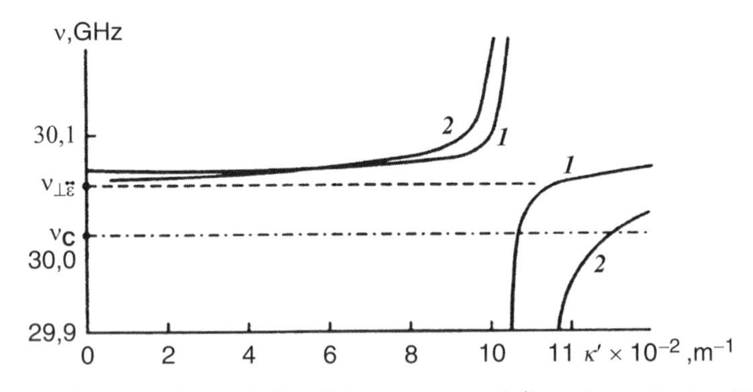

Fig. 3.33 The dispersive characteristics of phase constants $(\kappa_y')_{LM_t}$ for conducting film crystals $1 - h_2 = 10^{-4}\,\mathrm{m}$, $2 - h_2 = 5 \cdot 10^{-4}\,\mathrm{m}$, with parameters $h_1 = h_3 = 5 \cdot 10^{-3}\,\mathrm{m}, \varepsilon_{1,2} = 16, \varepsilon_3 = 1$, the frequency of relaxation $\nu_\tau = 0$, the concentration of electrons $n_e = 10^{16}\,\mathrm{m}^{-3}$, $H_0 = 852\,\mathrm{kA/m}$, the cyclotron frequency $\nu_c = 3 \cdot 10^{10}\,\mathrm{Hz}$ and $\kappa_\varepsilon' \approx 25 \cdot 10^2\,\mathrm{m}^{-1}$

LM_t^1 and LM_t^2 waves. The displacement of frequency $\nu_{\perp \overleftrightarrow{\varepsilon}}$ in the range of frequencies above ν_c near which the phase constants of these waves have strongly dispersive regions, is subject to the concentration of charge carriers n_e.

With due account of the relaxation frequency ($\nu_\tau = 10^6$Hz) the course of phase constants $(\kappa_y')_{LM_t^1}$ and $(\kappa_y')_{LM_t^2}$, in comparison with an idealized crystal with $\nu_\tau = 0$ considered above, practically has not changed. In Fig. 3.34 the dependences $1 - (\kappa_y')_{LM_t^1}$, $2 - (\kappa_y'')_{LM_t^1}$, $3 - (\kappa_y')_{LM_t^2}$, $4 - (\kappa_y'')_{LM_t^2}$ for a structure with parameters $\nu_c = 3 \cdot 10^{10}$Hz, $h_1 = h_3 = 5 \cdot 10^{-4}$m, $h_2 = 5 \cdot 10^{-5}$m, $\varepsilon_{1,2} = 16$, $\varepsilon_3 = 1$, $\nu_\tau = 10^6$Hz, $n_e = 10^{16}$m^{-3}, $H_0 = 852$kA/m are shown. In the fields of frequency $\nu_{\perp \overleftrightarrow{\varepsilon}} > \nu_c$ the dispersion of phase constants is close to zero, and at $\kappa_y' \approx 11.3 \cdot 10^2$m^{-1} a sharp increase of the course of dispersive dependences $(\kappa_y')_{LM_t^1}$ and $(\kappa_y')_{LM_t^2}$ takes place.

For a LM_t^1-wave in the field of frequencies $\nu_{\perp \overleftrightarrow{\varepsilon}}$ a sharp increase in the amplitude constant $\left|(\kappa_y'')_{LM_t^1}\right|$ takes place. For a LM_t^2 wave a selective increase in attenuation and the value of $\left|(\kappa_y''(\nu_{\perp \overleftrightarrow{\varepsilon}})_{LM_t^2}\right|_{max}$ is found.

At increasing losses ($\nu_\tau = 10^9$ Hz) the dispersive characteristics of LM_t^1 and LM_t^2 waves converge on frequency $\nu_{\perp \overleftrightarrow{\varepsilon}}$ in the field of wave numbers $\kappa_y' \approx 11.2 \cdot 10^2$m^{-1}.

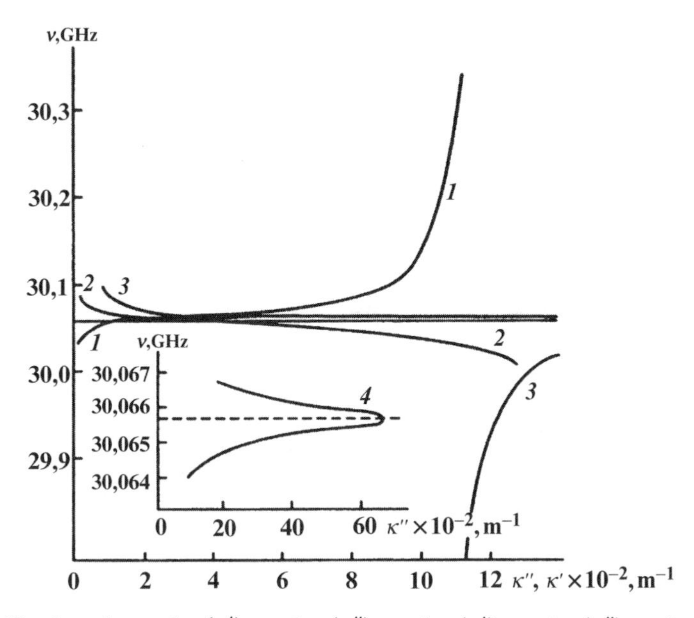

Fig. 3.34 The dependences $1 - (\kappa_y')_{LM_t^1}$, $2 - (\kappa_y'')_{LM_t^1}$, $3 - (\kappa_y')_{LM_t^2}$, $4 - (\kappa_y'')_{LM_t^2}$ for a structure with parameters $\nu_c = 3 \cdot 10^{10}$Hz, $h_1 = h_3 = 5 \cdot 10^{-4}$m, $h_2 = 5 \cdot 10^{-5}$m, $\varepsilon_{1,2} = 16$, $\varepsilon_3 = 1$, $\nu_\tau = 10^6$Hz, $n_e = 10^{16}$m^{-3}, $H_0 = 852$kA/m

3.4.3 A Bigyrotropic Structure in an External Magnetic Field

For a layered structure containing a tangentially-magnetized bigyrotropic film $(\overline{H}_0 \| 0Z)\overline{H}_0$, two cases are possible:

1. The gyrotropy of the layer due to $\overleftrightarrow{\mu}$, has a basic influence on the dispersive characteristics of LE_t waves, and $\overleftrightarrow{\varepsilon}$ has an influence close to $\nu_{\perp \overleftrightarrow{\varepsilon}}$.
2. The gyrotropy of the layer due to $\overleftrightarrow{\varepsilon}$ has a basic influence on the dispersive characteristics of LM_t waves, and $\overleftrightarrow{\mu}$ has an influence close to $\nu_{\perp \overleftrightarrow{\mu}}$.

The closeness of the characteristic frequencies:

- The frequency of cross-section resonance $\nu_{\perp \overleftrightarrow{\mu}}$ of the medium with $\overleftrightarrow{\mu}$
- The frequency of cross-section resonance $\nu_{\perp \overleftrightarrow{\varepsilon}}$ of the medium with $\overleftrightarrow{\varepsilon}$
- The cyclotron frequency ν_c of the medium with $\overleftrightarrow{\varepsilon}$

is determined by the parameters of such structures, namely, the concentration of charges n_e, their effective weights m^*, the lattice constant ε_L, the frequency of relaxation ν_τ for $\overleftrightarrow{\varepsilon}_2$; and also by the parameter of ferromagnetic losses α, dielectric permeability of layers, magnetization M_S for $\overleftrightarrow{\mu}_2$.

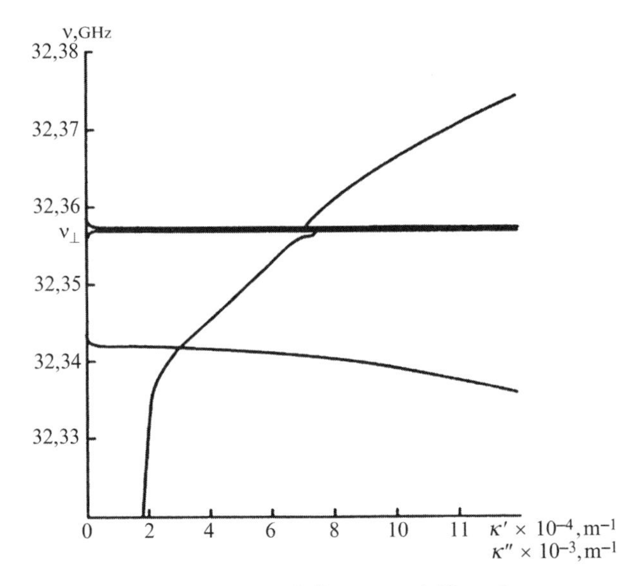

Fig. 3.35 The dispersive dependences $1 - (\kappa'_y)_{LE_t}$, $2 - (\kappa''_y)_{LE_t}$ for a structure with parameters $\nu_H = 3 \cdot 10^{10}\,\text{Hz}$, $\nu_{\perp \overleftrightarrow{\mu}} = 32.357 \cdot 10^9\,\text{Hz}$, $\nu_{\perp \overleftrightarrow{\varepsilon}} = 30.066 \cdot 10^9\,\text{Hz}$, $\nu_c = 3 \cdot 10^{10}\,\text{Hz}$, $\nu_\tau = 10^6\,\text{Hz}$, $h_1 = 5 \cdot 10^{-4}\,\text{m}$, $h_2 = 10^{-5}\,\text{m}$, $h_3 = 0.5\,\text{m}$, $\varepsilon_{1,2} = 16$, $\varepsilon_3 = 1$, $\frac{m^*}{m_0} = 0.5$, $\frac{\nu_p}{\sqrt{\varepsilon}} = 2 \cdot 10^9\,\text{Hz}$, $n_e = 10^{16}\,\text{m}^{-3}$, $4\pi M_S = 0.176\,\text{T}$

In Fig. 3.35 the dispersive dependences $1-(\kappa_y')_{LE_t}$, $2-(\kappa_y'')_{LE_t}$ for a structure with parameters $\nu_{\mathrm{H}}=3\cdot10^{10}\,\mathrm{Hz}$, $\nu_{\perp\overset{\leftrightarrow}{\mu}}=32.357\cdot10^{9}\,\mathrm{Hz}$, $\nu_{\perp\overset{\leftrightarrow}{\varepsilon}}=30.066\cdot10^{9}\,\mathrm{Hz}$, $\nu_c=3\cdot10^{10}\,\mathrm{Hz}$, $\nu_\tau=10^{6}\,\mathrm{Hz}$, $h_1=5\cdot10^{-4}\,\mathrm{m}$, $h_2=10^{-5}\,\mathrm{m}$, $h_3=0.5\,\mathrm{m}$, $\varepsilon_{1,2}=16$, $\varepsilon_3=1$, $\frac{m^*}{m_0}=0.5$, $\frac{\nu_p}{\sqrt{\varepsilon}}=2\cdot10^{9}\,\mathrm{Hz}$, $n_e=10^{16}\,\mathrm{m}^{-3}$, $4\pi M_S=0.176\,\mathrm{T}$ are shown. In the frequency field of cross-section resonance $\nu_{\perp\overset{\leftrightarrow}{\mu}}$ there is an insignificant perturbation of the dispersion of phase constant $(\kappa_y')_{LE_t}$ as a small region with an abnormal dispersion to whose middle part there corresponds a highly selective increase in the amplitude constant $\left|(\kappa_y'')_{LE_t}\right|_{\max}$. A sharp increase in the amplitude constant of this wave is observed at tuning-out by $-15\,\mathrm{MHz}$, below the frequency $\nu_{\perp\overset{\leftrightarrow}{\mu}}$.

For LM_t waves in such structures a highly selective attenuation of signal in the field of splitting of a LM_t wave into a LM_t^{F1} and LM_t^{F2} wave will be observed.

3.5 Conclusions

1. *LE* and *LM* waves have been simulated in tangentially and normally-magnetized (by an external field H_0) three-layer structures with losses, shielded with metal surfaces for cases when the middle layer represents a ferrite with a permeability μ or a conducting crystal with a permeability ε, or a bigyrotropic medium with tensors $\overset{\leftrightarrow}{\mu}$ and $\overset{\leftrightarrow}{\varepsilon}$.
2. The dispersions of *LE* and *LM* waves have been investigated, as near the resonant (ν_{H} and $\nu_\perp$ for a ferrite, $\nu_{\perp\overset{\leftrightarrow}{\mu}}$ and $\nu_{\perp\overset{\leftrightarrow}{\varepsilon}}$ for a bigyrotropic medium) and the cyclotron frequencies (ν_c for conducting crystals), and at significant tuning-out from them, including the bands of the existence of strongly delayed electromagnetic waves of the dipole–dipole interaction, i.e. magnetostatic spin waves of various types.
3. It is shown that the dispersive dependences of *LE* and *LM* waves in such structures near the resonant frequencies are strongly influenced by metal screens. Depending on the distance between the screens, determining the critical frequency (cutoff frequency) of the structure ν_{cr} and signal frequencies ν, selective attenuation, amplification of attenuation at $\nu\gg\nu_{cr}$ as well as selective signal transmission at $\nu\ll\nu_{cr}$ can be observed, determined by the parameter of ferromagnetic losses α . These effects are upset by frequencies from each other and determined by various types of waves.
4. In the prelimit mode ($\nu\gg\nu_{cr}$) of tangentially-magnetized structures the selective attenuation is related to the interference mechanism of selective attenuation of signals at interaction between LE_t (fast) and LE_t (slow) waves, and for the central frequency $\nu<\nu_\perp$ equality of the amplitude wave constants and the antiphased nature of phase wave constants are characteristic.
5. In the beyond-cutoff mode $\nu\ll\nu_{cr}$ of tangentially-magnetized structures, selective transmission of signals on frequency $\nu>\nu_\perp$ is caused, depending on the

range of frequencies, by competition between LE_t^S and LE_t^F waves localized on the opposite surfaces of the ferrite film:

- In the long-wave part of the UHF range ($\nu \leq 3\,\text{GHz}$) – by LE_t^S – waves, for which $(\kappa_y'')_{LE_t^S} \ll (\kappa_y'')_{LE_t^F}$.
- In the middle-wave parts of the UHF range ($\nu \approx 15$–$20\,\text{GHz}$) – equally by LE_t^S and LE_t^F waves, for which $(\kappa_y'')_{LE_t^S} \approx (\kappa_y'')_{LE_t^F}$.
- In the short-wave part of the UHF range and in the EHF range (on frequencies 20–30 GHz) – *by* LE_t^F waves only, for which $(\kappa_y'')_{LE_t^F} \ll (\kappa_y'')_{LE_t^S}$, and LE_t^S – and LE_t^F – waves.

6. It is shown that in structures based on weakly dissipative ferrite films the spectrum of excited and propagating waves falls outside the bounds defined by the MSW approximation, and, with advance into the EHF range this shift increases.
7. The borders of applicability of the MSW approximation in the UHF and EHF – ranges have been determined, from which is follows that for tangentially-magnetized structures in a range of frequencies $\nu_{\text{H}} = 2\,\text{GHz}$, $\kappa_{b.t}' \geq 10^3\text{m}^{-1}$, that corresponds to the traditionally accepted restriction, in a range of $\nu_{\text{H}} \approx 30\,\text{GHz}$, $\kappa_{b.t}' \geq 2 \cdot 10^4\text{m}^{-1}$, and for $\nu_{\text{H}} \approx 60\,\text{GHz}$, $\kappa_{b.t}' \geq 10^5\text{m}^{-1}$, that essentially diverges with the existing point of view and specifies the correctness of the description of wave processes in magneto-ordered structures in the short-wave range within the limits of the electrodynamic approach only, both near the resonant frequencies and in the band of MSW existence.
8. For LM_t waves in tangentially-magnetized ferrite-dielectric structures in the prelimit mode ($\nu \gg \nu_{cr}$) on frequency $\nu \leq \nu_{\perp}$ selective attenuation of a signal due to features of HF-permeability $\mu_{\perp}$ has been revealed.
9. The dispersive characteristics of *LM* waves in normally-magnetized ferrite-dielectric structures with losses in the UHF and EHF ranges were investigated.
10. It is shown that for LM_n waves the existence of both slow LM_n^S waves passing in a limit into SSLMSW, and fast LM_n^F waves having phase constants essentially smaller, in comparison with a wave in free space is characteristic in the UHF range.
11. For LM_n waves in the field of frequencies $\nu \gg \nu_{\text{H}}$ characteristic are:
    - The presence of regions with normal and abnormal dispersions of phase constants.
    - The presence of points of inflection on the dispersive characteristics of phase constants with an extremely high steepness.
    - Small amplitude constants in the passband, and $\left|(\kappa_y'')_{LM_n^F}\right|_{\min} < \left|(\kappa_y'')_{LM_n^S}\right|_{\min}$.
12. The interference mechanism of interaction between LM_n^S and LM_n^F waves with close dispersions of phase constants in the field of frequencies $\nu > \nu_{\text{H}}$ provides selective signal attenuation over the ferrite length $L \approx \frac{n\pi}{\Delta\kappa_y'}$, $n = 1, 2, \ldots$.

13. It was revealed that for LM_n waves in the millimeter range:

    – The value of permeability of the ferrite-dielectric structure ($\varepsilon = 1$–14) practically does not influence the dispersion of phase constant.
    – With increasing the dielectric permeability of both the ferrite and basis the value of selective attenuation on frequency $\nu > \nu_{\mathrm{H}}$ decreases.
    – On frequency ν_{H} there is a spasmodic increase of signal attenuation due to the action of losses in a bilaterally-metallized structure.
    – On frequencies $\nu < \nu_{\mathrm{H}}$ the amplitude constant has its minimum value.

14. It is shown that in normally-magnetized bilaterally-metallized ferrite-dielectric structures an LM wave in the field of frequencies $\nu \geq \nu_{\mathrm{H}}$ split into two waves, for one of which $\kappa'_y < \kappa'_0$, that allows one to define it as a fast LM_{n}^F wave, and for another tending to the boundary frequency of SSLMSW is characteristic at high values of κ'_y, that allows one to define it as a slow LM_{n}^S wave. In the field of splitting of dispersive characteristics at tuning-out from ν_{H} above and below by frequencies there is a raised steepness of the phase constants $(\kappa'_y)_{LM_{\mathrm{n}}^F}$ and $(\kappa'_y)_{LM_{\mathrm{n}}^S}$. In the field of frequencies ν_{H} both waves have a raised dispersion of their phase constants and the maximum attenuation of signal.
15. For planar waveguides with ferrite-dielectric filling with layered structures at normal magnetization there exist LM waves which:

    – Fall outside the ferromagnetic resonance ν_{H} in the form of two fast and one slow waves: LM_{n}^{F1}, LM_{n}^{F2}, and LM_{n}^S.
    – In the beyond-cutoff mode $(\nu \ll \nu_{cr})$ on frequency $\nu < \nu_{\mathrm{H}}$ a selective attenuation of signal due to the increase of the amplitude constants of fast LM_{n}^F waves is observed.
    – In the prelimit mode $(\nu \gg \nu_{cr})$ on frequency $\nu < \nu_{\perp}$ there is a selective attenuation of signal due to the increase of the amplitude constant of LM_{n}^F-waves within a narrow band of frequencies.

16. It is shown that the lower border of the MSW approximation for the phase constants of slow LM_n^S waves is $\kappa'_{b.n} \approx 2 \cdot 10^3 \mathrm{m}^{-1}$ in the field of frequencies $\nu_{\mathrm{H}} > 15$–$20\,\mathrm{GHz}$ and $\kappa'_{b.n} \approx 3 \cdot 10^3 \mathrm{m}^{-1}$ for $\nu_{\mathrm{H}} \approx 3\,\mathrm{GHz}$, however, calculation of the amplitude constants in the field of $\kappa' < \kappa'_{b.n}$ gives correct results for the UHF and EHF ranges only in the developed electrodynamic approach.
17. For planar waveguides containing ferrite-spinels the above tregularities are kept in tangentially and normally-magnetized structures, the band of frequencies extends, and under equal conditions the lower boundary frequency of the MSW approximation increases (for $\nu_{\mathrm{H}} = 3 \cdot 10^{10}\,\mathrm{Hz}$, $\kappa'_{b.n} \approx 5 \cdot 10^4 \mathrm{m}^{-1}$).
18. For planar waveguides containing layered structures with a conducting crystal (Gross' model) tangentially-magnetized by an external field, in the field of frequency $\nu_{\perp \overset{\leftrightarrow}{\varepsilon}}$ there exist fast LM_t^F waves, for which $(\kappa'_y)_{LM_t} < \kappa'_0$, and the dispersive characteristics of phase constant splits into two branches, LM_t^{F1} and LM_t^{F2}, and their repulsion is more for thicker films, and the displacement of frequency $\nu_{\perp \overset{\leftrightarrow}{\varepsilon}}$ in a range above ν_c is related with the concentration of charge carriers in the crystal.

19. In the presence of dissipation in a conducting crystal (the frequency of relaxation $\nu_\tau \approx 10^6$ Hz) for LM_t^{F1} and LM_t^{F2} waves the course of the dispersions of phase constants does not change, in the field of frequency $\nu_{\perp \overleftrightarrow{\varepsilon}} > \nu_c$ the dispersion of phase constants of these waves is close to zero, and at $\kappa_y' \approx 11.3 \cdot 10^2 \mathrm{m}^{-1}$ on frequency $\nu_\mathrm{H} \approx 3 \cdot 10^{10}$ Hz a sharp increase takes place. For a LM_t^{F2} wave in the field of frequency $\nu_{\perp \overleftrightarrow{\varepsilon}}$ a selective increase in the amplitude constant takes place. At increasing losses (ν_τ), the repulsion of the dispersive characteristics of LM_t^{F1} and LM_t^{F2} waves on frequency $\approx \nu_{\perp \overleftrightarrow{\varepsilon}}$ decreases.
20. For a planar waveguide containing a layered structure with a bigyrotropic film, tangentially-magnetized on frequency $\nu_{\perp \overleftrightarrow{\varepsilon}}$, for the phase constant of LM_t wave there is a small region with an abnormal dispersion, to which a highly selective attenuation of signal corresponds, and $\left|(\kappa_y'')_{LE_t}\right|_{\max}^{b/g} \gg \left|(\kappa_y'')_{LE_t}\right|_{\max}^{fc/c}$ and $(\Delta\nu_{3dB})^{b/g} \ll (\Delta\nu_{3dB})^{fc/c}$, where the superscripts "b/g" is bigyrotropic $(\overleftrightarrow{\mu}, \overleftrightarrow{\varepsilon})$, "$f$" is ferrite $(\overleftrightarrow{\mu})$, "$c/c$" is a conducting crystal $(\overleftrightarrow{\varepsilon})$ relate to possible media.

Chapter 4
Research of Dispersions of Weakly and Strongly Delayed Waves in Layered Structures

Laboratory breadboard models are described and the results of our experimental research of various types of surface and solid waves in layered structures on the basis of ferrite plates and films in a range of frequencies of 3–140 GHz are reported. Experimental data are compared with the results of theoretical analysis. Research was made on structures containing yttrium–iron garnet, spinels, barium hexaferrite. Special sensors of resonant frequencies (fields) and experimental methods have been developed for research of the dispersions of waves near the resonant frequencies. Waveguide and strip lines were used as converters.

4.1 Experimental Breadboard Models and Measurement Technique

Experimental studies were made on various types of waveguide and strip lines. Laboratory breadboard models of a waveguide type were devices (Fig. 4.1) on waveguides of standard and decreased cross-sections: (*a*) a module with a regulable crossrunner of its line for research of the influence of metal screens; (*b*) a module for research of the properties of connected structures; (*c*) a module for research of the properties of a slotted-guide converter and FDMT; (*d*) a module for research of the properties of FDMT and transmission line with a lossy jacket, (*e, f*) waveguide-to-beyond-cutoff modules. The ferrite was magnetized by an external field tangentially ($\overline{H}_0||0Z$) or normally ($\overline{H}_0||0X$).

Laboratory constructions of a strip type (Fig. 4.2) were of various topologies and mutual orientations of the entrance (input) and target (output) converters in the form of microstrip (Fig. 4.2*a, b*), slot (Fig. 4.2*c, d*) or coplanar (Fig. 4.2*e*) lines. Layered structures on the basis of ferrite films and plates were arranged in parallel (Fig. 4.2*a–c*) and orthogonally, along and across (Fig. 4.2*d, e*) the plane of converters. For tangentially magnetized structures with parallel orientation to the converter on the basis of a strip line, breadboard models of a special construction (Fig. 4.2) were used.

A.A. Ignatiev, *Magnetoelectronics of Microwaves and Extremely High Frequencies in Ferrite Films*.
DOI: 10.1007/978-0-387-85457-1_5, © Springer Science + Business Media, LLC 2009

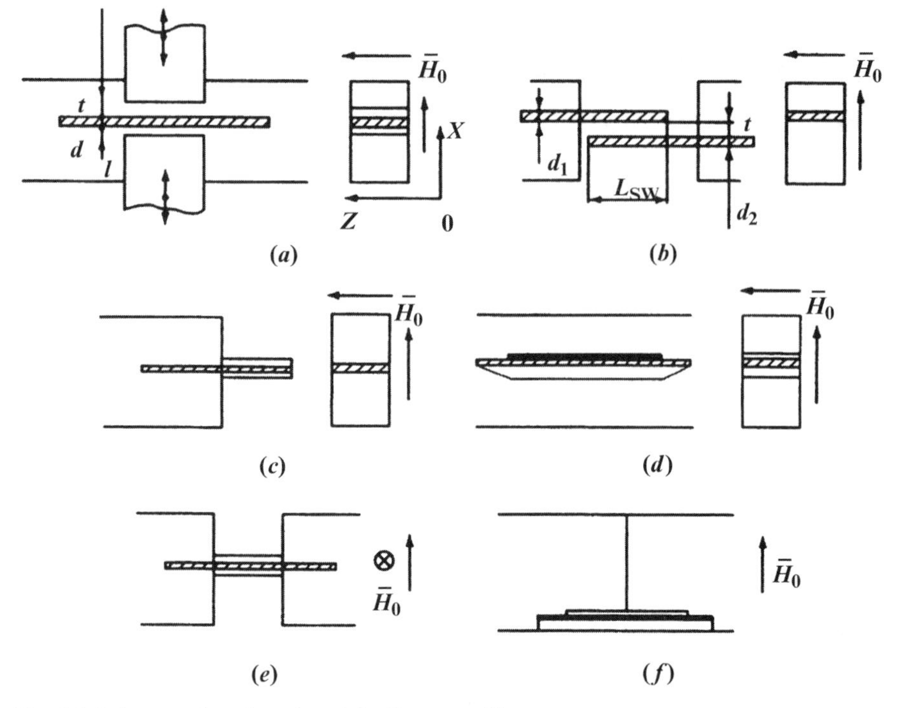

Fig. 4.1 Laboratory breadboard models of a waveguide type

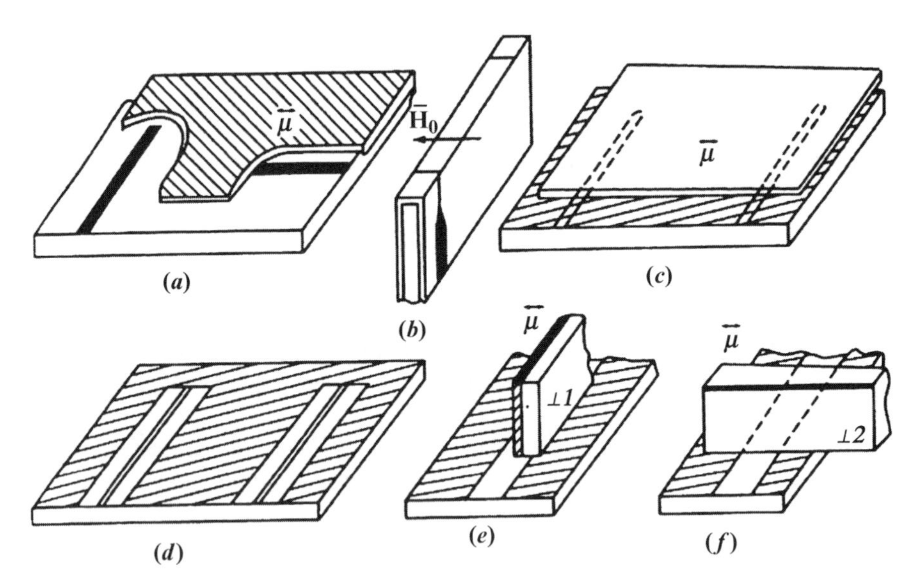

Fig. 4.2 Laboratory constructions of a strip type

Studies were made on industrial and pilot models of ferrite structures, included high-quality ferrites of YIG, spinels, barium hexeferrite. Ferrite plates had sizes of the order of $(0.50 \div 0.75) \times 3.5 \times 15\text{mm}^3$, a film thickness of 10^{-4}–$10^{-5}\,\text{m}^{-1}$ on dielectric bases of a thickness $(3–5) \cdot 10^{-4}\,\text{m}^{-1}$ of GGG, magnesium oxide, aluminum–garnet with $tg\delta \leq 10^{-4}$. Saturation magnetization was $4\pi M_S \cong (14–40)$ kA/m, fields of cubic anisotropy $H_A \cong 3\,\text{kA/m}$, monoaxial $H_s \cong 2\,\text{MA/m}$, a line width FMR $\Delta H_{||,\perp} \geq 8\,\text{A/m}$.

Experimental studies were made in a range of frequencies 3–140 GHz. In a range of frequencies 3–80 GHz standard panoramic measuring instruments of SWRe and attenuation of R2-44, R2-67, R2-66, R2-65, R2-68, R2-69 types, and non-standard installations RW-12, RW-13, RW-15, RW-16 on the basis of return wave lamps (Fig. 4.3) were used. In a range of frequencies 120–140 GHz an installation was designed on the basis of a G-144 *W* type. For raised power level $P = (3–5)\ W$ experimental samples of generator of backward-wave tube of M-type of the millimeter range were used.

The external magnetic field $\bar{H}_0 \leq 1.75\,\text{MA/m}$ was set up with laboratory EM-1 electromagnets with a diameter of their pole clamps 25 and 80 mm and a LM-01 with a size of its tips $60 \times 120\,\text{mm}^2$ and a diameter 100 mm. The electromagnets were energized from a current 8 A 620.34C × Э stabilizer and a voltage UIP-1 stabilizer. The magnetic induction was controlled with a measuring instrument, PIE. MG P-1 teslameter with a nominal error $\frac{\Delta B_0}{B_0} \cong 1–2\%$. To establish reference points by frequency and field, special sensors (Chapter 6) with an error $\frac{\Delta\nu_0}{\nu_0} \leq (10^{-2}–10^{-3})\%$ were used at an external field $\overline{H}_0 \cong 1.75\,\text{MA/m}$.

Time of signal delay was measured on an installation (Fig. 4.4) containing a mm-range klystron generator (1) (G3-37), a modulating pulse generator (2) *s* (G5-15), a waveguide highway with a coupler (3) and an UHF detector (4), a delay line (5), a high-frequency A3-29 (6), a measuring MZ-26 instrument of time intervals (7).

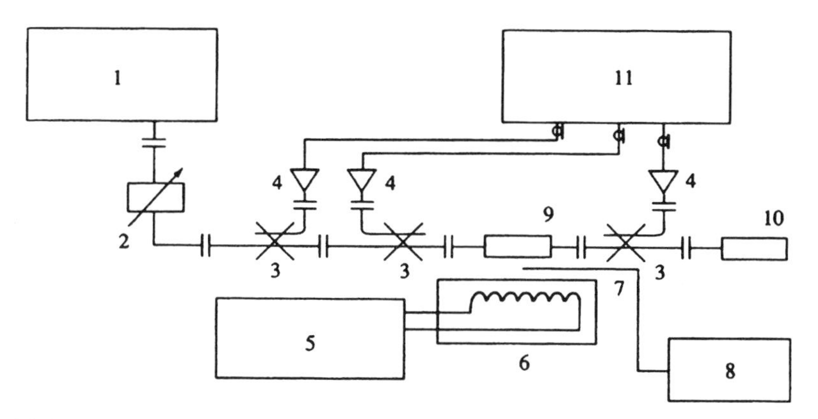

Fig. 4.3 Block-diagram of aggregate for investigation in frequency ranges 3–140 GHz

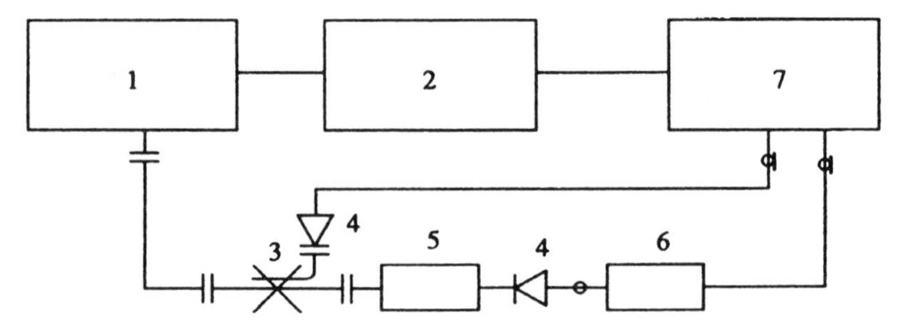

Fig. 4.4 Aggregate for investigation of delay time

AFC parameters of the transmitted and reflected from the ferrite-dielectric structures signals (their appearance, shape, and irregularity) were investigated. Current frequency in AFC was measured with half-accuracy wavemeters (MAW) with an error 0.02%. For the wavemeters built in panoramic measuring instruments the regular error did not exceed 0.4%. Attenuation measurements attenuation were carried out on panoramic installations according to standard techniques and schemes with an error $(1 \div 1.5)\%$.

Dispersive characteristics were examined by the frequency of interference maxima of AFC ν_n subject to their exact binding to the resonant frequency. The delay time of signal was determined by the front shift of a delayed UHF pulse with respect to the basic one. The measurement errors of time intervals did not exceed ± 5 ns.

Experimentally investigated were:

- The dispersive characteristics of phase constants $\Omega(\kappa') = \frac{\omega(\kappa')}{\omega_{\mathrm{H}}}$, $\nu(\kappa')$ and $\Omega(\kappa' d), \nu(\kappa' d), \Omega_n(n)$ and $\nu_n(n)$ where $\omega_{\mathrm{H}} = \gamma H_{0i}$ is the frequency of ferromagnetic resonance, γ the gyromagnetic ratio, H_{0i} the internal magnetic field, d the thickness of the ferrite layer (film), n the number of spin-wave resonance.
- The dispersive characteristics of amplitude constants $\Omega(\kappa'') = \frac{\omega(\kappa'')}{\omega_{\mathrm{H}}}$, $\nu(\kappa'')$, $\Omega(\kappa' d)$ and $\nu(\kappa' d)$, $K_{los} = 8.68\kappa'' L$, K_{los}–in dB, L is the length of the structure, K_{car} the carryover factor, $K_{car}(dB) = 10 \lg \frac{P_{out}(\nu)}{P_{inp}(\nu)}$.
- The dispersion of GDT of signal $\tau(\nu) = \frac{L}{U_{gr}(\nu)}; U_{gr}(\nu) = \frac{d\omega(\kappa)}{d\kappa}$ being the group wave speed in the structure.

4.2 Waves in Ferrite Plate Structures

At the first stage of our theoretical and experimental research basic attention was given to studying the dispersive characteristics of various types of magnetostatic spin waves in layered structures on the basis of weakly and strongly anisotropic ferrite plates in the millimeter range. Theoretical analysis was based on the MSW approximation [8]. From the position of an electrodynamic approach developed in our works, theoretical dependences for MSW will be limited by small wave numbers

κ' and $\kappa' d$. In some cases, instead of normalization on $\omega_H = 2\pi\nu_H$ accepted in the MSW approximation we shall use that on the lower boundary frequency $\nu_{H0}(\nu_{H0} < \nu_H)$ of the signal spectrum, determined at a typical level of sensitivity of the standard measuring equipment $(45 \div 50)$ dB mW.

Theoretical analysis in the MSW approximation gives: if a layered structure contains a ferrite plate symmetrically loaded with metal screens over equithick dielectric layers ($t = \ell$) and magnetized by an external field $\overline{H}_0$, then the following types of waves can propagate in the direction, perpendicular to $\overline{H}_0$ ($\kappa'_y \neq 0$):

- In a band of frequencies $\nu_H < \nu < \sqrt{\nu_H(\nu_H + \nu_M)}$: space spatial MSW (SSLMSW), which dispersions weakly depend on the influence of the metal screens.
- In a band of frequencies $\sqrt{\nu_H(\nu_H + \nu_M)} < \nu < \nu_H + \frac{1}{2}\nu_M$: space surface MSW or SSMSW (Eshbach-Damon's waves), corresponding to the case of a free ferrite plate ($\frac{t}{d} = \frac{\ell}{d} = \infty$).
- In a band of frequencies $\nu_H + \frac{1}{2}\nu_M < \nu < \nu_H + \nu_M$: return surface MSW (RSMSW) corresponding to the case of metal screen arrangement near the ferrite layer ($\infty \leq \frac{t}{d} \leq 0$).
- Along the field $\overline{H}_0$ in a band of frequencies $\nu_H < \nu < \sqrt{\nu_H(\nu_H + \nu_M)}$: return spatial MSW (RSLMSW).

In Fig. 4.5 the dispersive characteristics $\Omega(\kappa' d)$ of SSLMSW-1, SSMSW-2 and RSMSW-3-8 in such a structure ignoring losses ($\alpha = 0$) in the millimeter range are presented for $t = \ell$: $1 - t/d = 0$; $2 - t/d = \infty$; $3 - t/d = 5.0$; $4 - t/d = 2.0$; $5 - t/d = 0.25$; $6 - t/d = 0.10$; $7 - t/d = 0.05$; $8 - t/d = 0.01$; $9 - t/d = 0$; $4\pi M_S = 0.176\text{T}$; $\overline{H}_0 = 0.995\,\text{MA/m}$. The dotted lines depict such regions of dispersions of RSMSW, SSMSW and SSLMSW for which calculation the electrodynamic approach is required. For RSMSW with $t/d < 0.25$ the results of the MSW approximation and of LE_t^S waves practically coincide in a range of frequencies $\nu_H = 3 \cdot 10^{10}\,\text{Hz}$.

Unlike the known result [8], given $t/d \ll 1$, practically in the whole range of frequency change in the passband $\sqrt{1 + \Omega_M} < \Omega < 1 + \Omega_M$ SMSW can exist as return ones, $\Omega_M = \frac{\omega_M}{\omega_H} = \frac{4\pi\gamma M_s}{\gamma H_{0i}}$. Simultaneously, at the same Ω_M, but lower frequencies, SLMSW characteristic of normal magnetization of the ferrite plate on its other sides can be excited. From Fig. 4.5 it follows that the dispersion of SMSW strongly depends on t/d while that of SLMSW practically does not depend on the presence of metal screens ($t/d = 0$) is resulted. With the metal screens approaching the ferrite plate the RSMSW passband extends and, in the limit, comes nearer to the band of SSLMSW.

Account of the influence of ferromagnetic loss parameter α in the MSW approximation gives the following results. In Fig. 4.6 the dispersive characteristics of phase constants κ'_y of RSMSW ($t/d = \infty$, curves 1–4) and SSMSW ($t/d = 1$, curves 5–8) for several loss parameters are presented: $1, 5 - \alpha = 0$; $2, 6 - \alpha = 0.1$; $3, 7 - \alpha = 0.15$; $4, 8 - \alpha = 0.2$.

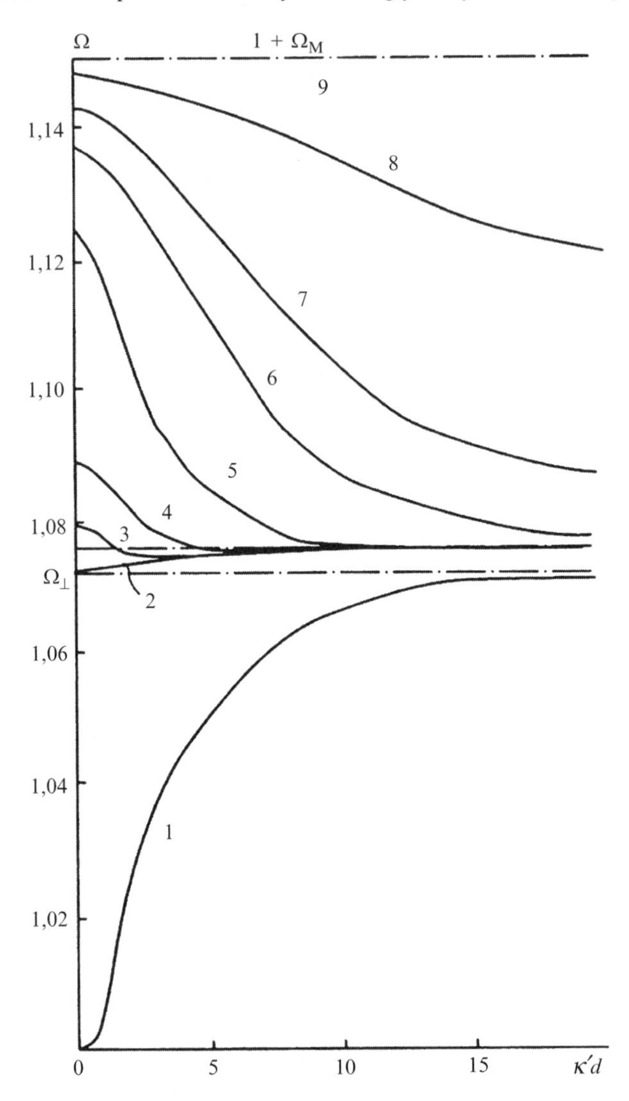

Fig. 4.5 The dispersive characteristics $\Omega(\kappa'd)$ of SSLMSW-1, SSMSW-2 and RSMSW-3–8 in such a structure ignoring losses ($\alpha = 0$) in the millimeter range are presented for $t = \ell$: $1 - t/d = 0$; $2 - t/d = \infty$; $3 - t/d = 5.0$; $4 - t/d = 2.0$; $5 - t/d = 0.25$; $6 - t/d = 0.10$; $7 - t/d = 0.05$; $8 - t/d = 0.01$; $9 - t/d = 0$; $4\pi M_S = 0.176\,\mathrm{T}$; $H_0 = 0.995\,\mathrm{MA/m}$

With incrattenuation of the parameter of losses α the boundary frequency of SMSW $\Omega = 1 + \frac{1}{2}\Omega_M$ for κ'_y decreases, and the dispersion κ'_y strongly changes in the field of $\kappa'_y d < 2$. Subject to the data of Section 3.1 for the lower border of the MSW approximation in the given range of frequencies we have $\kappa'_{b.t} > 2 \cdot 10^5\,\mathrm{m}^{-1}$ (Fig. 3.17*a*). Then. For the dispersions of SMSW with $\kappa'd > 0.1$ we have $d > 5 \cdot 10^{-6}\,\mathrm{m}$, that corresponds to thin ferrite films.

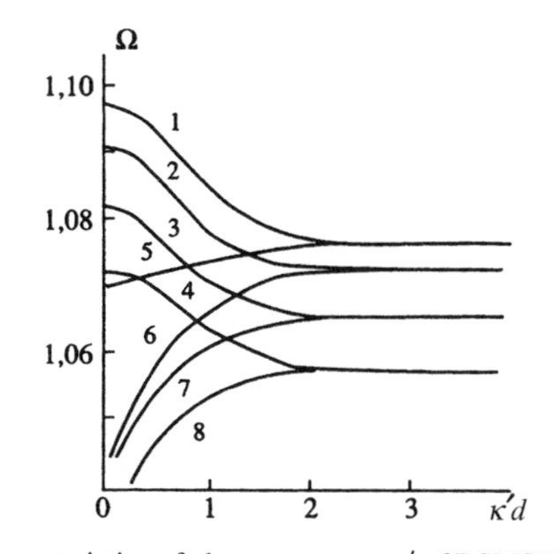

Fig. 4.6 The dispersive characteristics of phase constants κ'_y of RSMSW ($t/d = \infty$, curves 1–4) and SSMSW ($t/d = 1$, curves 5–8) for several loss parameters as are presented: $1,\ 5 - \alpha = 0$; $2,\ 6 - \alpha = 0.1$; $3,\ 7 - \alpha = 0.15$; $4,\ 8 - \alpha = 0.2$

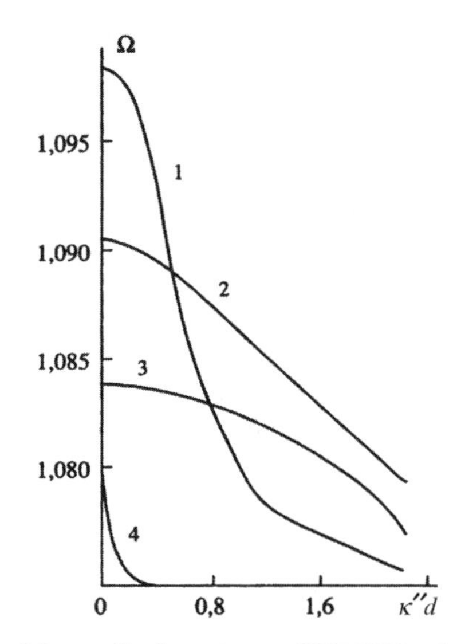

Fig. 4.7 The dependences of the amplitude constants of RSMSW on losses α for $t/d = 1$ ($1 - \alpha = 3 \cdot 10^{-2}$; $2 - \alpha = 1 \cdot 10^{-1}$; $3 - \alpha = 7.5 \cdot 10^{-2}$; $4 - \alpha = 1 \cdot 10^{-3}$)

For RSMSW the dispersion κ'_y practically does not change, and with growing losses α it displaces towards lower frequencies. At further metal screens approaching to the ferrite ($t/d < 1$) their influence is even weak. In Fig. 4.7 the dependences of the amplitude constants of RSMSW on losses α for $t/d = 1(1 - \alpha = 3 \cdot 10^{-2}$; $2 - \alpha = 1 \cdot 10^{-1}$; $3 - \alpha = 7.5 \cdot 10^{-2}$; $4 - \alpha = 1 \cdot 10^{-3})$ are shown.

With frequency reduction the amplitude constant of RSMSW grows, rather sharply at small losses ($\alpha = 10^{-3}$).

Theoretically investigated were also SMSW in structures with various kinds of loadings of the ferrite with metal screens, namely:

- Structures with a unilateral operating screen ($\ell/d = \infty$, and t/d to be changed)
- Structures with unilateral metallization ($\ell/d = 0$, and t/d to be varied)
- Structures with bilaterial metallization ($\ell/d = const$, $t/d = const$)

In Fig. 4.8 the dispersions of SMSW in such structures are presented for $\kappa_y' > 0$ and $\kappa_y' < 0$:

(a) At $\ell/d = \infty$: $1 - t/d = 0$; $2 - t/d = 0.05$; $3 - t/d = 0.1$, $4 - t/d = 0.2$, $5 - t/d = 0.5$, $6 - t/d = 1$, $7 - t/d = 2$, $8 - t/d = 10$
(b) At $\ell/d = 0$: $9 - t/d = 0.05$, $10 - t/d = 0.1$
(c) At $t/d = 0.1$: $11 - \ell/d = 0.1$, $12 - \ell/d = 0.5$, $13 - \ell/d = 1$, $14 - \ell/d = 2$

It is obvious that in bilaterally-metallized structures there exist SMSW with normal, abnormal, and normal–abnormal (combined) dispersions. At reversal of the magnetic field $\overline{H}_0$ in structures with asymmetrical loading of the ferrite with metal screens it is possible to get dispersive characteristics of RSMSW with various values of κ' and steepness $d\omega/d\kappa'$ for waves extending on various surfaces of the ferrite film in opposite directions. For RSMSW the value of group speed and dispersion laws $U_p(\nu)$ and $U_{gr}(\nu)$ vary over a wide range.

In Fig. 4.9 the dispersive characteristics $\Omega(\kappa')$ of connected SMSW are presented. In free there are no external metal screens $\ell_1/t = \ell_2/t = \infty$ structures from

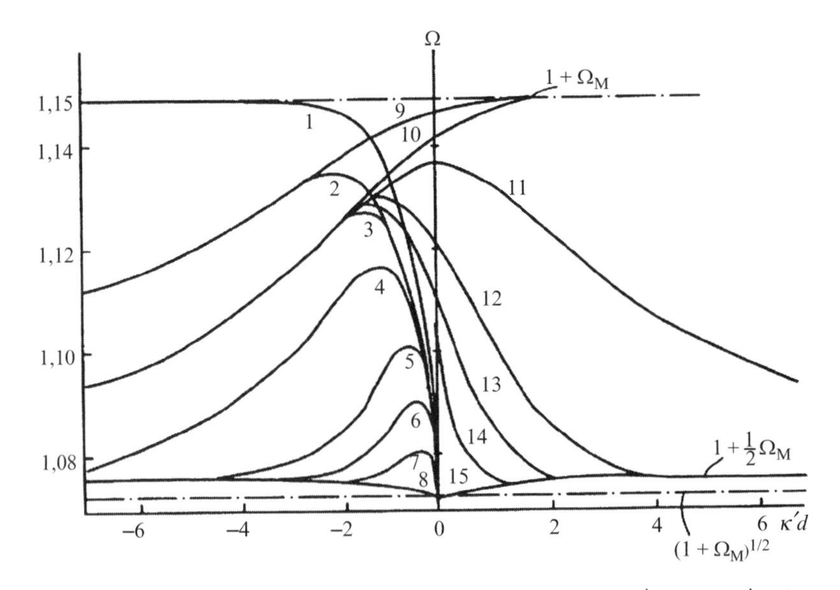

Fig. 4.8 The dispersions of SMSW in such structures are presented for $\kappa_y' > 0$ and $\kappa_y' < 0$:
(*a*) At $\ell/d = \infty$: $1 - t/d = 0$; $2 - t/d = 0.05$; $3 - t/d = 0.1$, $4 - t/d = 0.2$, $5 - t/d = 0.5$, $6 - t/d = 1$, $7 - t/d = 2$, $8 - t/d = 10$
(*b*) At $\ell/d = 0$: $9 - t/d = 0.05$, $10 - t/d = 0.1$
(*c*) At $t/d = 0.1$: $11 - \ell/d = 0.1$, $12 - \ell/d = 0.5$, $13 - \ell/d = 1$, $14 - \ell/d = 2$

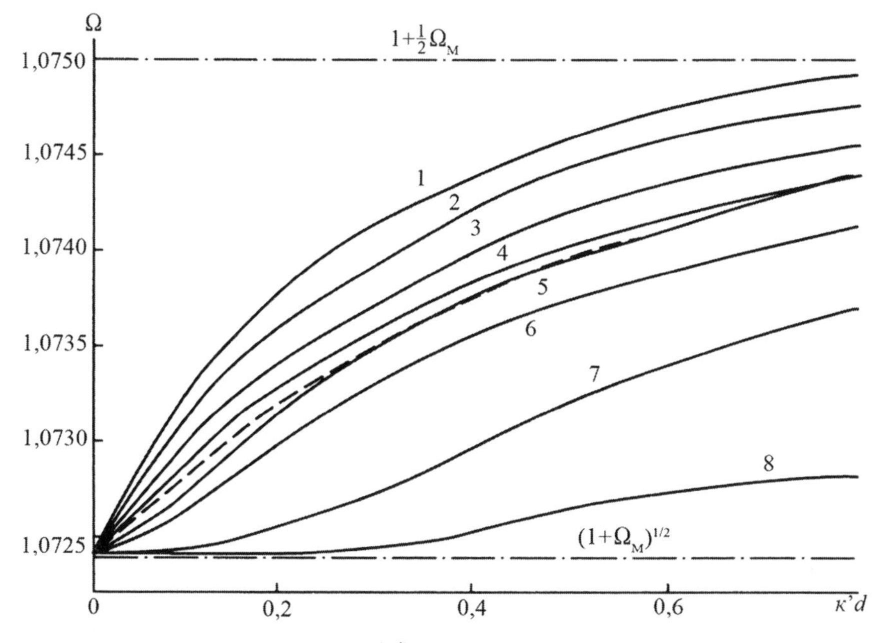

Fig. 4.9 The dispersive characteristics $\Omega(\kappa')$ of connected SMSW are presented

two layers of ferrite with thickness d_1 and d_2, shared a clearance $t(d_1 = d_2 = d)$, $\ell_1/d = \ell_2/d = \infty$, ℓ_1 and ℓ_2 is a distance between external surfaces of layers of ferrite d_1 and d_2 up to the corresponding metal screens, including dispersion characteristics of fast (1–4) and slow (5–8) vogues: 1, $8 - t/d = 0.2$; 2, $7 - t/d = 1$; 3, $6 - t/d = 5$; 4, $5 - t/d = 10$ at $\Omega_\mathrm{M} = 0.15$, a dotted line – wave of Eshbah-Daimon's. The terms ≪fast and slow modes≫ are introduced for Eshbach-Damon's wave.

In Fig. 4.10 the dispersions $\Omega(\kappa' d)$ of connected SMSW in bilaterally-metallized structures are presented at $\ell_1/d = \ell_2/d = \ell/d$ and $t/d = 0.2:\ 1 - \ell/d = 0.250$; $2 - \ell/d = 0.250$; $3 - \ell/d = 0.125$; $4 - \ell/d = 0.05$; $5 - \ell/d = 0.01$ at $\Omega_M = 0.15$ and the binding area $L_c = 6 \cdot 10^{-3}\,\mathrm{m}$, the points designate the experimental values of dispersions $\Omega(\kappa' d)$. One can see that, unlike free two-connected structures (Fig. 4.9), the character of dispersions $\Omega(\kappa' d)$ has changed. Fast modes have an abnormal dispersion $\Omega(\kappa' d)$ in such structures and belongs to a band of frequencies $1 + \frac{1}{2}\Omega_\mathrm{M} < \Omega < 1 + \Omega_\mathrm{M}$.

In Fig. 4.11 typical oscillograms of the experimentally observed spectra of waves traversing a symmetric layered ferrite-dielectric structure with parameters $t/d = \ell/d$: a – 0.013, b – 0.060, c – 0.250, for $\overline{H}_0 = 0.97\mathrm{MA/m}$ are shown. As frequency grows, two AFC corresponding to SSLMSW and RSMSW were consistently observed. Solid waves are excited more effectively and extend with smaller losses than surface ones. For SSLMSW with growth of frequency the distance between the resonances continuously decreases, for RSMSW an inverse relationship takes place. The character of irregularities of these AFC and their arrangement in various frequency ranges qualitatively agree with calculation.

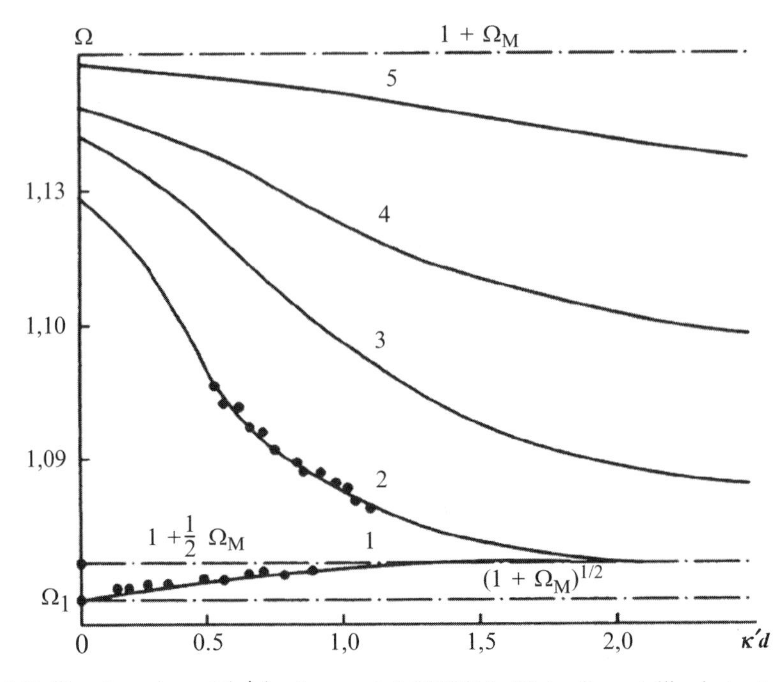

Fig. 4.10 The dispersions $\Omega(\kappa'd)$ of connected SMSW in bilaterally-metallized structures are presented at $\ell_1/d = \ell_2/d = \ell/d$ and $t/d = 0.2$: $1 - \ell/d = 0.250$; $2 - \ell/d = 0.250$; $3 - \ell/d = 0.125$; $4 - \ell/d = 0.05$; $5 - \ell/d = 0.01$ at $\Omega_M = 0.15$ and the binding area $L_{SW} = 6 \cdot 10^{-3}$ m, the points designate the experimental values of dispersions $\Omega(\kappa'd)$

The strong irregularity of AFC is related to reflections from the end-walls of the ferrite plate, which forms a spin-wave resonator. From the practical point of view such an irregularity is undesirable, however, at physical research it is rather useful. By the frequency of resonances it is possible to judge about the dispersions of phase and group speeds $(U_p(\nu))$ and $U_{gr}(\nu))$. It is necessary to reveal the mechanism of "false" resonances in AFC, appearing due to various re-reflections of waves in the device and converter, and either to eliminate them or to consider at treatment of experimental data.

Later researches have shown basic importance of the accuracy of registration of the external magnetic field $\overline{H}_0$ in experimental data treatment. The determination error of $\overline{H}_0$ should not be higher than that of registration of:

- The frequencies of spin-wave resonances ν_n in the wave spectrums
- The characteristic resonant frequencies of structures ν_{H} – for normally-magnetized structures, $\nu_{\perp}$ for tangentially- and ν_{φ} – for obliquely-magnetized structures, φ being the angle of the field $\overline{H}_0$ inclination to the plane of the ferrite

The measurement accuracy of the mentioned frequencies in the millimeter range is provided with rather simple external MAW wavemeters at a level not worse than 0.02%, but widespread measuring instruments of magnetic induction at $B_0 \approx 1\mathrm{T}$

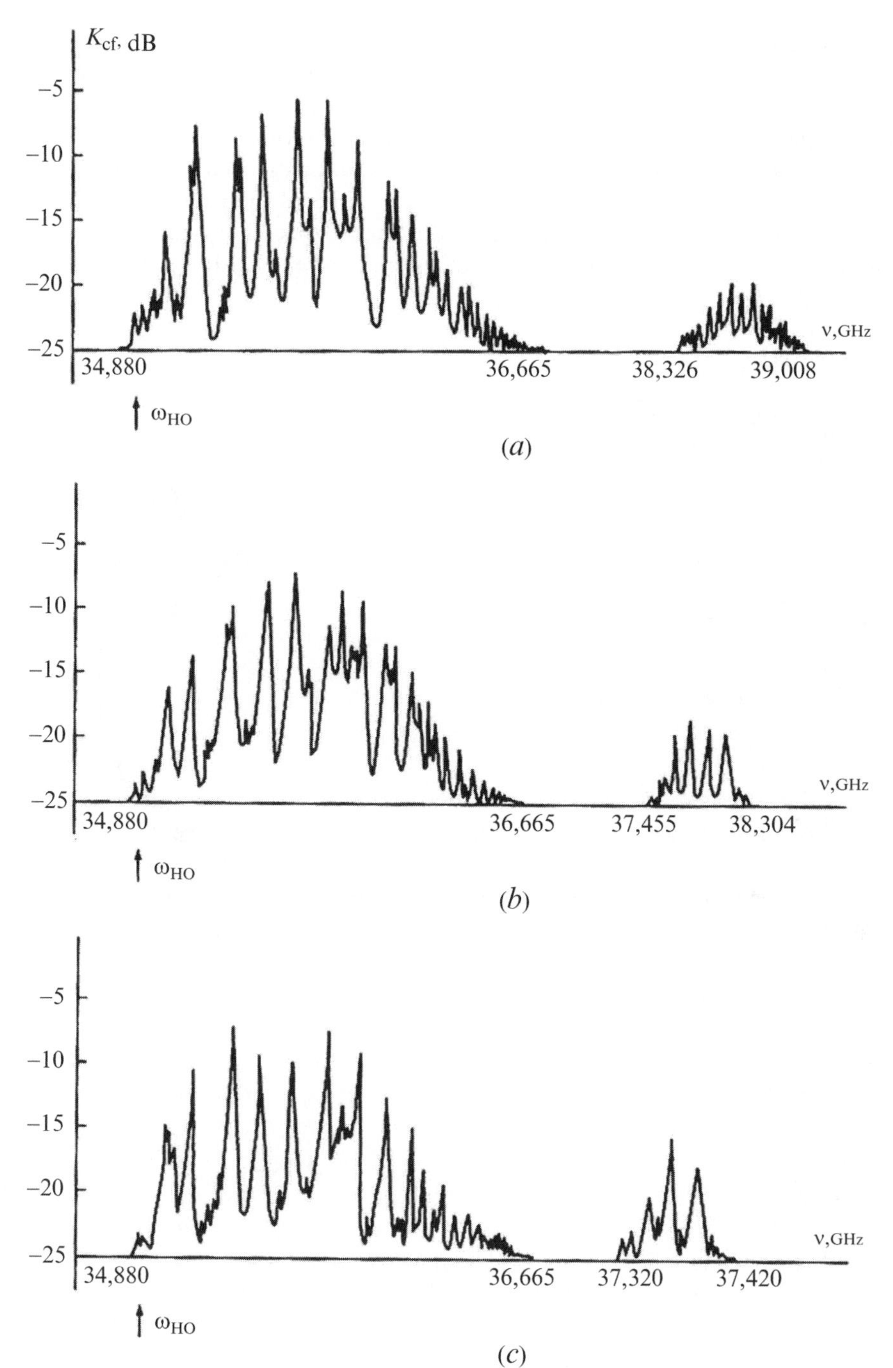

Fig. 4.11 Typical oscillograms of the experimentally observed spectra of waves traversing a symmetric layered ferrite-dielectric structure with parameters $t/d = \ell/d$: (*a*) 0.013, (*b*) 0.060, (*c*) 0.250, for $\overline{H}_0 = 0.97\,\mathrm{MA/m}$ are shown

have an error of the order of 1–2%. This difference of measurement accuracies of frequencies and fields is kept at advance from the millimeter range into the centimeter one, because both the accuracy of frequency measurement and that of measurement of fields $\overline{H}_0$ smaller, in comparison with the millimetric range, raise. These circumstances do not allow the value of an external magnetic field $\overline{H}_0$ and one of the characteristics of resonant frequencies of the structure (ν_{H}, $\nu_{\perp}$ or ν_{φ}) to be correctly associated, and development of special devices for their registration with an adequate accuracy is required. Figure 4.12 shows the dispersive

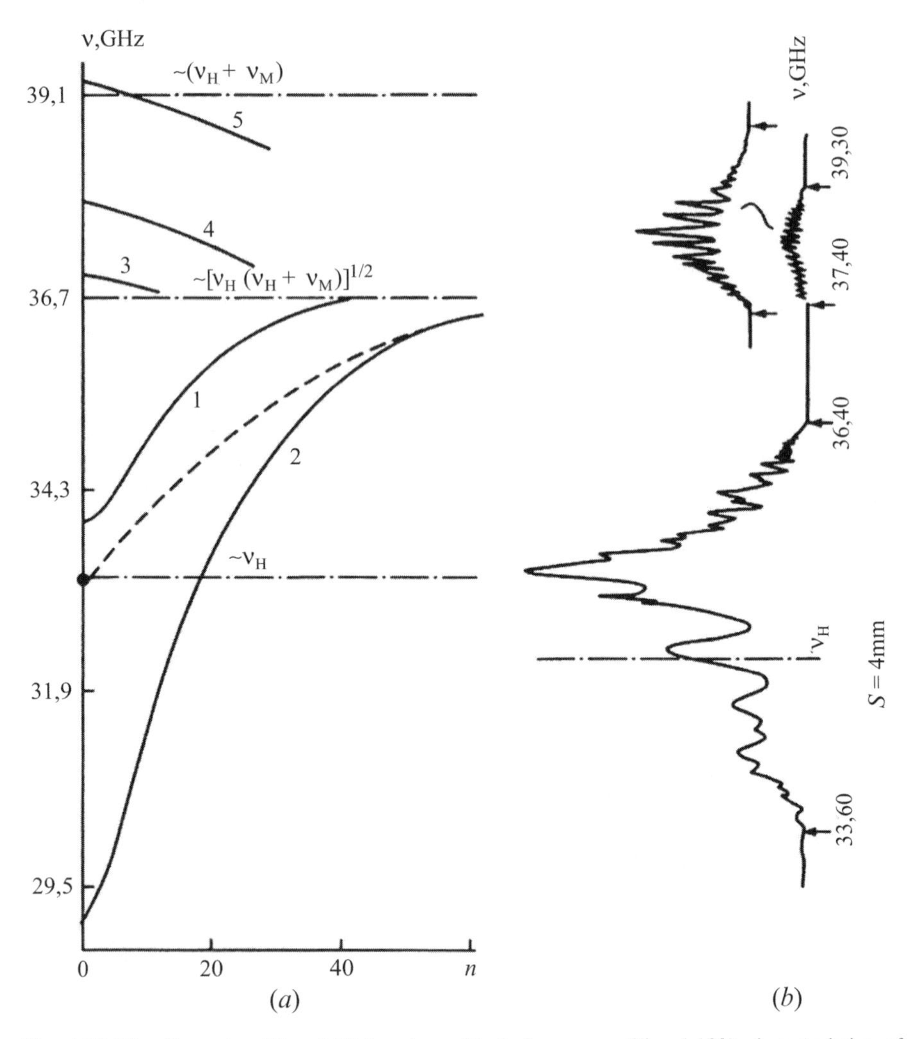

Fig. 4.12 The dispersive (Fig. 4.12*a*) and amplitude-frequency (Fig. 4.12*b*) characteristics of electromagnetic waves ν_n, including their magnetostatic branches in a ferrite-dielectric layered structure with symmetric loading ($t/d = \ell/d$), containing a YIG plate ($4\pi M_s = 0.176\mathrm{T}$) with sizes $0.5 \times 3.2 \times 14\,\mathrm{mm}$

(Fig. 4.12a) and amplitude-frequency (Fig. 4.12b) characteristics of electromagnetic waves ν_n, including their magnetostatic branches in a ferrite-dielectric layered structure with symmetric loading $(t/d = \ell/d)$, containing a YIG plate $(4\pi M_s = 0.176\,\mathrm{T})$ with sizes $0.5 \times 3.2 \times 14\,\mathrm{mm}$. In a wave range $\nu < \nu_\perp$ the dispersive characteristics of space spatial waves (SSLW) are located at various sizes of the ferrite's jut from its beyond-cutoff section: $1 - S = 4 \times 10^{-4}\,\mathrm{m}$; $2 - S = 8 \times 10^{-4}\,\mathrm{m}$. In a wave range $\nu > \nu_\perp$ – the dispersive characteristics of RSW for parameters t/d : $3 - 1.5 \times 10^{-1}$; $4 - 2 \times 10^{-2}$; $5 - 1 \times 10^{-2}$. The ferrite jut size S essentially influences the amplitude and band characteristics of the structure. For $S = 4 \times 10^{-4}\,\mathrm{m}$ the dispersive characteristics ν_n (curve 1) lays in the wave range $\nu_\mathrm{H} < \nu < \nu_\perp$, and for $S = 8 \times 10^{-4}\,\mathrm{m}$ it essentially falls outside the frequency ν_H. Estimations made with due account of demagnetization factors [8] give a wave range of the existence of SSLW $\Delta\nu = \nu_\perp - \nu_\mathrm{H} \cong 3.5\,\mathrm{GHz}$, that essentially less than that observed experimentally. This dependence evidently shows the role of electrodynamic effects in the millimeter range. The dotted line in Fig. 4.12*a* shows the qualitative course of the dispersion of forward solid magnetostatic spin waves.

In Fig. 4.13*a*, *b* the experimental dependences $\Omega_{\mathrm{H0}}(\kappa' d)$ and $\Omega_{\mathrm{H0}}(U_{gr})$ for SSLW in a transmission line made of a ferrite plate in a range of frequencies of 30–40 GHz are given, where $\Omega = \frac{\omega}{\omega_\mathrm{H}}$, ω_H is the lower boundary frequency of the wave spectrum. The dotted line depicts the course of dependences $\Omega(\kappa' d)$ and $\Omega(U_{gr})$, obtained in the MSW approximation [8], where $\Omega = \frac{\omega}{\omega_\mathrm{H}}$, ω_H is the frequency of ferromagnetic resonance. Discrepancy of the frequencies ω_H and ω_{H0} $(\omega_\mathrm{H} > \omega_{\mathrm{H0}})$ can be established only in the case of direct measurement of H_0 with a raised accuracy by two- or three orders exceeding that of Hall sensors or special sensors of resonant frequencies (Chapter 5). The dependences $\Omega_0(\kappa' d)$ and $\Omega(\kappa' d)$, together with $\Omega_0(U_{gr})$ and $\Omega(U_{gr})$ show how the instrumental error of measurement of raised H_0 values together with the dependence of passband on the length S of FDMT lead to incorrect treatment of experimental results. In fact, the noted quite satisfactory agreement between the theoretical and experimental dispersive dependences suggests an idea about the magnetostatic nature of waves under study in the mm range.

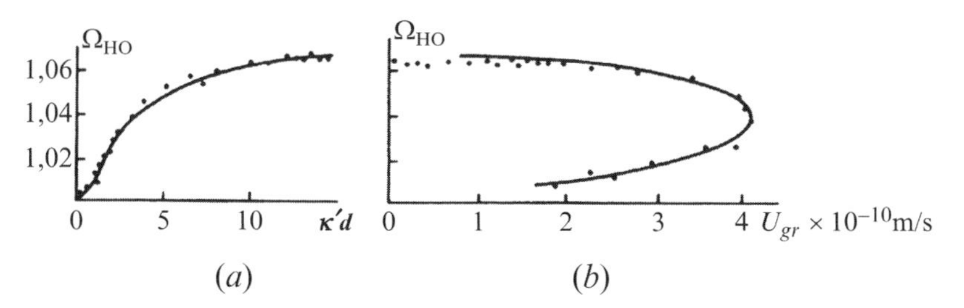

Fig. 4.13 The experimental dependences $\Omega_{\mathrm{H0}}(\kappa' d)$ and $\Omega_{\mathrm{H0}}(U_{gr})$ for SSLW in a transmission line made of a ferrite plate in a range of frequencies of 30–40 GHz are given, where $\Omega = \frac{\omega}{\omega_\mathrm{H}}$, ω_H is the lower boundary frequency of the wave spectrum

Our researches have shown that for return surface waves the MSW approximation in the MMR quite well describes the dispersive characteristics of waves in the structure in the beyond-cutoff mode, where waves with a surface character of the distribution of HF field extend.

In Fig. 4.14 the experimental dependences $\Omega_{\mathrm{H0}}(\kappa' d)$ (Fig. 4.14*a*) and $\Omega_{\mathrm{H0}}(U_{gr})$ (Fig. 4.14*b*) for RSMSW in a tangentially-magnetized ferrite-dielectric structure with $t/d = \ell/d$ are presented with parameters.

$1 - t/d = 0.013$; $2 - t/d = 0.066$; $3 - t/d = 0.107$; $4 - t/d = 0.250$; $\nu_{\mathrm{H}} = 33.99\,\mathrm{GHz}$; $d = 7.5 \times 10^{-4}\,\mathrm{m}$; $L = 6 \times 10^{-3}\,\mathrm{m}$; $H_{0i} = 994.718$ kA/m; $H_{0i} = 964.479$ kA/m; $4\pi M_S = 0.176\,\mathrm{T}$. The *dotted lines* depict the dependences in the MSW approximation. As follows from the theory [8], RSMSW are located in a wave range $1 + \Omega_{\mathrm{M}} < \Omega < (1 + \Omega_{\mathrm{M}})^{1/2}$, they have abnormal dispersions of phase constants which parameters are controlled by means of metal screens over a wide range. On removal of metal screens from the ferrite layer the upper boundary frequency of RSMSW reduces, the experimentally observable wave range extends, and the value of group speed U_{gr} raises.

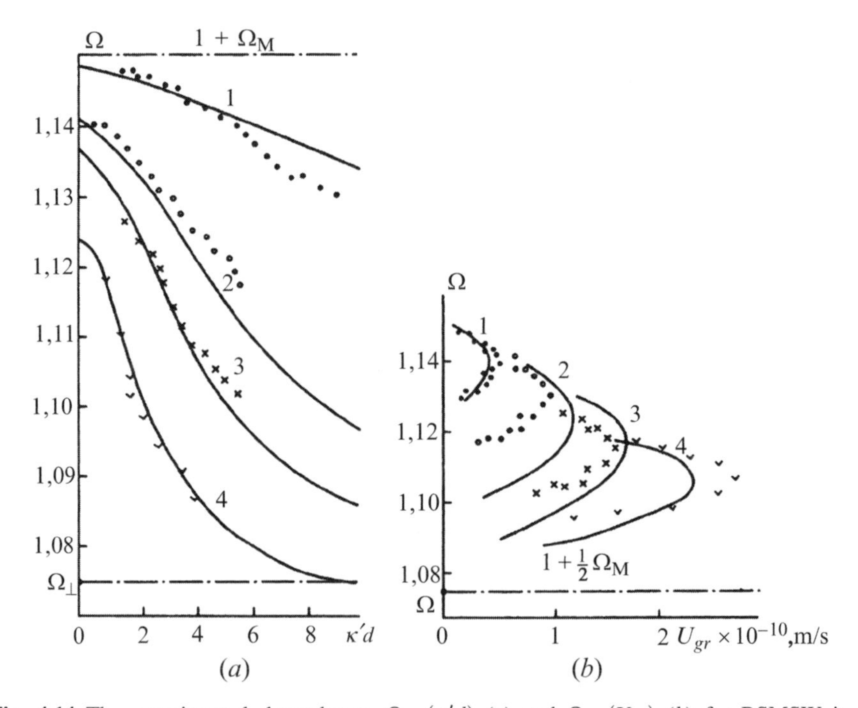

Fig. 4.14 The experimental dependences $\Omega_{\mathrm{H0}}(\kappa' d)$ (*a*) and $\Omega_{\mathrm{H0}}(U_{gr})$ (*b*) for RSMSW in a tangentially-magnetized ferrite-dielectric structure with $t/d = \ell/d$ are presented with parameters: $1 - t/d = 0.013$; $2 - t/d = 0.066$; $3 - t/d = 0.107$; $4 - t/d = 0.250$; $\nu_{\mathrm{H}} = 33.99\,\mathrm{GHz}$; $d = 7.5 \times 10^{-4}\,\mathrm{m}$; $L = 6 \times 10^{-3}\,\mathrm{m}$; $\mathrm{H}_{0i} = 994.718\,\mathrm{kA/m}$; $\mathrm{H}_{0i} = 964.479\,\mathrm{kA/m}$; $4\pi M_S = 0.176\,\mathrm{T}$. The *dotted lines* depict the dependences in the MSW approximation

Let's note that physical experiments to study the dispersions of RSMSW in bilaterally-metallized layered structures with parameters $t/d \ll 1$ and $\ell/d \ll 1$, representing most interest from the point of view of formation of transmission lines with preset dispersive characteristics, can be made on ferrite plates only.

In Fig. 4.15*a*, *b* the theoretical and experimental dependences of phase $\Omega(\kappa' d)$ (Fig. 4.15*a*) and group $\Omega(U_{gr})$ (Fig. 4.15*b*) speeds of RSMSW in bilaterally-metallized layered structures are presented: $1 - t/d = \ell/d = 0.24$; $2 - t/d = 0.24$, $\ell/d = 0.49$; $3 - t/d = 0.73$, $\ell/d = 0.24$; $4 - t/d = 0.24$, $\ell/d = 0.11$; $L = 7.2 \times 10^{-3}$ m; $\Omega_M = 0.15$.

In Fig. 4.15*c* typical experimental dependences of introduced transfer losses on SSLMSW-1 and RSMSW-2-5 in the millimeter range are presented: $1 - d = 3 \times 10^{-3}$ m; $L = 7.1 \times 10^{-3}$ m; $2 - t/d = 0.250$; $3 - t/d = 0.107$; $4 - t/d = 0.066$;

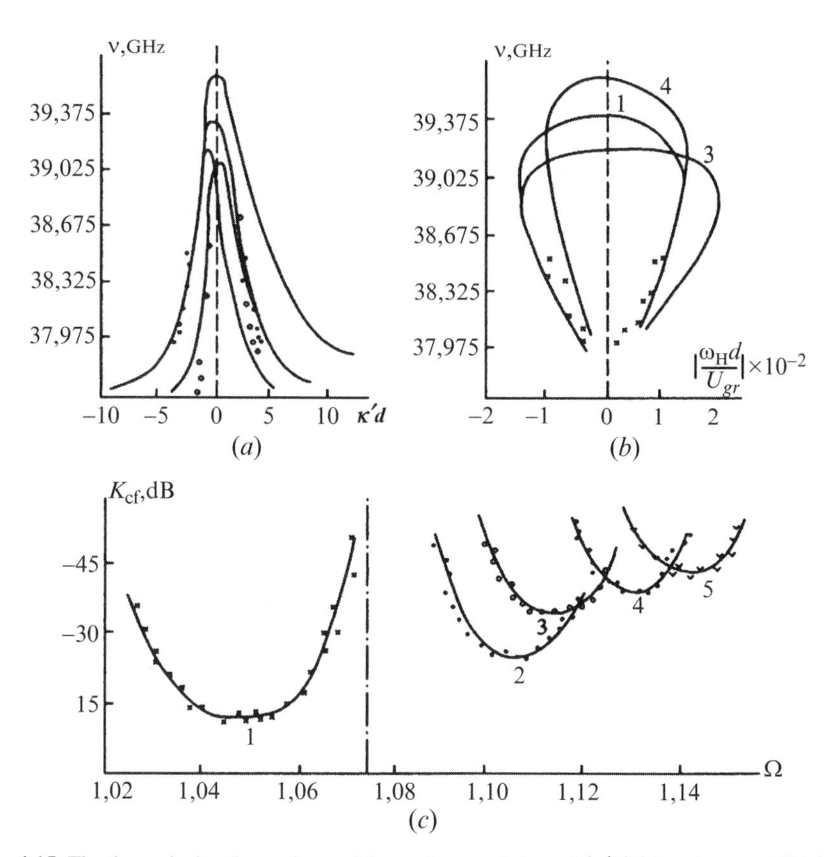

Fig. 4.15 The theoretical and experimental dependences of phase $\Omega(\kappa' d)$ (*a*) and group $\Omega(U_{gr})$ (*b*) speeds of RSMSW in bilaterally-metallized layered structures are presented: $1 - t/d = \ell/d = 0.24$; $2 - t/d = 0.24$, $\ell/d = 0.49$; $3 - t/d = 0.73$, $\ell/d = 0.24$; $4 - t/d = 0.24$, $\ell/d = 0.11$; $L = 7.2 \times 10^{-3}$ m; $\Omega_M = 0.15$ and experimental dependences of introduced transfer losses (*c*) on SSLMSW-1 and RSMSW-2-5 in the millimeter range are presented: $1 - d = 3 \times 10^{-3}$ m; $L = 7.1 \times 10^{-3}$ m; $2 - t/d = 0.250$; $3 - t/d = 0.107$; $4 - t/d = 0.066$; $5 - t/d = 0.013$, $d = 7.0 \times 10^{-3}$ m; $L = 7.1 \times 10^{-3}$ m, $\Omega_M = 0.15$

$5 - t/d = 0.013$, $d = 7.0 \times 10^{-3}$ m; $L = 7.1 \times 10^{-3}$ m, $\Omega_M = 0.15$. It is obvious that the introduced transfer losses for SSLMSW are by 10–30 dB lower than those for RSMSW.

For SSLMSW in a range of frequencies $\nu_H = 3.4 \times 10^{10}$ GHz on a structure with parameters $d = 3.0 \times 10^{-3}$ m; $L = 1.4 \times 10^{-2}$ m, $\Omega_M = 0.15$, the delay time of signal $\tau \cong 3.48 \times 10^{-2}$ µs at losses $(K_{los})_{\min} \cong 5$ dB that, according to the known relation $K(\text{dB}/\mu\text{s}) = 76.4\Delta H(\text{Oe})$, for the investigated crystals gives a FMR line width $\Delta H_{\perp} \cong 2$ Oe. For RSMSW in crystals with $d = 1.0 \times 10^{-3}$ m; $L = 1.4 \times 10^{-2}$ m; $\Omega_M = 0.15$, the delay time of signal is $\tau \cong 5 \times 10^2$ µs at $(K_{los})_{\min} \cong 25$ dB, that gives $\Delta H_{||} \cong 0.5$ Oe. These values of parameters $\Delta H_{||,\perp}$ should be considered as upper bounds as the total signal transfer losses include the transformation losses at input and output.

Excitation of electromagnetic waves in strongly anisotropic barium hexaferritte monocrystals was experimentally investigated. Waveguide-beyond-cutoff modules with a basis waveguide channel of a cross section 3.6×1.8 were used. In Fig. 4.16 are: the signal AFC at output of the module (Fig. 4.16*a*) and the dependences of resonant frequencies (Fig. 4.16*b*):

- For the orientation of the external field $\overline{H}_0$ along the axis of easy magnetization $\bar{c}$, which is perpendicular to the plane of the crystal ($\overline{H}_0 || \bar{c}$– curve 1)
- For the orientation of the field $\overline{H}_0$ orthogonally to the axes $\bar{c}$($\overline{H}_0 \perp \bar{c}$-curve 2)

For comparison, the dependence $\nu_{H0}(H_0)$ for a normally-magnetized YIG film (curve 3) is shown. It is obvious that for achievement of the same range of frequencies at using solid and film barium hexaferrite crystals (BHF), fields by 0.995 MA/m lower than those for weakly anisotropic ferrites are required. One of the major factors limiting the BHF application in lengthy transfer lines of the millimeter range, is a wide enough line of FMR in such structures, which by five to seven times can surpass the FMR line width for weakly anisotropic YIG ferrites.

The results obtained at our research of the dispersive characteristics of various types of waves in layered structures on the basis of solid ferrite plates, determine the expediency of studying wave processes in high-quality (with a low parameter of ferromagnetic losses α) epitaxial ferrite films and layered structures on their basis in the millimeter range of radiowaves.

4.3 Waves in Ferrite Film Structures

On passing to ferrite films:

- The influence of demagnetization factors in wide and rather lengthy tangentially-magnetized films ($W \gg d, L \gg d$, W the width, L the length, d– the thickness of the film) decreases and their influence for normally-magnetized films extremely increases, that gives the maximum separation of resonant frequencies $\nu_{\perp}$ and ν_H, and $\nu_{\perp} - \nu_H \cong \gamma M_s$.

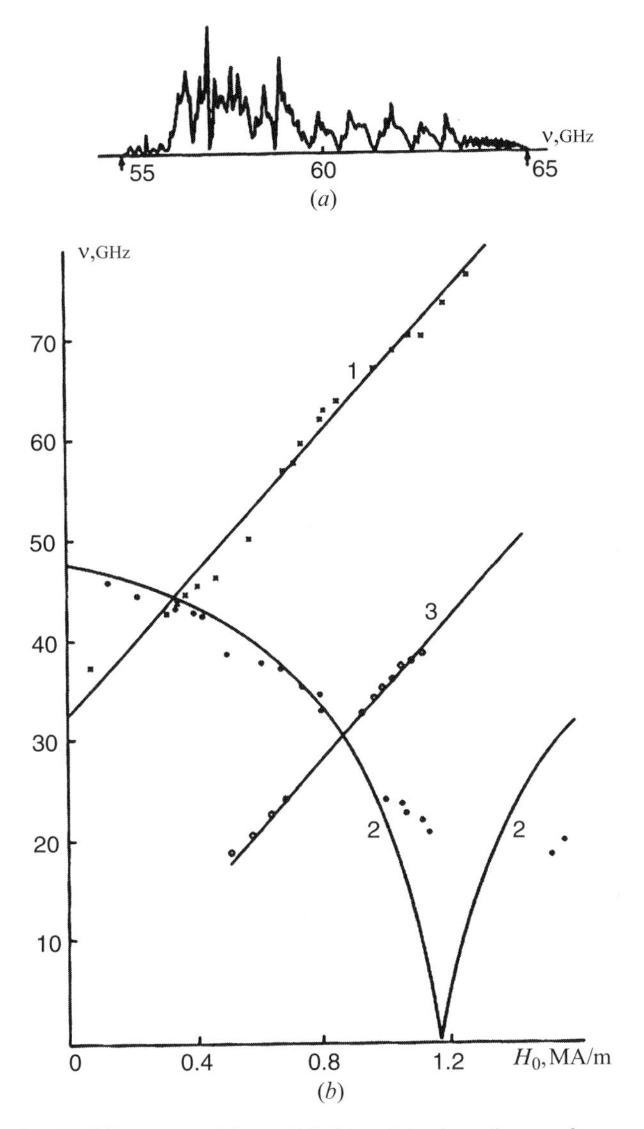

Fig. 4.16 The signal AFC at output of the module (*a*) and the dependences of resonant frequencies (*b*):
– For the orientation of the external field $\overline{H}_0$ along the axis of easy magnetization $\bar{c}$, which is perpendicular to the plane of the crystal ($\overline{H}_0 || \bar{c}$ – curve 1)
– For the orientation of the field $\overline{H}_0$ orthogonally to the axes $\bar{c}$ ($\overline{H}_0 \perp \bar{c}$ – curve 2)

- The area of internal field uniformity extends (at $WL = \text{const}$, boundary fields are within the limits of $3d$).
- The group speed of waves $(U_{gr} \sim d^{-1})$ decreases.
- The volume and weight of ferrite decrease, and the efficiency of magnetostatic wave excitation reduces.

These circumstances, on the one hand, allow wave spectra to be divided well enough by frequency ranges, and, on the other hand, require search for effective converters.

Our researches made on waveguide-beyond-cutoff devices have shown that at switching from plates to films and reduction of the thickness of ferrite films a sharp reduction of the spectrum band of excited waves and its localization close to $\nu_{\perp}$ (tangential) or to ν_{H} (normal magnetization of the structure) are observed.

4.3.1 Tangentially Magnetized Structures

In Fig. 4.17 the spectra of a signal passing through a waveguide-beyond-cutoff device with a ferrite (YIG) films of various thickness are presented: a – $d = 8 \times 10^{-5}$ m; b – $d = 7 \times 10^{-5}$ m; c – $d = 4.6 \times 10^{-5}$ m; d – $d = 2.5 \times 10^{-5}$ m.

For each film the size of its jut from the beyond-cutoff section was taken optimal and corresponded to the minimum of introduced transfer losses. It is obvious that at reduction of the thickness d of the ferrite film the passband contracts, its irregularity decreases, and the spectrum is localized near the frequency $\nu_{\perp}$. A major factor here is the increase of uniformity of exciting HF fields within the ferrite layer in the crack shorting the walls of the waveguide.

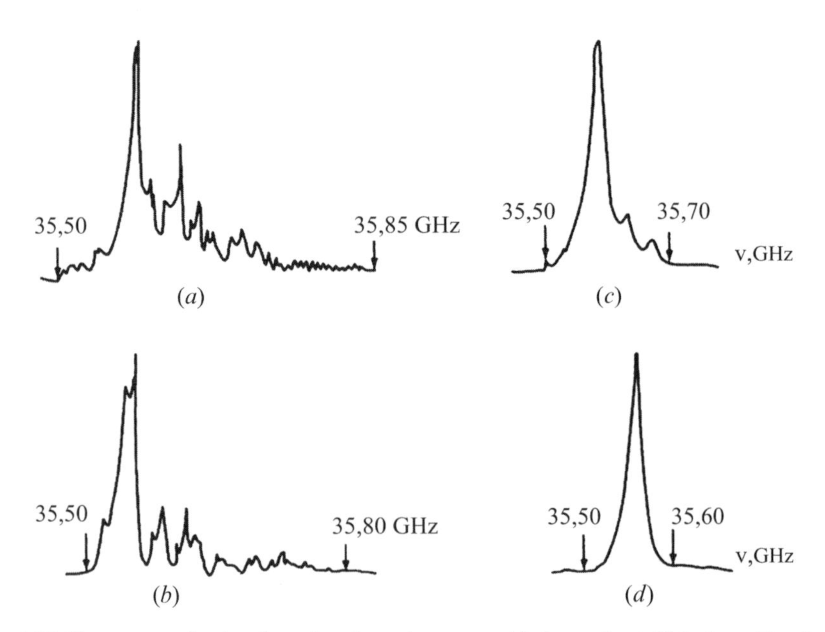

Fig. 4.17 The spectra of a signal passing through a waveguide-beyond-cutoff device with a ferrite (YIG) films of various thickness are presented: (*a*) $d = 8 \times 10^{-5}$ m; (*b*) $d = 7 \times 10^{-5}$ m; (*c*) $d = 4.6 \times 10^{-5}$ m; (*d*) $d = 2.5 \times 10^{-5}$ m

Inherently, a waveguide-beyond-cutoff device with ferrite is a converter with orthogonal orientation of its structure to the exciting plane, and with the width of the crack W_a and the thickness of the ferrite layer d the following cases are possible:

- For ferrite plates the width of the active zone $W_a \approx d \approx W_{SL}$, the heterogeneity of exciting internal HF fields within the ferrite layer raises, the spectrum of wave numbers $\Delta\kappa'$ and the wave band $\Delta\nu(\kappa')$ excited (transmitted) by the converter extend, and the mode will be broadband.
- For ferrite films $W_a \ll W_{SL}$, that reduces the heterogeneity of exciting internal HF fields, reduces the spectrum $\Delta\kappa'$ and the wave range $\Delta\nu(\kappa')$, so the mode of the converter will be narrow-band.

The part of the ferrite with the length S bulging out of the beyond-cutoff section of the waveguide acts as a FDT matching HF fields in the waveguide with the HF-fields in the ferrite-dielectric structure in the beyond-cutoff section. In the plane of the exciting slot the fields in the FDR with the length S are coupled with those of the structure in its beyond-cutoff section with the factor of coupling depending on the parameters of FDR, the slot width, its filling with a layered structure, the distance between the screens.

The properties of FDT and SGC were investigated theoretically and experimentally for ferrite plates (Section 5.1) and films (Section 5.2). Researches were made in the following directions:

- Theoretical and experimental analysis of the influence of metal screens and the parameters of structure on wave dispersion, treatment of the dispersive part of the problem and the characteristics of an excitation and transfer signal (the problem of signal excitation in a structure by a converter with orthogonal orientation)
- Experimental study of the influence of FDT parameters on the characteristics of a transfer signal

Figure 4.18*a*, *c* shows the experimental dependences of AFC signals on the output of a waveguide with a ferrite-dielectric structure located in a section with a changing distance a between the narrow walls of the waveguide in two ranges of frequencies: $a - \nu_\perp \approx 9\,\text{GHz}$ and $b - \nu_\perp \approx 27\,\text{GHz}$ for a YIG film with $d = 3.5 \times 10^{-6}\,\text{m}$, $t/d = \ell/d = 5 \times 10^{-4}\,\text{m}$.

It is obvious that at changing a the middle section of the waveguide passes consecutively from the prelimit mode ($\nu \gg \nu_{cr}$) to the limit one ($\nu \approx \nu_{cr}$) and then to the beyond-cutoff ($\nu \ll \nu_{cr}$) mode for a signal of frequency ν.

In the limit mode ($\nu \gg \nu_{cr}$) selective attenuation of a signal on frequency $\nu < \nu_\perp$ ($\nu_\perp$ is marked by a dotted line and was determined by specially designed sensors) is observed.

In the saturated mode ($\nu \approx \nu_{cr}$):

- A displacement of the central frequency $\nu_\perp^1$ of selective attenuation towards lower frequencies and expansion of the wave range $\Delta\nu_{3dB}$ are observed.
- On frequency $\nu_\perp^1 > \nu_\perp$ reduction of the selective attenuation of a system transparency signal in the wave range $\Delta\nu_2$ is observed.

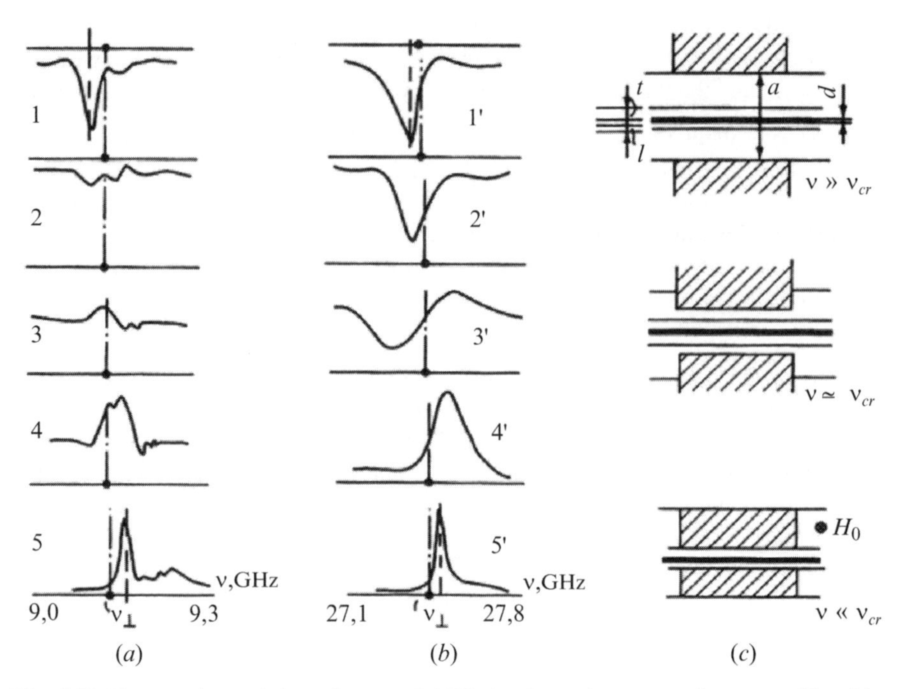

Fig. 4.18 The experimental dependences of AFC signals on the output of a waveguide with a ferrite-dielectric structure located in a section with a changing distance a between the narrow walls of the waveguide in two ranges of frequencies: $a - \nu_{\perp} \approx 9\,\text{GHz}$ and $b - \nu_{\perp} \approx 27\,\text{GHz}$ for a YIG film with $d = 3.5 \times 10^{-6}\,\text{m}$, $t/d = \ell/d = 5 \times 10^{-4}\,\text{m}$

- The signal level on tuned-out by $(3 \div 5)\ \Delta\nu^1{}_{\perp}$ and $\Delta\nu^2{}_{\perp}$ from $\nu_{\perp}$ frequencies, i.e., the level of barrage, decreases on the average, that is due to amplification of the mismatch between the waveguides with various cross sections.

In the beyond-cutoff mode $(\nu \ll \nu_{cr})$ the transparency of the electrodynamic system amplifies, the central frequency of a transmitted signal $v_{\perp}^{1}$ decreases and tends to $v_{\perp}$, the level of barrage reaches its maximum value.

The following features are characteristic:

- The central frequency of selective signal attenuation $v_{\perp}^{1}$ does not coincide with the central frequency of signal passage $v_{\perp}^{2}$.
- The wave range of signal attenuation at a level 3 dB Δv_{3dB} is wider than that of passage Δv.
- At transition from the UHF range into the EHF one, expansion of the signal attenuation. Δv_{3dB} and passage Δv bands is observed.

Thus, two most typical modes have been revealed, namely:

- The limit one, at which metal screens are rather distant (t/d and $\ell/d \gg 1$) from the ferrite layer and the frequency of a signal is $\nu \ll \nu_{cr}$.
- The beyond-cut off one, at which metal screens are close to the ferrite layer (t/d and $\ell/d \ll 1$) and $\nu \ll \nu_{cr}$.

4.3.2 *Pre-limit Mode* $(\nu \gg \nu_{cr})$

Research of the selective signal attenuation mode on frequency $\nu_{\perp}^{2} < \nu_{\perp}$ was made on waveguide breadboard models (Fig. 4.18). Most simply a physical experiment on studying the wave properties of FDT is carried out in the mid the centimeter range and in the long-wave part of the millimeter range, with the sizes of the ferrite-dielectric structure (length S and width W_S) protruding into a rectangular waveguide to be varied.

In Fig. 4.19 experimental dependences of the AFC of waves excited in FDT on S in the reflection mode are presented. First we consider a feature of selective signal attenuation on frequencies close to $\nu_{\perp}$. From experiment it follows that:

- At $S \approx 1.5 \times 10^{-3}$ m on frequency $\nu_{\perp}^{2} > \nu_{\perp}$ selective attenuation (oscillogram 1) is registered, whose value increases with increasing S, and continuously decreases and tends to $\nu_{\perp}$ (oscillograms 1, 2).
- At $S \approx 5.5 \times 10^{-3}$ m the value of signal attenuation on $\nu_{\perp}^{2}$ reaches the maximum value (oscillogram 3) on frequency $\nu_{\perp}^{2} < \nu_{\perp}$.
- At further increasing S (oscillograms 4, 5) reduction of selective attenuation on $\nu_{\perp}^{2}$ and an insignificant lowering of frequency $(\nu_{\perp}^{2} < \nu_{\perp})$ are observed.
- At $S > 12.5 \times 10^{-3}$ m selective attenuation on $\nu_{\perp}^{2}$ increases again (oscillograms 6 and 7).
- Further the dependence of attenuation on S has a character close to periodic one.

Let's analyze AFC in a wave range from ν_1 up to ν_2 (Fig. 4.19). Experimental research shows that in a wave range $\nu_{\perp}^{2} < \nu < \nu_2$ at increasing S a change of the AFC irregularity is observed, and two kinds of dependences are seen:

- The irregularity which depends on S, and at increasing S α reduction of the frequencies of detunings between resonances is observed, that is related to the waves extending along FDT $(\kappa_y' \neq 0)$.
- The irregularity independent of S which and is related to oscillatory processes across of FDT (along the layer width w) and $\kappa_z' \neq 0$.

Besides, from experimental data (Fig. 4.19) it follows that in a wave range $\nu_{\perp}^{2} < \nu < \nu_2$ only the irregularity of AFC varies, while the average level of signal attenuation enveloping by resonances depends on S very weakly. In a wave range $\nu_{\perp} < \nu < \nu_2$ with growing S a monotonous growth of more broadband, in comparison with selective attenuation on frequency ν, signal with the central frequency $\nu_{\perp 2}$ is observed.

Analysis of experimental data in the centimeter and millimeter ranges shows that near the resonant frequency $\nu_{\perp}$ in a wave range $\nu_{\perp}^{2} < \nu < \nu_2$ there are the forward surface waves (LE_t^S waves) reaching the lower boundary frequency $\nu_{\perp}^{2}$ $(\nu_{\perp}^{2} < \nu_{\perp})$. Close to $\nu_{\perp}$ and, in particular, in bands of frequencies $\nu_{\perp}^{2} < \nu < \nu_{\perp}$ and $\nu_{\perp} < \nu < \nu_{\perp}^{2}$, this wave, as theoretical analysis shows, has a dispersion close to that of fast LE_t^{F1} and LE_t^{F2} waves in such structures.

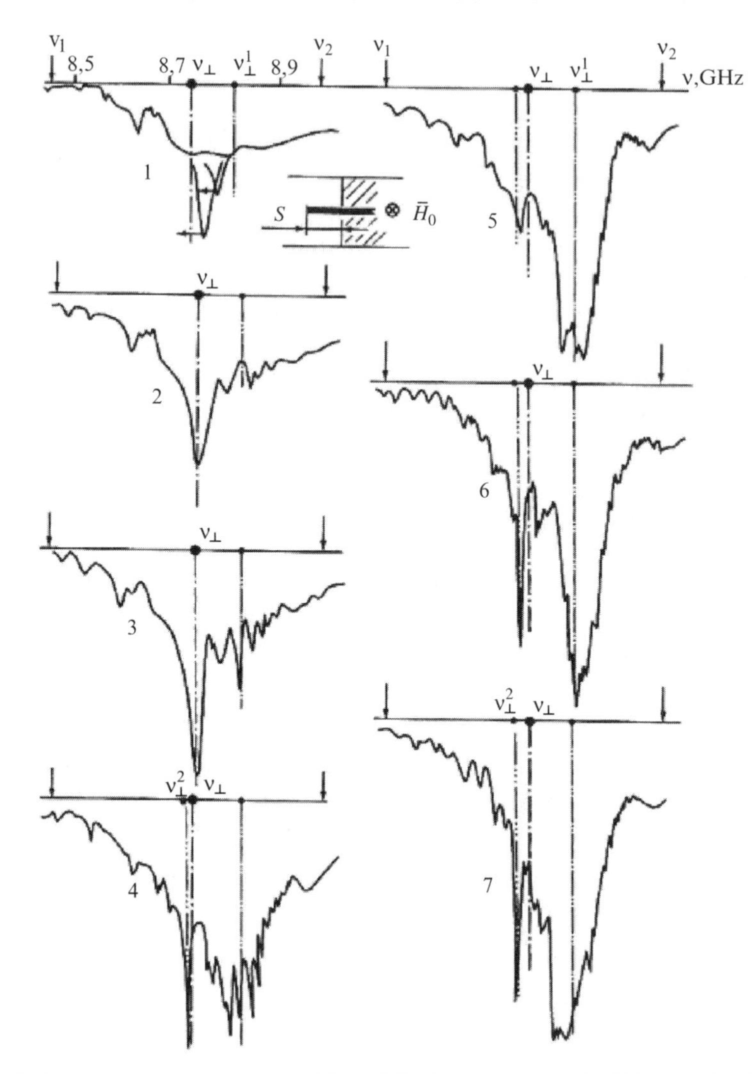

Fig. 4.19 Experimental dependences of the AFC of waves excited in FDT on S in the reflection mode

For adequate interpretation of the observable spectrum of LE_t^S, LE_t^{F1}, and LE_t^{F2}-waves near the resonant frequency the following is required:

– To theoretically compare the phase and amplitude constants of these waves and their dispersions, to compare their power fluxes and AFC signals, and to develop most essential criteria for their distinction by frequency ranges of existence in the common or different wave range, the level of selective signal attenuation, AFC shapes and the steepness of their slopes, etc.

- To experimentally confirm the criteria of effective excitation and propagation of these waves through most essential dependences, e.g., interference attenuation in the prelimit mode, selective attenuation and transmission of a signal in various modes determined by the ratio of the frequency of signal ν and the critical frequency ν_{cr}.

This complex of researches will be considered by stages.

The results of studying the dispersive properties of fast and slow waves in various modes are discussed.

Let's present the results of our theoretical analysis of H_{n0} waves in weakly dissipative structures screened by metal walls. In Figs. 4.20 and 4.21 the dispersive characteristics $\nu(\kappa)$ of H_{n0} waves in flat ferrite-dielectric waveguides near the frequency $\nu_{\perp}$ in two ranges of frequencies $\nu_{\mathrm{H}} = 8.9\,\mathrm{GHz}$ and $\nu_{\mathrm{H}} = 160\,\mathrm{GHz}$ are shown. In Fig. 4.20*a*, *e* the dependences of phase constants for various parameters of ferromagnetic losses are given: $a - \alpha = 10^{-5}$; $b - \alpha = 3 \times 10^{-4}$; $c - \alpha = 10^{-3}$; $d - \alpha = 10^{-2}$; $e - \alpha = 2 \times 10^{-2}$; $f - \alpha = 3 \times 10^{-2}$ at $\nu_{\mathrm{H}} = 8.9 \times 10^{10}\,\mathrm{Hz}$, $4\pi M_S = 0.176\,\mathrm{T}$. Figure 4.21*a*–*d* presents similar dependences: $a - \alpha_{||} = 2 \times 10^{-5}$; $b - \alpha_{||} = 5 \times 10^{-5}$; $c - \alpha_{||} = 5 \times 10^{-4}$; $d - \alpha_{||} = 5 \times 10^{-3}$ at $4\pi M_S = 0.176\,\mathrm{T}$.

Let's consider consistently the characteristics $\nu(\kappa')$ for $\nu_{\mathrm{H}} = 8.9\,\mathrm{GHz}$ (Fig. 4.20*a*–*d*):

- In an idealized structure (*a*) at $\alpha_{||} 10^{-5} (\Delta H_{||} \approx 2.39\,\mathrm{A/m})$ there exist two forward surface waves (SSW_1^* and SSW_2^*), which in the limit (at $\kappa' \to \infty$) pass into Eshbach-Damon's wave, and a SFW, whose dispersion with growing κ' asymptotically tends to the dispersive characteristics of a wave in a waveguide with dielectric filling κ_2' i.e, in the field of $\nu_{\perp}$ the mode is two-wave one (SSW_1^* or SSW_2^* and SFW).
- With growing ferromagnetic losses (*b*) at $\alpha_{||} = 3 \times 10^{-4} (\Delta H_{||} \approx 76.4\,\mathrm{A/m})$ the dispersion of SFW in the field of $\nu_{\perp}$ experiences amplification of its steepness and has an inflexion point, SSW* in the limit passes into SSMSW, and the mode is two-wave one (SSW* and SFW).
- At $\alpha_{||} = 1 \times 10^{-3} (\Delta H_{||} \approx 0.25\ \mathrm{kA/m})$ SFW and SSW* acquire (*c*) regions with abnormal dispersions in the field of frequencies $\nu < \nu_{\perp}$ (SSW* and space wave).
- At $\alpha_{||} = 1 \times 10^{-2} (\Delta H_{||} \approx 2.54\ \mathrm{kA/m})$ the dispersive characteristic of SFW_2 approaches the dispersion of SSW* in the field of frequencies $\nu < \nu_{\perp}$.
- At $\alpha_{||} = 2 \times 10^{-2} (\Delta H_{||} \approx 5\ \mathrm{kA/m})$ in the field of frequencies $\nu < \nu_{\perp}$ there appears a point of splitting of the SFW dispersion into two branches, namely, SFW_1 and SFW_2 (*e*), and in the field of frequencies below this point of splitting the mode is two-wave one for fast waves (SFW_1 and SFW_2).
- At $\alpha_{||} = 3 \times 10^{-2} (\Delta H_{||} \approx 7.6\ \mathrm{kA/m})$ the dispersion $\nu(\kappa')$ is close to that described in [8], and the mode is single-wave one as only SFW exists.

For $\nu_{\mathrm{H}} = 160\,\mathrm{GHz}$ (Fig. 4.21*a*–*d*) at $\alpha_{||} \leq 5 \cdot 10^{-5} (\Delta H_{||} \approx 2.2\,\mathrm{kA/m})$ the two-wave mode (SSW* and SFW) is also observed.

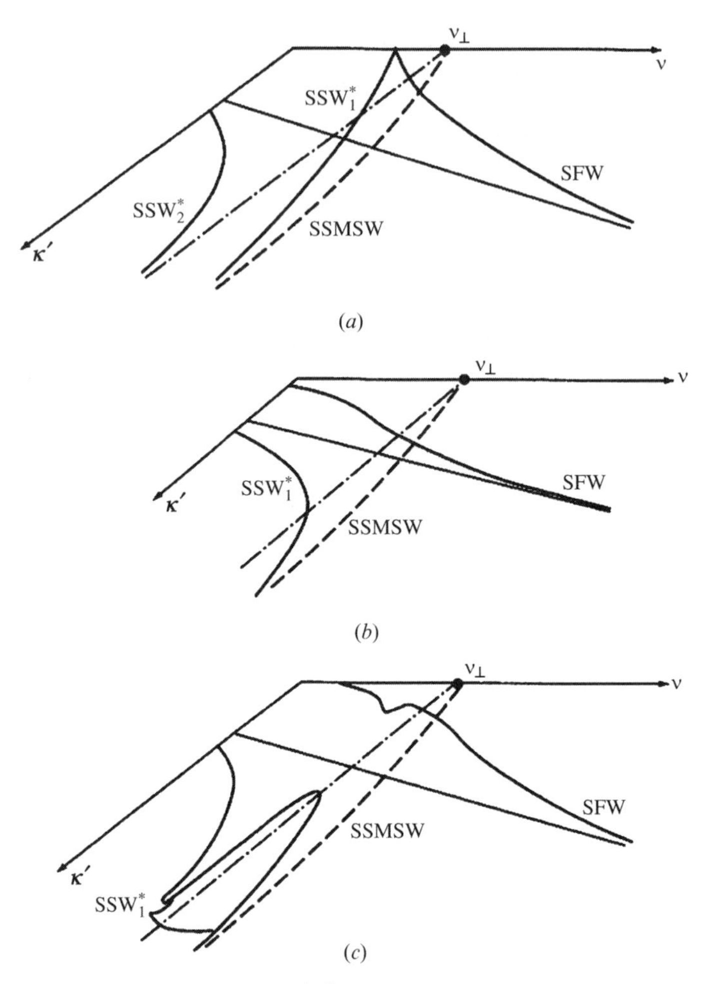

Fig. 4.20 The dispersive characteristics $\nu(\kappa')$ of H_{n0} waves in flat ferrite-dielectric waveguides near the frequency $\nu_{\perp}$ in range of frequencies $\nu_{\mathrm{H}} = 8.9\,\mathrm{GHz}$ and $\nu_{\mathrm{H}} = 160\,\mathrm{GHz}$ are shown. In (*a–e*) the dependences of phase constants for various parameters of ferromagnetic losses are given: (*a*) $\alpha = 10^{-5}$; (*b*) $\alpha = 3\times10^{-4}$; (*c*) $\alpha = 10^{-3}$; (*d*) $\alpha = 10^{-2}$; (*e*) $\alpha = 2\times10^{-2}$; (*f*) $\alpha = 3\times10^{-2}$ at $\nu_{\mathrm{H}} = 8.9\times10^{10}\,\mathrm{Hz}$, $4\pi M_S = 0.176\,\mathrm{T}$. Figure 4.21a–d presents similar dependences: (*a*) $\alpha_{||} = 2\times10^{-5}$; (*b*) $\alpha_{||} = 5\times10^{-5}$; (*c*) $\alpha_{||} = 5\times10^{-4}$; (*d*) $-\alpha_{||} = 5\times10^{-3}$ at $4\pi M_S = 0.176\,\mathrm{T}$

Figure 4.22 presents the qualitative dependences to illustrate: the course of $\kappa'(\nu)$ and $\kappa'(\Theta)$ for SFW and SSW* (two-wave mode) in weakly dissipative layered structures (*a*); the course of dependences of signal attenuation in structures with a length L (*b*); attenuation of a signal in an antiphase-balanced waveguide circuit at phase inversion on the ferrite-dielectric structure (*c*) and localizations of SFW and SSW* on opposite surfaces of the structure (*d*). In Fig. 4.22*a* the following dispersions are shown:

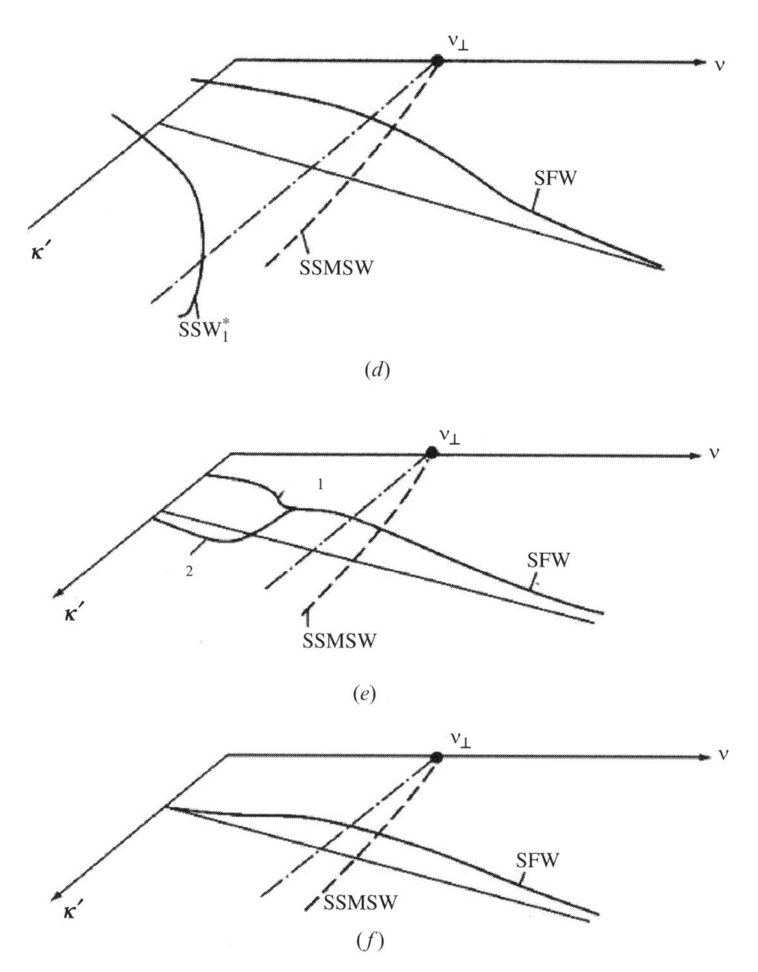

Fig. 4.20 Continued

$1-\kappa'_{SSW^*}$; $2-\kappa''_{SSW^*}$; $3-\kappa'_{SFW}$; $4-\kappa''_{SFW}$; $5-\kappa'_{\varepsilon}$ in a structure with $\alpha < 10^{-3}$–10^{-4}. Figure 4.22*b* gives selective signal attenuation: $1-K_{SFW}$; $2-K_{SSW^*-SFW}$; $3-K_{SSW^*}$. In Fig. 4.22*c* is selective transmission of a signal (the factor of transfer K_{cf}) at phase inversion on the ferrite: $1-K_{cf.SFW}$; $2-K_{cf.SSW^*-SFW}$; $3-K_{cf.SSW^*}$. In Fig. 4.22*d* localization of the HF fields of SSW* and SFW in a planar waveguide with ferrite is shown.

From Fig. 4.22 it follows that the dispersive characteristics for phase constants of SFW and SSW* do not intercept, and for the amplitude ones have a range of interception on frequency $\nu_0 = \nu_\perp^2 < \nu_\perp$, for which

$$\Delta\kappa' L = (\kappa'_{SSW^*} - \kappa'_{SFW})L = (1+2n)\pi, \qquad n = 0,1,2,\ldots,$$
$$\kappa''_{SSW^*}(\nu) = \kappa''_{SFW}(\nu_0),$$

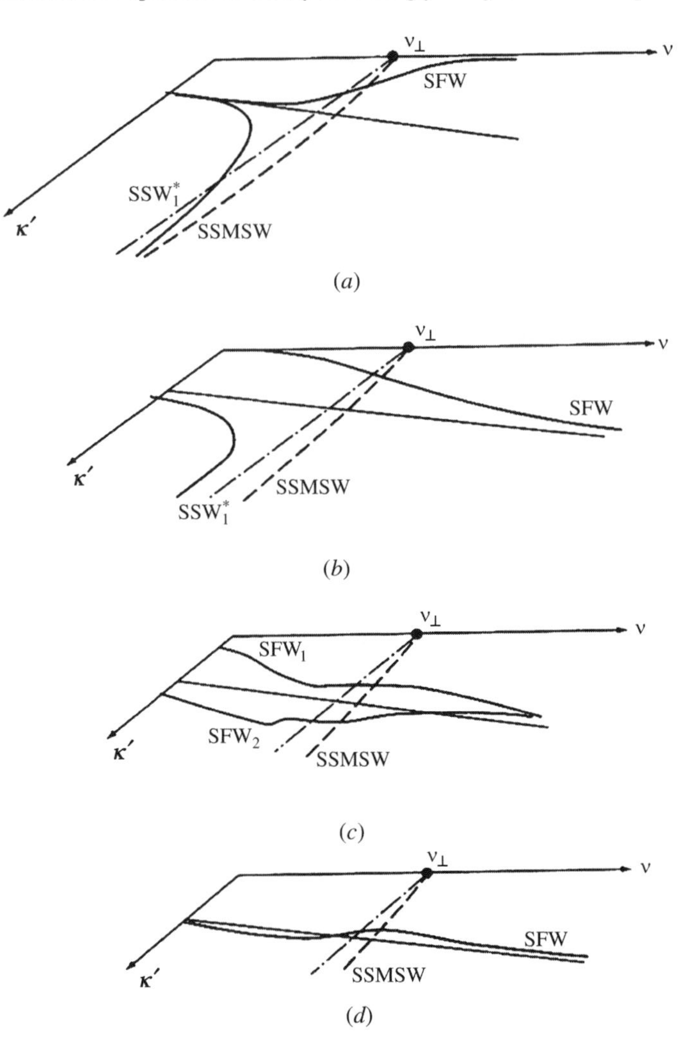

Fig. 4.21 The dependences of phase constants for various parameters of ferromagnetic losses are given: (*a*) $\alpha_{||} = 2 \times 10^{-5}$; (*b*) $\alpha_{||} = 5 \times 10^{-5}$; (*c*) $\alpha_{||} = 5 \times 10^{-4}$; (*d*) $\alpha_{||} = 5 \times 10^{-3}$ at $4\pi M_S = 0.176\,\mathrm{T}$

that leads to interference signal attenuation along the length of the structure L. On frequencies $\nu < \nu_0$ and $\nu > \nu_0$, i.e., below and above the resonance with its frequency ν_0, selective signal attenuation related to the implementation of similar conditions for phase constants will be observed:

- For SSW $\kappa'_{SSW*}L \cong (1+2n)\pi$ and for SFW $\kappa'_{SFW}L \cong (1+2n)\pi$. It will be the first $(n = 0)$ maxima of signal attenuation on frequencies ν_{01} and ν_{02} respectively. For higher orders $(n = 1,\ 2,\ \ldots)$ the amplitude and phase constants of waves sharply increase. These circumstances determine the opportunity of

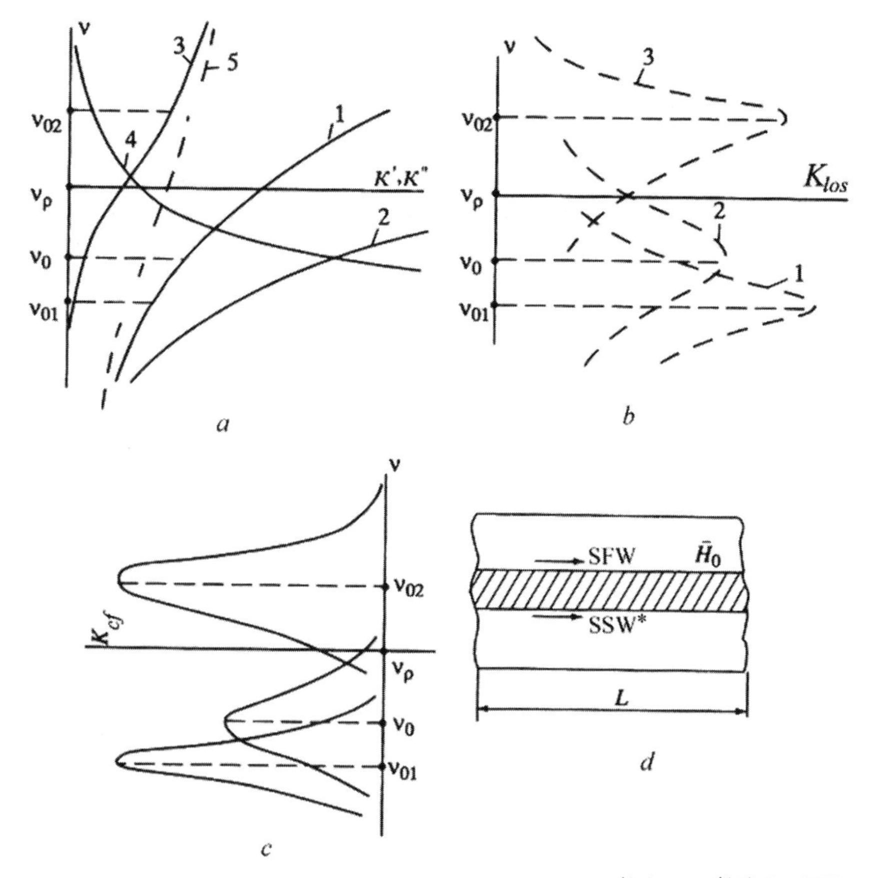

Fig. 4.22 The qualitative dependences to illustrate: the course of $\kappa'(\nu)$ and $\kappa'(\Theta)$ for SFW and SSW* (two-wave mode) in weakly dissipative layered structures (*a*); the course of dependences of signal attenuation in structures with a length L (*b*); attenuation of a signal in an antiphase-balanced waveguide circuit at phase inversion on the ferrite-dielectric structure (*c*) and localizations of SFW and SSW* on opposite surfaces of the structure (*d*)

observation of AFC with three characteristic maxima of attenuation in a passing signal near the frequency $\nu_\perp$ for high-quality ferrites ($\alpha < 10^{-3}$–10^{-4}) for a waveguide with the ferrite-dielectric structure (Fig. 4.22*b*) or passage of a signal (Fig. 4.22*c*) in an antiphase balanced circuit at satisfying the conditions of phase inversion of waves on the structure. The values of transmission losses $K(\nu_{01})$, $K(\nu_0)$, $K(\nu_{02})$ and their minimum values depending on the length of the structure $L - K_{\min}(\nu_{01})$, $K_{\min}(\nu_0)$, $K_{\min}(\nu_{02})$, and also the steepness of the AFC dependences $K(\kappa''(\nu_{01}))$, $K(\kappa''(\nu_0))$, $K(\kappa''(\nu_{02}))$ determine amplitude and phase constants.

- For SSW* $\kappa'_{SSW^*}(\nu_{01})$, $\kappa''_{SSW^*}(\nu_{01})$.
- For SFW – $\kappa'_{SFW}(\nu_{02})$, $\kappa''_{SFW}(\nu_{02})$.

– For frequency ν_0, on which SSW* and SFW coexist – $\Delta\kappa'(\nu_0)$ and $\kappa''(\nu_0)$ and also the parameters of ferromagnetic losses near one and the other surfaces of the ferrite film (Chapter 6).

In Fig. 4.23 you see experimental AFC for a ferrite-dielectric structure, on which phase inversion in an antiphase balanced circuit is carried out with parameters: $d = 30.5 \times 10^{-6}\,\text{m}$, $h = 2 \times 10^{-3}\,\text{m}$, $4\pi M_S = 0.176\,\text{T}$, $H_0 = 929.9\,\text{kA/m}$, $L = 4 \times 10^{-3}\,\text{m}$.

Table 4.1 collects key parameters for this experiment and constants of SFW and SSW* near the resonant frequency of the structure.

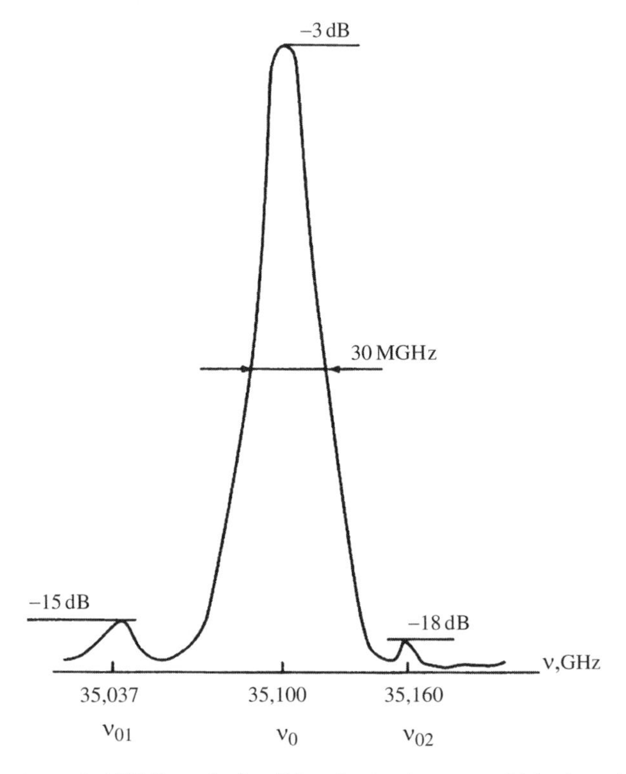

Fig. 4.23 Experimental AFC for a ferrite-dielectric structure, on which phase inversion in an antiphase balanced circuit is carried out with parameters: $d = 30.5 \times 10^{-6}\,\text{m}$, $h = 2 \times 10^{-3}\,\text{m}$, $4\pi M_S = 0.176\,\text{T}$, $H_0 = 929.9\,\text{kA/m}$, $L = 4 \times 10^{-3}\,\text{m}$

Table 4.1 Key parameters and constants of SFW and SSW* near the resonant frequency of the structure for case, which is showed on Fig. 4.32

Wave, frequency	$K_{cf}(\nu_{01}, \nu_0, \nu_{02})$, dB	$\Delta\nu_{3dB}(\nu_{01}, \nu_0, \nu_{02})$, MHz	$\kappa' \cdot 10^{-2}$, m^{-1} $\Delta\kappa \cdot 10^{-2}$, m^{-1}	$\kappa'' \cdot 10^{-2}$, m^{-1}
SSW*, ν_{01}	−15	15	=16	4.32
SFW, ν_{02}	−18	7.5	=16	5.18
SSW*–SFW, ν_0	−3	30	=16	0.86

The specified technique of substantiation of the existence of two-wave modes in high-quality ferrite layers was used for the first time.

The dispersions of amplitude and phase constants of SSW* within the whole range of their existence were investigated with using of stripline converters. In a broadband operating mode ($W_a/d \leq 1$), where W_a is the width of the stripline converter, d the thickness of the ferrite film, the widest (by frequency band) wave spectrum was studied while in the narrow-band mode (at $W_a/d \gg 1$) the wave spectrum close to $\nu_\perp$ was These results are considered below.

For separation of the resonances related to RSLW with $\kappa_z \neq 0$, and SFW and SSW* resonances with $\kappa_y \neq 0$ in the wave range $\nu_1 < \nu_0 < \nu_2$ in the AFC of a signal, experiments were made to change the width of the film W at $S = const$, which usually corresponded to the mode $\kappa''_{\max}(\nu_0)$ providing control binding with respect to $\nu_\perp$. In Fig. 4.24 are presented the AFC of a signal for structures of various width: a – $W = 7 \times 10^{-3}$ m, b – $W = 5 \times 10^{-3}$ m, c – $W = 2 \times 10^{-3}$ m, d – $W = 1 \times 10^{-3}$ m. At reduction of W observed are:

- An equidistant spectrum of RSLW located below the frequency $\nu_\perp$ in a band $\nu_1 < \nu < \nu_\perp$.
- The irregularity of the RSLW spectrum related to the width of the ferrite film (the size along field $\overline{H}_0$) is of equidistant character, and the distance between resonances is $\Delta\nu_n \sim W^{-1}$.
- A monotonous reduction of amplitude in the band $\nu_1 < \nu < \nu_2$ and primary localization of the SSW* spectrum in the field of frequencies $\nu_\perp < \nu < \nu_2$. The latter feature is subject, first of all, to the reduction of the volume of ferrite in HF fields of excitation and the decrease in the efficiency of RSLW excitation.

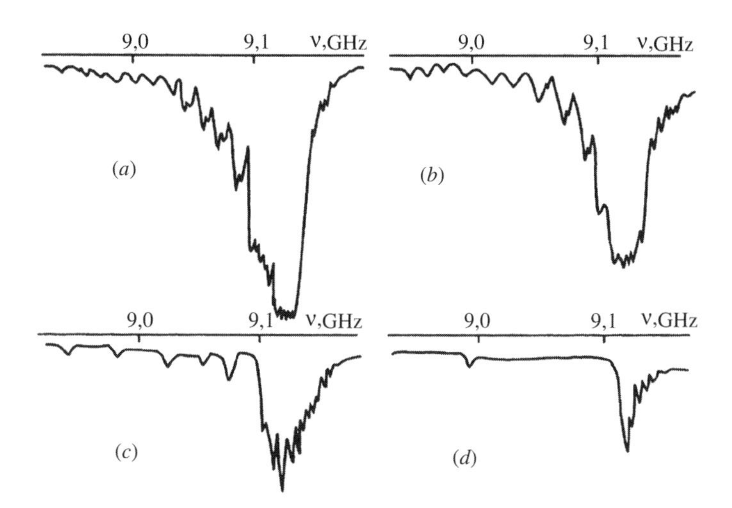

Fig. 4.24 The AFC of a signal for structures of various width: (*a*) $W = 7 \times 10^{-3}$ m, (*b*) $W = 5 \times 10^{-3}$ m, (*c*) $W = 2 \times 10^{-3}$ m, (*d*) $W = 1 \times 10^{-3}$ m

To prove that the two-wave mode (the existence of both SFW and SSW*) manifests itself in the UHF and EHF ranges only in high-quality ferrite-dielectric structures with $\alpha < 10^{-4}$, experiments have been made:

- On films before and after their processing with diamond paste
- At heating of these films

In both the cases the parameter α increased. At heating of the films due to the dependence $M_S \sim T^{-1}$ a sharp drop of frequency ν_0 was observed. At mechanical treatment of the surface of a ferrite film, which in its initial condition had 13 or −14 class of precision, the magnetization M_S did not change, and the central frequency of selective signal attenuation or passage (determined by the conditions of experiment in the transmission mode in a waveguide, or in the mode of phase inversion in a balancing circuit) was within the limits of initial $\nu_\perp$. Figure 4.25 show the AFC of signals on structures of various thickness: $d = 15.8 \times 10^{-4}$ m (Fig. 4.25*a*) and $d = 20 \times 10^{-4}$ m (Fig. 4.25*b*) before processing with diamond paste and after it Fig. (4.25*c*, *d*), respectively. It is obvious that at increasing α the irregularity of AFC is eliminated and the level of signal attenuation decreases, and the character of AFC corresponds to the one-wave mode described in the literature [5–8] at $\alpha \approx 10^{-2}$. In such structures no interference mechanism of attenuation on frequency was observed. In balancing circuits only single low-durable resonance on frequency ν_0 was registered.

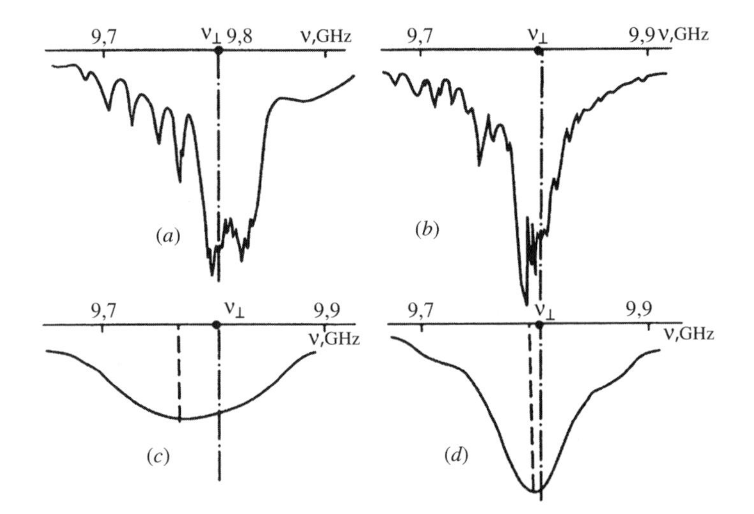

Fig. 4.25 The AFC of signals on structures of various thickness: $d = 15.8 \times 10^{-4}$ m (*a*) and $d = 20 \times 10^{-4}$ m (*b*) before processing with diamond paste and after it (*c, d*)

4.3.3 Beyond-Cutoff Mode

Researches were executed in a range of frequencies $\nu_{\mathrm{H}} = 3$–$80\,\mathrm{GHz}$ and were directed to studying the mechanism of selective signal passage in a ferrite-dielectric structure with closely located screens (t/d and $\ell/d \leq 1$) for $\nu < \nu_{cr}$.

Theoretical analysis in the H_{n0} wave approximation in such structures has shown that selective passage (the transparency of the structure) of a signal in BCM this mode is related:

- In the long-wave part of the UHF range ($\nu < 10\,\mathrm{GHz}$) – to forward surface waves, for which $(\kappa''_{\min})_{SSW^*} \ll (\kappa''_{\min})_{SFW}$
- In the middle parts of the UHF range ($\nu \approx 10$–$15\,\mathrm{GHz}$) – to SSW* and SFW, for which $(\kappa''_{\min})_{SSW^*} \cong (\kappa''_{\min})_{SFW}$
- In the short-wave part of the UHF and EHF ranges ($\nu \approx 15$–$20\,\mathrm{GHz}$) – only to surface wave, for which $(\kappa''_{\min})_{SSW^*} \gg (\kappa''_{\min})_{SFW}$

Let's remind that SSW* and SFW are localized on different surfaces of the ferrite layer, this requires checking at experimental researches of the direction of magnetic field H_0. This manifests itself most essentially in layered structures with an asymmetrical arrangement of metal screens ($t/d \neq \ell/d$).

In Fig. 4.26 the dispersive characteristics $\kappa'_y(\nu)$ and $\kappa''_y(\nu)$ for SFW (1, 5, 9) and (3, 7, 11) and SSW* (2, 6, 10) and (4, 8, 12), respectively, of a bilaterally-metallized ferrite-dielectric structure in э р the beyond-cutoff mode are presented for $a = \ell + d + t = 1 \times 10^{-3}\,\mathrm{m}$, $d/a = 2 \times 10^{-2}$, $4\pi M_S = 0.176\mathrm{T}$ for: $a - \alpha = 5 \times 10^{-4}$, $H_0 = 194.95\,\mathrm{kA/m}$; $b - \alpha = 5 \times 10^{-5}$, $H_0 = 835.49\,\mathrm{kA/m}$; $c - \alpha = 5 \times 10^{-5}$, $a = 5 \times 10^{-4}\,\mathrm{m}$, $H_0 = 2.387\,\mathrm{MA/m}$. In Fig. 4.27 the dependences $\kappa'_{\min}(\nu_0)$ for SSW* (1) and SFW (2), and in Fig. 4.28 the passbands $\Delta\nu_{3dB}(\nu_0)$ for the same waves (the dotted line depicts the full passband dependence of Eshbach-Damon's waves) are shown.

Similar dependences have been obtained for LE_t^S and LE_t^F waves as well (Chapter 3). At experimental research of the properties of ferrite-dielectric structures basic attention was given to studying the characteristics of a transmission signal, namely, the passbands $\Delta\nu_{3dB}$ and introduced losses K_{los}. Out research has been made in a range of frequencies of 3–60 GHz on an extensive number of industrial and pilot YIG models.

In Fig. 4.29 are shown the experimental (1, 2) and theoretical (3, 4) dependences $\Delta\nu_{3dB}$ on the thickness of YIG film d in two frequency ranges: 1, 3 – $\nu_{\mathrm{H}} = (8$–$10)$ GHz; 2, 4 – $\nu_{\mathrm{H}} = (30$–$40)$ GHz; theoretical dependences 3 – $\nu_{\perp} = 32\,\mathrm{GHz}$, $\Delta H_{||} = 398\,\mathrm{A/m}$; 4 – $\nu_{\perp} = 9\,\mathrm{GHz}$, $\Delta H_{||} = 95\,\mathrm{A/m}$. Agreement is quite satisfactory. According to [28], at advance into the millimeter range the value of $\Delta H_{||}$ for monocrystal YIG spheres linearly increases with frequency and reaches $\Delta H_{||} \cong (159$–$318)\,\mathrm{A/m}$.

The selective properties of the structure, depending on the position of the ferrite layer between the screens h_1/a, are shown in Fig. 4.30: 1 – $\Delta\nu_{3dB}$ – theory; 2 – $\Delta\nu_{3dB}$ – experiment; 3 – K_{los} – theory; 4 – K_{los} – experiment. The experimental dependence $\Delta\nu_{3dB}(h)$ is in accord with the theoretical one. The dependence

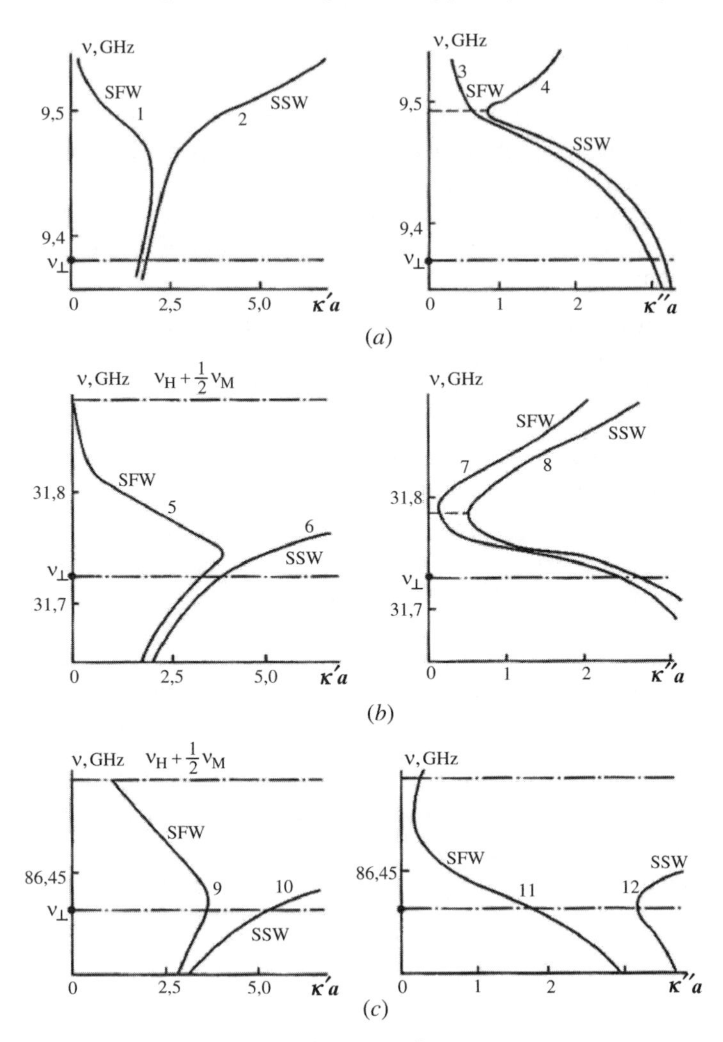

Fig. 4.26 The dispersive characteristics $\kappa_y'(\nu)$ and $\kappa_y''(\nu)$ for SFW (1, 5, 9) and (3, 7, 11) and SSW* (2, 6, 10) and (4, 8, 12), respectively, of a bilaterally-metallized ferrite-dielectric structure in BCM the beyond-cutoff mode are presented for $a = \ell + d + t = 1 \times 10^{-3}$ m, $d/a = 2 \times 10^{-2}$, $4\pi M_S = 0.176$ T for: (*a*) $\alpha = 5 \times 10^{-4}$, $H_0 = 194.95$ kA/m; (*b*) $\alpha = 5 \times 10^{-5}$, $H_0 = 835.49$ kA/m; (*c*) $\alpha = 5 \times 10^{-5}$, $a = 5 \times 10^{-4}$ m, $H_0 = 2.387$ MA/m

$\Delta\nu_{3dB}(h)$ coincides with the calculated one in an interval $h_1/a = 0.15$–0.80. At $h_1/a < 0.15$ the experimental value of $\Delta\nu_{3dB}$ is finite while the theoretical one calculated in the H_{n0} wave approximation, $\Delta\nu_{3dB}$ increases.

In Fig. 4.31*a* are presented the theoretical (1) and experimental (2) dependences $\Delta\nu_{3dB}$ on the length of the beyond-cutoff section, and in Fig. 4.31*b* are the calculated AFC at L_{bcs}/a: 1 – 1, 2 – 2, 3 – 8 for $d = 3 \times 10^{-5}$ m, $a = 1 \times 10^{-3}$ m, $4\pi M_S = 0.176$ T, $h_1/a = 5 \times 10^{-4}$ m. Such dependences can be used in the millimeter range for reduction of the signal passband.

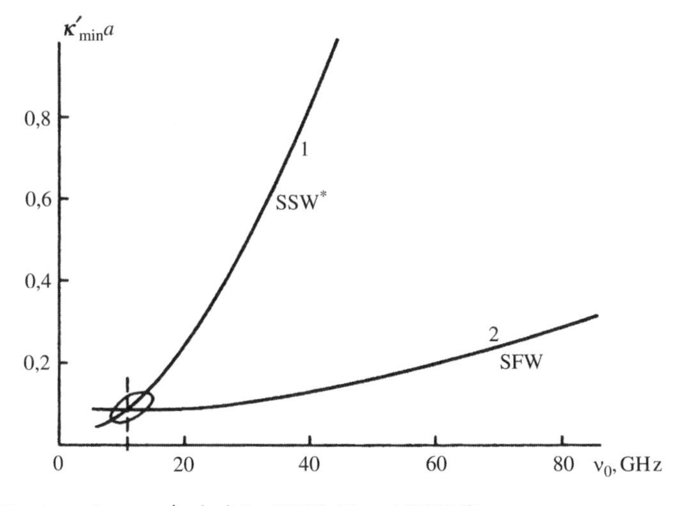

Fig. 4.27 The dependences $\kappa'_{\min}(\nu_0)$ for SSW* (1) and SFW (2)

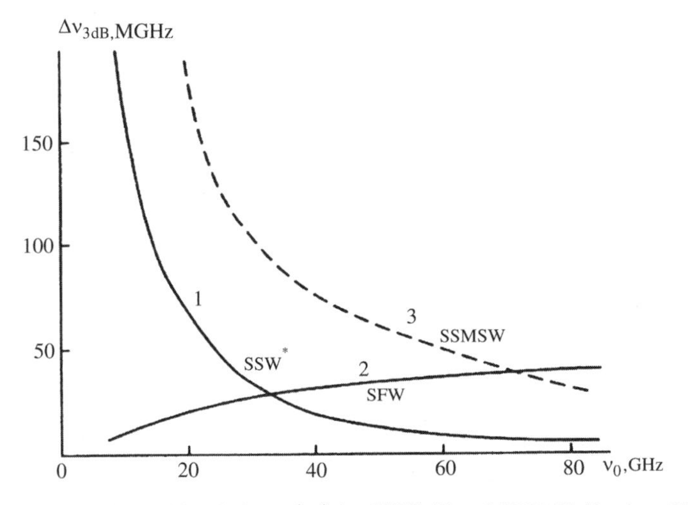

Fig. 4.28 Dependence the passbands $\Delta\nu_{3dB}(\nu_0)$ for SSW* (1) and SFW (2) (the dotted line depicts the full passband dependence of Eshbach-Damon's waves) are shown

Figure 4.32 shows the experimental dependences $\Delta\nu_{3dB}$ on frequency ν for YIG films with their thickness: $1 - h = 10.2 \times 10^{-6}$ m; $2 - h = 2 \times 10^{-5}$ m. It is obvious that for various structures the dependence $\Delta\nu_{3dB}$ is linear. The change of the steepness of the dependence $\Delta\nu_{3dB}(\nu)$ is due to different ferromagnetic losses in our samples films.

In Fig. 4.33 theoretical and experimental dependences of the dispersions of phase constants of surface MSW in symmetrical (1) and asymmetrical (2) structures on the basis of films of various thickness are presented: $1 - d = 3 \times 10^{-5}$ m; $t/d = \ell/d = 50$;

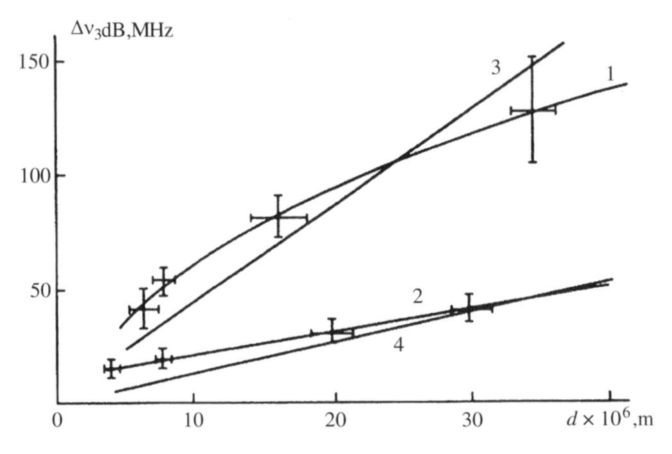

Fig. 4.29 The experimental (1, 2) and theoretical (3, 4) dependences $\Delta\nu_{3dB}$ on the thickness of YIG film d in two frequency ranges: 1, 3 – $\nu_{\mathrm{H}} = (8\text{–}10)$ GHz; 2, 4 – $\nu_{\mathrm{H}} = (30\text{–}40)$ GHz; theoretical dependences 3 – $\nu_{\perp} = 32$ GHz, $\Delta H_{||} = 398$ A/m; 4 – $\nu_{\perp} = 9$ GHz, $\Delta H_{||} = 95$ A/m

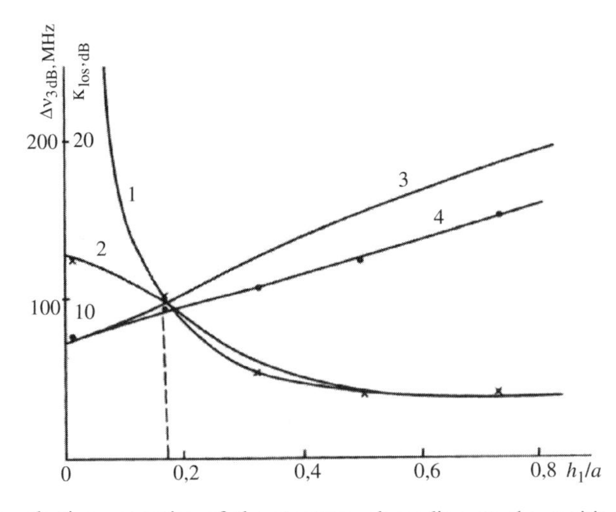

Fig. 4.30 The selective properties of the structure, depending on the position of the ferrite layer between the screens h_1/a: 1 – $\Delta\nu_{3dB}$ – theory; 2 – $\Delta\nu_{3dB}$ – experiment; 3 – K_{los} – theory; 4 – K_{los} – experiment

2 – $h = 7 \times 10^{-5}$ m; $t/d = 1$, $\ell/d = 7.1$; $4\pi M_S = 0.176$T; $\Omega_{\mathrm{M}} = 0.15$. In asymmetrical structures the experimentally observed wave spectra (2) are separated by frequencies. The different steepness of the dispersions $\Omega(\kappa' d)$ for $y > 0$ and $y < 0$ means different group speeds of SMSW.

In Fig. 4.34 the dispersive characteristics $\Omega(\kappa' d)$ SMSW in connected two-layer structures are presented: 1, 2 – $g/h = \ell/h = 0.07$; 3, 4 – $g/h = \ell/h = 0.125$, $4\pi M_S = 0.176$T, $\Omega_{\mathrm{M}} = 0.15$. One can see that SMSW with normal dispersions, belonging to the frequency band of the existence of Eshbach-Damon's

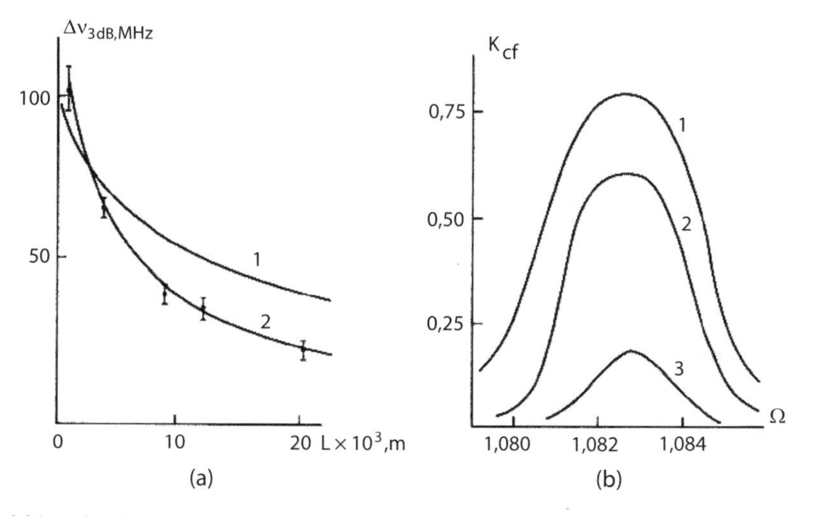

Fig. 4.31 *a* the theoretical (1) and experimental (2) dependences $\Delta\nu_{3dB}$ on the length of the beyond-cutoff section, and in Fig. 4.31*b* are the calculated AFC at L_{bcs}/a: 1–1, 2–2, 3–8 for $d = 3 \times 10^{-5}$ m, $a = 1 \times 10^{-3}$ m, $4\pi M_S = 0.176$T, $h_1/a = 5 \times 10^{-4}$ m

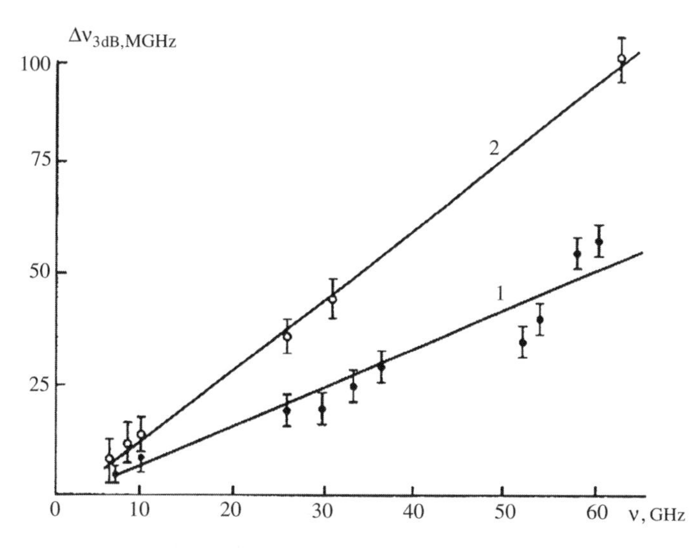

Fig. 4.32 The experimental dependences $\Delta\nu_{3dB}$ on frequency ν for YIG films with their thickness: $1 - h = 10.2 \times 10^{-6}$ m; $2 - h = 2 \times 10^{-5}$ m

waves, are practically independent of the g/h and ℓ/h ratios. For SMSW with abnormal dispersions a strong dependence of the frequency band and its dispersion on parameters g/h and ℓ/h takes place, and with its reduction the dispersion of RSMSW goes above by frequencies with respect to $\nu_{\perp}$.

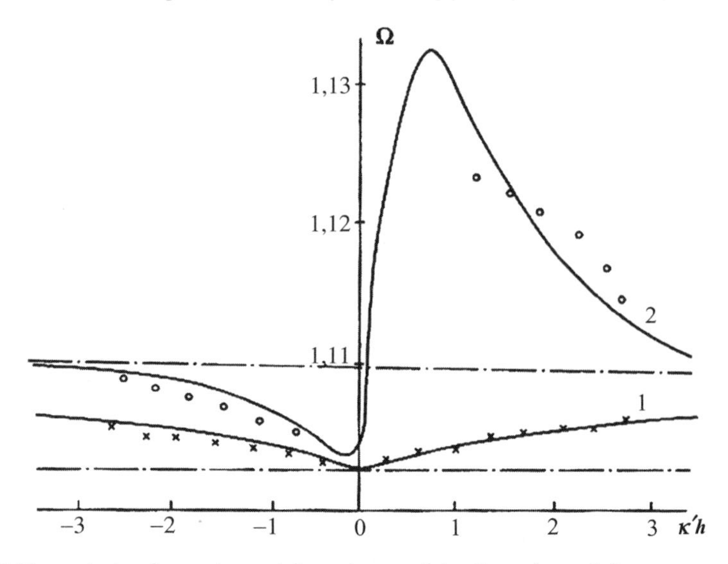

Fig. 4.33 Theoretical and experimental dependences of the dispersions of phase constants of surface MSW in symmetrical (1) and asymmetrical (2) structures on the basis of films of various thickness are presented: $1 - d = 3 \times 10^{-5}$ m; $t/d = \ell/d = 50$; $2 - h = 7 \times 10^{-5}$ m; $t/d = 1$, $\ell/d = 7.1$; $4\pi M_S = 0.176$T; $\Omega_M = 0.15$

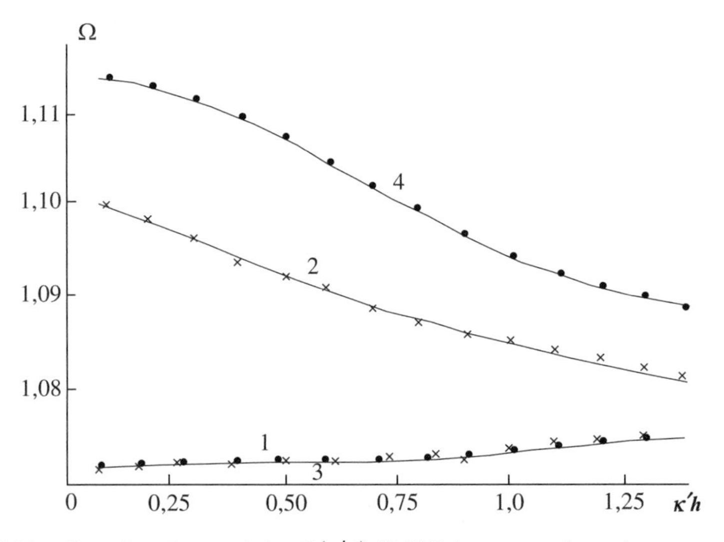

Fig. 4.34 The dispersive characteristics $\Omega(\kappa' d)$ SMSW in connected two-layer structures are presented: $1, 2 - g/h = \ell/h = 0.07$; $3, 4 - g/h = \ell/h = 0.125$, $4\pi M_S = 0.176$T, $\Omega_M = 0.15$

Our experimental research of the dispersive characteristics of SSW*, SSLW, and RSLW in the UHF and EHF ranges has shown that:

- The dispersive dependences $\Omega(\kappa)$ fall outside the borders $\nu_{\perp}$ for SSW* and RSLW and outside ν_{H} for SSLW.
- This falling outside the borders $\nu_{\perp}$ and ν_{H} at advance into the millimeter range gets stronger.

In Fig. 4.35 the experimental dispersive dependences $\nu_{H}(n)$ in a tangentially-magnetized structure with parameters: $d = 35 \times 10^{-6}$ m, $t = 10^{-4}$ m, $W_{MSL} = 5 \times 10^{-5}$ m for two directions of magnetic field: $1 - H_0 \uparrow\uparrow 0Z$; $2 - H_0 \uparrow\downarrow 0Z$ are presented. Frequency $\nu_{\perp}$ (dotted line) was determined experimentally by means of specially developed sensors. The various course of the dependences ν_n can be due to:

- The crystallographic anisotropy field $\overline{H}_A$, including its cross-section gradient by thickness of the film $\nabla_x H_A$
- The cross-section gradient of saturation magnetization $\nabla_x M_S$

In Part III methods of experimental determination of $\overline{H}_A$, $\nabla_x H_A$, $\nabla_x M_S$ and other parameters of ferrite film structures s are discussed.

Another type of waves which can be excited at tangential magnetization, is RSLW extending along the field H_0. The dispersions of RSLW in tangentially-linear magnetized ferrite-dielectric structures for various mutual orientations of the radiating and reception converters (microstrip converter – MSC) were investigated.

In Fig. 4.36 the dispersive dependence ν_H for RSLW in the millimeter range is presented for $d = 45 \times 10^{-6}$ m, $t = 10^{-4}$ m, $W_{MSL} = 5 \times 0^{-5}$ m, $4\pi M_S = 0.176$ T, $H_0 = 822.6$ kA/m. Figure 4.37 shows a similar dependence for Li-Zn spinel: $d = 13 \times 10^{-6}$ m, $t = 10^{-4}$ m, $W_{MS} = 5 \times 10^{-5}$ m, $4\pi M_S = 0.5$ T. The reader sees that the increase in magnetization M_S leads to expansion of the band of RSLW existence.

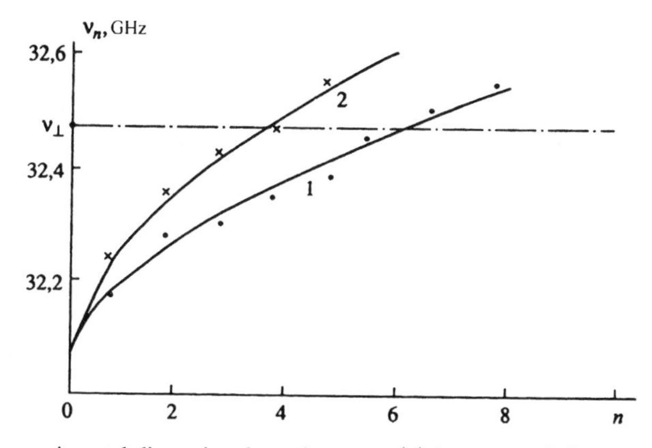

Fig. 4.35 The experimental dispersive dependences $\nu_{H}(n)$ in a tangentially-magnetized structure with parameters: $d = 35 \times 10^{-6}$ m, $t = 10^{-4}$ m, $W_{MS} = 5 \times 10^{-5}$ m for two directions of magnetic field: $1 - H_0 \uparrow\uparrow 0Z$; $2 - H_0 \uparrow\downarrow 0Z$

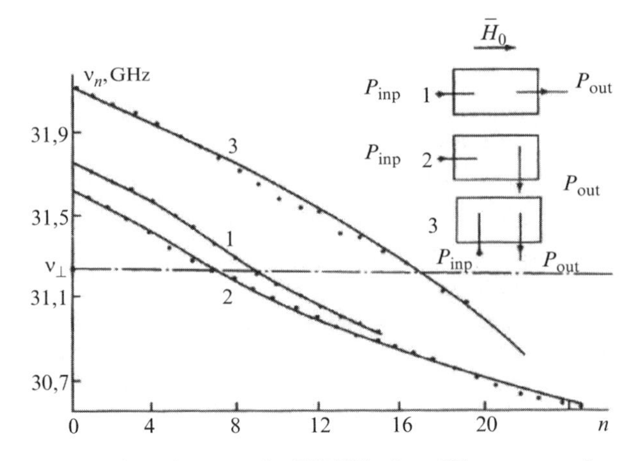

Fig. 4.36 The dispersive dependence ν_H for RSLW in the millimeter range is presented for $d = 45 \times 10^{-6}$ m, $t = 10^{-4}$ m, $W_{MSL} = 5 \times 10^{-5}$ m, $4\pi M_S = 0.176$T, $H_0 = 822.6$kA/m

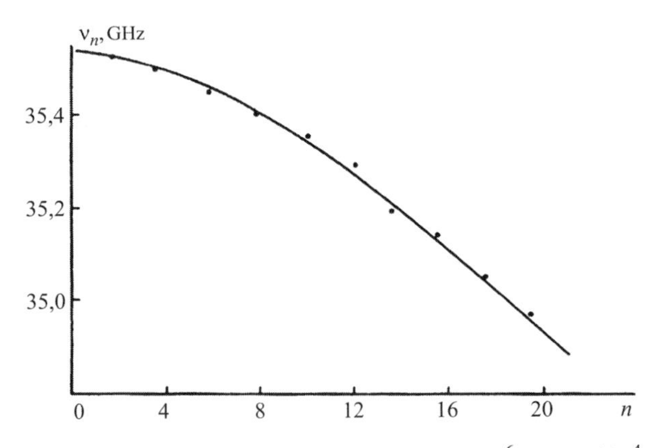

Fig. 4.37 Dispersion dependence for Li–Zn spinel: $d = 13 \times 10^{-6}$ m, $t = 10^{-4}$ m, $W_{MSL} = 5 \times 10^{-5}$ m, $4\pi M_S = 0.5$T

It follows from the results of our research of SSW* and RSLW in tangentially-magnetized high-quality structures with $\alpha < 10^{-4}$ that the dispersive dependences ν_n in the millimeter range fall essentially outside the bounds:

- SSW* exist both in the frequency area $\nu_\perp \leq \nu < \nu_H + \frac{1}{2}\nu_M$, and at $\nu < \nu_\perp$
- SSLW exist both in the frequency area $\nu_H \leq \nu < \nu_H(\nu_H + \nu_M)$, and at $\nu > \nu_\perp$

For SSW* in single layers and SSW* in two-connected structures the opportunity to control wave dispersions both in symmetric and asymmetrical layered structures is shown.

4.3.4 Normally Magnetized Layered Structures

Waves in normally-magnetized layered structures on the basis of ferrite films in narrow-band operating modes of converters are localized near the frequency ν_H, and in broadband ones they exist both in a band of frequencies $\nu_H \leq \nu < \nu_\perp$ and below ν_H.

To estimate the upper bound frequency and values $(\kappa'_y)_{\max}$ for SSLW, structures with ferrite films of various thickness were investigated.

In Fig. 4.38 the experimental dispersive characteristics $\Omega_{0H}(\kappa' d)$ for SSLMSW in normally-magnetized structures on the basis of ferrite films of various thickness are shown: $1 - d = 1.8 \times 10^{-5}\,\text{m}$, $2 - d = 7 \times 10^{-5}\,\text{m}$ at $W_{MSL} = 8 \times 10^{-6}\,\text{m}$, $H_0 = 954.84\,\text{kA/m}$, $4\pi M_S = 0.176\,\text{T}$. Experiment has shown that SSLW in ferrite film structures exist in a range of wave numbers $\kappa' \leq 10^5\,\text{m}^{-1}$.

A study of the dispersive characteristics of SSLW with the aid of sensors of resonant frequency ν_H was made in the centimeter and millimeter ranges. In Fig. 4.39 the dispersion of phase constant ν_n for a unilaterally-metallized ferrite film $\nu_H = 8.47 \times 10^9\,\text{Hz}$, $d = 45 \times 10^{-6}\,\text{m}$, $\ell = 10^{-4}\,\text{m}$, $\varepsilon_d = \varepsilon_\ell = 14$, $W_{MSL} = 3 \times 10^{-5}\,\text{m}$, $H_0 = 380.74\,\text{kA/m}$, $4\pi M_S = 0.176\,\text{T}$, $\Delta H_{||} = 47.4\,\text{A/m}$ is shown. It is obvious that SSLW fall essentially outside the frequencies ν_H in the range $\nu < \nu_H$. The dotted line depicts the course of the dispersion of SSLMSW. The divergence of the MSW dependence shows an incorrectness of these calculations within the limits of the MSW approximation near frequency ν_H at tuning-out less than $\pm 2.5\%$.

Similar results have been obtained in the centimeter range in narrow-band operating modes of various types of converters.

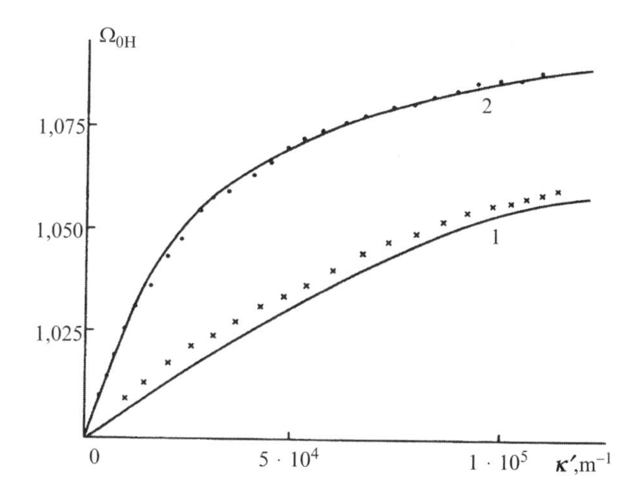

Fig. 4.38 The experimental dispersive characteristics $\Omega_{0H}(\kappa' d)$ for SSLMSW in normally-magnetized structures on the basis of ferrite films of various thickness are shown: $1 - d = 1.8 \times 10^{-5}\,\text{m}$, $2 - d = 7 \times 10^{-5}\,\text{m}$ at $W_{MSL} = 8 \times 10^{-6}\,\text{m}$, $H_0 = 954.84\,\text{kA/m}$, $4\pi M_S = 0.176\,\text{T}$

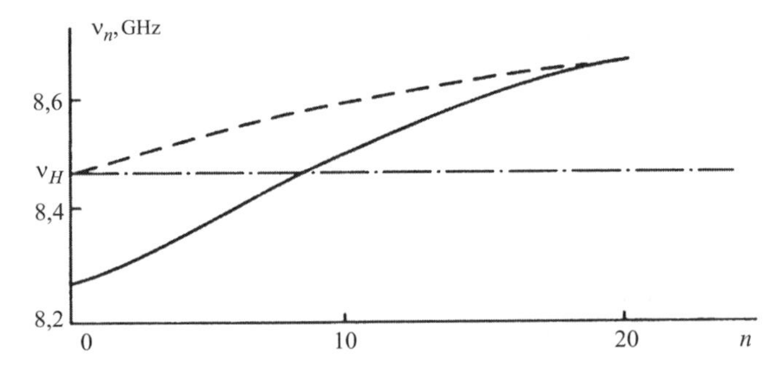

Fig. 4.39 The dispersion of phase constant ν_n for a unilaterally-metallized ferrite film $\nu_H = 8.47 \times 10^9\,\text{Hz}$, $d = 45 \times 10^{-6}\,\text{m}$, $\ell = 10^{-4}\,\text{m}$, $\varepsilon_d = \varepsilon_\ell = 14$, $W_{MS} = 3 \times 10^{-5}\,\text{m}$, $H_0 = 380.74\,\text{kA/m}$, $4\pi M_S = 0.176\,\text{T}$, $\Delta H_{||} = 47.4\,\text{A/m}$

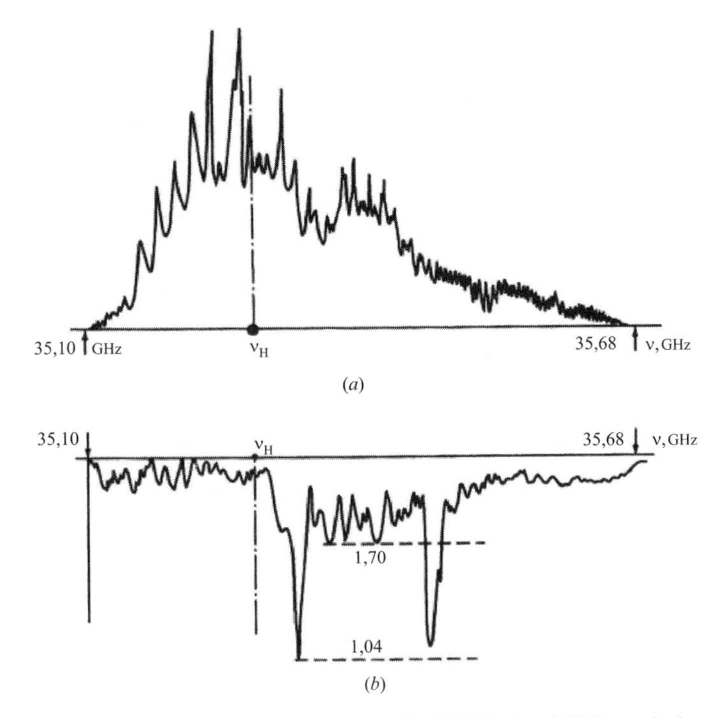

Fig. 4.40 In the millimeter range presented are AFC for SSLW (*a*) and SWR_e at the input (*b*) of the transmission line on the basis of a ferrite film $\nu_H = 35.6 \times 10^9$ Hz, $d = 40 \times 10^{-6}$ m, $t = 3 \times 10^{-4}$ m, $\Delta H_{||} = 119.4\,\text{A/m}$, $W_{MSL} = 10^{-5}$ m, $4\pi M_S = 0.176\,\text{T}$, $H_0 = 1.152\,\text{MA/m}$

In Fig. 4.40 for SSLW in the millimeter range presented are AFC (Fig. 4.40*a*) and SWR_e at the input (Fig. 4.40*b*) of the transmission line on the basis of a ferrite film $\nu_H = 35.6 \times 10^9\,\text{Hz}$, $d = 40 \times 10^{-6}\,\text{m}$, $t = 3 \times 10^{-4}\,\text{m}$, $\Delta H_{||} = 119.4\,\text{A/m}$, $W_{MSL} = 10^{-5}\,\text{m}$, $4\pi M_S = 0.176\,\text{T}$, $H_0 = 1.152\,\text{MA/m}$. The dot-and-dash line marks the frequency ν_H measured by means of own sensors. It is obvious that in the field of ν_H on SWR_e (Fig. 4.40*b*) there are no prominent features. In Fig. 4.41 the experimental dispersion of phase constant ν_H for the given transmission line is shown. The dotted line depicts the course of SSLMSW. One can see that at advance into the millimeter range (Fig. 4.39) the move of the dispersive characteristic ν_n into the field of frequencies $\nu < \nu_H$ has essentially amplified. At tunings-out $\pm 1.7\%$ from ν_H the usage of the MSW-approximation is incorrect.

Experimental research near the frequency ν_H was made, as above, by means of a balancing waveguide circuit, in which phase inversion was carried out in the ferrite-dielectric structure located on the wide wall of the waveguide. The external field $\overline{H}_0$ was oriented normally to the wide wall of the waveguide.

Figure 4.42 gives the AFC of a signal at the output of such a circuit for a film with $d = 35 \times 10^{-6}$ m, $l = 5 \times 10^{-3}$ m, $H_0 = 1.172$ MA/m, $4\pi M_S = 0.176$ T.

One can see that on frequency ν_{H0} of interference attenuation of SSLW and SFW, as well as for tangentially-magnetized structures considered above, the minimum attenuation of a signal ($K_{los} \cong -4$ dB) is observed, and

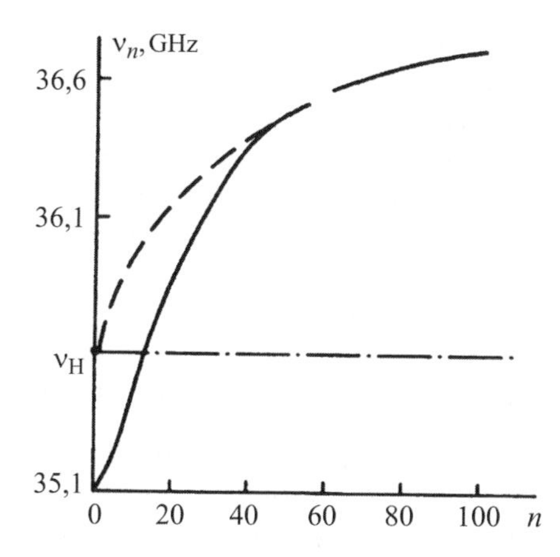

Fig. 4.41 The experimental dispersion of phase constant ν_n for transmission line

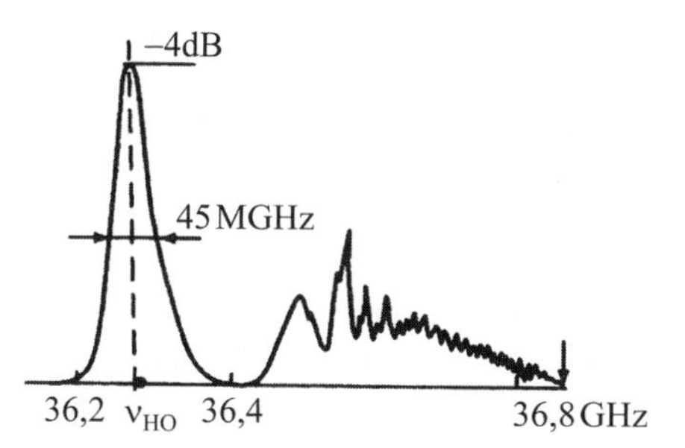

Fig. 4.42 The AFC of a signal at the output of a circuit with film $d = 35 \times 10^{-6}$ m, $l = 5 \times 10^{-3}$ m, $H_0 = 1.172$ MA/m, $4\pi M_S = 0.176$ T

$$\kappa'_{SFW}(V_{\mathrm{H0}}) = \kappa'_{SSW}(V_{\mathrm{H0}}),$$
$$\Delta\varphi = \Delta\kappa' L = (1+2n)\pi, \qquad n = 0,1,2,\ldots$$

Above the frequency ν_{H0} an incised AFC is observed, related to the dispersion of SSLW in the structure.

Thus, our research of the dispersive characteristics of various types of waves in tangentially and normally magnetized layered structures on the basis of ferrite films has confirmed the basic conclusions of our theoretical analysis and shown the expedience of their using in the millimeter range. Waves near the resonant frequencies can provide the design of effective selective devices and delay lines with low introduced losses, reconstructed by frequency and approaching waveguide devices by their parameters (first of all, introduced attenuation and selectivity).

4.4 Conclusions

1. Theoretical and experimental studies were made of the dispersions of electromagnetic waves, including strongly delayed – magnetostatic waves in films and rectangular waveguides with layered structures, including structures with unilateral metallization, containing ferrite plates and films with various magnetization.
2. Comparison of the MMR dispersions of forward and return surface and solid waves shows that in the widest band of frequencies there are return surface forward and return solid waves, with the propagation losses of surface waves being by 15–20 dB/cm higher than that of solid waves.
3. The dispersive characteristics of forward surface, space and return solid waves fall essentially outside the frequency limits given by the MSW approximation:

   – In the field of frequencies $\nu < \nu_{\perp}$ for forward surface and $\nu > \nu_{\perp}$ for return solid waves at tangential magnetization.
   – In the field of frequencies $\nu < \nu_{\mathrm{H}}$ for forward solid waves at normal magnetization.

4. The conclusion of our theoretical analysis that in the prelimit mode $(\nu \gg \nu_{cr})$ in layered structures on the basis of weakly dissipative $(\alpha < 10^{-4})$ ferrites, selective processes near the resonant frequency (ν_{H} – at normal, $\nu_{\perp}$ – at tangential or ν_{φ} – at arbitrary magnetization) are determined by interference attenuation on a certain length of the structure of fast and slow electromagnetic waves with close dispersion laws of phase constants and equal values of the amplitude constants of these waves is confirmed.
5. The conclusion of our theoretical analysis is confirmed that in the beyond-cutoff mode $(\nu \ll \nu_{cr})$ in layered structures on the basis of weakly dissipative ferrites $(\alpha < 10^{-4})$ the effect of the transparency of the beyond-cutoff waveguide, observable in the field of frequencies $\nu > \nu_{\mathrm{H}}$ at normal, $\nu > \nu_{\perp}$ – at tangential, and $\nu > \nu_{\varphi}$ – at arbitrary magnetization is related to selective passage of fast waves through the beyond-cutoff section.

6. The processes of selective signal attenuation and passage are most effective near the resonant frequency of a screened layered structure with ferrite and so are promising for the creation of waveguide devices for filtration – single- and multichannel TF and LHPF, filters-preselectors with a low level of introduced losses and an expanded dynamic range of power levels in MMR, and also for the purposes of testing of the parameters of film ferrites.
7. The broadband modes of excitation and propagation of various types of waves in flat layered structures, including strongly delayed– magnetostatic waves (in frequency bands essentially exceeding those of MSW), determined in the MSW approximation, are promising for the creation of transmission lines with preset dispersion laws of their phase and amplitude constants of wave propagation and can be used in operated delay lines, transversal filters, receivers of IFM and signal analysis.
8. Surfaces MSW at tangential magnetization and solid MSW at normal one are promising for nondestructive control of the parameters of film structures in MMR in the presence of a raised spatial resolution.

Chapter 5
Electromagnetic Wave Excitation by Waveguide and Stripe-Line Converters

Properties of waveguide and strip-line converters with parallel and orthogonal orientations of their layered structures (on the basis of ferrite plates and films) to the exciting plane in both centimeter and millimeter ranges of radiowaves are investigated. Theoretical analysis for *LE* and *LM* waves has allowed features of electromagnetic wave excitation in ferrite-dielectric structures with losses to be revealed, the characteristics of such converters in the presence of metal screens from both sides of the structure in the prelimit ($\nu \gg \nu_{cr}$) and beyond-cutoff ($\nu \ll \nu_{cr}$) modes in structures with unilateral metallization to be studied. Characteristics of the converters were calculated in both near and far zones of radiation. The transmission factors in structures with a reasonable level of ferromagnetic losses are calculated. Features of excitation of various waves, of their propagation and reception, both near the resonant frequency of a structure and at tuning-out from it are investigated.

5.1 Theoretical Analysis of Waveguide Converters Characteristics

Let's consider the results of our theoretical analysis of electromagnetic wave excitation in flat waveguides with ferrite-dielectric structures by an extraneous electric current in the limit and beyond-cutoff modes, and this process in ferrite-dielectric structures with bilateral metallization. Our computer program for *LE* and *LM* waves in the complex plane of wave numbers $\kappa = \kappa' - j\kappa''$ included:

- A dispersive part of the problem for characteristic waves
- An amplitude part of the problem
- Calculation of partial capacities in the layers of the structure
- Calculation of the total power flux in the structure

At excitation of an electromagnetic wave in the structure with losses the total power flux P_0 brought to the active zone of the converter, is distributed into three components

A.A. Ignatiev, *Magnetoelectronics of Microwaves and Extremely High Frequencies in Ferrite Films*.
DOI: 10.1007/978-0-387-85457-1_6,

$$P_0 = P_{TL} + j(P_{los.R} + P_{los.B}),$$

where $P_{los.R}$ is the power of losses at transformation (the power of losses in the near zone of radiation), $P_{los.B}$ the reactive power (in the near zone of radiation), P_{TL} the power transferred by the wave in the transmission line on the basis of ferrite (the power in the distant zone of radiation).

For a passage device with an input and output converters the power of a signal on the output is

$$P_{out} = P_0 + j[(P_{los.R} + P_{los.B})_{en.con.} + (P_{los.R} + P_{los.B})_{out.con.}],$$

where the subscripts mean "*inp. con.* – input converter" and "*out. con.* – output converter".

The transmission factor by power in such a device is

$$K_{car}(dB) = 10\ \lg\left\{1 + j\left[\left(\frac{P_{los.R}}{P_0} + \frac{P_{los.B}}{P_0}\right)_{inp.con.} + \left(\frac{P_{los.R}}{P_0} + \frac{P_{los.B}}{P_0}\right)_{out.con.}\right]\right\}.$$

The following values can be experimentally measured:

- Transmission factor $K_{los.} = \frac{P_{out.}}{P_{inp.}}$ considering transformation losses in the input and output elements (converters) and distribution losses in the transmission line
- The modulus of the reflection coefficient of power from the converter

$$|\mathrm{G}| = K_{car.} = \frac{P_0 - P_{tl}}{P_0}$$

- The phase of reflection coefficient $\varphi = arctg\frac{P_{los.B}}{P_{los.R}}$ considering the reactive power of the converter

Let's analyze some properties of MSL at a tangential field ($\bar{H}_0 \| 0X$) and a normal field $\bar{H}_0 \| 0X$ of magnetization of a ferrite film.

5.1.1 MSL with a Tangentially Magnetized Layered Ferrite Film Structure

The strip properties of converters are mainly determined by the width of the microstrip line (the aerial of electromagnetic waves radiated into a layered structure) – W_{MSL}.

Begin our analysis, as well as at consideration of dispersive properties in Chapter 3, from a hypothetical case of electromagnetic wave excitation in bilaterally-metallized ferrite-dielectric structures with their layers' penetrability $\varepsilon_{1,2,3} = 1$. In Fig. 5.1 dependencies of the loss factors on transformation of active power $K_{los.R} = \frac{P_{los.R}(\nu)}{P_0(\nu)}$ in the near zone of MSL with a varying width W_{MSL} are shown: 1–10^{-5}m, 2–10^{-4}m, 3–10^{-3}m at $\nu_{\mathrm{H}} = 3 \cdot 10^{10}$Gz, $\alpha_{\|} = 10^{-4}$, $h_1 = h_3 = 5 \cdot 10^{-4}$m, $h_2 = 25 \cdot 10^{-6}$m, $4\pi M_S = 0.176$T. In the field of frequencies $\nu < \nu_{\perp}$ the

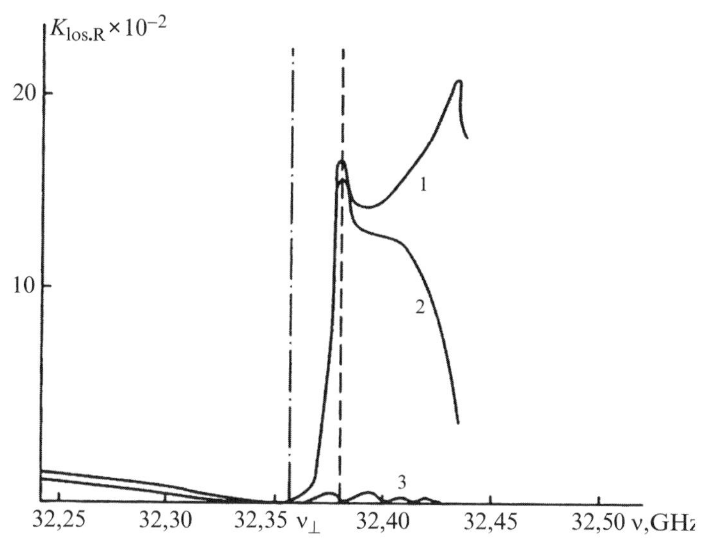

Fig. 5.1 Dependencies of the loss factors on transformation of active power $K_{los.R} = \frac{P_{los.R}(\nu)}{P_0(\nu)}$ in the near zone of MSL with a varying width W_{MSL} are shown: 1–10^{-5}m, 2–10^{-4}m, 3–10^{-3}m at $\nu_H = 3 \cdot 10^{10}$Hz, $\alpha_\| = 10^{-4}$, $h_1 = h_3 = 5 \cdot 10^{-4}$m

value of $K_{los.R}$ changes weakly and practically does not depend on the width of the microstrip W_{MSL}. At a frequency $\nu \to \nu_\perp$ from below the dependence $K_{los.R}$ monotonically decreases. In the field of frequency $\nu_\perp \leq \nu \leq \nu_1$ the value of $K_{los.R}$ sharply increases, and the course of the dependencies is practically independent of $W_{MSL} \cong (10\text{–}100) \times 10^{-6}$m. For frequencies $\nu > \nu_1$ the course of the dependencies $K_{los.R}$ is different. At $W_{MSL} = 10^{-3}$m the character of the dependence $K_{los.R}$ is pulsing, and its zeros are determined by the function $\frac{\sin \kappa'_y W_{MSL}/2}{\kappa'_y W_{MSL}/2}$ describing the basic harmonic of decomposition of the extraneous electric current j_E uniformly distributed over the microstrip, and the zeros of this function are $\kappa'_y(\nu) = \frac{2n\pi}{W}$, $n = 0,\ 1,\ 2, \ldots$. For $W_{MSL} = 10^{-3}$m, $\kappa'_y(\nu) = 2n\pi \cdot 10^3 \text{m}^{-1}$. The value of W_{MSL} renders the strongest influence on the course of the dependence $K_{los.R}$ in a frequency area $\nu > \nu_\perp$, that is related to changes of the propagation conditions of LE_t wave in the beyond-cutoff waveguide.

In Fig. 5.2 dependencies of the conversion coefficient of reactive radiation power $K_{los.B} = \frac{P_{los.B}}{P_0}$ in the near zone of the analyzed MSL with a varying width W_{MSL} are presented: 1–10^{-5}m, 2–10^{-4}m, 3–10^{-3}m. As well as above (Fig. 5.1), in the field of frequencies $\nu < \nu_\perp$ the course of the dependence $K_{los.B}$ weakly depends on the value of W_{MSL}. At $\nu \to \nu_\perp$ from below the courses of the dependencies $K_{los.B}$ for $W_{MSL} = 10^{-5}$m (1) and $W_{MSL} = 10^{-4}$m (2) differ, that is due to a change of the dispersive characteristics of LM_t and LE_t waves. On frequencies $\nu \approx \nu_\perp$ the value of $K_{los.B}$ is negligible, and on $\nu \geq \nu_\perp$ it reaches the minimum value for $W_{MSL} \cong (10\text{–}100) \cdot 10^{-6}$m in a narrow band. For $W_{MSL} \cong 10^{-3}$m the dependence $K_{los.B}$ has a pulsing character, and its zeros correspond to those of $K_{los.R}$ (Fig. 5.1).

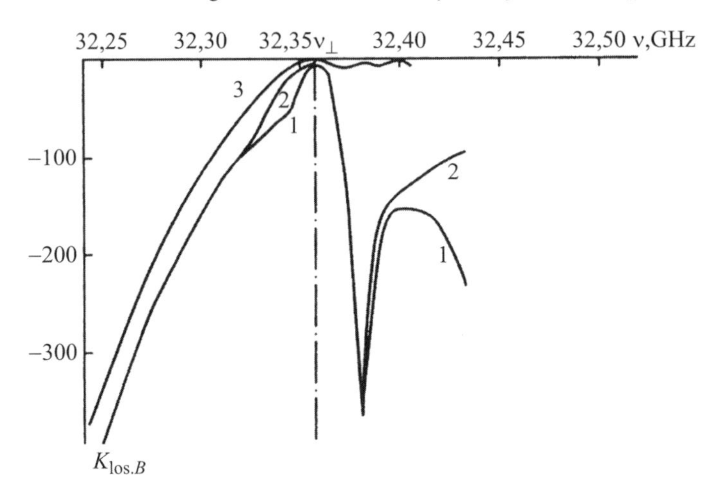

Fig. 5.2 Dependencies of the conversion coefficient of reactive radiation power $K_{los.B} = \frac{P_{los.B}}{P_0}$ in the near zone of the analyzed MSL with a varying width W_{MSL} are presented: 1–10^{-5}m, 2–10^{-4}m, 3–10^{-3}m

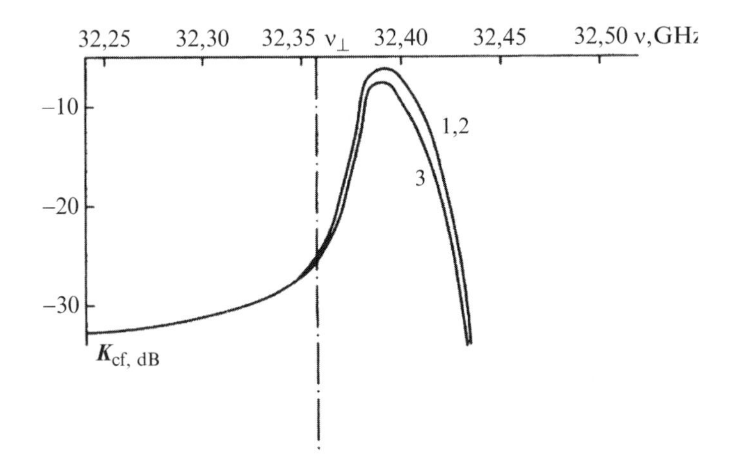

Fig. 5.3 Theoretical dependencies of the transmission factor by power of a wave in the structure excited with MSL in the distant zone, K_{cf}, for various W_{MSL} are presented: 1–10^{-5}m, 2–10^{-4}m, 3–10^{-3}m at a distance $L = 10^{-7}$m from the longitudinal axis of the converter

On Fig. 5.3 theoretical dependencies of the transmission factor by power of a wave in the structure excited with MSL in the distant zone, K_{cf}, for various W_{MSL} are presented: 1–10^{-5}m, 2–10^{-4}m, 3–10^{-3}m at a distance $L = 10^{-7}$m from the longitudinal axis of the converter. The character of this AFC is asymmetric. At changing W_{MSL} by three orders of magnitude the course of K_{cf} practically did not change. The central frequency ν_1^* of the AFC envelope determines detuing with respect to $\nu_\perp$ and corresponds to the minimum value of the amplitude wave constant κ_y'' in the beyond-cutoff mode of a flat waveguide. The weak dependence of the passband of the TL $\Delta\nu_{3dB}$ on the width of MSL allows waveguide-beyond-cutoff devices to be used for control of the parameters of magnetic films in the centimeter and millimeter ranges. Attenuation of a signal outside of the passband in beyond-cutoff waveguides with a FDLS can be increased by means of reduction of the gap size h_1 and h_3, the lengths of the transmission line L, that allows reconstructable filters with a raised blocking level to be designed in the millimeter range. The central frequency ν_1, for which $(K_{los.R})_{\max}$, and a wave range $\Delta\nu_{3dB}$ can be measured in a mode of signal reflection, that allows such devices to be used for purposes of diagnostics of the dissipative and magnetic parameters of ferrite films and determination of the resonant frequencies of FDLS, the corresponding internal magnetic fields, and the quantities determining them (Chapter 6).

Our account of the real values of the dielectric penetrability of FDLS layers has yielded the following results. In Fig. 5.4 dependencies $K_{los.R}$ in a MSL containing

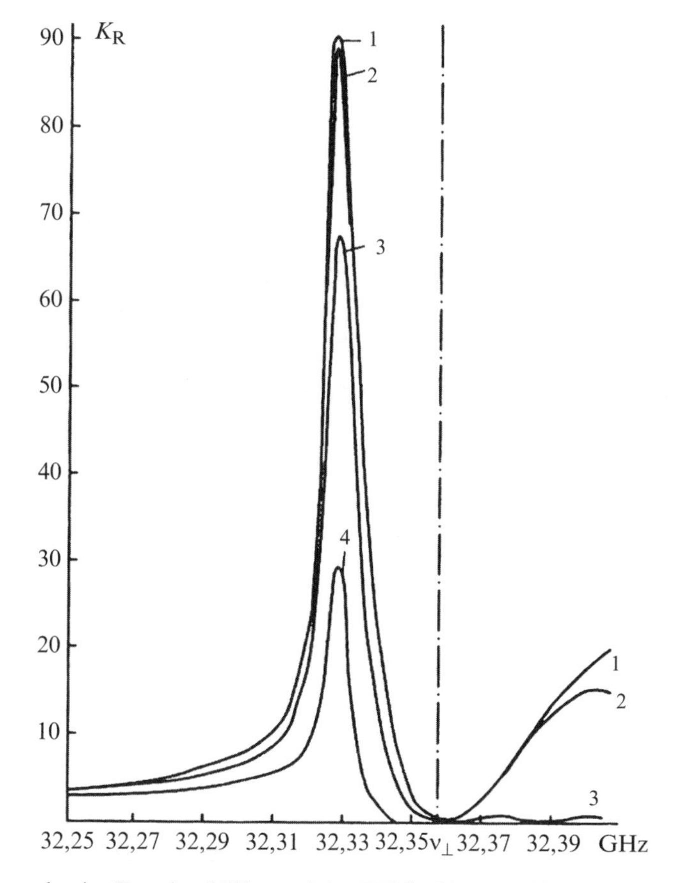

Fig. 5.4 Dependencies $K_{los.R}$ in a MSL containing FDLS with $\varepsilon_{1,2} = 14$, $\varepsilon_3 = 1$, for various W_{MSL} are presented: 1–10^{-5}m, 2–10^{-4}m, 3–$5\cdot10^{-4}$m, 4–10^{-3}m at $\nu_H = 3\cdot10^{10}$Hz, $\alpha_\parallel = 10^{-4}$, $h_1 = h_3 = 5\cdot10^{-4}$m, $h_2 = 25\cdot10^{-6}$m, $4\pi M_S = 0.176$T

FDLS with $\varepsilon_{1,2} = 14$, $\varepsilon_3 = 1$, for various W_{MSL} are presented: 1–10^{-5} m, 2–10^{-4}m, 3–$5 \cdot 10^{-4}$m, 4–10^{-3} m at $\nu_{\mathrm{H}} = 3 \cdot 10^{10}$Gz, $\alpha_{||} = 10^{-4}$, $h_1 = h_3 = 5 \cdot 10^{-4}$m, $h_2 = 25 \cdot 10^{-6}$m, $4\pi M_S = 0.176$T. The main maximum $K_{los.R}$ is displaced to a frequency range $\nu < \nu_\perp$, and the correction frequency with respect to $\nu_\perp$ is equal to $\Delta\nu = -28 \cdot 10^6$ Hz (for $\varepsilon_{1,2,3} = 1$, in Fig. 5.1, $\Delta\nu = +22 \cdot 10^6$ Hz). Side maxima $K_{los.R}$ are clearly seen. The value of W_{MSL} renders the most essential influence on the values of the major and side maxima $K_{los.R}$. In Fig. 5.5 dependencies of the major (5) and side (6) maxima on W_{MSL} are shown. In Fig. 5.6 the dependence of tuning–out of the first high-frequency side maximum with respect to $\nu_\perp$ on W_{MSL} of the analyzed MSL is presented.

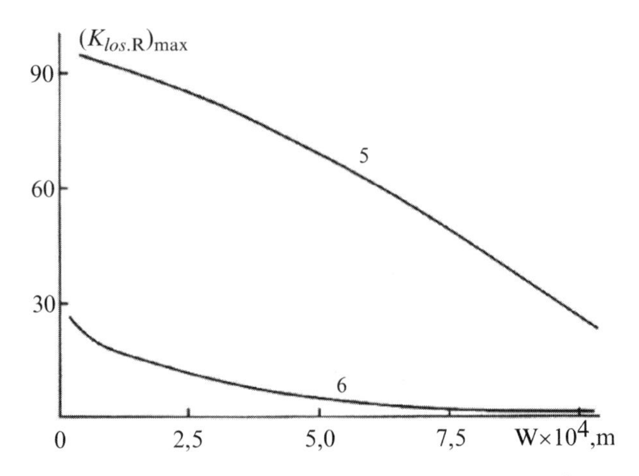

Fig. 5.5 Dependencies of the major (5) and side (6) maxima on W_{MSL} are shown

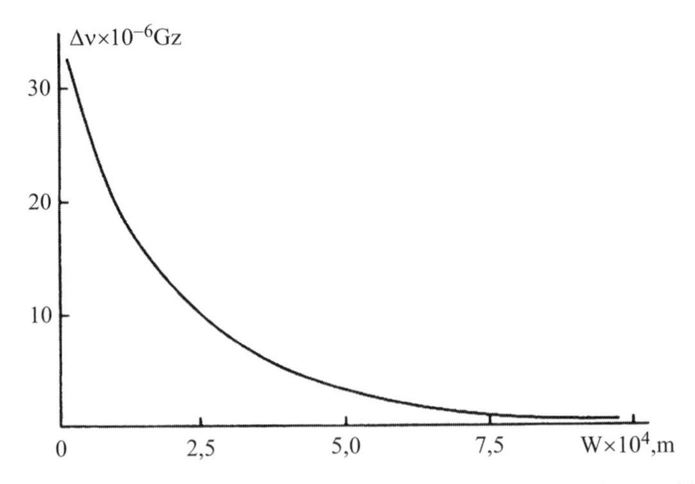

Fig. 5.6 The dependence of tuning-out of the first high-frequency side maximum with respect to $\nu_\perp$ on W_{MSL} of the analyzed MSL is presented

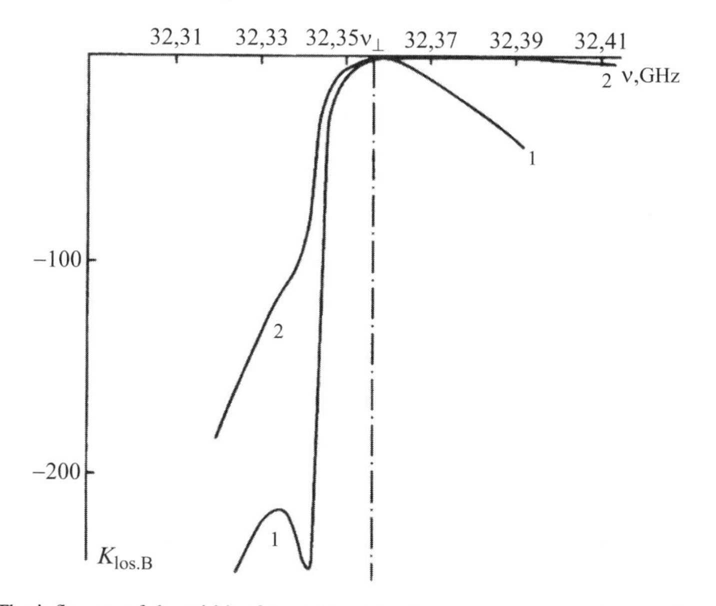

Fig. 5.7 The influence of the width of the microstrip W_{MSL} on the course of $K_{los.B}$ for a structure with $\varepsilon_{1,2} = 14, \varepsilon_3 = 1$: $1 - W_{MSL} = 10^{-5}$ m, $2 - W_{MSL} = 10^{-3}$ m

Figure 5.7 illustrates the influence of the width of the microstrip W_{MSL} on the course of $K_{los.B}$ for a structure with $\varepsilon_{1,2} = 14$, $\varepsilon_3 = 1$: $1 - W_{MSL} = 10^{-5}$m, $2 - W_{MSL} = 10^{-3}$m. In comparison with a hypothetical case of $\varepsilon_{1,2,3} = 1$ (Fig. 5.2), the account of dielectric permeability $\varepsilon_{1,2} = 14$ leads to a deformation of the course of $K_{los.B}$:

- In the field of frequencies $\nu < \nu_2 < \nu_\perp$ sharp reduction of $K_{los.B}$ is observed.
- The selective reduction is displaced from the frequency range $\nu_1 > \nu_\perp$ to $\nu_2 < \nu_\perp$, and a sharp reduction of $|K_{los.B}(\nu_2)|_{max}$ is observed for $W_{MSL} \cong 10^{-5}$ m while for wider microstrips it is not shown at all.
- In the field of frequencies $\nu > \nu_\perp$ an increase of $K_{los.B}$ is observed, and for wide microstrips ($W_{MSL} = 10^{-3}$m) $K_{los.B} \to 0$.

Figure 5.8 shows the dependencies of K_{cf} for FDLS with a length $L = 10^{-2}$m at W_{MSL} : 1–10^{-5}m, 2–10^{-3}m. The central frequency ν_2^* corresponding to the minimum of introduced losses, unlike the previous case (Fig. 5.3), was displaced to a range below $\nu_\perp$, but does not coincide with the central frequency ν_2, for which $|K_{los.R}(\nu_2)|_{max}$ and $|K_{los.B}(\nu_2)|_{max}$. This speaks for the selective processes of electromagnetic wave excitation in the near and far zones of radiation being upset by frequencies, that is due to the change of the distances between the conducting coverings of MSL. In the near zone there are two half-space, one is above the exciting microstrip of a height $h_2 + h_3$, and the other one is under it, of a height h_1. In the distant zone the distance between the external metal coverings is $h_1 + h_2 + h_3$. At

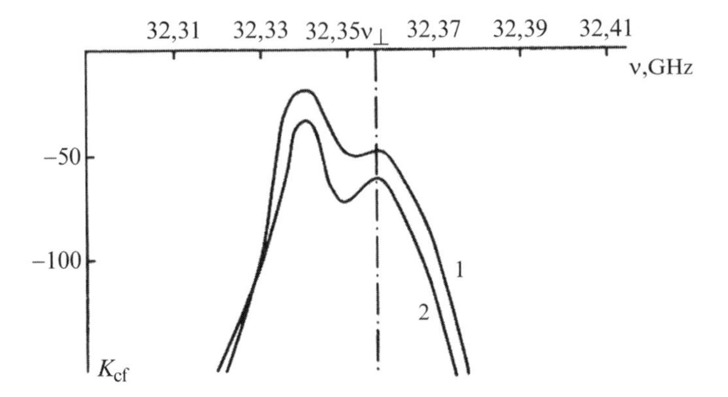

Fig. 5.8 The dependencies of K_{cf} for FDLS with a length $L = 10^{-2}$m at W_{MSL}: 1–10^{-5} m, 2–10^{-3} m

increasing W_{MSL} the passband of the converter in the near zone (Fig. 5.4) decreases, that is related to the increase in the uniformity of exciting current, and leads to a reduction of the band of frequencies of K_{cf} in the distant zone. The difference between the passband of the converter in both near and far zones decreases with increasing W_{MSL} (cf. curves 1 and 4 in Fig. 5.4 with curves 1 and 2 in Fig. 5.8).

Let's consider the properties of MSL when the metal screens are arranged far enough from the ferrite layer and the structure is in its prelimit mode $(\nu \gg \nu_{cr})$.

In Figs. 5.9 and 5.10 dependencies of $K_{los.R}(\nu)$ are presented and $K_{los.B}(\nu)$ at wave excitation in a flat waveguide with a tangentially magnetized ferrite-dielectric structure at various W_{MSL}: $1-(10 \div 100)\cdot 10^{-6}$m, 2–$10^{-3}$ m for parameters $\nu_{\mathrm{H}} = 3\cdot 10^{10}$Hz, $\alpha_{\|} = 10^{-4}$, $h_1 = h_3 = 3.6\cdot 10^{-3}$m, $h_2 = 25\cdot 10^{-6}$m, $\varepsilon_{1,2,3} = 1$, $4\pi M_S = 0.176$T.

The selective reduction of $K_{los.R}(\nu_{\perp})$ and the increase of $K_{los.B}(\nu_{\perp})$ at the frequency $\nu_{\perp}$, and the course of these dependencies at $\nu > \nu_{\perp}$ and $\nu < \nu_{\perp}$ practically does not depend on the width W_{MSL} of the microstrip. Analysis has shown that the basic influence on the passband and value of $K_{los.R}(\nu_{\perp})$ and $K_{los.B}(\nu_{\perp})$ is rendered by ferromagnetic losses in the ferrite film. In Fig. 5.11 the dependence $K_{cf}(\nu)$ in the analyzed structure at a distance $L = 10^{-2}$m from the longitudinal axis of the converter is shown. The selective attenuation of a signal on frequency $\nu_{\perp}$ in the prelimit mode $(\nu \gg \nu_{cr})$ of a flat waveguide can be used for design of sensors of resonant frequencies (fields) and for diagnostics of ferrite film structures in the millimeter range. Note that the passband of MSL in the near zone (Fig. 5.9) exceeds approx. twice the band of rejection at the same level in the distant zone (Fig. 5.10), that should be considered at design of rejection devices for filtration of both reflective and transmission types and at development of diagnostic methods of ferromagnetic losses in ferrite film structures.

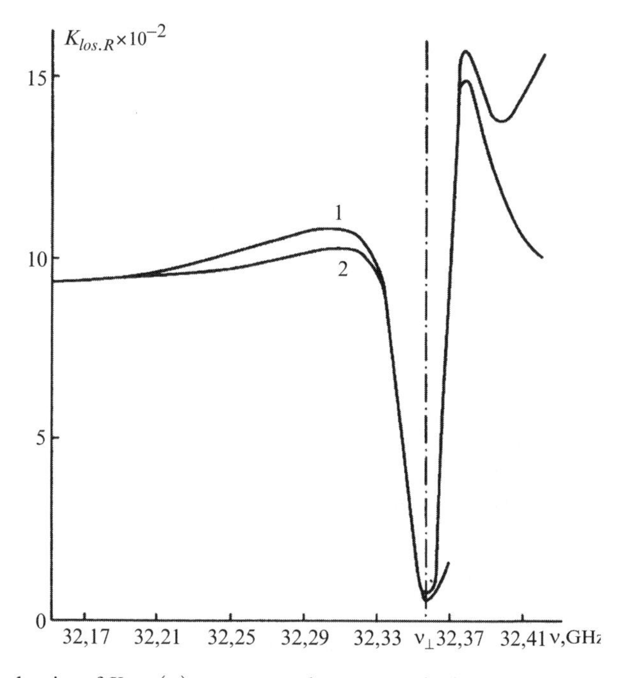

Fig. 5.9 Dependencies of $K_{los.R}(\nu)$ are presented at wave excitation in a flat waveguide with a tangentially magnetized ferrite-dielectric structure at various W_{MSL}: $1-(10\div100)\cdot10^{-6}\,\text{m}$, $2\text{–}10^{-3}\,\text{m}$ for parameters $\nu_{\text{H}} = 3\cdot10^{10}\,\text{Hz}, \alpha_{\|} = 10^{-4}$, $h_1 = h_3 = 3,\ 6\cdot10^{-3}\,\text{m}$, $h_2 = 25\cdot10^{-6}\,\text{m}$, $\varepsilon_{1,2,3} = 1$, $4\pi M_S = 0.176\,\text{T}$

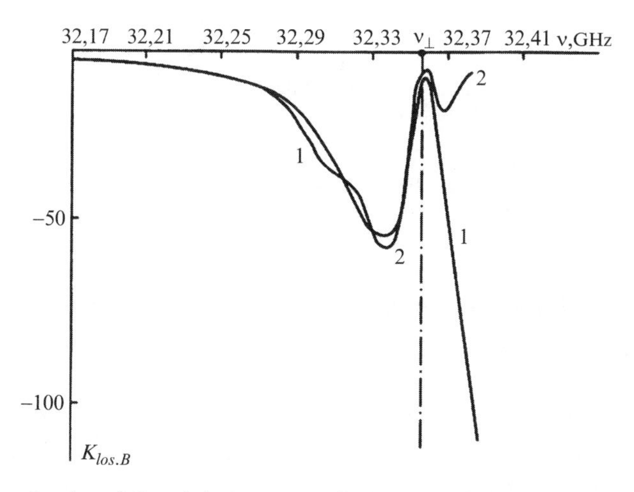

Fig. 5.10 Dependencies of $K_{los.B}(\nu)$ are presented at wave excitation in a flat waveguide with a tangentially magnetized ferrite-dielectric structure at various W_{MSL}: $1-(10\div100)\cdot10^{-6}\,\text{m}$, $2-10^{-3}\,\text{m}$ for parameters $\nu_{\text{H}} = 3\cdot10^{10}\,\text{Hz}, \alpha_{\|} = 10^{-4}$, $h_1 = h_3 = 3,\ 6\cdot10^{-3}\,\text{m}$, $h_2 = 25\cdot10^{-6}\,\text{m}$, $\varepsilon_{1,2,3} = 1$, $4\pi M_S = 0.176\,\text{T}$

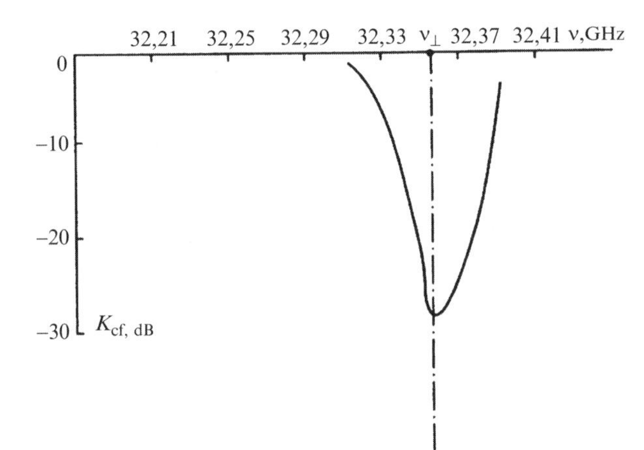

Fig. 5.11 The dependence $K_{cf}(\nu)$ in the analyzed structure at a distance $L = 10^{-2}$ m from the longitudinal axis of the converter is shown

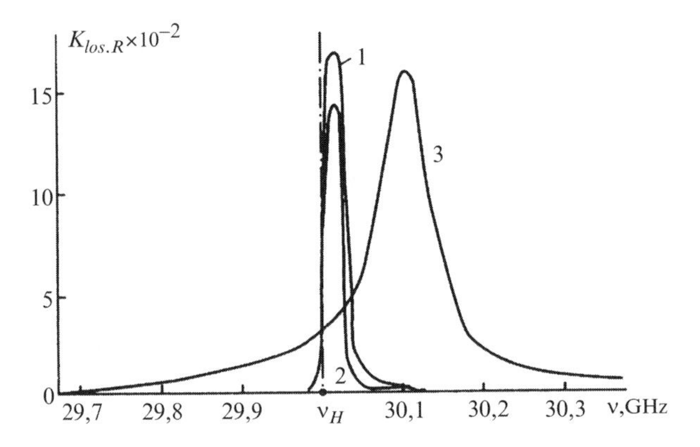

Fig. 5.12 The dependencies $K_{los.R}$ for MSL at normal magnetization with various W_{MSL} and h_2: 1 – $W_{MSL} = 10^{-5}$ m, $h_2 = 10^{-5}$ m, 2 – $W_{MSL} = 10^{-3}$ m, $h_2 = 10^{-5}$ m, 3 – $W_{MSL} = 10^{-5}$ m, $h_2 = 5 \cdot 10^{-5}$ m at $\nu_H = 3 \cdot 10^{10}$ Hz, $\alpha_\perp = 10^{-4}$, $h_1 = h_3 = 5 \cdot 10^{-4}$ m, $\varepsilon_{1,2,3} = 1$, $4\pi M_S = 0.176$ T

5.1.2 MSL with Ferrite-Dielectric Structure at Normal Magnetization

In Figs. 5.12 and 5.13 the dependencies $K_{los.R}$ and $K_{los.B}$ for MSL at normal magnetization with various W_{MSL} and h_2: 1 – $W_{MSL} = 10^{-5}$ m, $h_2 = 10^{-5}$ m, 2 – $W_{MSL} = 10^{-3}$ m, $h_2 = 10^{-5}$ m, 3 – $W_{MSL} = 10^{-5}$ m, $h_2 = 5 \cdot 10^{-5}$ m at $\nu_H = 3 \cdot 10^{10}$ Hz, $\alpha_\perp = 10^{-4}$, $h_1 = h_3 = 5 \cdot 10^{-4}$ m, $\varepsilon_{1,2,3} = 1, 4\pi M_S = 0.176$ T are presented. At increasing thickness of the ferrite film h_2 the passband of MSL in the near zone of radiation extends and the central frequency ν_{H0} is displaced towards higher frequencies with respect to ν_H. $K_{los.B}(\nu)$ behave similarly with a selective reduction at frequencies $\nu_{H0} > \nu_H$.

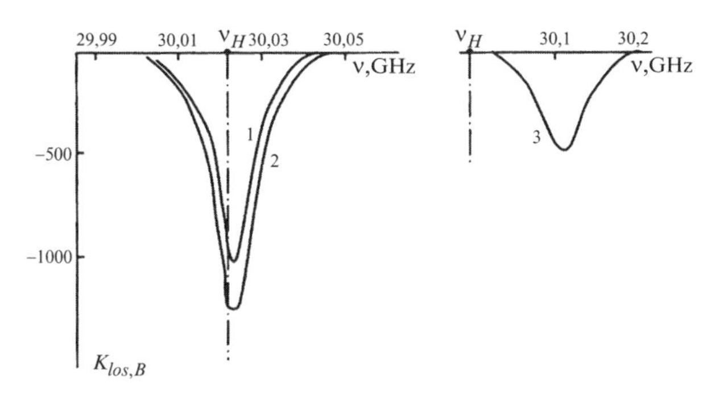

Fig. 5.13 The dependencies $K_{los.B}$ for MSL at normal magnetization with various W_{MSL} and h_2: 1 – $W_{MSL} = 10^{-5}$ m, $h_2 = 10^{-5}$ m, 2 – $W_{MSL} = 10^{-3}$ m, $h_2 = 10^{-5}$ m, 3 – $W_{MSL} = 10^{-5}$ m, $h_2 = 5 \cdot 10^{-5}$ m at $\nu_{\mathrm{H}} = 3 \cdot 10^{10}$Hz, $\alpha_{\perp} = 10^{-4}$, $h_1 = h_3 = 5 \cdot 10^{-4}$m, $\varepsilon_{1,2,3} = 1$, $4\pi M_S = 0.176$T

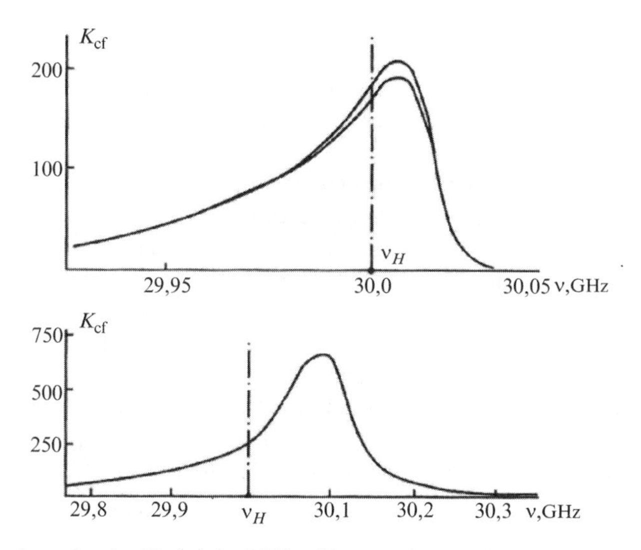

Fig. 5.14 The dependencies $K_{cf}(\nu)$ for MSL with a varying W_{MSL} and the thickness of the ferrite film h_2 are shown: 1 – $W_{MSL} = (1 \div 100) \cdot 10^{-6}$m, $h_2 = 10^{-5}$ m, 2 – $W_{MSL} = 10^{-3}$ m, $h_2 = 10^{-5}$ m, 3 – $W_{MSL} = 10^{-5}$ m, $h_2 = 5 \cdot 10^{-5}$ m for $L = 10^{-2}$ m and the parameters specified above

In Fig. 5.14 the dependencies $K_{cf}(\nu)$ for MSL with a varying W_{MSL} and the thickness of the ferrite film h_2 are shown: 1 – $W_{MSL} = (1 \div 100) \cdot 10^{-6}$m, $h_2 = 10^{-5}$m, 2 – $W_{MSL} = 10^{-3}$m, $h_2 = 10^{-5}$m, 3 – $W_{MSL} = 10^{-5}$m, $h_2 = 5 \cdot 10^{-5}$m for $L = 10^{-2}$m and the parameters specified above. We note the following features. The passband $\Delta\nu_{3dB}$ for MSL with $W_{MSL} = (10 \div 500) \cdot 10^{-6}$m does not depend on the width of the microstrip and practically coincides with that for $K_{cf}(\nu)$. For $W_{MSL} \cong 10^{-3}$m the passband by K_{cf} decreases by 10%, and for $K_{los.R}(\nu)$ (curve 3, Fig. 5.12) (thick ferrite films) it increases. Earlier such features for the band properties of MSL with normally magnetized structures have not been discussed in the literature.

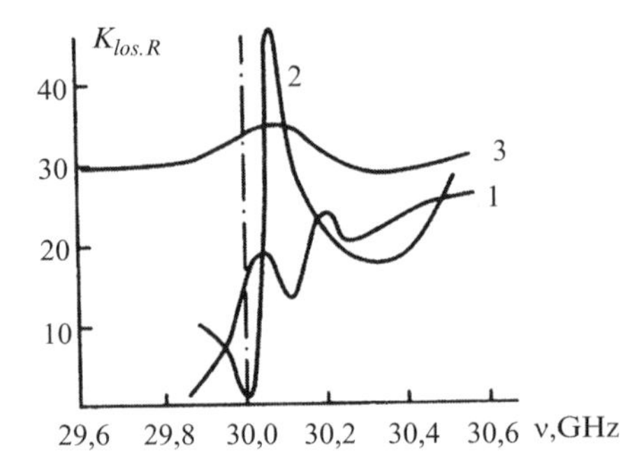

Fig. 5.15 The dependencies $K_{los.R}(\nu)$ in MSL with a normally-magnetized structure are presented: 1 – a fast LM_n^F wave in the structure with parameters $\varepsilon_{1,2} = 14$, $\varepsilon_3 = 1$; 2 – a slow LM_n^S wave, $\varepsilon_{1,2,3} = 1$; 3 – slow LM_n wave, $\varepsilon_{1,2} = 14$, $\varepsilon_3 = 1$ at $\nu_H = 3 \cdot 10^{10}$Hz, $\alpha_\perp = 10^{-4}$, $h_1 = h_3 = 5 \cdot 10^{-4}$m, $h_2 = 5 \cdot 10^{-5}$m, $W_{MSL} = 10^{-5}$ m, $4\pi M_S = 0.176$T

In layered structures on the basis of thick ferrite films ($h_2 > (50 \div 100) \cdot 10^{-6}$m) at normal magnetization of a bilaterally-metallized layered structure the role of dielectric modes for LM waves ($LM_n^{F_1}$ and $LM_n^{F_2}$ waves) near the resonant frequencies enhances. In Fig. 5.15 the dependencies $K_{los.R}(\nu)$ in MSL with a normally-magnetized structure are presented: 1 – a fast LM_n^F wave in the structure with parameters $\varepsilon_{1,2} = 14$, $\varepsilon_3 = 1$; 2 – a slow LM_n^S wave, $\varepsilon_{1,2,3} = 1$; 3 – slow LM_n wave, $\varepsilon_{1,2} = 14$, $\varepsilon_3 = 1$ at $\nu_H = 3 \cdot 10^{10}$Hz, $\alpha_\perp = 10^{-4}$, $h_1 = h_3 = 5 \cdot 10^{-4}$m, $h_2 = 5 \cdot 10^{-5}$m, $W_{MSL} = 10^{-5}$m, $4\pi M_S = 0.176$T. For the LM_n^S wave on frequency $\nu > \nu_H$ a selective increase of $K_{los.R}(\nu)$ takes place, and at increasing the dielectric permeability of the base of MSL (ε_1) and the ferrite film (ε_2) its maximum value decreases, and the average level on the tuned frequencies sharply increases. For a LM_n^F wave in the field of frequencies $\nu > \nu_H$ there is a double "splash" of $K_{los.R}(\nu)$. At tuning out from ν_H towards higher frequencies the value of $(K_{los.})_{LM_n^F}$ increases and tends to a constant value at tuning out by $\frac{\Delta\nu}{\nu_H} > 2\%$. Near the resonant frequency ν_H the losses for active power transformation in the near zone of MSL are essentially higher in the field of frequencies $\nu < \nu_H$ for LM_n^F waves, and in the field of frequencies $\nu > \nu_H (K_{los.R})_{LM_n^F} \cong (K_{los.R})_{LM_n^S}$.

In Fig. 5.16 theoretical dependencies of losses for reactive power transformation in the near zone of MSL $K_{los.B}(\nu)$ are presented: 1 – a fast LM_n^F wave with $\varepsilon_{1,2} = 14$, $\varepsilon_3 = 1$; 2 – a slow LM_n^S wave with $\varepsilon_{1,2,3} = 1$; 3 – a slow LM_n wave in the structure with $\varepsilon_{1,2} = 14$, $\varepsilon_3 = 1$ and the other parameters specified above. For a LM_n^F wave in the field of the frequency corresponding to the selective reduction of $K_{los.B}$ (Fig. 5.15) we have a "splash" in $K_{los.B}(\nu)$ which decreases with growing penetrability $\varepsilon_{1,2}$.

In Figs. 5.17–5.19 the dependencies $K_{cf}(\nu)$ in the distant zone of MSL at a distance $L = 10^{-2}$m from the longitudinal axis of the converter are depicted, where Fig. 5.17 – a LM_n^F wave with $\varepsilon_{1,2} = 14$, $\varepsilon_3 = 1$; Fig. 5.18 – a LM_n^S wave with $\varepsilon_{1,2,3} = 1$; Fig. 5.19 – a LM_n^S wave, $\varepsilon_{1,2} = 14$, $\varepsilon_3 = 1$. In real structures the value of K_{cf} has, as calculation shows, a weak maximum on frequency ν_H for a LM_n^F

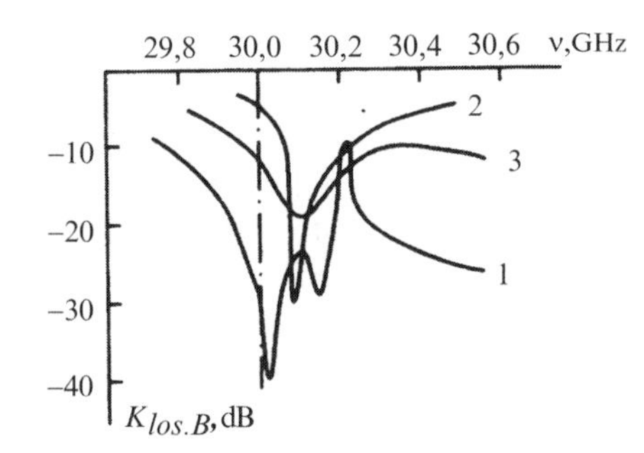

Fig. 5.16 Theoretical dependencies of losses for reactive power transformation in the near zone of MSL $K_{los.B}(\nu)$ are presented: 1 – a fast LM_n^F wave with $\varepsilon_{1,2} = 14,\ \varepsilon_3 = 1$; 2 – a slow LM_n^S wave with $\varepsilon_{1,2,3} = 1$; 3 – a slow LM_n wave in the structure with $\varepsilon_{1,2} = 14,\ \varepsilon_3 = 1$

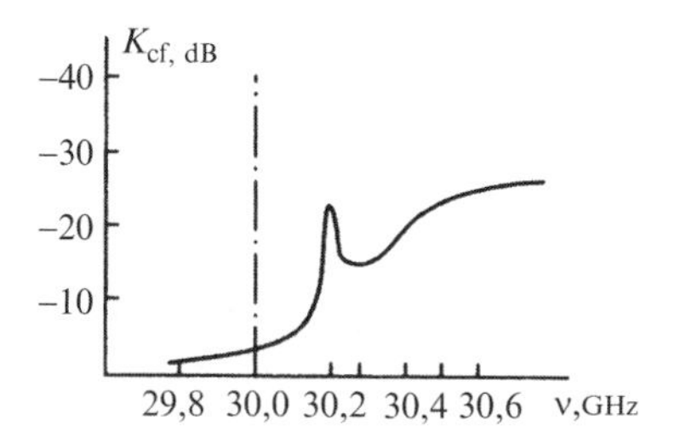

Fig. 5.17 The dependencies $K_{cf}(\nu)$ in the distant zone of MSL at a distance $L = 10^{-2}$m from the longitudinal axis of the converter are depicted, where a LM_n^F waves with $\varepsilon_{1,2} = 14,\ \varepsilon_3 = 1$

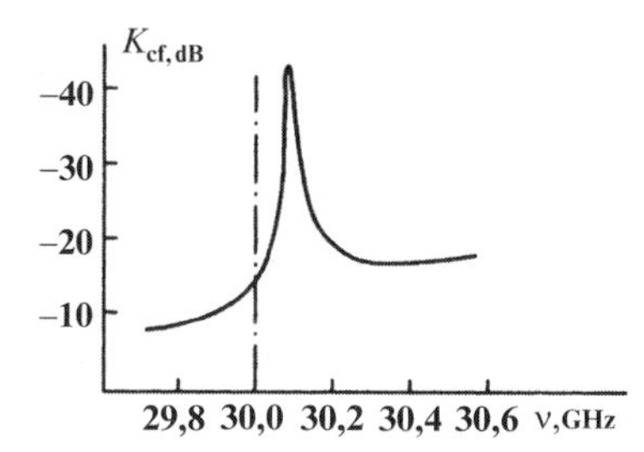

Fig. 5.18 The dependencies K_{cf} (ν) in the distant zone of MSL at a distance $L = 10^{-2}$m from the longitudinal axis of the converter are depicted for LM_n^S wave with $\varepsilon_{1,2,3} = 1$

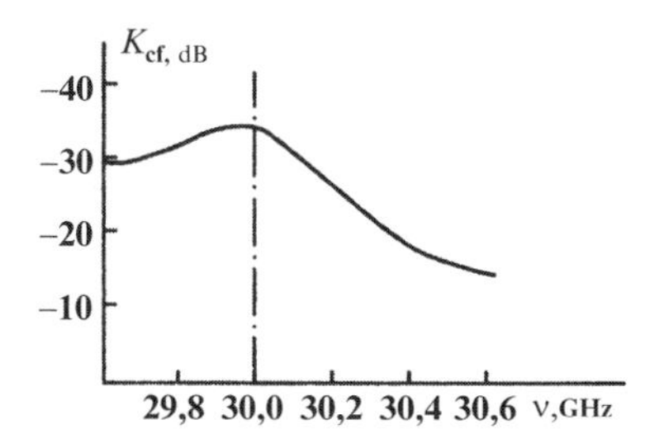

Fig. 5.19 The dependencies $K_{cf}(\nu)$ in the distant zone of MSL at a distance $L = 10^{-2}$m from the longitudinal axis of the converter are depicted for LM_n^S wave, $\varepsilon_{1,2} = 14, \varepsilon_3 = 1$

wave, and the dielectric mode prevails on frequencies tuned out above ν_{H} by more than 2%.

For the first time the properties of converters with normal and tangential magnetization of FDLS with losses for *LE* and *LM* waves were investigated. An essential exit of the spectrum of excited waves, in both near and distant zones of radiation was found, which builds up at advance into the short-wave part of the centimeter range and into the millimeter range of radiowaves. A complex character of the dependencies of AFC of excited waves near the resonant frequencies ν_{H} and $\nu_{\perp}$ of FDLS is shown, tuning-up frequencies for structures with losses in the prelimit and beyond-cutoff modes have been determined. Losses by transformation of active and reactive power in the near zone and transmission factor in the distant zone of converters have been calculated.

5.2 Experimental Research of Waveguide Converter Characteristics

For research of excitation, reception, and propagation of various types of waves in layered structures on the basis of ferrite plates and films the following waveguide devices were used:

- A beyond-cutoff waveguide with a ferrite-dielectric layered structure (FDLS), having parts of the ferrite protruding from the beyond-cutoff part into the bringing and tailrace waveguides (Fig. 4.1*c*, *e*)
- A rectangular waveguide with FDLS, containing an absorbing covering on some part of the ferrite (Fig. 4.1*d*)

These devices, WC, contain a part of the ferrite layer or of the ferrite-dielectric structure, being at the maximum of HF magnetic fields of excitation and coordinating water the fields of the bringing waveguide with those in the structure on input and the fields in the transmission line which can be metallized layered structures or structures with absorbing coverings and the fields in these structures with those of the tailrace waveguide on output. The free part of the plate or ferrite-dielectric structure will be called (Fig. 4.1) a FDT.

In waveguide-beyond-cutoff devices (Fig. 4.1*c*, *e*) in the plane of the shorting wall we have a slotted-guide ferrite-dielectric converter (SGC) with a layered ferrite-dielectric structure located orthogonally to the localization plane of HF fields of excitation.

Before passing to consideration of the results of our experimental research of various types of waveguide and strip converters, we make some remarks.

The following parameters of converters were experimentally investigated:

- Dispersive characteristics $\kappa'(\nu)$ and $\kappa''(\nu)$ of excited and extending waves in FDLS
- Power conversion coefficients (the power P_0 brought to the converter is converted into the power of a wave in FDLS $P_{\parallel,\perp}$) at various orientations of the layer $K_{\parallel,\perp} = \frac{P_{\parallel,\perp}}{P_0} = 1 - \frac{(P_{ref})_{\parallel,\perp}}{P_0}$, where $(P_{ref})_{\parallel,\perp}$ is the level of reflected power on the absorption frequency, $P_{\parallel,\perp} = P_0 - (P_{ref})_{\parallel,\perp}$, $K_{\parallel,\perp}(dB) = 10\lg\left[1 - \frac{(P_{ref})_{\parallel,\perp}}{P_0}\right]$
- Wave band $\Delta\nu_{\parallel,\perp}$ by absorption at a level $\frac{P_0 - P_{ref})_{\parallel,\perp}}{2}$
- Wave bands $\Delta\nu_{0\parallel}$, $\Delta\nu_{0,\perp}$ by level P_0

transmission factor $K_{\parallel,\perp} = (K_{\parallel,\perp})_{inp} \cdot (K_{cf})_{\parallel,\perp}(K_{\parallel,\perp})_{out}$, $K_{\parallel,\perp}(dB) = 10[\lg(K_{\parallel,\perp})_{inp} + \lg(K_{cf})_{\parallel,\perp} + \lg(K_{|,\perp})_{out}](K_{cf})_{\parallel,\perp} = \left(\frac{P_{out}}{P_{inp}}\right)$, where $(P_{inp})_{\parallel,\perp}$ is the level of power in the distant zone of the input converter; $(P_{out})_{\parallel,\perp}$ the level of power in the distant zone of the output converter, i.e. – the output FDLS power.

Let's consider the results of our research of the properties of WC and FDT in the millimeter range.

First experiments to study excitation of various types of waves, including magnetostatic ones, in layered structures on the basis of YIG monocrystal plates in the millimeter range were made on WBCC as SGC with FDT.

In Fig. 5.20 experimental dependencies of the transmission factor $K_\parallel$ of a beyond-cutoff waveguide with FDLS, having FDT (see Fig. 4.1*c*, *e*), on various frequencies of the spectrum of target signal ν_1, ν_0 and ν_2 on the size of the protruding part S are shown: $1 - K_\parallel(\nu_0)$, ν_0 is the frequency corresponding to the minimum size of introduced losses, $\nu_0 = 35.75 \cdot 10^{10}$ Hz; 2, $3 - K_\parallel(\nu_1)$ and $K_\parallel(\nu_2)$, ν_1 and ν_2 are the average frequencies from the lower and upper half passbands of AFC signal

$$\nu_1 = \frac{\nu_0 - \nu_{10}}{2}, \quad \nu_2 = \frac{\nu_{20} - \nu_0}{2},$$

where ν_{10} is the lower, ν_{20} the upper boundary frequency of the spectrum of target signal; $\nu_1 = 34.14 \cdot 10^{10}$ Hz; $\nu_2 = 38.75 \cdot 10^{10}$ Hz. YIG monocrystals with sizes $d = 5 \cdot 10^{-4}$ m, $c = 3.2 \cdot 10^{-3}$ m, $L = 7 \cdot 10^{-3}$ m, $H_0 = 1.156$ MA/m, $4\pi M_S = 0.176$ T were used. For $K_\parallel(\nu_0)$, $K_\parallel(\nu_1)$, and $K_\parallel(\nu_2)$ there are such values of S at which transformation is most effective, but $\left|K_\parallel(\nu_0)\right|_{\min}$ and $\left|K_\parallel(\nu_1,\ \nu_2)\right|_{\min}$ do not coincide, that speaks for a dependence of the passband of such a converter from S. The value $\left|K_\parallel(\nu_0)\right|_{\min}$ is observed at $S \cong 4 \cdot 10^{-4}$ m, that, subject to the dielectric permeability of ferrite ($\varepsilon = 14$), corresponds to $S \cong \lambda/4$, λ being the wavelength in ferrite. On frequencies $\nu_{1,2}$ tuned out from ν_0, more complex dependencies on S are observed. The transmission factors on the tunes-out frequencies $K_\parallel(\nu_1)$ also $K_\parallel(\nu_2)$ have rather an expressed dependence on S, and for the average value of $\overline{K_\parallel} = \frac{K_\parallel(\nu_1) + K_\parallel(\nu_2)}{2}$ the periodicity by S is also close to $\lambda/4$.

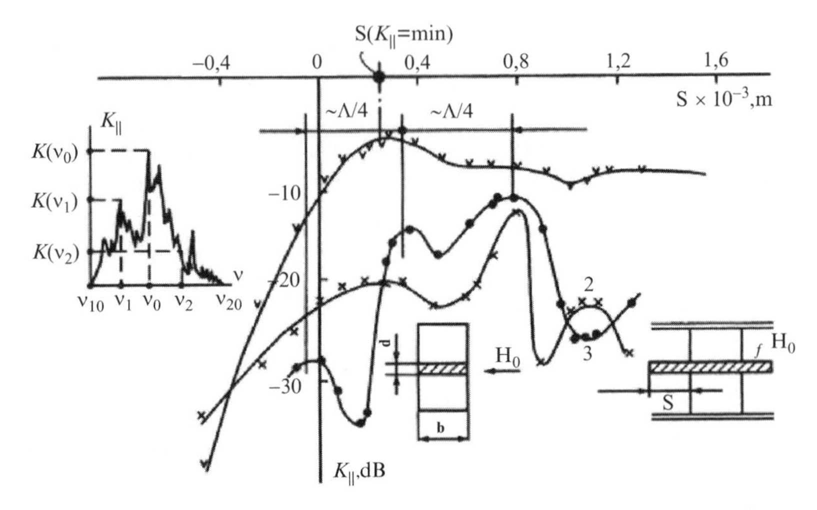

Fig. 5.20 Experimental dependencies of the transmission factor $K_{\|}$ of a beyond-cutoff waveguide with FDLS, having FDT (see Fig. 4.1*c*, *e*), on various frequencies of the spectrum of target signal ν_1, ν_0 and ν_2 on the size of the protruding part S are shown: $1 - K_{\|}(\nu_0), \nu_0$ is the frequency corresponding to the minimum size of introduced losses, $\nu_0 = 35.75 \cdot 10^{10}$ Hz; $2, 3 - K_{\|}(\nu_1)$ and $K_{\|}(\nu_2), \nu_1$ and ν_2 are the average frequencies from the lower and upper half passbands of AFC signal

Let's note that the distinction of the values of $\left|K_{\|}(\nu_0)\right|_{\min}$, $\left|K_{\|}(\nu_1)\right|_{\min}$ and $\left|K_{\|}(\nu_2)\right|_{\min}$ provides control of the passband $\Delta\nu$ of the spectrum of excited waves in such devices, that should be considered at treatment of experimental dependencies of wave dispersions.

In Fig. 5.21 experimental dependencies of the conversion coefficient of FDT on S are presented: (Fig. 5.21) for YIG films of a varying thickness: $1 - d = 9 \cdot 10^{-6}$ m; $2 - d = 22 \cdot 10^{-6}$ m, and the dependencies of the passbands at a level of $3\,\text{dB} - \Delta\nu_{3dB}$ and by the basis $-\Delta\nu_{bas}$ (Fig. 5.21): 1, $2 - d = 9 \cdot 10^{-6}$ m; 3, $4 - d = 28 \cdot 10^{-6}$ m; 5, $6 - d = 22 \cdot 10^{-6}$ m; $\Omega_M = 0.14$.

In Fig. 5.22 theoretical (curves 1, 5) and experimental (curves 2, 3, 4) dependencies of the minimum transfer losses are presented: 1, $2 - K_{\min}(\nu_0)$ on the central frequency $\nu_0 \cong \nu_{\perp}$, and the passbands: 4, $5 - \Delta\nu_{3dB}$, and the level of out-of-band attenuation: $3 - K(\nu)$ at tuning out from ν_0 by $\pm\Delta\nu \cong (3\text{–}5)\Delta\nu_{3dB}$ from the length of the layered structure with an absorbing covering. The course of the experimental dependencies $K_{\min}(\nu_0)$ and $\Delta\nu_{3dB}$ on L qualitatively agree with the theoretical ones. The average steepness is $\Delta K_{\min}/\Delta L \cong 0.5\,\text{dB/mm}$ and $\Delta K/\Delta L \cong 2\,\text{dB/mm}$.

It was experimentally revealed that in the presence of FDT at an edge of the film before the absorbing covering on input and behind the covering at the opposite edge of the film on output, the transfer losses in the device generally decrease. In this connection the length of FDT was experimentally selected

$$S = \frac{L - L_n}{2}$$

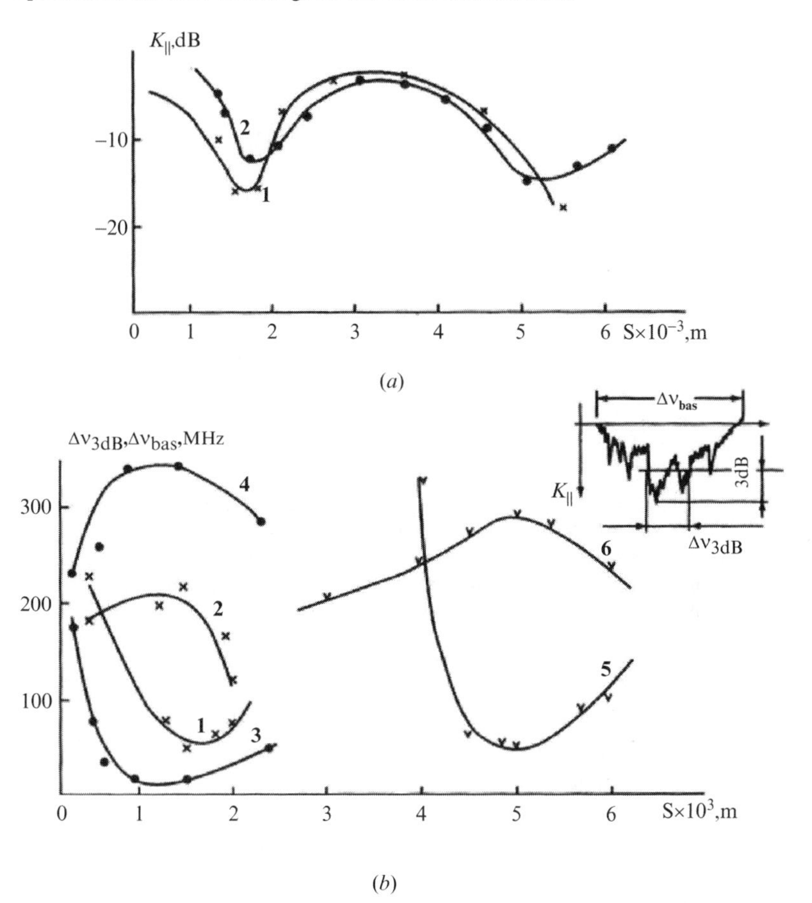

Fig. 5.21 Experimental dependencies of the conversion coefficient of FDT on S are presented: (a) for YIG films of a varying thickness: $1-d=9\cdot10^{-6}$ m; $2-d=22\cdot10^{-6}$ m, and the dependencies of the passbands at a level of 3 dB – $\Delta\nu_{3dB}$ and by the basis – $\Delta\nu_{bas}$ (b): $1,2-d=9\cdot10^{-6}$ m; $3,4-d=28\cdot10^{-6}$ m; $5,6-d=22\cdot10^{-6}$ m; $\Omega_M=0.14$

in devices with an absorbing covering. In Fig. 5.23 typical dependencies $K_{\min}(\nu_0)$ of the L_n/L ratio in a range of frequencies of 30–40 GHz for YIG films are presented for thickness: $1-d=34.5\cdot10^{-6}$m; $2-d=19.8\cdot10^{-6}$m, $4\pi M_S=0.176$T, $H_0=927.435$ kA/m. The characteristic break on the dependence $K_{\min}/(L_n/L)$ determines the length S for FDT, the steepness of the dependence $\Delta K_{\min}/\Delta(L_n/L)$ is related to ferromagnetic losses. Figure 5.24 compares the course of similar dependencies for Li–Zn spinels: $1-d=13\cdot10^{-6}$m; $2-d=20\cdot10^{-6}$m, $3-d=7\cdot10^{-6}$m, $4\pi M_S=0.35$T, $H_0=865.08$ kA/m. For spinels the curves $K/(L_n/L)$ go by above an order of magnitude, more abruptly than for YIG, in the same range of frequencies ($\nu_\perp=35$ GHz).

The presence of FDT facilitates the conditions of coordination of the bringing waveguide with the ferrite-dielectric structure.

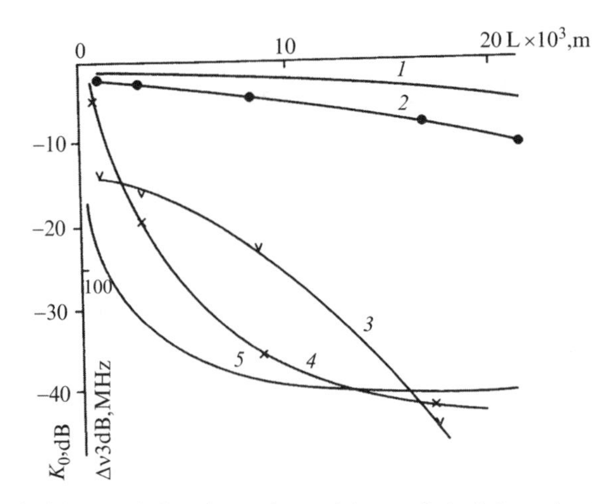

Fig. 5.22 Theoretical (curves 1, 5) and experimental (curves 2, 3, 4) dependencies of the minimum transfer losses are presented: $1,2 - K_{\min}(\nu_0)$ on the central frequency $\nu_0 \cong \nu_\perp$, and the passbands: $4,5 - \Delta\nu_{3dB}$, and the level of out-of-band attenuation: $3 - K(\nu)$ at tuning out from ν_0 by $\pm\Delta\nu \cong (3-5)\Delta\nu_{3dB}$ from the length of the layered structure with an absorbing covering

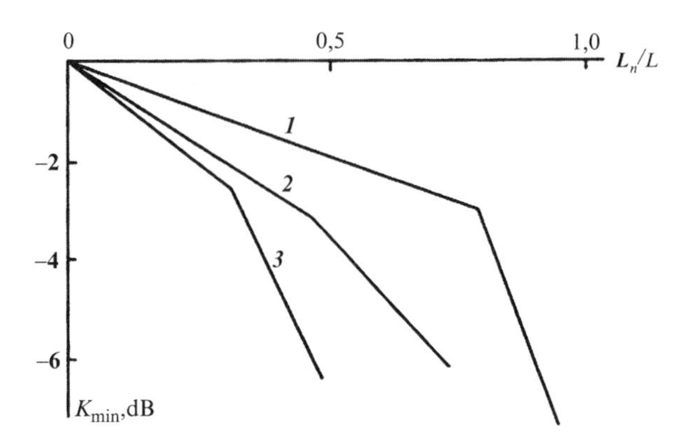

Fig. 5.23 Dependencies $K_{\min}(\nu_0)$ of the L_n/L ratio in a range of frequencies of 30–40 GHz for YIG films are presented for thickness: $1 - d = 34.5 \cdot 10^{-6}$m; $2 - d = 19.8 \cdot 10^{-6}$ m, $4\pi M_S = 0.176$ T, $H_0 = 927.435$ kA/m

Thus, the properties of waveguide-slotted converters and ferrite-dielectric matching transformers on the basis of solid ferrite plates and films in the centimeter and millimeter ranges were investigated for the first time. It is shown that waveguide converters with a ferrite films are most promising for design of electronic wavemeters, sensors of resonant frequencies of low-and-high-pass and transmitting filters at low and high levels of continuous and pulse power, diagnostic devices in the UHF and EHF ranges.

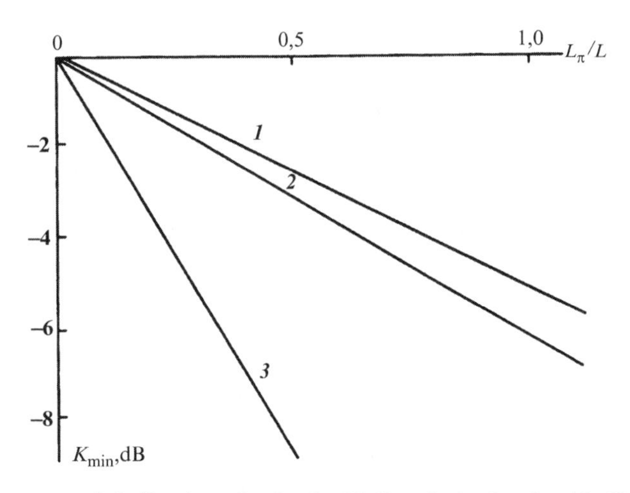

Fig. 5.24 The course of similar dependencies for Li–Zn spinels: $1-d=13\cdot10^{-6}$m; $2-d=20\cdot10^{-6}$m, $3-d=7\cdot10^{-6}$m, $4\pi M_S=0.35\,\mathrm{T}$, $H_0=865.08\,\mathrm{kA/m}$

5.3 Converters on Microstrip Lines

Unlike WC, strip converters:

- Are bi-directional, i.e. excite waves with $\kappa' > 0$ and $\kappa' < 0$ in the directions perpendicular to the longitudinal axis of a converter.
- Have an enhanced localization of WC fields of excitation and currents, and permit variation of the degree of their heterogeneity in wider limits.
- Provide variation of the characteristic resistance over a wide range at small cross-section sizes.
- Allow using of a planar technology and application in integrated circuits of various levels.

Strip converters, before our research, had not been applied for excitation and reception of waves in layered structures on the basis of ferrite films in the short-wave part of the UHF and EHF ranges, that required theoretical and experimental research.

The increased level of HF fields localization near planar aerials required studying features of excitation and reception of surface and solid, fast and slow waves, including magnetostatic ones, with various dispersions in layered ferrite-dielectric structures located at some distance from them.

Another problem is concerned with the irregularity of AFC waves excited in ferrite-dielectric structures of finite sizes. Prior to our research any treatment of the influence of losses in structures and, first of all, in a ferrite layer, on the properties of converters and their characteristics had been problematic also; features of excitation in structures of the millimeter range, both near the resonant frequencies and at significant tuning out from it, and, first of all, in the field of existence of the dipole–dipole interaction or magnetostatic waves were to be revealed. The basic type of converters widely used in various regions of the UHF – range is converters of various designs of symmetric and asymmetrical MSL, and their various modifications.

One- and multielement converters provide excitation and reception of various types of electromagnetic waves in wide and narrow bands of frequencies.

Until recently the most essential error at treatment of experimentally observed AFC of various types of waves in structures with tangential and normal magnetizations was due to the absence of exact conformity between the beginning of the spectrum of waves at a preset level of sensitivity of measuring equipment and the values of external and internal magnetic fields.

Control over the phase and amplitude constants of waves, their dispersions, the resistance of radiation near the resonant frequencies is associated with serious technical difficulties in an experimental plan. The absence of electrodynamic approaches to calculation of various types of converters from a united methodical approach had complicated the development of theoretical models adequate to experimental data and considering most essential features of film ferrite-dielectric structures in the UHF and EHF ranges. Until recently the point of view was widespread that near the resonant frequencies (no problem of this frequency falling inside the passband was discussed) the MSW – approximation failed, and the signal itself was treated as an "electromagnetic pickup". This resulted in, more likely, qualitative, frequently – very "exact" agreement between theoretical and experimental data. And the more strict model and mathematical apparatus are used (self-coordination in the MSW approximation, exchange interaction, etc.), the better agreement between the theory and experiment is. However, attempts to calculate and design effective devices on the basis of ferrite films in the short-wave part of the centimeter range and in the millimeter ranges are practically unknown to the authors. Though for broadband operating modes of converters there still is an opportunity to have experiment fit the mode of an approximate MSW theory, for narrow-band modes near the resonant frequencies essential divergences are observed.

5.3.1 Microstrip Line Converters

A study was made of layered structures on the basis of YIG films and Li–Zn-spinels in the UHF and EHF ranges. For MSL passing from the broadband mode to the narrow-band one is possible by means of changing the width of the strip. To characterize the mode of the converter it is convenient to use the ratio of the strip width W_{MSL} to the thickness of the ferrite layer d: W_{MSL}/d. For $W_{MSL}/d \gg 1$ the mode is narrow-band, and the spectrum of waves excited in such a structure lays near the resonant frequency, and the passband $\Delta\nu/\nu \leq 0.3\%$. For $W_{MSL}/d \leq 1$ the mode is broadband, and the spectrum of excited waves covers a frequency range below and above the resonant frequency and includes MSW branches, the passband making hundreds MHz or few GHz.

Figure 5.25 shows the AFC of excited waves a in normally magnetized ferrite-dielectric structure in two ranges of frequencies $\nu_{\mathrm{H}} = 8$ and 27 GHz in a broadband operating mode of converters with parameters: $W_{MSL}/d = 1$; $d = 3 \cdot 10^{-5}$ m; $h = 5 \cdot 10^{-4}$ m; $L_a = 4 \cdot 10^{-3}$ m; $\varepsilon_h = 14$; $\varepsilon_d = 14$; $4\pi M_S = 0.176$ T; $a - H_0 = 380.74$ kA/m;

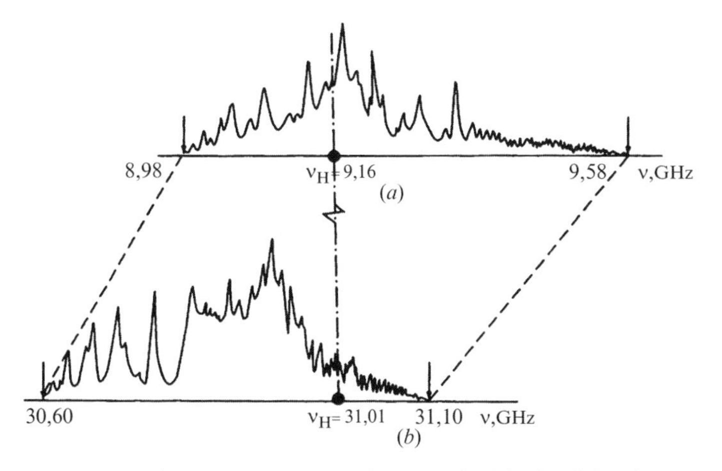

Fig. 5.25 The AFC of excited waves a in normally magnetized ferrite-dielectric structure in two ranges of frequencies $\nu_{\mathrm{H}} = 8$ and 27 GHz in a broadband operating mode of converters with parameters: $W_{MSL}/d = 1$; $d = 3 \cdot 10^{-5}\,\mathrm{m}$; $h = 5 \cdot 10^{-4}\,\mathrm{m}$; $L_a = 4 \cdot 10^{-3}\,\mathrm{m}$; $\varepsilon_h = 14$; $\varepsilon_d = 14$; $4\pi M_S = 0.176\,\mathrm{T}$; (*a*) $H_0 = 380.74\,\mathrm{kA/m}$; (*b*) $H_0 = 928.35\,\mathrm{kA/m}$. The dot-and-dash line marks frequency ν_{H} measured by means of our developed sensors

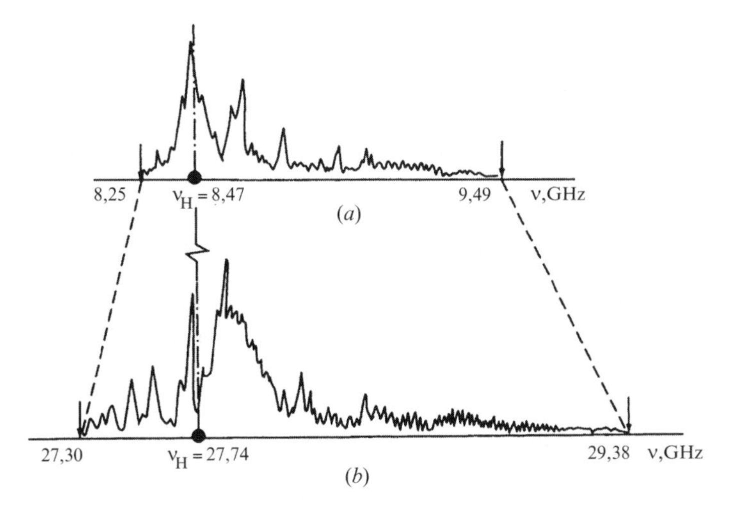

Fig. 5.26 The AFC of waves excited in a tangentially magnetized structures with similar parameters: (*a*) $H_0 = 199.54\,\mathrm{kA/m}$; (*b*) $H_0 = 813.99\,\mathrm{kA/m}$

b $- H_0 = 928.35\,\mathrm{kA/m}$. The dot-and-dash line marks frequency ν_{H} measured by means of our developed sensors. It is obvious that at advance into the mm range the wave spectrum extends both in the field of frequencies $\nu < \nu_{\mathrm{H}}$ and in the field of frequencies $\nu > \nu_{\mathrm{H}}$.

Figure 5.26 presents the AFC of waves excited in a tangentially magnetized structures with similar parameters: $a - H_0 = 199.54\,\mathrm{kA/m}$; $b - H_0 = 813.99\,\mathrm{kA/m}$. Unlike normally-magnetized structures, at advance into the millimeter range the

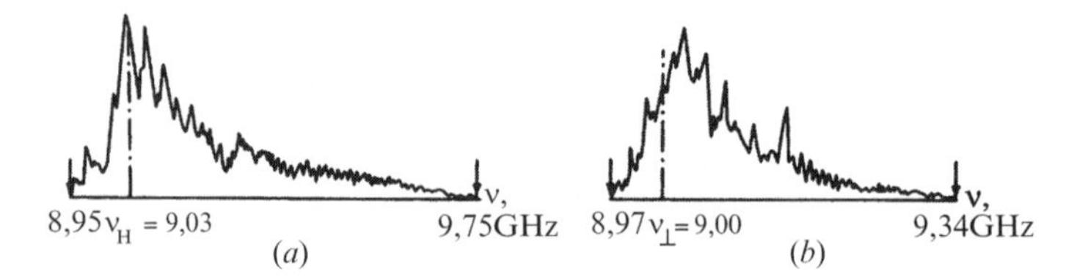

Fig. 5.27 The AFC of waves excited in the narrow-band mode of MSL in layered structures on the basis of ferrite films at normal (*a*) and tangential (*b*) magnetization: (*a*) $H_0 = 141.66\,\text{kA/m}$; (*b*) $H_0 = 195.20\,\text{kA/m}$; $d = 25 \cdot 10^{-6}\,\text{m}$; $4\pi M_S = 0.176\,\text{T}$; $W_{MSL}/d \gg 1$

wave spectrum in the field of frequencies $\nu < \nu_\perp$ extends, but in the field of frequencies $\nu > \nu_\perp$ it is contracted.

In Fig. 5.27 are presented typical AFC of waves excited in the narrow-band mode of MSL in layered structures on the basis of ferrite films at normal (Fig. 5.27*a*) and tangential (Fig. 5.27*b*) magnetization: a – $H_0 = 141.66\,\text{kA/m}$; b – $H_0 = 195.20\,\text{kA/m}$; $d = 25 \cdot 10^{-6}\,\text{m}$; $4\pi M_S = 0.176\,\text{T}$; $W_{MSL}/d \gg 1$. One can seen that in the narrow-band operating mode of MSL the wave spectrum falls outside ν_H and $\nu_\perp$ as well,and the passbands of signal are $\nu_\text{H} - \nu_{\text{H}1} \cong \nu_\perp - \nu_{\perp 1}$ and $\nu_{\text{H}2} - \nu_\text{H} \cong \nu_{\perp 2} - \nu_{\perp 1}$.

In Fig. 5.28 is presented a typical AFC of a signal (Fig. 5.28*a*) in the broadband mode of excitation of a wave spectrum in a normally-magnetized structure in a range of frequencies of 30–40 GHz, and SWR_e (Fig. 5.28*b*) of the input of MSL as well. The dot-and-dash line designates frequency ν_H, the dotted line depicts the range of the lower border frequency $\kappa'_{bor.\text{H}}$, from which the MSW approximation is valid for the given structure and range of frequencies. It is obvious that the most part of the passband of MSL (ca. 57–60%) is due to the excitation of waves described in the electrodynamic approach. The fraction of the passband of the waves described by the MSW approximation makes about 43–40% of the total passband. These results show incorrectness of application of the MSW approximation in the mm range within the total frequency passband of MSL.

The course of the dependence of SWR_e shows that on frequency ν_H no specific features are observed in the investigated breadboard model. Two characteristic minima of SWR_e in the field of frequencies $\nu < \nu_\text{H}$ are related to the conditions of reflection in TL of a specific breadboard model. They were not observed for other devices.

Expansion of the wave spectrum in the field of frequencies essentially far from the resonant frequency of the structure ν_H and reduction of the MSW fraction in the total wave spectrum at advance into the short-wave part of the centimeter range and in the millimeter ranges of radiowaves is related to strengthening of the vortex character of HF fields in a layered structure, and with an increase in the influence of the wings of the tensor $\overset{\leftrightarrow}{\mu}$ components at broadening of the ferromagnetic resonance lines ΔH.

The loss coefficients for transformation of the active $K_{los.R}(\nu)$ and reactive $K_{los.B}(\nu)$ power of radiation in the near zone of radiation, and the transmission

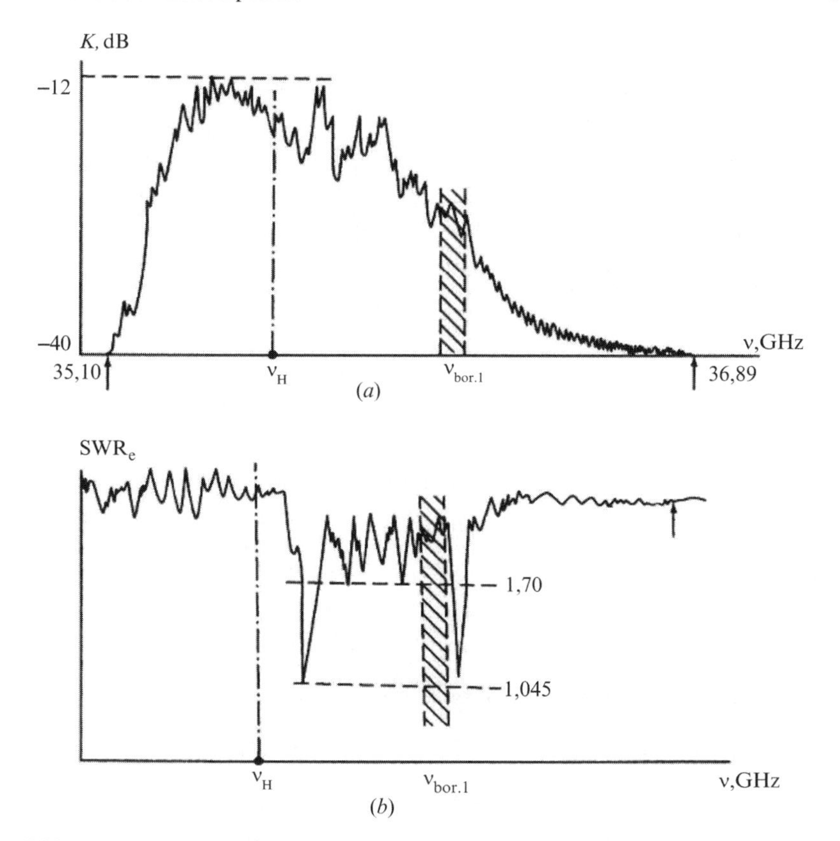

Fig. 5.28 The AFC of a signal (*a*) in the broadband mode of excitation of a wave spectrum in a normally-magnetized structure in a range of frequencies of 30–40 GHz, and SWR$_e$ (*b*) of the input of MSL as well. The dot-and-dash line designates frequency ν_H, the dotted line depicts the range of the lower border frequency $\kappa'_{bor.H}$, from which the MSW approximation is valid for the given structure and range of frequencies

factor $K_{los}(\nu)$ of a wave extending in FDLS in the distant zone are key parameters of converters. Reflective (Fig. 5.29*a*) and straight-through passage (Fig. 5.29*b*) breadboard models were applied for their estimation. In transmission lines and converters with losses the HF of power are complex. The input power P_0 in a device of the straight-through passage type is generally

$$\dot{P}_0 = -(\dot{P}_{ref})_{inp} + \dot{P}^{\pm}_{los} + \dot{P}^{\pm} - (\dot{P}_{ref})_{out}, \tag{5.1}$$

where the superscripts $\pm$ are concerned with waves with $\kappa'_y > 0$ and $\kappa'_y < 0$, respectively;

$$\dot{P}_0 = Re\dot{P}_0 + j\mathrm{Im}\dot{P}_0, Re\dot{P}_0 = P_{0R}, \mathrm{Im}\dot{P}_0 = P_{0B},$$

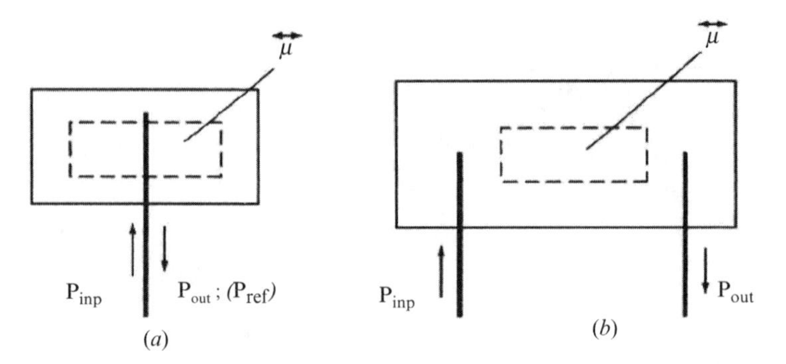

Fig. 5.29 Reflective (*a*) and straight-through passage (*b*) breadboard models

$(\dot{P}_{ref})_{inp,out}$ is the power reflected from the border of FDLS from input and output outside of the working band of frequencies of waves excited in the structure; $\dot{P}^{\pm}_{los} = Re\dot{P}^{\pm}_{los} + j\mathrm{Im}\dot{P}^{\pm}_{los}$ the power describing losses for transformation in the near zone of radiation;

$$Re\dot{P}^{\pm}_{los} = \dot{P}^{\pm}_{los.R}, \mathrm{Im}\dot{P}^{\pm}_{los} = \dot{P}^{\pm}_{los.B},$$

$\dot{P}^{\pm}_{los.R}$ the active part of the power of radiation generating heat;

$\dot{P}^{\pm}_{los.B}$ – the reactive part of the power of radiation describing attenuation of a signal (tail from the converter) due to mismatch with the bringing line; $\dot{P}^{\pm} = Re\dot{P}^{\pm} + j\mathrm{Im}\dot{P}^{\pm}$ the power of a wave in the distant zone of radiation;

$$Re\dot{P}^{\pm} = P^{\pm}_{R};\ \mathrm{Im}\dot{P}^{\pm} = P^{\pm}_{B}.$$

In usual practice the level of reflected power can be accepted for the initial one from which other powers are measured. In regular bringing lines $P^{\pm}_{0R} \gg P^{\pm}_{0B}$, in the distant zone of radiation $P^{\pm}_{los.R} \gg P^{\pm}_{los.B}$. By normalization by P_R in Eq. (5.1), in view of these remarks, we have

$$1 \cong K_{los.R} + jK_{los.B} + K_{cf}, \tag{5.2}$$

where $K_{los.R} = \frac{P_{los.R}}{P_{0R}}$ is the loss coefficient for transformation of active power of radiation to the near zone; $K_{los.B} = \frac{P_{los.B}}{P_{0R}}$ the loss coefficient by transformation of reactive power of radiation to the near zone; $K_{cf} = \frac{P^{\pm}_{R}}{P_{0R}}$ the transmission (attenuation) of signal in the distant zone of radiation.

Usually theoretical analysis uses running values $P_{los.R.B.}(W/m)$ normed by a unit of the active length of the converter L_a(m). Then $P_{los.R.B.}(W) = P_{los.R.B.}L_a$, where

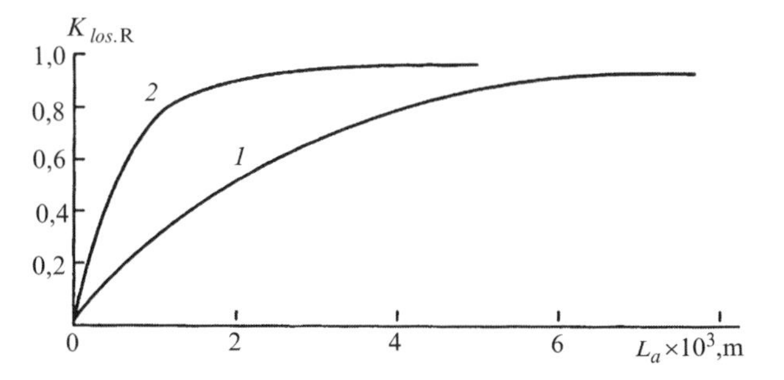

Fig. 5.30 Theoretical and experimental dependencies of $K_{los.R}(\nu)$ of MSL with a normally magnetized FDLS (experiment on passage under the scheme in Fig. 5.29b) from the active length L_a for YIG films of a varying thickness d are presented: 1–$3 \cdot 10^{-6}$ m; 2–$17 \cdot 10^{-6}$ m; at $W_{MSL} = 5 \cdot 10^{-5}$ m; $h_1 = 5 \cdot 10^{-4}$ m; $\varepsilon_1 = 14$; $4\pi M_S = 0.176$T; $\Omega_M = 0.22$

L_a characterizes the length of the ferrite covering of the microstrip. In the reflection mode the value is measured

$$\dot{G} = |G|e^{j\varphi},$$

where $|G| = \frac{P_{ref}}{P_{0R}}$; $|P_{ref}| = \sqrt{(P_{los.R})^2 + (P_{los.B})^2}$; $\varphi = arctg\frac{P_{los.B}}{P_{los.R}}$.

For $P_{los.R} \gg P_{los.B}$, $P_{ref} \cong P_{los.R}, \varphi \cong 0$. In the passage mode $K_{cf} = \frac{P_R(L_a)}{P_{0R}}$ is measured.

In Fig. 5.30 theoretical and experimental dependencies of $K_{los.R}(\nu)$ of MSL with a normally magnetized FDLS (experiment on passage under the scheme in Fig. 5.29*b*) from the active length L_a for YIG films of a varying thickness d are presented: 1–$3 \cdot 10^{-6}$ m; 2–$17 \cdot 10^{-6}$ m; at $W_{\text{MSL}} = 5 \cdot 10^{-5}$ m; $h_1 = 5 \cdot 10^{-4}$ m; $\varepsilon_1 = 14$; $4\pi M_S = 0.176$T; $\Omega_M = 0.22$.

For each value of W_{MSL}/d there is a certain value of L_a at which $K_{los.R} \to 1$. The same value L_a determines the width of the ferrite layer, which can influence the dispersive characteristics of a transmission line. The change limits of the parameter W_{MSL}/d are determined by conditions of electrodynamic coordination of the bringing strip line, MSL, layered structure on the basis of ferrite, and technological opportunities as well.

In Fig. 5.31 experimental dependencies of the loss coefficients for transformation of both active $K_{los.R} - 1$ and reactive $K_{los.B} - 2$ power of radiation in MSL in the reflection mode (Fig. 5.29*a*) for parameters $\nu_H = 25.25 \cdot 10^{10}$ GHz, $W_{MSL}/d = 2$, $H_0 = 0.85$ MA/m, $Z_0 = 70$ Ohm are shown. At increasing parameter W_{MSL}/d both $|K_{los.R}|$ and $|K_{los.B}|$ decrease.

Our research has shown that the parameter W_{MSL}/d renders an essential influence on the passband of the converter. In Fig. 5.32 theoretical and experimental dependencies of the passband of frequencies $\Delta\nu_{3dB}$ of MSL on the W_{MSL}/d ratio on frequencies $\nu_H = 20$–25 GHz are presented. It is obvious that when $W_{MSL}/d \leq 1$ a broadband operating mode of MSL takes place, and when $W_{MSL}/d \geq 10$ the mode is narrow-band. The value of the W_{MSL}/d ratio can change due to both a variation

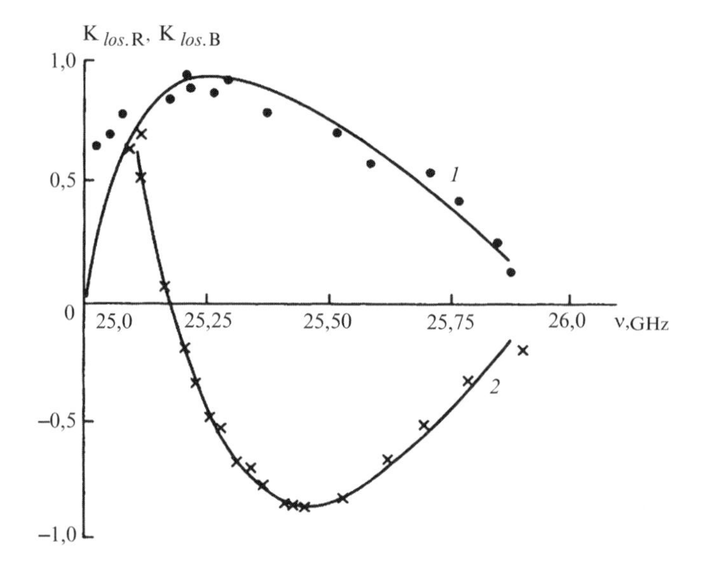

Fig. 5.31 Experimental dependencies of the loss coefficients for transformation of both active $K_{los.R} - 1$ and reactive $K_{los.B} - 2$ power of radiation in MSL in the reflection mode (Fig. 5.29*a*) for parameters $\nu_{\mathrm{H}} = 25.25 \cdot 10^{10}\,\mathrm{GHz}$, $W_{MSL}/d = 2$, $H_0 = 0.85\,\mathrm{MA/m}$, $Z_0 = 70\,\mathrm{Ohm}$

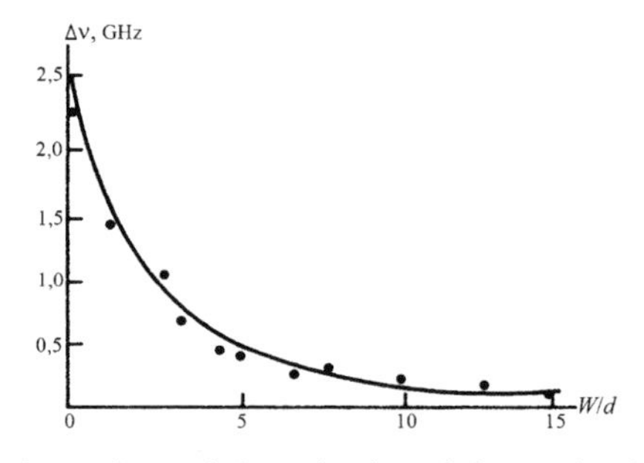

Fig. 5.32 Theoretical and experimental dependencies of the passband of frequencies $\Delta\nu_{3dB}$ of MSL on the W_{MSL}/d ratio on frequencies $\nu_{\mathrm{H}} = 20$–$25\,\mathrm{GHz}$

of the width W_{MSL} of the conducting strip of the converter and the thickness of the ferrite layer d, and W_{MSL} determines, first of all, the width of the spectrum of wave numbers ($\Delta k = 2n\pi/W_{MSL}$, $n = 1, 2, \ldots$) or spatial-temporal harmonics $\kappa_n(\nu_n)$.

Our experimental research has shown that the frequency passband of MSL depends also on the thickness of the dielectric base h_1. In Fig. 5.33 experimental dependencies of $\Delta\nu_{3dB}$ on ℓ for a varying width of the strip W_{MSL} are presented: 1–$20 \cdot 10^{-6}$m; 2–$50 \cdot 10^{-6}$m; 3–$100 \cdot 10^{-6}$m at $d = 17 \cdot 10^{-6}$m, $4\pi M_S = 0.176\,\mathrm{T}$, $\Omega_{\mathrm{M}} = 0.22$. It is obvious that the lesser the thickness of the dielectric base h_1 and the lesser the width of the strip W_{MSL}, the wider the passband $\Delta\nu_{3dB}$ is.

Transfer losses in the distant zone of radiation were investigated in devices of the straight-through type (see Fig. 5.29*b*). In Fig. 5.34 theoretical dependencies of transfer losses – 1, 2, calculated in the MSW approximation (1) and under the

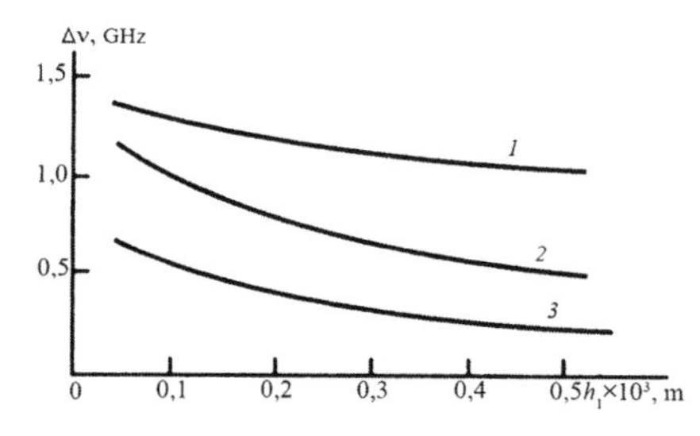

Fig. 5.33 Experimental dependencies of $\Delta\nu_{3dB}$ on ℓ for a varying width of the strip W_{MSL} are presented: 1–$20\cdot10^{-6}$ m; 2–$50\cdot10^{-6}$ m; 3–$100\cdot10^{-6}$ m at $d=17\cdot10^{-6}$ m, $4\pi M_S=0.176$T, $\Omega_M=0.22$. It is obvious that the lesser the thickness of the dielectric base h_1 and the lesser the width of the strip W_{MSL}, the wider the passband $\Delta\nu_{3dB}$ is

electrodynamic approach (2) developed in work, and an experimental (3) dependence are shown. In the MSW approximation curve 1 was determined by the relationship

$$K_{MSW}=1-\frac{1-\frac{Z_{rad}}{Z_{MSL}}}{1+\frac{Z_{rad}}{Z_{MSL}}}.$$

The total transfer losses measured experimentally include the losses at forward (at input) transformation of an electromagnetic wave from the bringing lines into a wave in the line on the basis of a ferrite-dielectric structure, and the loss for return transformation at input, and the losses for distribution of a wave in the structure. Therefore, the theoretical dependence of total transfer losses in the MSW approximation contains the level of losses of a signal in the bringing elements in the line, then the theoretical dependence is $-K_T=K_{MSW}+A$. At comparison with the MSW approximation at the first stages of our research the beginning of the experimentally observed AFC was accepted for ν_H, the frequency of ferromagnetic resonance, that, as follows from the last data, displaces all the spectrum towards lower frequencies by 30–40% from the total frequency band of SSLW in the short-wave part of the centimeter range. In the electrodynamic approach the spectrum of excited and extending waves falls outside the limits ν_H.

5.4 Converters on the Basis of Slot and Coplanar Strip Lines

Converters on the basis of SL have a number of basic differences from MSL:

- The presence of a metal screen in the exciting plane allows dispersive characteristics at direct metallization of ferrite $h_1/d=0$ and at any distance to be formed (for MSL $h_1/d\geq5\cdot10^{-5}$ m, that is connected with certain technological restrictions, on the one hand, and with electrodynamic requirements, on the other hand).
- The volume character of HF fields and their increased ellipticity allows various types of waves in layered structures, both with parallel (traditional) and

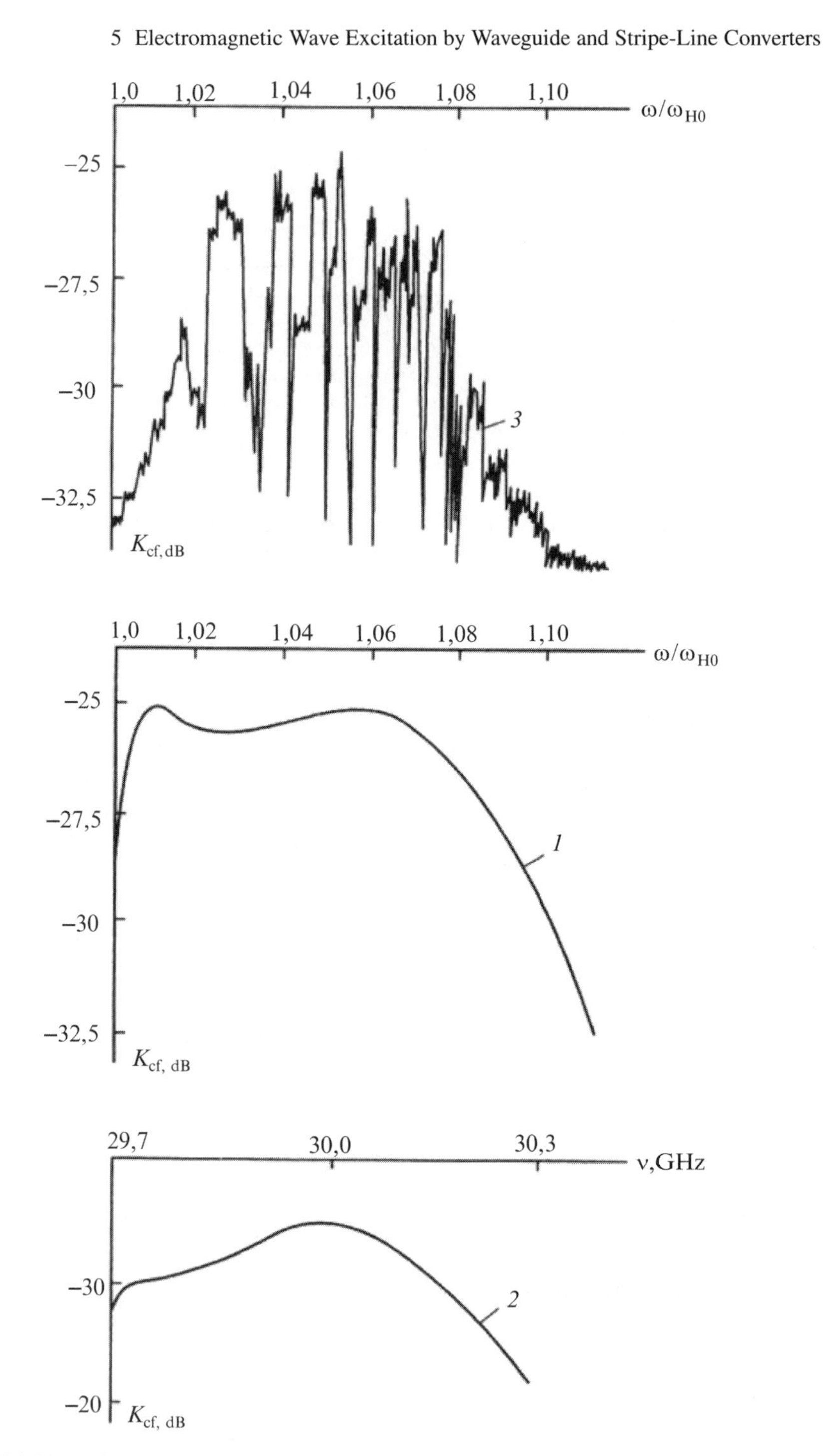

Fig. 5.34 Theoretical dependencies of transfer losses – 1, 2, calculated in the MSW approximation (1) and under the electrodynamic approach (2) developed in work, and an experimental (3) dependence are shown

orthogonal (new) orientations of the structures to the exciting plane to be effectively excited.
- Strong sagging of HF fields outside of the slot provides an opportunity of effective excitation of electromagnetic waves in structures located on a metal surface outside of the slot of excitation, that eliminates any additional heterogeneity in the region of interface of the strip for MSL or slots for SL with a ferrite transmission line.

Converters on the basis of CL, combining the advantages of SL, allow twofold reduction of the active length L_a, in comparison with CL at an inphase type of the exciting wave to be provided, that leads to reduction of the width of the transmission line, and at an antiphase type of the exciting wave – to excite waves in the structure with a π shift, that at certain geometrical sizes of CL and in a certain range of frequencies presumes realizing unidirectional properties, as in waveguide converters.

For SL, along with parallel, the orthogonal orientation of FDLS to the plane of the converter is effective as well. Therefore, below we shall distinguish:

- Parallel orientation of FDLS to SL (a subscript $\|$ to designate the corresponding parameters of converters)
- Orthogonal orientation of FDLS to the exciting plane at an arrangement of the structure along the longitudinal axis of the slot (subscript $\perp 1$)
- Orthogonal orientation of the structure at its arrangement in the plane perpendicular to the longitudinal axis of the slot line (subscript $\perp 2$)

Before passing to consideration of the properties of SL and CL, we shall result results of our research of the dispersive characteristics of waves excited by SL in the structures represented in Fig. 5.35: *a* – for excitation of SSLW in normally magnetized structures with parallel orientation of the ferrite layer; *b*, *c* – for simultaneous excitation of RSLW and SSW in tangentially magnetized structures with orthogonal orientation of the ferrite layer; *d* – for excitation of RSW in a tangentially magnetized structure with orthogonal orientation of the ferrite layer in the presence of a metallized covering with a sample in the form of a segment with a radius $R = 3 \cdot 10^{-2}$ m and height $h_R > \lambda$, λ being the wavelength in SL.

Epitaxial YIG films with the thickness $d = (7\text{–}17) \cdot 10^{-6}$ m and the size $(1 \cdot 1.5) \cdot 10^{-4}$ m^2, and a volume monocrystal of YIG with the thickness $d = 6 \cdot 10^{-4}$ m, the size $(1 \cdot 1.5) \cdot 10^{-4}$m^2 with magnetization $4\pi M_S = 0.176$T were used.

In Fig. 5.36 theoretical (curve) and experimental (points) dispersions $\Omega(\kappa d)$ with indication of the serial numbers of backs-wave resonances for SL with a volume crystal are presented: *a* – SSLW; *b* – RSLW; *c* – SSW; *d* – RSW for a ferrite with $d = 6 \cdot 10^{-4}$m, $4\pi M_S = 0.176$ T, $\Omega_{\mathrm{M}} = 0.245$.

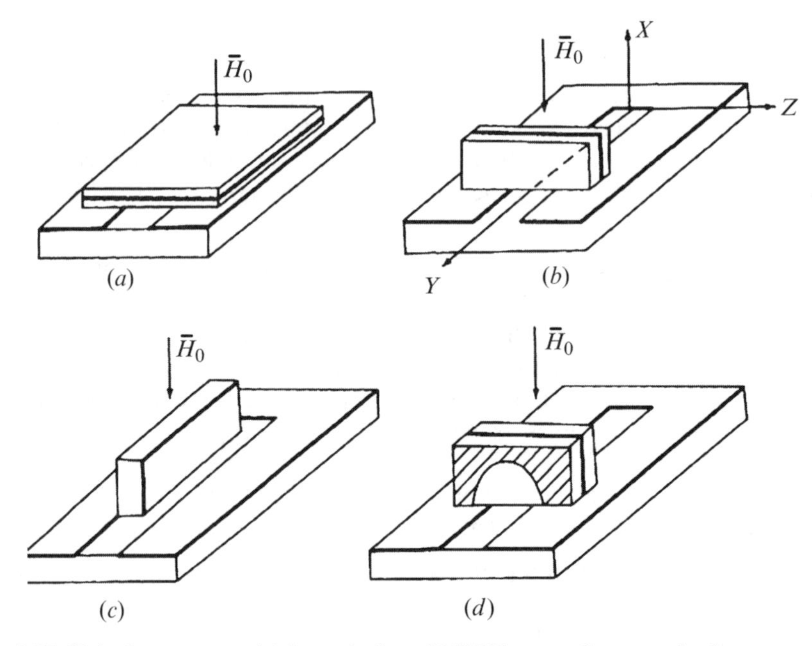

Fig. 5.35 SL in the structures: (*a*) for excitation of SSLW in normally magnetized structures with parallel orientation of the ferrite layer; (*b, c*) for simultaneous excitation of RSLW and SSW in tangentially magnetized structures with orthogonal orientation of the ferrite layer; (*d*) for excitation of RSW in a tangentially magnetized structure with orthogonal orientation of the ferrite layer in the presence of a metallized covering with a sample in the form of a segment with a radius $R = 3 \cdot 10^{-2}$ m and height $h_R > \lambda$, λ being the wavelength in SL

Let's note that in the structures with orthogonal orientation (Fig. 5.35*b–d*) simultaneous excitation of the following can be observed:

- RSLW and SSW, having, as a common frequency area of existence close to $\nu \cong \nu_\perp$, and belonging to various frequency ranges, at $\nu > \nu_\perp$ and $\nu < \nu_\perp$ and extending in mutually-perpendicular directions
- RSLW and RSW, belonging to various frequency ranges and extending in mutually-perpendicular directions, that can be used for design of devices of frequency separation of signals

5.4.1 Properties of SL at Parallel Orientation of the Ferrite Layer to the Converter Plane

In Fig. 5.37 typical oscillograms of the wave spectrum excited by SL in a normally magnetized structure (K_{los}) for various parameters W_{SL}/d and L_a in two ranges of frequencies are presented: $a - \nu_H = 9.08\,\text{GHz}$; $H_0 = 398.08\,\text{kA/m}$; $b - \nu_H = 31.7\,\text{GHz}$, $H_0 = 1.041\,\text{MA/m}$; $1 - W_{SL}/d = 15$, $L_a = 3.5 \cdot 10^{-3}\,\text{m}$; $2 - W_{SL}/d = 20$,

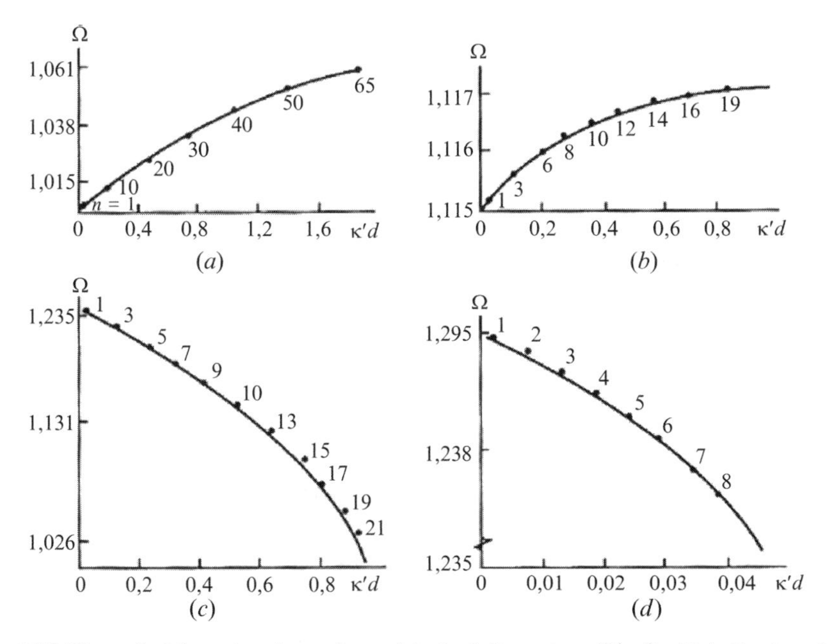

Fig. 5.36 Theoretical (curve) and experimental (points) dispersions $\Omega(\kappa d)$ with indication of the serial numbers of backs-wave resonances for SL with a volume crystal are presented: (*a*) SSLW; (*b*) RSLW; (*c*) SSW; (*d*) RSW for a ferrite with $d = 6 \cdot 10^{-4}$ m, $4\pi M_S = 0.176$T, $\Omega_M = 0.245$

$L_a = 3.5 \cdot 10^{-3}$ m; $3 - W_{SL}/d = 35$, $L_a = 3.2 \cdot 10^{-3}$ m; $4 - W_{SL}/d = 70$, $L_a = 3.5 \cdot 10^{-3}$ m; $5 - W_{SL}/d = 15$, $L_a = 11.8 \cdot 10^{-3}$ m; $6 - W_{SL}/d = 17.5$, $L_a = 11.8 \cdot 10^{-3}$ m; $7 - W_{SL}/d = 29$, $L_a = 11.8 \cdot 10^{-3}$ m; $8 - W_{SL}/d = 50$, $L_a = 5.1 \cdot 10^{-3}$ m; $4\pi M_S = 0.176$T, $\varepsilon_h = 5$, $\varepsilon_d = 14$. It is obvious that with growth of W_{SL}/d the wave spectrum is localized close to ν_H and the mode of SL is narrow-band. At transition into the millimeter range the frequency range of the wave spectrum, for both $\nu < \nu_H$ and $\nu > \nu_H$ extends and the required value of the active length of the converter L_a increases.

In Fig. 5.38 typical oscillograms of the wave spectra excited by SL in normally magnetized structures (K_{los}), depending on distance ℓ between the plane of the converter and the ferrite layer for two frequency ranges $a - \nu_H = 9.08$ GHz and $b - \nu_H = 31.7$ GHz are presented: $1 - l/d = 0$; $2 - l/d = 4$; $3 - l/d = 8$; $4 - l/d = 0$; $5 - l/d = 2$; $6 - l/d = 6$; $4\pi M_S = 0.176$T, $\varepsilon_h = 5$, $\varepsilon_d = 14$. The spectrum of excited waves still falls outside the borders ν_H. At increasing l a reduction of the efficiency of wave excitation in the structure and the passbands of SL is observed, that is connected with the reduction of the amplitude of excitation fields.

In Fig. 5.39 experimental dependencies of the transformation factor $K_{\|}$ for SSLW excited in a short-circuited SL ($W_{SL} = 3.5 \cdot 10^{-4}$ m, $\ell = 5 \cdot 10^{-4}$ m, $\varepsilon = 5$) on the active length of the converter L_a for various values of the W_{SL}/d ratio are shown for $\Omega_M = 0.245$: $1 - W_{SL}/d = 14$; $2 - W_{SL}/d = 23$; $3 - W_{SL}/d = 50$, and $K_{\|}$ for SSLW excited in MSL ($W_{MSL} = 5 \cdot 10^{-4}$ m, $\ell = 5 \cdot 10^{-4}$ m, $\varepsilon = 14$) at

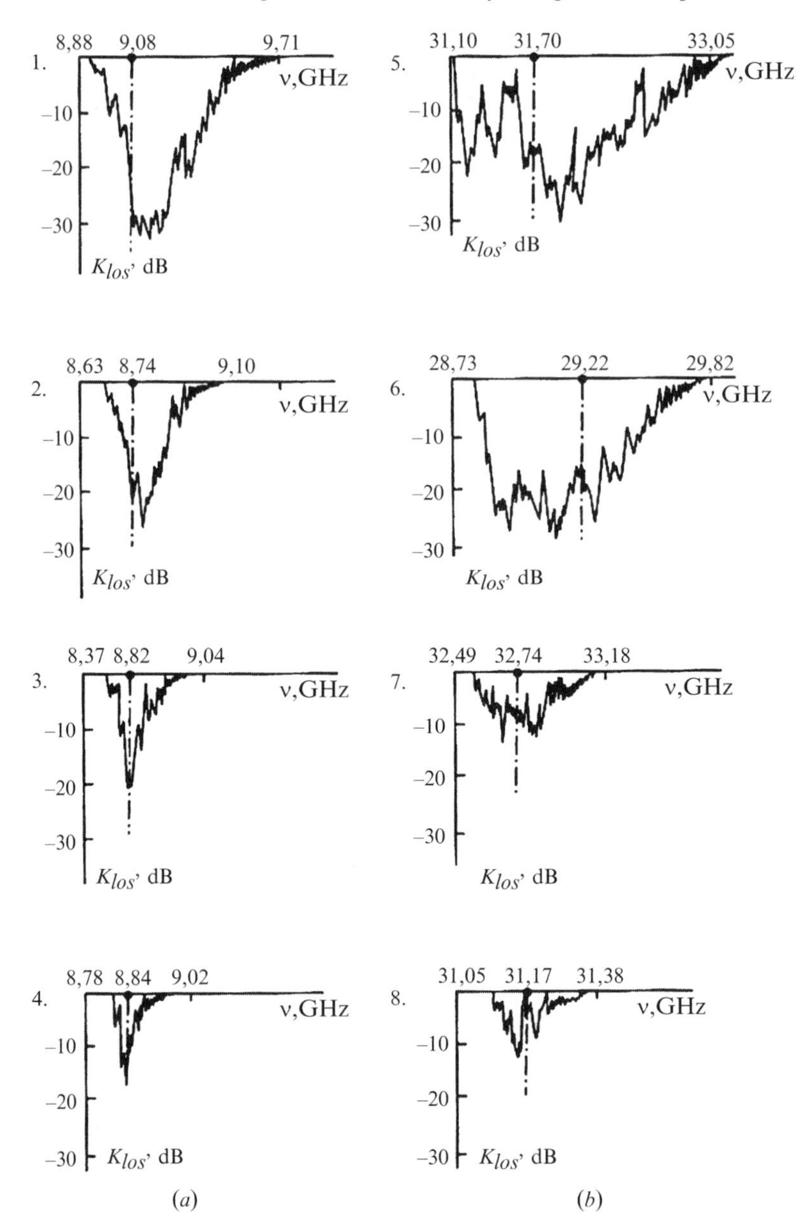

Fig. 5.37 Typical oscillograms of the wave spectrum excited by SL in a normally magnetized structure (K_{los}) for various parameters W_{SL}/d and L_a in two ranges of frequencies are presented: (*a*) $\nu_H = 9.08\,\text{GHz}$; $H_0 = 398.08\,\text{kA/m}$; (*b*) $\nu_H = 31.7\,\text{GHz}$, $H_0 = 1.041\,\text{MA/m}$; $1 - W_{SL}/d = 15$, $L_a = 3.5 \cdot 10^{-3}\,\text{m}$; $2 - W_{SL}/d = 20$, $L_a = 3.5 \cdot 10^{-3}\,\text{m}$; $3 - W_{SL}/d = 35$, $L_a = 3.2 \cdot 10^{-3}\,\text{m}$; $4 - W_{SL}/d = 70$, $L_a = 3.5 \cdot 10^{-3}\,\text{m}$; $5 - W_{SL}/d = 15$, $L_a = 11.8 \cdot 10^{-3}\,\text{m}$; $6 - W_{SL}/d = 17.5$, $L_a = 11.8 \cdot 10^{-3}\,\text{m}$; $7 - W_{SL}/d = 29$, $L_a = 11.8 \cdot 10^{-3}\,\text{m}$; $8 - W_{SL}/d = 50$, $L_a = 5.1 \cdot 10^{-3}\,\text{m}$; $4\pi M_S = 0.176\,\text{T}$, $\varepsilon_h = 5$, $\varepsilon_d = 14$

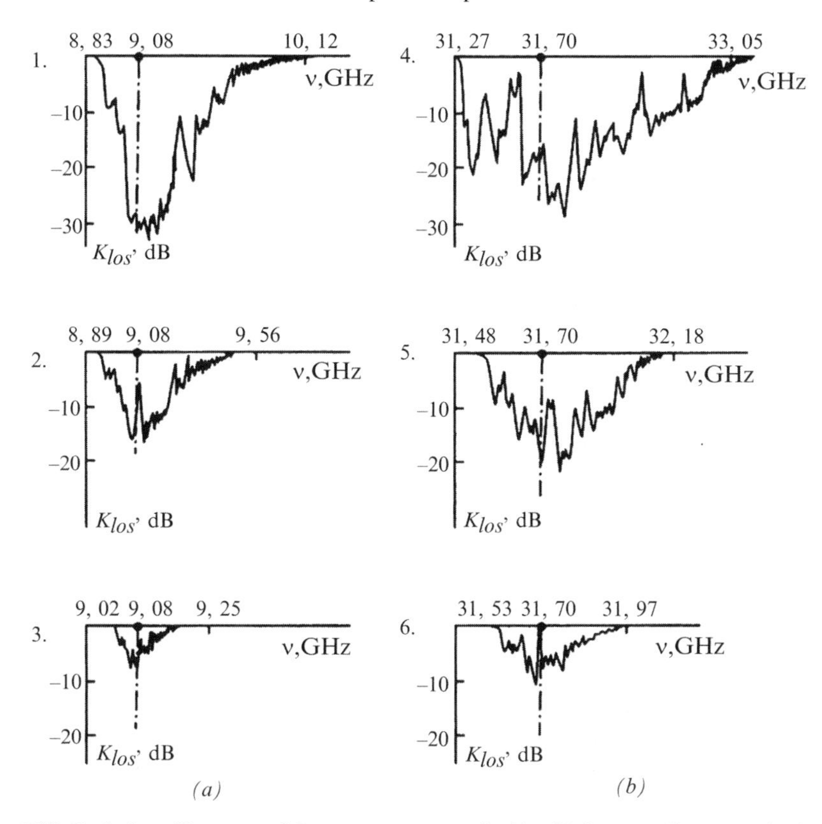

Fig. 5.38 Typical oscillograms of the wave spectra excited by SL in normally magnetized structures (K_{los}), depending on distance ℓ between the plane of the converter and the ferrite layer for two frequency ranges (*a*) $\nu_{\mathrm{H}} = 9.08\,\mathrm{GHz}$ and (*b*) $\nu_{\mathrm{H}} = 31.7\,\mathrm{GHz}$ are presented: $1 - l/d = 0$; $2 - l/d = 4$; $3 - l/d = 8$; $4 - l/d = 0$; $5 - l/d = 2$; $6 - l/d = 6$; $4\pi M_S = 0.176\,\mathrm{T}$, $\varepsilon_h = 5$, $\varepsilon_d = 14$

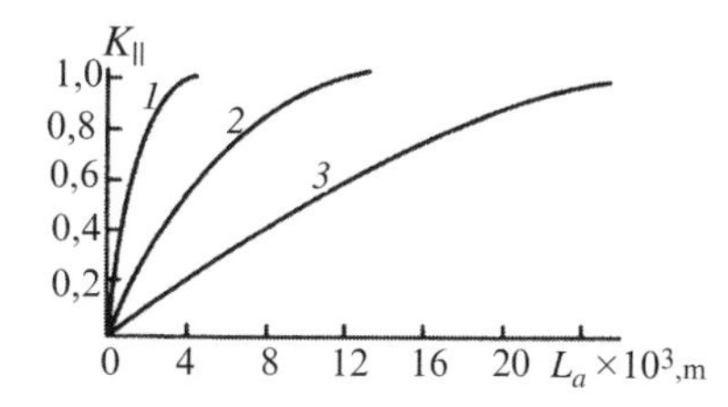

Fig. 5.39 Experimental dependencies of the transformation factor $K_{\|}$ for SSLW excited in a short-circuited SL ($W_{SL} = 3.5 \cdot 10^{-4}\,\mathrm{m}$, $\ell = 5 \cdot 10^{-4}\,\mathrm{m}$, $\varepsilon = 5$) on the active length of the converter L_a for various values of the W_{SL}/d ratio are shown for $\Omega_{\mathrm{M}} = 0.245$: $1 - W_{SL}/d = 14$; $2 - W_{SL}/d = 23$; $3 - W_{SL}/d = 50$, and $K_{\|}$ for SSLW excited in MSL ($\mathrm{W_{MSL}} = 5 \cdot 10^{-4}\,\mathrm{m}$, $\ell = 5 \cdot 10^{-4}\,\mathrm{m}$, $\varepsilon = 14$) at $\Omega_{\mathrm{M}} = 0.22$; $4 - W_{SL}/d = 3$; $5 - W_{SL}/d = 14$

$\Omega_M = 0.22$; $4 - W_{SL}/d = 3$; $5 - W_{SL}/d = 14$. One can see that with reduction of W_{SL}/d (reduction of W_{SL} or increase of d with W_{SL}, naturally, having major importance from the viewpoint of the influence on the spatial-temporal wave spectrum) the length of the active area of SL determining the value of conversion coefficient, sharply increases and at certain values W_{SL}/d $K_{||} \rightarrow 1$. In comparison with MSL, with other things being equal, the length of the active area of SL approximately by 1.5 times is longer, that speaks for the opportunity of using SL in the millimeter range.

In Fig. 5.40 dependencies of the conversion coefficients $K_{||}$ and passbands $\Delta\nu$ of SL in two frequency ranges $\nu_H = 6$–12 GHz (Fig. 5.40*a*, *b*) and $\nu_H = 35$–40 GHz (Fig. 5.40*c*, *d*) are presented. In Fig. 5.40*a*, *c*:

1 – Conversion coefficient by frequency $\nu_H - K_{||}(\nu_H)$
2 – Conversion coefficient corresponding to $\frac{1}{2}\left[K_{||}(\nu_{H1})\right]_{max}$ for the passband $\Delta\nu_{H1} \in \nu_H - \nu_{H1}$, ν_{H1} being the lower boundary frequency of SL
3 – Conversion coefficient corresponding to $\frac{1}{2}\left[K_{||}(\nu_{H2})\right]_{max}$ for the passband $\Delta\nu_{H2} \in \nu_{H2} - \nu_H$, ν_{H2} being the upper boundary frequency SL, and in Fig. 5.40*b*, *d*, *e*:

(i) – Passband $\Delta\nu_{H1} = \nu_H - \nu_{H1}$
(ii) – Passband $\Delta\nu_{H2} = \nu_{H2} - \nu_H$

As the thickness of the ferrite film d grows, an increase in the transformation factor $K_{||}$ and the total passband $\Delta\nu_H = \nu_{H1} + \nu_{H2}$ is observed in the centimeter range. In the millimeter range saturation of the specified dependencies takes place.

In Fig. 5.41 dependencies of the specified parameters on the active length L_a for a short-circuited SL in ranges of frequencies $\nu_H = 8.35$ GHz (Fig. 5.41*a*, *b*) and $\nu_H = 32.71$ GHz (Fig. 5.41*c*, *d*) are shown. It is obvious that in the millimeter range the saturation boundaries of the dependence $K_{||}(L_a)$ move towards higher values of L_a.

In Fig. 5.42 the same dependencies on the distance h between the ferrite layer and the plane of SL are shown. Monotonous reduction of $K_{||}$ and $\Delta\nu$ with growth of h is seen.

The most essential influence on the passband of SL is rendered by the width of the slot W_{SL} of the converter. In Fig. 5.43 experimental dependencies of $\Delta\nu$ of SL and MSL on parameters W_{SL}/d and W_{MSL}/d for structures on the basis of YIG films with $\Omega_M = 0.245$ are shown for comparison: 1 – SL; 2 – MSL. You see that the passband of SL with growing W_{SL}/d decreases more slowly than the dependence for MSL does with growing W_{MSL}/d. It is due to different laws of dispersions of SSLW. Therefore, SL at equal factors of transformation are more broadband. The same passband $\Delta\nu$ at a preset thickness of the ferrite film is provided at $W_{SL} = (3 \div 4)\ W_{MSL}$. On the other hand, in SL with parallel orientation of the ferrite layer it is not possible to reach such narrow passbands as in MSL. For SL in the millimeter range:

– A narrow-band mode at $W_{SL}/d > 60$
– A broadband mode at $W_{SL}/d \leq 3 \div 5$

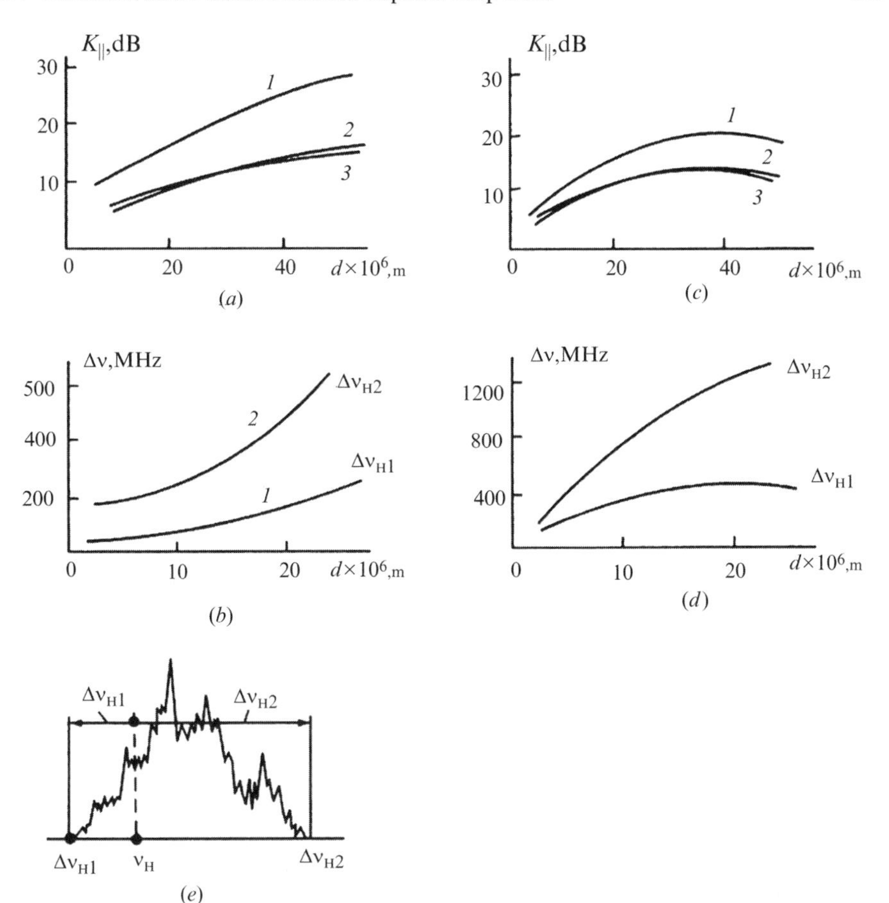

Fig. 5.40 Dependencies of the conversion coefficients $K_{\|}$ and passbands $\Delta\nu_{\mathrm{H}}$ of SL in two frequency ranges $\nu_{\mathrm{H}} = 6$–$12\,\mathrm{GHz}$ (*a, b*) and $\nu_{\mathrm{H}} = 35$–$40\,\mathrm{GHz}$ (*c, d*) are presented. In Fig. 5.40*a, c*:

1 – Conversion coefficient by frequency $\nu_{\mathrm{H}} - K_{\|}(\nu_{\mathrm{H}})$

2 – Conversion coefficient corresponding to $\frac{1}{2}\left[K_{\|}(\nu_{\mathrm{H1}})\right]_{\max}$ for the passband $\Delta\nu_{\mathrm{H1}} \in \nu_{\mathrm{H}} - \nu_{\mathrm{H1}}$, ν_{H1} being the lower boundary frequency of SL

3 – Conversion coefficient corresponding to $\frac{1}{2}\left[K_{\|}(\nu_{\mathrm{H2}})\right]_{\max}$ for the passband $\Delta\nu_{\mathrm{H2}} \in \nu_{\mathrm{H2}} - \nu_{\mathrm{H}}$, ν_{H2} being the upper boundary frequency SL, and in Fig. 5.40*b, d, e*:

(i) – Passband $\Delta\nu_{\mathrm{H1}} = \nu_{\mathrm{H}} - \nu_{\mathrm{H1}}$
(ii) – Passband $\Delta\nu_{\mathrm{H2}} = \nu_{\mathrm{H2}} - \nu_{\mathrm{H}}$

Thus, SL with parallel orientation of the ferrite layer are more promising for design of broadband devices with preset laws of dispersion. The dispersive characteristics of excited SSLW in normally magnetized layered structures suppose wider limits of variation, in comparison with MSL devices. Basic importance in SL devices in the millimeter range has the increase of the level of parasitic infiltration along the dielectric base of UHF power from input to output that requires additional measures to shield input from output of the device.

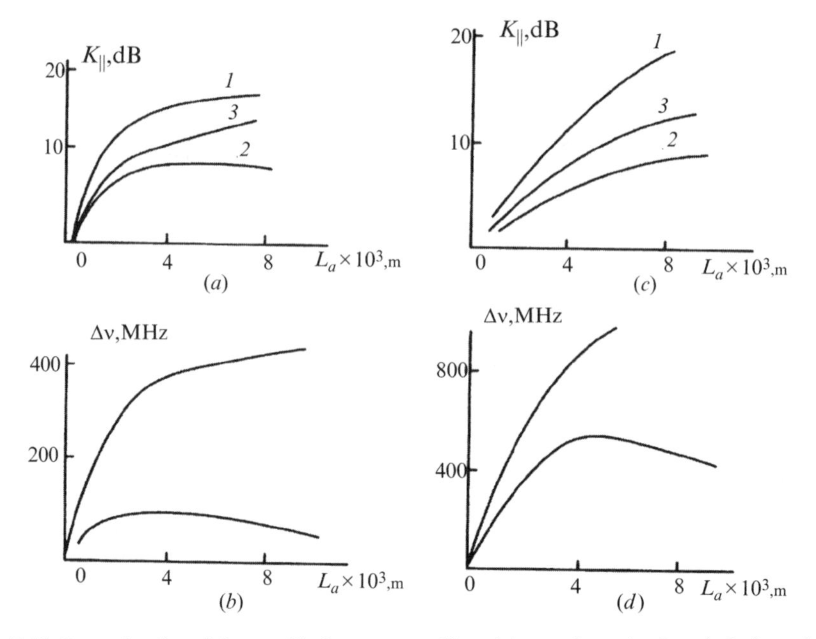

Fig. 5.41 Dependencies of the specified parameters $K_{\|}$ and $\Delta\nu$ on the active length L_a for a short-circuited SL in ranges of frequencies $\nu_H = 8.35\,\text{GHz}$ (*a, b*) and $\nu_H = 32.71\,\text{GHz}$ (*c, d*) are shown

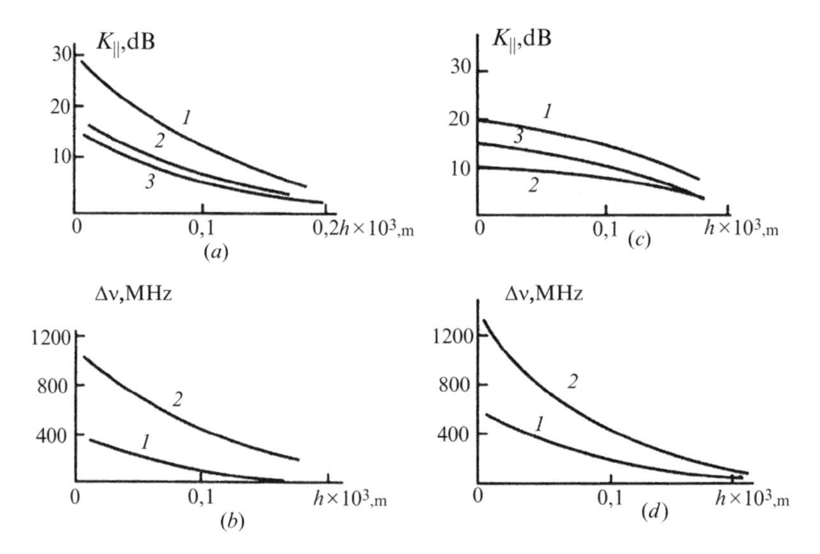

Fig. 5.42 The same dependencies on the distance h between the ferrite layer and the plane of SL are shown. Monotonous reduction of $K_{\|}$ and $\Delta\nu$ with growth of h is seen

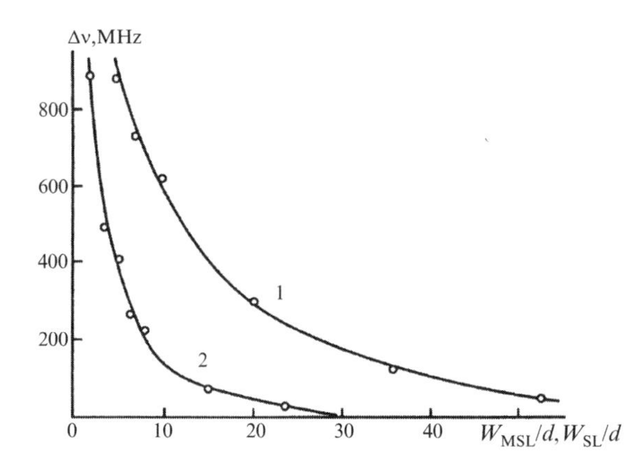

Fig. 5.43 Experimental dependencies of $\Delta\nu$ of SL and MSL on parameters W_{SL}/d and W_{MSL}/d for structures on the basis of YIG films with $\Omega_M = 0.245$ are shown for comparison: 1 – SL; 2 – MSL

5.4.2 Properties of SL and CL at Orthogonal Orientation of the Ferrite Layer to the Converter Plane

At orthogonal orientation of the ferrite layer to the converter plane the active zone area $S_\perp \ll S_{||}$ ($S_{||}$ is the area of the active zone at parallel orientation). So, at a typical thickness of the ferrite films $d = 10$–$20\,\mu\text{m}$, $a = 1\,\text{mm}$, $b = 1\,\text{mm}$ we get $S_\perp \cong (1\text{–}2) \cdot 10^{-2}\,\text{mm}^2$. Such values of $S_\perp$ and its high reproducibility (to a tolerance of tenth μm) can not be achieved by means of micron machining and chemical etching.

In Fig. 5.44*a–c* oscillograms of the AFC of signals in the reflection mode from SL with orthogonal orientation of the ferrite-dielectric structure located along the axis of the slot line, depending on the height of the ferrite film b are presented. It is obvious that near the frequency $\nu_\perp$ selective attenuation of a signal is observed, which depth practically does not depend on b.

In Fig. 5.45 the dependence of the transformation factor $K_{\perp 1}$ on the cross-section coordinate X_0 of the center of the ferrite layer for a short-circuited SL is presented at orthogonal orientation of FDS along the slot and tangential magnetization. The area above the slot of SL is shaded. Rather effective excitation $(K_{\perp 1} > 0.5)$ is observed at an arrangement of the film within the limits of the slot $x \leq (2\text{–}3)\ W_{SL}/2$. The metal surfaces of SL are shaded. High-frequency magnetic fields, as follows from Fig. 5.45, are strongly dispersed outside the limits of the slot.

In Fig. 5.46 the dependence of the conversion coefficient $K_{\perp 2}$ on the longitudinal coordinate of the ferrite layer z_0 for a short-circuited SL (Fig. 5.35*b*) is given at tangential magnetization (field H_0 is oriented to the plane of the converter). The most effective transformation $(K_{\perp 1} \to 1)$ takes place near the shorting coverings and repeats along the slot with a period close to $\lambda/2$ for $z_0 > 0$. Rather effective transformation is also observed at an arrangement of the ferrite layer on the conducting

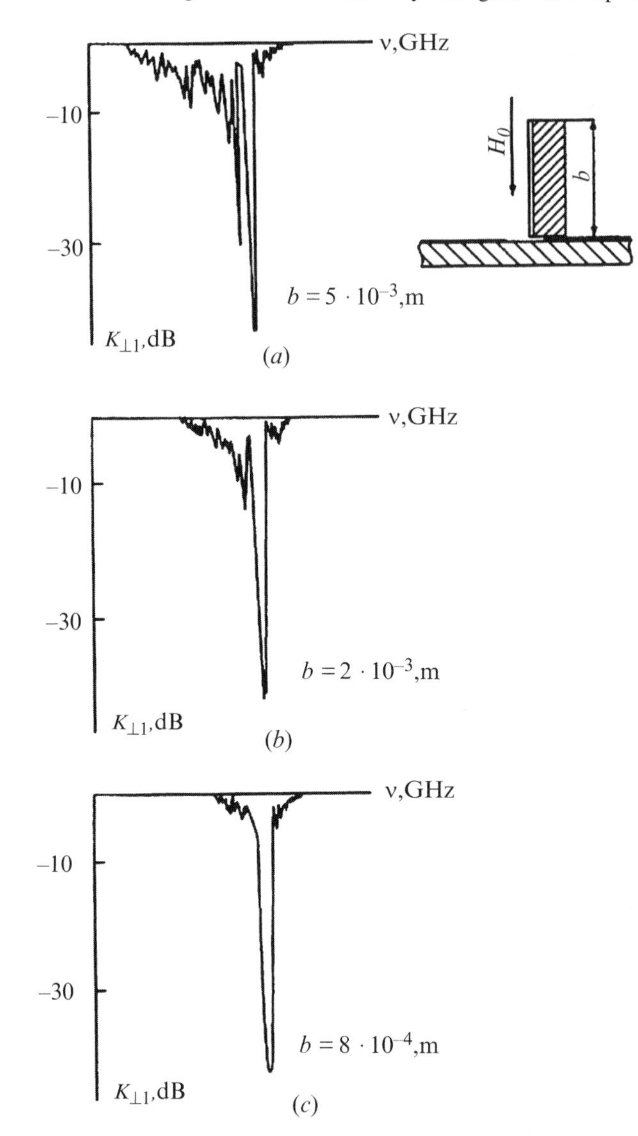

Fig. 5.44 Oscillograms of the AFC of signals in the reflection mode from SL with orthogonal orientation of the ferrite-dielectric structure located along the axis of the slot line, depending on the height of the ferrite film b are presented

surface of the end face of the slot at $z_0 < 0$. It speaks for strong sagging of HF fields of SL for the end face of the slot. The given circumstance should be considered at arrangement of SL near other converters and HF elements.

The essential sagging of HF fields outside the limits of the slot allows one to reach $K_{\perp 1} \leq 0.8$ and $K_{\perp 2} \leq 0.8$ at direct arrangement of the ferrite film on the metal surface of SL, that can be used in various devices of the millimeter range.

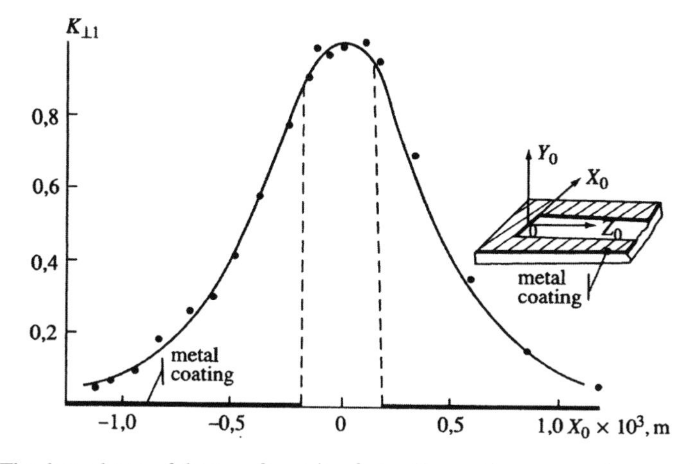

Fig. 5.45 The dependence of the transformation factor $K_{\perp 1}$ on the cross-section coordinate X_0 of the center of the ferrite layer for a short-circuited SL is presented at orthogonal orientation of FDS along the slot and tangential magnetization

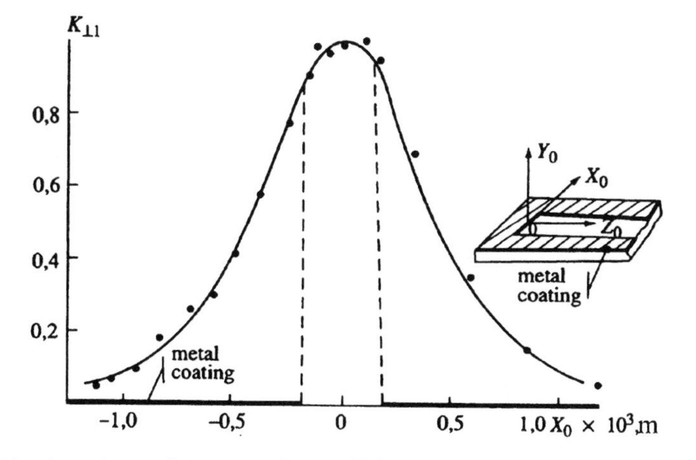

Fig. 5.46 The dependence of the conversion coefficient $K_{\perp 2}$ on the longitudinal coordinate of the ferrite layer z_0 for a short-circuited SL (Fig. 5.35*b*) is given at tangential magnetization (field H_0 is oriented to the plane of the converter)

The resulted dependencies allow positions of the ferrite layer in SL for effective excitation of various types of waves to be determined.

Transition to extremely small active zones $S_{\perp}$ allows one to provide, depending on the kind of loading of ferrite with metal screens, excitation of forward surface and return volume fast and close waves in free and unilaterally metallized ferrite-dielectric structures. In ferrite plates return surface waves can be excited in the

Table 5.1 The efficiency of excitation of forward fast and surface waves in a converter on the basis of CL at an even type of a wave in CL, the orthogonal orientation of the ferrite layer and its tangential magnetization by field H_0 at $W_{CL} = 50\mu$ m, $S_{CL} = 350\mu$ m, $h/d = 0.125\%140$, d = 30μ m in a range of frequencies of 30–40 MHz

Frequency ν_0, GHz	32.646	32.894	33.868	34.992	35.634	36.864	38.742
Conversion coefficient, $K_{\parallel}$	0.95	0.96	0.97	0.94	0.93	0.95	0.94
Frequency band width $\Delta\nu$, MHz	24	23	25	25	26	27	30

presence of metal screens at distances $t/d \cong l/d \cong 0.25$. The dependencies of the passband of SL $\nu_{\perp 1}$ and $\nu_{\perp 2}$ for various orientations of FDS were experimentally investigated. It was revealed that $\nu_{\perp 1} \cong \nu_{\perp 2}$ and the value of the passband in high-quality films with $\alpha < 10^{-4}$ is mainly determined by the thickness of the layer.

The efficiency of excitation of forward fast and surface waves in a converter on the basis of CL at an even type of a wave in CL, the orthogonal orientation of the ferrite layer and its tangential magnetization by field H_0 at $W_{CL} = 50\,\mu$m, $S_{CL} = 350\,\mu$m, $h/d = 0.125 \div 140$, $d = 30\,\mu$m in a range of frequencies of 30–40 GHz is illustrated in Table 5.1. Reorganization of the central frequency of absorption ν_0 was carried out by changing the external field H_0.

At orthogonal orientation of the ferrite layer (or layered structures on its basis) to the plane of the converter narrow frequency bands of effective excitation of various types of waves can be achieved, as well as in MSL with a wide exciting strip $(W/d \gg 1)$ at parallel orientation of the ferrite layer to the converter. On the other hand, in SL and CL such contradictory requirements as sharp reduction of the area of the active zone and increase in the effective thickness of the ferrite layer in the positive direction of axis $0X$ (Fig. 5.35) can be combined.

The properties of converters on the basis of slot and coplanar lines with parallel and orthogonal orientations of FDLS to the slot (coplanar) lines in the centimeter and millimeter ranges by means of of specially designed sensors of resonant frequencies were investigated for the first time. An exit of the spectra of electromagnetic waves outside the resonant frequencies of FDLS, which enhances at promotion into the millimeter range was found. It is shown that in such devices it is possible to carry out variation of the laws of dispersions of excited waves in FDLS in wider limits due to an extremely close arrangement of metal coverings in the plane of the ferrite film. At orthogonal orientation of FDLS to the slot line, extremely narrow-band modes of excitation and propagation of electromagnetic waves in the structures, close to the narrow-band mode of MSL, are realized, they are practically independent of the thickness of the ferrite film.

Converters on the basis of SL and CL are promising for design of sensors of resonant frequencies and electronic wavemeters, one- and multichannel TF and LHPF, delay lines in the millimeter range.

5.5 Conclusions

1. The properties of various types of wave and strip converters of electromagnetic waves in FDLS on the basis of weakly dissipative ($\alpha < 10^{-4}$) of ferrite films were theoretically and experimentally investigated in the centimeter and millimeter ranges for the first time.
2. The essential exit outside the boundary frequencies defined by the MSW approximation, of the spectra of waves excited in both near and distant zones of radiation, which increases at promotion into the millimeter range, was found out.
3. The narrow-band and broadband modes of excitation and propagation of electromagnetic waves in bilaterally metallized flat waveguides containing layered structures with various magnetization on the basis of ferrite films were investigated. A complex character of the dependencies of tuning-out of the central frequencies of excited and extending waves on the dielectric permeability, losses in layers, modes of the structure is shown.
4. Waveguide-slot converters are classified among converters with orthogonal orientation of FDLS to the exciting plane with the slot and provide effective excitation and propagation with small losses of forward and return surface and volume waves in a band of frequencies essentially wider than by 100% surpassing the band of existence of forward surface and volume magnetostatic waves connected with the increased heterogeneity of regional internal fields of the plate at using of plates made of weakly dissipative ferrite; for return surface waves at bilateral metallization of the plate, agreement with the MSW approximation was obtained. At using ferrite films, effective narrow-band excitation and propagation of waves on the frequencies close to the resonant ones is observed, that is connected with the increase of the uniformity of exciting HF fields within the limits of the ferrite film.
5. Ferrite-dielectric matched transformers in the form of open places of the ferrite film provide control over the passband and losses for transformation in a wide range due to changing their length at control over interference interaction of fast and slow waves with their close laws of dispersions, which reduces the criticality of waveguide-slot converters in the mm range.
6. Waveguide-slot converters are unidirectional and promising for construction of electronic wavemeters, sensors of resonant frequencies, one- and multichannel TF and LHPF for lowered and increased levels of continuous and pulse power, devices for diagnostics of ferrite films in the millimeter range.
7. Converters on the basis of MSL are most universal and provide effective excitation and reception of signals on various types of waves, in both narrow-band modes near the resonant frequencies and broadband modes with a band of the wave spectrum essentially surpassing that given by the MSW approximation.
8. Converters on the basis of slot and coplanar lines provide effective excitation, at both parallel and orthogonal orientation of a signal, and its transfer in the narrow-band and broadband modes at one normal direction of the external field to the plane of the slot line.

9. Strong sagging of HF fields outside the limits of the slot and outside the shorting plane provide rather effective excitation of various types of waves at an arrangement of the layered structure directly on the metal surface or at a small distance from it, that allows the range of variation of the laws of dispersions of various types of excited and accepted waves to be essentially expanded.
10. At orthogonal orientation of FDLS to the slot line, extremely narrow-band modes of excitation and reception due to the improved uniformity of HF fields of excitation within the limits of the ferrite film are realized.
11. The most promising scope of waveguide-slot and strip converters of various types in the centimeter and millimeter ranges was outlined.
12. Selective processes of attenuation and passage of signals on various types of electromagnetic waves, in both broadband and narrow-band modes, are correctly described in the centimeter and millimeter ranges in the electrodynamic calculation only, the fraction of strongly delayed magnetostatic spin waves at promotion into the millimeter range essentially decreases.

Part III
Methods and Devices for High-Frequency Parameter Control of Ferrite Films

The scientific problems to be discussed here are related to the development of methods and devices for control of the high-frequency and static parameters of ferrite-dielectric structures, permanent external magnetic fields, and permanent extending waves in suvh structures. The development of magnetoelectronic devices of the millimeter range will be considered also.

It is necessary to control the high-frequency parameters – (the line width of ferromagnetic resonance, the parameter of ferromagnetic losses, the penetrability tensor components, the resonant frequency of the structure and material, amplitude and phase constants, their differential characteristics) in ferrite films near their resonant frequencies in the EHF range.

Such static parameters as (saturation magnetization, an internal magnetic field, its distribution over the thickness of a film and its area, the field of crystallographic anisotropie, increased values of an external constant magnetic field and its heterogeneity in small working gaps) should be controlled with an essentially higher accuracy and spatial resolution, in comparison with the known methods and devices.

Solution of these problems, the development of new devices, and the reached parameters in the millimetric range were determined by the level of theoretical developments and physical investigations of wave excitation and propagation in magneto-ordered film structures.

Chapter 6
High-Frequency Control Methods and Devices

Considered are methods for control of high-frequency parameters of layered structures on the basis of ferrite films on the principles of:

- Selective excitation of a signal in the near zone of a strip line converter in the prelimiting $(\nu \gg \nu_{cr})$ and post-limiting $(\nu \ll \nu_{cr})$ modes
- Interference attenuation of fast and slow waves in weakly-dissipative layered structures
- Transparency of a post-limit waveguide with a layered structure on the basis of a ferrite film near its resonant frequency
- Transparency in layered ferrite-dielectric structures with an absorbing covering
- Transparency in antiphased balanced waveguide and strip bridges at phase inversion on a ferrite-dielectric structure

Investigations of the microwave and UHF properties of layered structures on the basis of ferrite films allowed requirements to the parameters determining the quality of films in the millimeter range to be formulated. These are, first, the value of ferromagnetic losses (the line width of ferromagnetic resonance ΔH and the parameter of ferromagnetic losses α), their transverse gradients by film thickness $\nabla_x \Delta H$, $\nabla_x \alpha$, alnd the transverse gradient of saturation magnetization $\nabla_x M_x$ and the field of anisotropy $\nabla_x \Delta H_A$ (A means anisotropy).

6.1 Control on the Basis of Selective Signal Excitation in the Near Zone of Radiation and Interference Attenuation of Fast and Slow Waves

Our theoretical and experimental investigations of excitation in the near zone of radiation and propagation in the far zone of electromagnetic waves in flat waveguides partially filled with weakly dissipative structures on the basis of ferrite films have revealed some features re;ated to arrangement of metal walls with respect to a ferrite film. With walls distant from the ferrite by $h_{1,3} \geq \frac{\lambda_{cr}}{2}$, where λ_{cr} is the critical

A.A. Ignatiev, *Magnetoelectronics of Microwaves and Extremely High Frequencies in Ferrite Films*.

DOI: 10.1007/978-0-387-85457-1_7,

wave length in a flat waveguide with FDLS (a ferrit-dielectric layered_ structure), the converter and transmission line mode is the prelimiting one $(\nu \gg \nu_{cr})$. For such a mode, selective reduction of active power emission on a frequency near the resonant frequency of the structure is characteristic. For sufficiently closely spaced screens at $h_{1,3} \leq \frac{\lambda_{cr}}{2}$ the converter and transmission line mode is the post-limiting one $(\nu \ll \nu_{cr})$, at which a selective increase of the active emission power on a frequency exceeding the resonant one of the structure takes place. Various combinations of screen arrangement are possible as well:

- The prelimiting mode for the converter in the near emission zone and the post-limit mode of the transmission line
- The post-limit mode of the converter in the near emission zone and the post-limit mode of the transmission line

Below we shall consider a case of signals reflected from the converters in different modes. At contact positioning of the ferrite film on the strip line converter in the reflection mode the following quantities can be measured:

- Trapping depth K_{los}
- Stopband $\Delta\nu_{los}$
- Central frequency ν_0
- Tuning-out $\Delta\nu$ from the resonant frequency of an idealized structure ν_r (r means resonance)

Comparison of these parameters of AFC with calculation data allows one to determine:

- Width of ferromagnetic resonance line ΔH and parameter of ferromagnetic losses α
- Amplitude and phase constants on the center frequency – $\kappa''(\nu_0)$ and $\kappa'(\nu_0)$
- Internal magnetic field corresponding to the resonant frequency $H_{0i} \sim \nu_0$

At changing the signal frequency ν and the value of constant magnetic field H_o we can determine the functional dependences $\Delta H(\nu)$, $\alpha(\nu)$, $\Delta\nu(\nu)$, $k'(\nu)$, and $\kappa''(\nu)$, being of great importance for advancing investigations and development into the millimeter range. The analysis area S_a is determined by the strip line converter (antenna) dimensions – $S_a = W_a L_a$ (W_a – antenna width, L_a – antenna length) covered with a ferrite film. The technology allows achieving $W_a \cong 10^{-5}\,\mathrm{m}$ and $L_a \cong (1 \div 10) \cdot 10^{-3}\,\mathrm{m}$, that corresponds to $S_a \cong (1\text{–}10) \cdot 10^{-8}\,\mathrm{m}^2$. From calculation data and the sensitivity of the used measuring equipment the limits of parameters registration of ferrite-dielectric structures can be obtained, requirements for the converter demensions of the diagnostic sensor are formulated. So, for standard panoramic measuring instruments SWR$_e$ and attenuation in the millimetric range, registration of such parameters a $K_{los} \leq (40\text{–}50)\,\mathrm{dB}$, $\Delta\nu \geq (1\text{–}3) \cdot 10^6\,\mathrm{Hz}$, $\frac{\Delta\nu}{\nu} \geq 0.1\%$, for external MAW $\frac{\Delta\nu}{\nu} \approx 10^{-3}\%$ is possible, that provides control of $\Delta H \geq 20\,\mathrm{A/m}$, $\alpha > 10^{-5}$, $\frac{\Delta\nu_0}{\nu_0} \approx 10^{-3}\%$, $\frac{\Delta H_{0i}}{H_{0i}} \approx 10^{-3}\%$.

Let's note that such a type of lines and measuring sensors were applied to measure the AFC parameters of reflected signals, however, correct processing of experimental data can be conducted only at electrodynamic calculation of electromagnetic wave excitation in the structures in view of some major factors for film structures.

In some cases, estimation of the specified above parameters for FDLS magnetized by a field H_0 at an arbitrary angle φ to the structure plane is required. This approach allows this problem to be solved as well.

Let's consider an opportunity of using features of electromagnetic waves propagation in FDLS in the far zone of the converter for diagnostic purposes. An FDT has similar properties. Theoretical analysis has shown that in a flat waveguide with metal screens at a distance $h_{1,3} \geq \frac{\lambda_{cr}}{2}$, in the prelimiting mode $(\nu \gg \nu_{cr})$ in weakly dissipative FDLS the following waves can be excited by outside sources and extended:

- Space fast and space surface waves, and return slow spatial waves at tangential magnetization
- Space fast and space slow spatial waves at normal magnetization

In obliquely magnetized structures (field $\overline{H}_0 \perp \kappa$ and H_0 directed at an angle φ to the ferrite plane) can be excited and extended:

- Space fast, space slow semi-surface and return slow spatial waves at magnetizations close to oblique $(0 \leq \varphi < 45°)$
- Space fast and space slow semi-spatial waves at magnetizations close to normal $(45° \leq \varphi < 90°)$

For direct coherent waves with close dispersion laws of their phase constants in such structures interference is possible. The reduction factor is:

- At the central frequency ν_0

$$K_\varphi(\nu_0) = 8.68\ \kappa''_\varphi(\nu_0)L + 20\lg\left|\cos\frac{\Delta\kappa'_\varphi L}{2}\right| \tag{6.1a}$$

- At frequencies $\nu > \nu_0$ *and* $\nu < \nu_0$ for $0 \leq \varphi < 45°$

$$K_\varphi = 8.68 \cdot (\pm\kappa''_{\varphi SSSW} \mp \kappa''_{FW})L + 20\lg\left|\cos\frac{\Delta\kappa'_{\varphi,t} L}{2}\right|; \tag{6.1b}$$

for $45° \leq \varphi < 90°$

$$K_\varphi = 8.68 \cdot (\pm\kappa''_{\varphi SSSLW} \mp \kappa''_{FW})L + 20\lg\left|\cos\frac{\Delta\kappa'_{\varphi,n} L}{2}\right|, \tag{6.1c}$$

where for $0 \leq \varphi < 45°$:

- At $\nu = \nu_0$

$$\kappa''_\varphi = \kappa''_{\varphi SSSW} = \kappa''_{\varphi FW}, \quad \Delta\kappa'_{\varphi,t} = \left|\kappa'_{\varphi SSSW} - \kappa'_{\varphi FW}\right|$$

– At $\nu > \nu_0$

$$\Delta\kappa'_{\varphi,t} = |\kappa'_{\varphi SSSW}(\nu > \nu_0) - \kappa'_{\varphi FW}(\nu > \nu_0)|$$

– At $\nu < \nu_0$

$$\Delta\kappa'_{\varphi,t} = |\kappa'_{\varphi SSSLW}(\nu < \nu_0) - \kappa'_{\varphi FW}(\nu < \nu_0)|$$

and for $45^\circ \leq \varphi < 90^\circ$:

– On $\nu = \nu_0$

$$\kappa''_{\varphi} = \kappa''_{\varphi SSSLW} = \kappa''_{\varphi FW}, \quad \Delta\kappa'_{\varphi,n} = |\kappa'_{\varphi SSSLW} - \kappa'_{\varphi FW}|$$

– At $\nu > \nu_0$

$$\Delta\kappa'_{\varphi,n} = |\kappa'_{\varphi SSSLW}(\nu > \nu_0) - \kappa'_{\varphi FW}(\nu > \nu_0)|$$

– At $\nu < \nu_0$

$$\Delta\kappa'_{\varphi,n} = |\kappa'_{\varphi SSSLW}(\nu < \nu_0) - \kappa'_{\varphi FW}(\nu < \nu_0)|$$

L is the length of the structure, the superscripts "+" and "−" correspond to the case of $\nu > \nu_0$, and the subscripts do to $\nu < \nu_0$.

The dispersions of direct waves (a) and envelope curves of interference attenuation (b) near the resonant frequencies $\nu_\perp$ and ν_H are shown in Fig. 6.1.

The dependences of transformation losses $K_{los.R}$, on L for loss parameters α: $1\text{–}2 \cdot 10^{-4}$, $2 - 1 \cdot 10^{-4}$ at $\nu_H = 1.9 \cdot 10^9\,\text{Hz}$, $h_2 = 42 \cdot 10^{-6}\,\text{m}$, $h_1 = h_3 = 1.029 \cdot 10^{-3}\,\text{m}$, $H_0 = 69.96\,\text{kA/m}$ are shown in Fig. 6.2. When the parameter of ferromagnetic losses increases the value of amplitude constant κ'' increases too, that leads to a proportional dependence of the dissipative term (the first one in Fig. 6.1*a*)*L*.

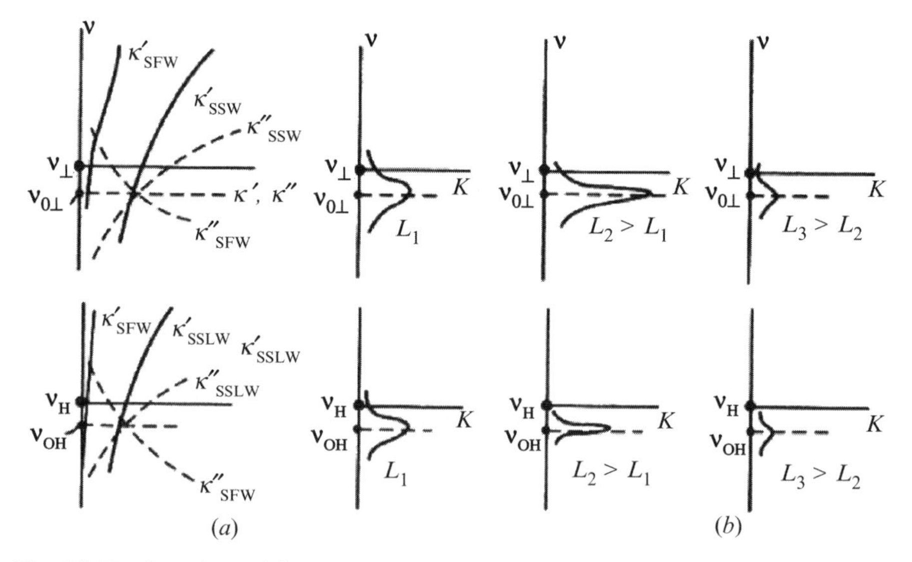

Fig. 6.1 The dispersions of direct waves (*a*) and envelope curves of interference attenuation (*b*) near the resonant frequencies $\nu_\perp$ and ν_H

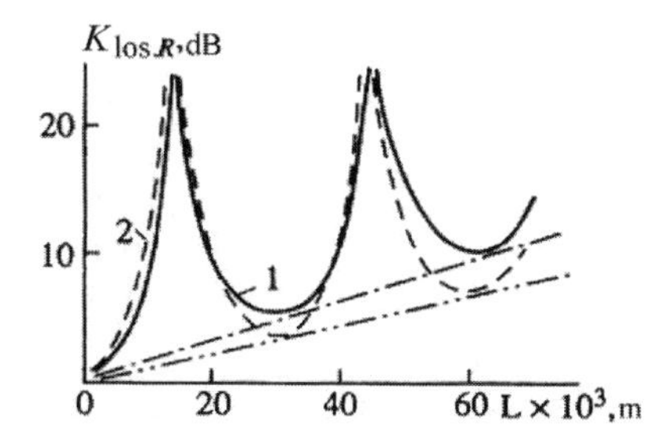

Fig. 6.2 The dependences of transformation losses $K_{los.R}$, on L for loss parameters α: 1–$2\cdot 10^{-4}$, $2-1\cdot 10^{-4}$ at $\nu_{\mathrm{H}} = 1.9\cdot 10^{9}\,\mathrm{Hz}$, $h_2 = 42\cdot 10^{-6}\,\mathrm{m}$, $h_1 = h_3 = 1.029\cdot 10^{-3}\,\mathrm{m}$, $H_0 = 69.96\,\mathrm{kA/m}$

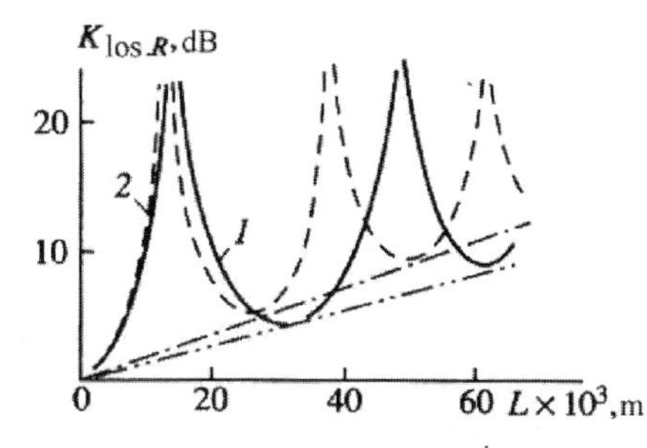

Fig. 6.3 The dependences $K_{los.R}$(L) for filling parameters $\frac{h_2}{a}$ for: 1–$2\cdot 10^{-3}$, 2–$2\cdot 10^{-2}$ at $\nu_{\mathrm{H}} = 1.9\cdot 10^{9}\,\mathrm{Hz}$, $a = 2\cdot 10^{-2}\,\mathrm{m}$, $h_1 = h_3$, $H_0 = 69.96\,\mathrm{kA/m}$, $4\pi M_S = 0.214\,\mathrm{T}$

The course of the dispersive characteristics of fast and slow waves near the resonant frequency of the structure, i.e., the length L at which Eq. (6.1) are satisfied, depends on the parameter of ferrite filling of a flat waveguide $\frac{h_2}{a}$, $a = h_1 + h_2 + h_3$.

The dependences $K_{los.R}(L)$ for filling parameters $\frac{h_2}{a}$ are shown in Fig. 6.3 for: 1–$2\cdot 10^{-3}$, $2-2\cdot 10^{-2}$ at $\nu_{\mathrm{H}} = 1.9\cdot 10^{9}\,\mathrm{Hz}$, $a = 2\cdot 10^{-2}\,\mathrm{m}$, $h_1 = h_3$, $H_0 = 69.96\,\mathrm{kA/m}$, $4\pi M_S = 0.214\,\mathrm{T}$.

When the resonant frequency increases the amplitude constant of waves κ'' and $\Delta\kappa'$ increase, that leads to reduction of the resonant length of the structure. The dependences $K_{los.R}(L)$ for various frequencies ν_{H} are shown in Fig. 6.4 for: 1–$153.27\cdot 10^{10}\,\mathrm{Hz}$ at $H_0 = 4.326\,\mathrm{MA/m}$, $2-1.93\cdot 10^{10}\,\mathrm{Hz}$ at $H_0 = 69.96\,\mathrm{kA/m}$ in a structure with parameters $\frac{h_2}{a} = 2\cdot 10^{-2}$, $h_1 = h_3$, $a = 2.1\cdot 10^{-2}\,\mathrm{m}$, $4\pi M_S = 0.214\,\mathrm{T}$.

From Figs. 6.2–6.4 such functional dependences follow:

- Steepness $|K_{los.R}(L)|_{\min}$ (dot-and-dash lines)
- "Periodicity" of following $|K_{los.R}(L)|_{\max}$ and $|K_{los.R}(L)|_{\min}$
- Variation range $|K_{los.R}(L)|_{\max}$ and $|K_{los.R}(L)|_{\min}$ which carries information about the parameters of fast and slow waves, high-frequency properties of a layered structure

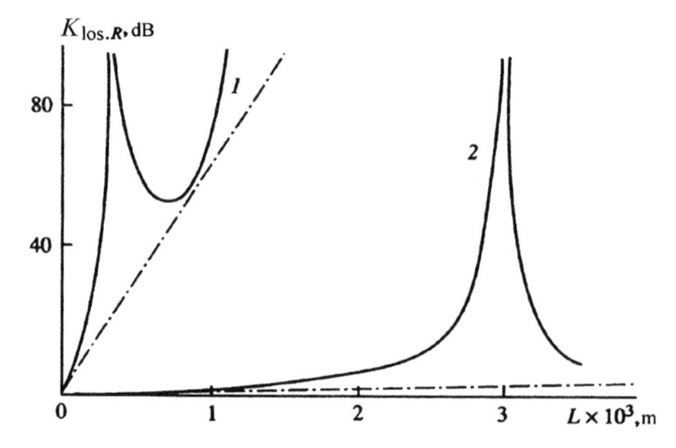

Fig. 6.4 The dependences $K_{los.R}(L)$ for various frequencies ν_H for: $1 - 153.27 \cdot 10^{10}\,\text{Hz}$ at $H_0 = 4.326\,\text{MA/m}$, $2 - 1.93 \cdot 10^{10}\,\text{Hz}$ at $H_0 = 69.96\,\text{kA/m}$ in a structure with parameters $\frac{h_2}{a} = 2 \cdot 10^{-2}$, $h_1 = h_3$, $a = 2.1 \cdot 10^{-2}\,\text{m}$, $4\pi M_S = 0.214\,\text{T}$

Measuring devices can be designed on rectangular waveguides, short-circuited wave guides, and strip lines. Devices of the short-circuited type operate by reflection $L_{ref} = 2L$, rectangular waveguides do by passage, and the length of the structure is L. The interference method provides rather a high sensitivity, and standard panoramic SWR_e and attenuation measuring instruments are used for registration of signal.

Waveguide devices of the reflective (*a*) and passage (*b*) types, a device on a slot line (*c*), a waveguide (*d*) and slot (*e*) measuring cells for nondestructive control, a variant of multilayered structure (*f*), cells for control of parameters of obliquely magnetizated FDLS with an opportunity of turning the waveguide (*g*) and structures (*h*), are shown in Fig. 6.5. Localization of the control area under nondestructive measurements was reached due to the application of external magnetic fields to fall down outside of the analysis zone. Measuring schemes for reflection (*a*) and passage are shown in Fig. 6.6*b*.

Let's consider basic parameters determined by the interference mechanism.

The amplitude constant κ'' on the central frequency ν_0 is determined from the average steepness of the experimental dependence $K_{los.R}(L)$, and

$$\kappa''(\nu_0) = \frac{K(L_2) - K(L_1)}{8.68(L_2 - L_1)}, \tag{6.2}$$

where $K(L_{1,2})$ is the loss factor by transformation in dB, $\kappa''(\nu_0)$ the amplitude constant in $\left[\frac{1}{L}\right]$.

The relative error is

$$\left|\frac{\Delta\kappa''}{\kappa''}\right| = \left|\frac{\Delta K_{los.R}(L_2) + \Delta K_{los.R}(L_{21})}{K_{los.R}(L_1) - \Delta K_{los.R}(L_2)}\right| + \left|\frac{\Delta L_1 + \Delta L_2}{L_1 - L_2}\right|, \tag{6.3}$$

where the symbol Δ hereinafter means the absolute error of a value.

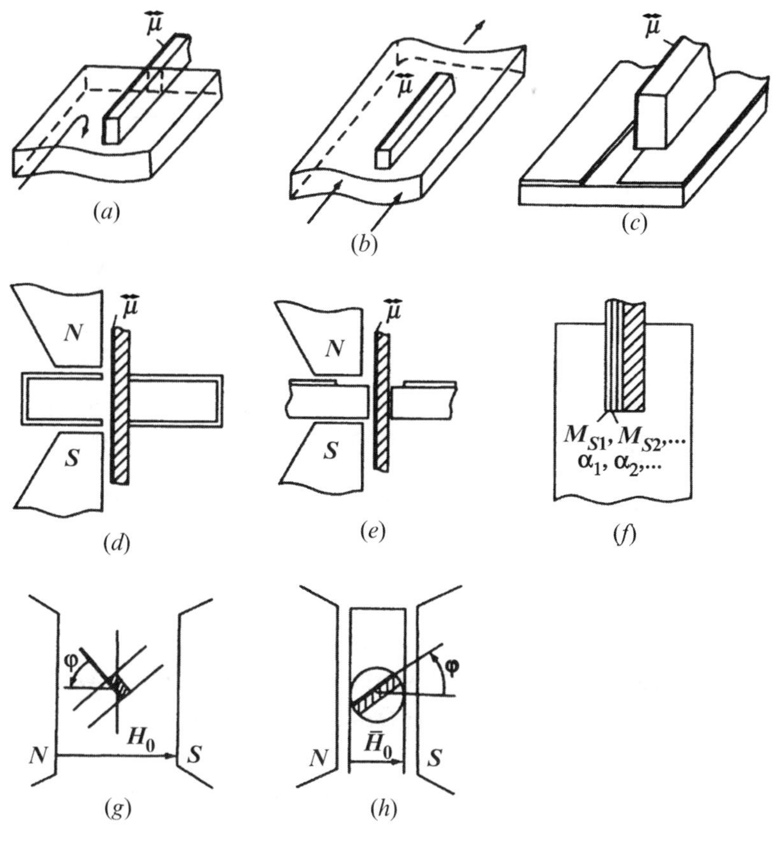

Fig. 6.5 Waveguide devices of the reflective (*a*) and passage (*b*) types, a device on a slot line (*c*), a waveguide (*d*) and slot (*e*) measuring cells for nondestructive control, a variant of multilayered structure (*f*), cells for control of parameters of obliquely magnetizated FDLS with an opportunity of turning the waveguide (*g*) and structures (*h*)

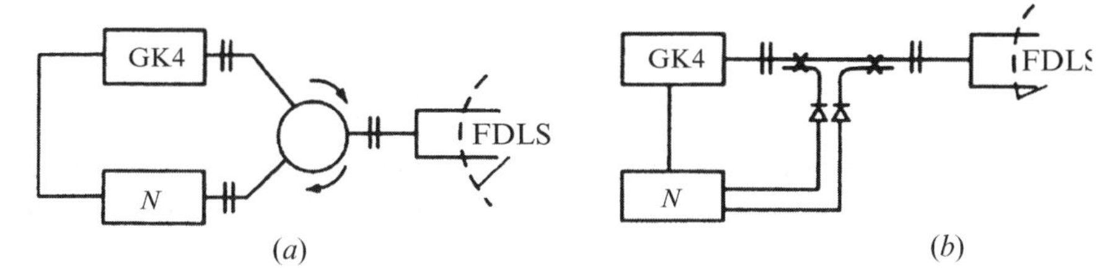

Fig. 6.6 Measuring schemes for reflection (*a*) and passage (*b*)

Table 6.1 The values of k in a range of frequencies $\nu_\perp = (12.7-17.5)$ GHz for two YIG films with a thickness $h_2 = 36\cdot10^{-6}$ m, $h_2 = 12\cdot10^{-6}$ m, $4\pi MS = 0.176$T

ν_0, GHz	Tangential magnetization			Normal magnetization		
	$\lvert K_{los.R}(L_2) - K_{los.R}(L_1)\rvert$, dB	(L_2-L_1), $\cdot10^2$ m	$\kappa''_\parallel$, m^{-1}	$\lvert K_{los.R}(L_2) - K_{los.R}(L_1)\rvert$, dB	(L_2-L_1), $\cdot10^2$ m	$\kappa''_\perp$, m^{-1}
12.7	3.60	2.3	73	0.8	4.4	21
17.5	0.85	2.0	49	1.2	4.8	29
12.7	0.3	3.5	10	1.0	4.8	24
17.5	0.8	5.6	16	0.9	5.2	20

At using of standard (R2-28, R2-65, R2-69) panoramic measuring instruments of SWR$_e$ and attenuation, and a laboratory MBS-2 microscope with a coordinate scale graduated to $2.5\cdot10^{-6}$m the value is $\frac{\Delta\kappa''}{\kappa''} \leq (3\text{–}5)\%$.

The values of κ'' in a range of frequencies $\nu_\perp = (12.7\text{–}17.5)$ GHz for two YIG films with a thickness $h_2 = 36\cdot10^{-6}$m, $h_2 = 12\cdot10^{-6}$m, $4\pi M_S = 0.176$T are shown in Table 6.1.

6.1.1 *Amplitude-Constant Difference of Slow and Fast Waves on Frequencies Tuned Out from the Center One*

At aninterference minimum from (6.1) for difference of amplitude constants we have:

– At tangential magnetization

$$|\pm\kappa''_{\varphi,SSW} \mp \kappa''_{\varphi,SFW}| \approx \left|\frac{K_{\varphi,\min}}{8.68\,L_{\min}}\right| \tag{6.4a}$$

– At normal magnetization

$$|\pm\kappa''_{\varphi,SSLW} \mp \kappa''_{\varphi,SFW}| \approx \frac{K_{\varphi,\min}}{8.68\,L_{\min}} \tag{6.4b}$$

At tuning out in the area of AFC boundaries (on frequencies $\nu_1 = \nu_0 + \Delta\nu$, $\nu_2 = \nu_0 - \Delta\nu$, $\Delta\nu = (3\text{–}5)\Delta\nu_{3dB}$, $\Delta\nu_{3dB}$ is the wave range by a level of 3 dB) we can neglect small corrections to amplitude constants in Eq. (6.4*a*). Then

– At tangential magnetization

$$\kappa''_{\varphi,SSW} \approx \left(\frac{K_{\varphi,\min}}{8.68\,L_{\min}}\right)_\parallel, \quad \nu > \nu_0 + \Delta\nu,$$

$$\kappa''_{\varphi,SFW} \approx \left(\frac{K_{\varphi,\min}}{8.68\,L_{\min}}\right)_\parallel, \quad \nu < \nu_0 - \Delta\nu \tag{6.5}$$

– At normal magnetization

$$\kappa''_{\varphi,SFW} \approx \left(\frac{K_{\varphi,\min}}{8.68\ L_{\min}}\right)_{\perp}, \quad \nu > \nu_0 + \Delta\nu,$$
$$\kappa''_{\varphi,SSLW} \approx \left(\frac{K_{\varphi,\min}}{8.68\ L_{\min}}\right)_{\perp}, \quad \nu < \nu_0 - \Delta\nu \tag{6.6}$$

The range of the measured dispersions of amplitude constants is subject to the sensitivity of the using equipment.

The relative error is

$$\left|\frac{\Delta\kappa''}{\kappa''}\right| = \left|\frac{\Delta K_{\min}}{K_{\min}} + \frac{\Delta L_{\min}}{L_{\min}}\right|. \tag{6.7}$$

The difference of the phase constants of slow and fast waves (phase disalignment on the central frequency) follows from Eq. (6.1*a*)

$$\Delta\kappa'(\nu_0) = \frac{2}{L}\arccos\left(10\frac{K - 8.68\ \kappa'' L}{20}\right), \tag{6.8}$$

where κ'' is found by the technique stated above.

The relative error is

$$\left|\frac{\Delta(\Delta\kappa')}{\Delta\kappa'}\right| = \left|\frac{\Delta L}{L}\left[\Delta K + 8.68(\Delta\kappa'' L + \kappa''\Delta L)\right]\frac{(K - 8.68\ \kappa'' L)10^A \ln A}{\arccos 10^A\sqrt{1-10^A}}\right|, \tag{6.9}$$

where $A = (K - 8.68\kappa'' L)/20$.

The values of $\Delta\kappa$ for frequencies $\nu_\perp = (12.7\text{–}17.5)\,\text{GHz}$ for two YIG films with $h_2 = 36\cdot 10^{-6}\,\text{m}$ and $h_2 = 12\cdot 10^{-6}\,\text{m}$, $4\pi M_S = 0.176\,\text{T}$ are shown in Table 6.2.

Dissipative factors such as the line width of FMR ΔH_φ and the parameter of ferromagnetic losses α_φ in an arbitrary magnetized FDLS can be estimated from comparison of theoretical and experimental dependences of the stopband $(\Delta\nu_{3dB})_{Th}$ (Th – theoretical) and $(\Delta\nu_{3dB})_{Ex}$ (Ex – experimental). For a symmetrically loaded

Table 6.2 The values of $\Delta\kappa$ for frequencies $\nu_\perp = (12.7\text{–}17.5)\,\text{GHz}$ for two YIG films with $h_2 = 36\cdot 10^{-6}\,\text{m}$ and $h_2 = 12\cdot 10^{-6}\,\text{m}$, $4\pi M_S = 0.176\,\text{T}$

$\nu_\perp$, *GHz*	Tangential magnetization				Normal magnetization			
	κ'', m^{-1}	$K_\parallel, dB$	$L\cdot 10^2, m$	$\Delta\kappa'_\parallel, m^{-1}$	κ'', m^{-1}	$K_\perp, dB$	$L\cdot 10^2, m$	$\Delta\kappa'_\perp, \text{m}^{-1}$
				$h_2 = 36\,\text{mc}\mu\text{m}$				
12.7	0.73	0.6	4.8	64	0.21	0.3	4.2	55
17.5	0.49	0.85	4.0	72	0.29	0.9	5.2	53
				$h_2 = 12\,\mu\text{m}$				
12.7	0.10	0.3	7.0	30	0.24	0.4	4.0	60
17.5	0.16	0.4	5.8	41	0.29	0.5	6.0	47

ferrite film from its two sizes, i.e., the bases at reversal of the external magnetic field $\overline{H}_0$ in tangentially magnetized structures the dissipative factors near both the surfaces of the film ΔH_1, α_1 and ΔH_2, α_2 can be determined nondestructively. For a normally magnetized FDLS we get the parameters $\Delta\overline{H}$ and $\bar{\alpha}$ belonging to the middle area of the ferrite film. A basic parameter tto characterize the ferrite film quality in the millimetric range (the transversal gradient dissipative factors $\nabla_x \Delta H$ and $\nabla_x \alpha$) can be determined from these data. The influence of shape anisotropic fields and crystallographic anisotropie is considered by means of special requirements to the dimensions of the investigated films and their orientations and discussed in Chapter 7.

6.2 Control by Transparency Effect of a Beyond-Cutoff Waveguide with Ferrite-Dielectric Filling Near Its Resonant Frequency

Our theoretical and experimental investigations of electromagnetic wave excitation (by outside sources) and propagation in flat and rectangular waveguides filled with ferrite film layered structures have found a number of features of signal passage with low losses on a frequency near the resonant one. The opportunity of calculation of the AFC envelope of a signal, losses by transformation in the near zone, and transfer losses in the distant zone of radiation in view of basic features of the millimetric range has allowed experimental results to be adequatly described and most effective ways of using such devices for parameter control of ferrite films to be found.

Electrodynamic self-consistent excitation of waveguides coupled by a small-sized spheric or ellipsoidal ferromagnetic resonator was investigated earlier [8]. Selective bridging of two waveguides separated by a below-cutoff section with a spheric ferromagnetic resonator was studied [8]. The parameters of a passing signal, namely, transfer, passband, losses on the center frequency and outside of the passband depended on the magnetization of the model, its losses, volume, type of coupling with waveguides. In paper [496] the results of electrodynamic calculation and some data of our experimental investigation of the filtering properties of a similar device in the centimetric range, in which a ferrite plate (instead of a sphere) was located in the middle part of the below-cutoff section, are presented.

Until recently the folloinf things were not considered from an electrodynamic point of view:

- Properties of below-cutoff waveguides with an extensive layered structure on the basis of ferrite film, having matching lugs of the structure from its below-cutoff section into the supplying and diverting waveguides, subject to field excitation in dissipative structures by outside electrical and magnetic currents both in the near and far zones of radiation in view of specific factors of the millimetric range
- Features of selective processes in FDT and slotted-guide converters with a cross-section outside magnetic current at orthogonal orientation of FDLS to the exciting plane of the slot

The theoretical approach developed in the monograph allows analyzing properties of such electrodynamic systems for the full basis of *LE* and *LM* waves in structures with random orientation of an external magnetic field $\overline{H}_0$ with due account of the impedances of shielding surfaces, a bigyrotropic structure with dissipation of its electric and magnetic parameters.

The basic conclusion of the our investigations is that in below-cutoff waveguides with extensive film weakly-dissipative ($\alpha < 10^{-4}$) FDLS a binary mechanism of bandpass filtering near the center frequency of transparency operates.

The first mechanism being basis for the millimeter range is connected with that fact that the passband of a slotted-guide transformer in the flat of aperture in the short-circuiting waveguide face substantially smaller than the passband of the ferrite-dielectric transformer of the pritrudinf part of FDLS. FDT has an extensive area of influence of outside HF fields with a length $L_{FDT} \geq \lambda/4$ and in the two-wave mode its factor of transformation losses of active power is related to interference of direct and slow electromagnetic waves, and $\kappa_{los.R}$ periodically increases with growing L_{FDT} at a growing (on the average) level of losses. The central frequency of selective signal attenuation in FDT is $\nu_{FDT} < \nu_0$, ν_0 being the resonant frequency of the loss-free structure, and the attenuation band of frequencies $\Delta\nu$ has the minimum value at $|K_{los.R}|_{\max}$ and the maximum one at $|K_{los.R}|_{\min}$, that is essentially depends on L_{FDT}. The passband of a slotted-guide transformer for FDLS with orthogonal orientation to the exciting plane and being together with the transformer in a below-cutoff regime ($\nu \ll \nu_{cr}$) or for a three-layer structure $\lambda \gg 2(h_1+h_2+h_3)$ is characterized by an increased homogeneity of exciting HF fields $W_{SGC} \gg h_2$, and its center frequency is $\nu > \nu_0$ for ferrite films. Therefore, the basis factor of signal filtering in waveguide – beyond-cutoff devices with FDT is the passband in the beyond-cutoff regime of SGC with orthogonal orientation of the ferrite film to the exciting plane.

The second mechanism of band-pass filtering in waveguide – beyond-cutoff electrodynamic systems with weakly dissipative ferrite films near the resonant frequency is connected with a decrease of the frequency band of the transmission factor at an increase of the length of the beyond-cutoff section L_{bcs} (bcs – beyond-cutoff section) or of the extent of the distant zone of the irradiator, underlie an increase of the steepness of the amplitude constant of wave extending in the amplitude beyond-cutoff section with a ferrite film $d\kappa''/d\omega$ at tuning out from the center frequency ν.

These two factors are supplemented with known properties of a beyond-cutoff waveguide. First of them is connected with a jump of the cross-sections of the supplying or tailrace and beyond-cutoff waveguides, that leads to their sharp mismatch over wave resistance and to an increase of the out-of-band barrage of a signal. The second property is connected with the dependence of attenuation introduced by the beyond-cutoff section in its length L_{bcs}. These circumstances allow obtaining high levels of the out-of-band barrage of a signal at simple technical designs, that is important when development of diagnostic devices and various types of filters for the millimeter range.

In Figs. 6.7–6.9 the dependences explaining physical features of transformation and distribution of HF power in waveguide – beyond-cutoff devices, containing

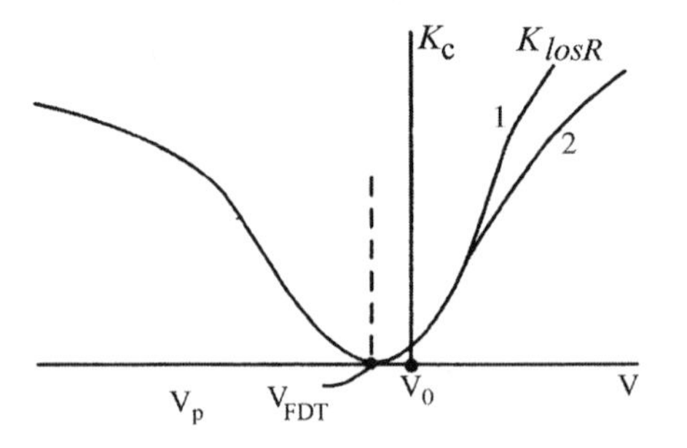

Fig. 6.7 The conversion coefficient of active capacity $K_{los.R}$ – curve 1 in the near region of the input FDT which is dispersed over a length $S \geq \lambda/4$, λ being the wavelength in the waveguide

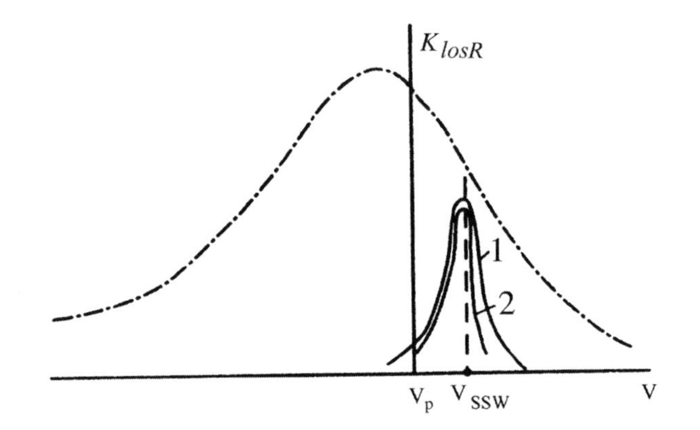

Fig. 6.8 The conversion coefficient of active power in the near region of SGC in the beyond-cutoff regime $K_{los.R} - 1$ is shown, and $(K_{los.R})_{MAW} = (K_H)_{MAW} + K_{car}$, where $(K_H)_{MAW}$ is the conversion coefficient of active power into heat in the near region; K_{car} the transfer factor of power in the distant region of the transformer, in the transmission line on FDLS in the beyond-cutoff waveguide

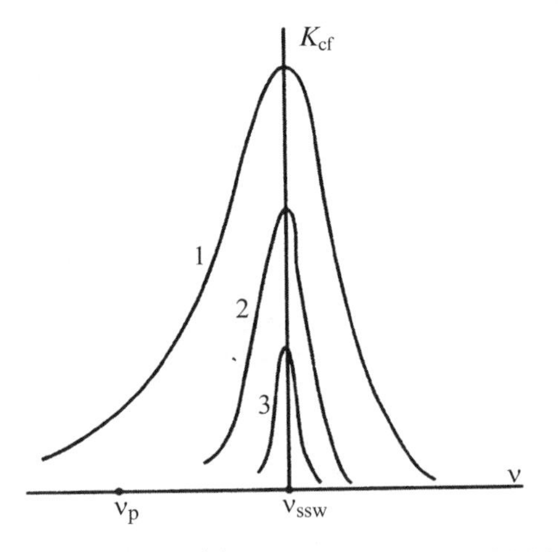

Fig. 6.9 The scaled-up regularities $K_{car}(\nu)$ in the distant region of SGC – 1 at a distance L_1 from the longitudinal axis of the slot, in a beyond-cutoff waveguide at various distances from the transformer $2 - L_2 > L_1$, $3 - L_3 > L_2 > L_1$

layered structures on the basis of a ferrite film near the resonant frequency are presented.

In Fig. 6.7 the conversion coefficient of active capacity $K_{los.R}$ – curve 1 in the near region of the input FDT which is dispersed over a length $S \geq \lambda/4$, λ being the wavelength in the waveguide, is shown, and

$$(K_{los.R})_{FDT} = (K_{con})_{FDT} + K_{car},$$

where K_{con} is the HF-heat conversion coefficient related to losses in FDT; K_{car} the carryover factor of active power accumulated in FDT into power on SGC input in the beyond-cutoff regime with an orthogonal oriented FDLS to the slot plane (curve 2).

In Fig. 6.8 the conversion coefficient of active power in the near region of SGC in the beyond-cutoff regime $K_{los.R} - 1$ is shown, and

$$(K_{los.R})_{SGC} = (K_{con})_{SGC} + K_{car},$$

where $(K_{con})_{SGC}$ is the conversion coefficient of active power into heat in the near region; K_{car} the transfer factor of power in the distant region of the transformer, in the transmission line on FDLS in the beyond-cutoff waveguide. The transfer factor on the output of FDT is shown by the dot-and-dash line. The central frequency for SGC in the beyond-cutoff region ν_{SGC} is located above the resonant frequency of the structure ν_r (r – resonance).

In Fig. 6.9 the scaled-up regularities $K_{car}(\nu)$ in the distant region of SGT – 1 at a distance L_1 from the longitudinal axis of the slot, in a beyond-cutoff waveguide at various distances from the transformer $2 - L_2 > L_1$, $3 - L_3 > L_2 > L_1$ are presented.

The output FDT is usually identical to the input one and at $L_{FDT} > \frac{\lambda}{4}$ its passband is essentially wider than the band of TL on FDLS in the beyond-cutoff region.

Therefore, for the given devices signal filtering in the near region of the input SGC is characteristic, first of all, and more narrow-band filtering depends on the length of the beyond-cutoff section. These circumstances have fundamental importance in the development of devices for diagnostics of ferrite fims and operated filters.

In Fig. 6.10 some design variants of measuring cells on beyond-cutoff waveguides are shown: (Fig. 6.10*a*) – on a waveguide of the conventional section, (Fig. 6.10*b*, *c*) – on waveguides of a shorter section. For non-destructive inspection over the wide walls of pull-on and diverting waveguides and in the side walls of the beyond-cutoff section through grooves with a size close to the thickness of the investigated FDLS are made. This leads to the appearance of stray coupling of dripping HF power from input to output on the dielectric base and directly on the slots in waveguides, that reduces the sensitivity and range of measured parameters.

The level of stray coupling decreased at due choice of the design of waveguides, for example, by using of pull-on and diverting waveguides of the same decreased section as in the beyond-cutoff section and the proper arrangement of HF absorbers (Fig. 6.10*b*). In the short-wave part of the centimeter range, HF fields are well localized within the range of beyond-cutoff waveguide with FDLS. Besides, outside of a waveguide the structure is in the prelimiting regime $(\nu \gg \nu_{cr})$, metal screens are

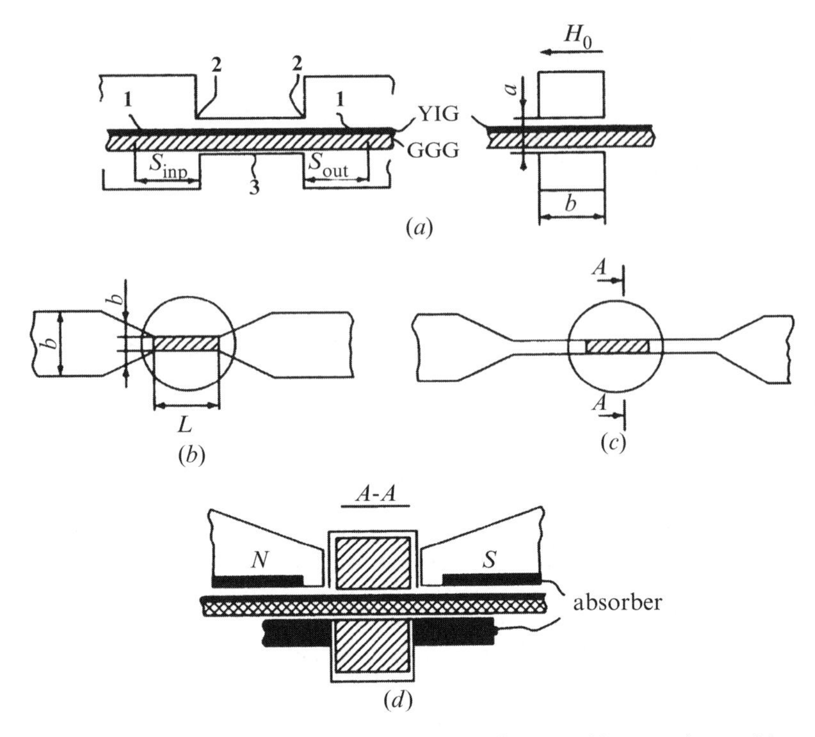

Fig. 6.10 Variants of measuring cells on beyond-cutoff waveguides are shown: (*a*) – on a waveguide of the conventional section, (*b, c*) – on waveguides of a shorter section

removed, and the wave processes in FDLS have a number of distinctive features – (the central frequencies are detuned, outside of the beyond-cutoff section a signal should be selectively attenuated in a wider band than passage in the beyond-cutoff section). In some cases, additional measures for localization of the analyzed area on FDLS in the form of non-uniform boundary magnetic fields (Fig. 6.10*b*, *c*, section *A-A*) were used.

The area under analysis in the beyond-cutoff section is $S_a = b_{bcs}L_{bcs}, b_{bcs}$ (a – analysis) – the width of the beyond-cutoff section, L_{bcs} – its length. Sagging of HF fields outside of the lateral walls of the beyond-cutoff section is $\Delta a_{bcs} \leq$ (1–2) $\cdot\, a_{bcs}, a_{bcs} = h_1 + h_2 + h_3 + h_4$, h_1 is thethickness of the dielectric base, h_2 – the thickness of the ferrite film, h_3 the thickness of the protective coat, h_4 – a technological gap for scanning. Then $S_a \approx (b_{bcs} + 2\Delta a_{bcs})L_{bcs}$.

In such cells the area under analysis can have various shape and represent a square $(b_{bcs} \times b_{bcs})$, a rectangular $(b_{bcs} \times L_{bcs})$, a filamentous area $(b_{bcs} \ll L_{bcs})$, and its value is $5 \cdot 10^{-7} \leq S_a(m^2) \leq 10^{-4}$. In the long-wave part of the microwave range instant analysis over practically the total area of the structure Ø (Ø – diameter) $\approx$ (45–76) $\cdot\, 10^{-3}$ m can be provided.

Good agreement of experimental results was observed at measurements of the same structure by a nondestructive method in a cell with lateral grooves and working elements made of the same FDLS, in a rectangular beyond-cutoff waveguide.

Let's discuss basic requirements and directions of increasing sensitivity and frequency resolution.

In Fig. 6.11*a* the dependences of introduced attenuation per length unit of the beyond-cutoff section $K(\mathrm{dB/m})$ are presented for various parameters of filling $\frac{h_2}{a}$ of the beyond-cutoff section with a ferrite layer: $1 - \frac{h_2}{a} = 10^{-2}$, $2 - \frac{h_2}{a} = 4 \cdot 10^{-2}$, $3 -$

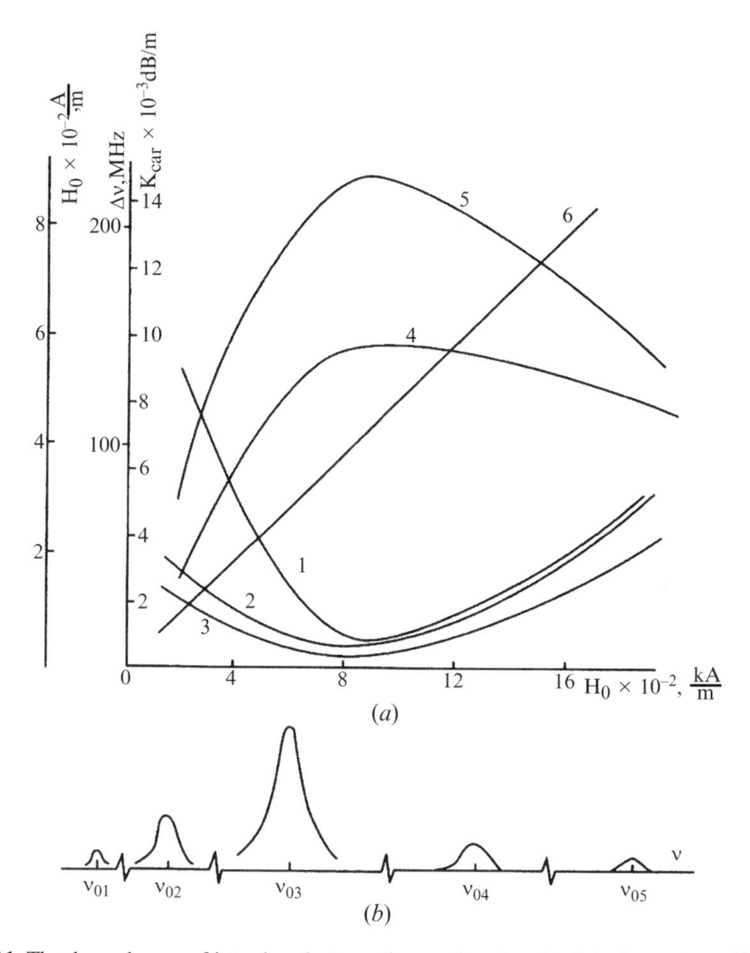

Fig. 6.11 The dependences of introduced attenuation per length unit of the beyond-cutoff section $K(\mathrm{dB/m})$ are presented for various parameters of filling $\frac{h_2}{a}$ of the beyond-cutoff section with a ferrite layer (*a*): $1 - \frac{h_2}{a} = 10^{-2}$, $2 - \frac{h_2}{a} = 4 \cdot 10^{-2}$, $3 - \frac{h_2}{a} = 6 \cdot 10^{-2}$; passbands $\Delta\nu_{3dB}(\mathrm{MHz}), 4 - \frac{h_2}{a} = 4 \cdot 10^{-2}$, $5 - \frac{h_2}{a} = 6 \cdot 10^{-2}$ for parameters $\alpha = 5 \cdot 10^{-4}, h_1 = h_3, a = 1 \cdot 10^{-3}\,\mathrm{m}, 4\pi M_S = 0.176\,\mathrm{T}$ and the dependence of line width of ferromagnetic resonance $\Delta H_{\|}$ – 6 on the magnetic field H_0 overlapping the range of resonant frequencies for YIG up to 70 GHz; the qualitative appearance of AFC of signals on the output of the device in this range of frequencies is shown (*b*)

$\frac{h_2}{a} = 6 \cdot 10^{-2}$; passbands $\Delta\nu_{3dB}$(MHz), $4 - \frac{h_2}{a} = 4 \cdot 10^{-2}$, $5 - \frac{h_2}{a} = 6 \cdot 10^{-2}$ for parameters $\alpha = 5 \cdot 10^{-4}, h_1 = h_3, a = 1 \cdot 10^{-3}\,\text{m}, 4\pi M_S = 0.176\,\text{T}$, and the dependence of line width of ferromagnetic resonance $\Delta H_{\|} - 6$ on the magnetic field H_0 overlapping the range of resonant frequencies for YIG up to 70 GHz. In Fig. 6.11b the qualitative appearance of AFC of signals on the output of the device in this range of frequencies is shown. For a ferrite film symmetrically loaded with layers of dielectrics with $h_1 = h_3$ in the near part of the millimetric range ($\nu_{\perp} \approx 30\,\text{GHz}$) at various parameters of filling, $\frac{h_2}{a}$ has a minimum of introduced losses for signal propagation, that provides an increase of sensitivity. In a higher range the influence of FMR line widening intensifies. For parameters of filling $\frac{h_2}{a} \approx (4\text{–}6) \cdot 10^{-2}$ the highest sensitivity in a range of frequencies of 9–70 GHz is realized. For such a structure the frequency resolution is highest in the microwave range, and in the EHF range it is higher at lower $\frac{h_2}{a}$ (for $a = const$ this means the usage of thinner ferrite films).

The simultaneous influence of filling of a beyond-cutoff waveguide $\frac{h_2}{a}$ and the position of the ferrite layer $\frac{h_1}{a}$ on the passband $\Delta\Omega = \frac{\Delta\nu_{3dB}}{\nu_{\perp}} - 1$, 2 and introduced attenuation on the central frequency $K_0 - 3$, 4 is illustrated by Fig. 6.12, where 1, $3 - \frac{h_2}{a} = 2 \cdot 10^{-2}$, 2, $4 - \frac{h_2}{a} = 4 \cdot 10^{-2}$, at $\nu_{\text{H}} = 3 \cdot 10^{10}\,\text{Hz}, \alpha = 10^{-4}, 4\pi M_S = 0.176\,\text{T}$.

From these dependences (Figs. 6.11, 6.12) one can see that the requirements for decreasing introduced losses and increasing frequency resolution are inconsistent.

In Fig. 6.13 dependences of the passband $\Delta\Omega - 1$ and introduced losses K at the central frequency on the width of a beyond-cutoff waveguide a_{bcs} for a struc-

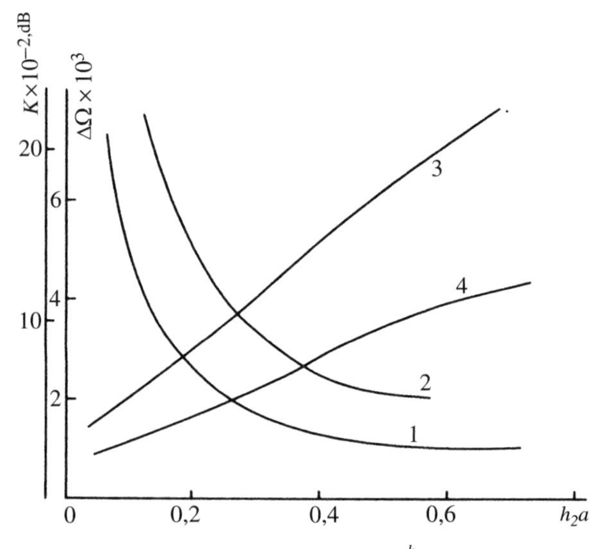

Fig. 6.12 Influence of filling of a beyond-cutoff waveguide $\frac{h_2}{a}$ and the position of the ferrite layer $\frac{h_1}{a}$ on the passband $\Delta\Omega = \frac{\Delta\nu_{3dB}}{\nu_{\perp}} - 1$, 2 and introduced attenuation on the central frequency $K_0 - 3$, 4 is illustrated by Fig. 6.12, where 1, $3 - \frac{h_2}{a} = 2 \cdot 10^{-2}$, 2, $4 - \frac{h_2}{a} = 4 \cdot 10^{-2}$, at $\nu_{\text{H}} = 3 \cdot 10^{10}\,\text{Hz}$, $\alpha = 10^{-4}, 4\pi M_S = 0.176\,\text{T}$

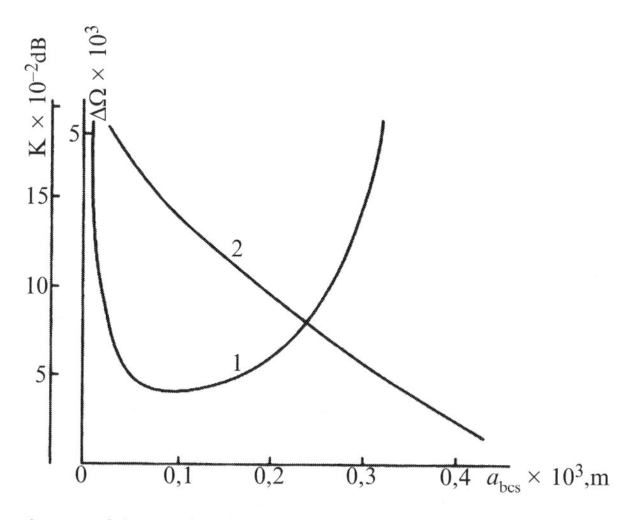

Fig. 6.13 Dependences of the passband $\Delta\Omega - 1$ and introduced losses K at the central frequency on the width of a beyond-cutoff waveguide a_{bcs} for a structure with parameters $\alpha = 10^{-4}$, $h_2 = 10^{-5}\,\mathrm{m}$, $h_1 = h_3$, $a = 1 \cdot 10^{-3}\,\mathrm{m}$, $H_0 = 835.48\,\mathrm{kA/m}$ $4\pi M_S = 0.176\,\mathrm{T}$

ture with parameters $\alpha = 10^{-4}$, $h_2 = 10^{-5}\,\mathrm{m}$, $h_1 = h_3$, $a = 1 \cdot 10^{-3}\,\mathrm{m}$, $H_0 = 835.48\,\mathrm{kA/m}$ $4\pi M_S = 0.176\,\mathrm{T}$ are shown. With growing a_{bcs} the value of K monotonously decreases, that is due toa weaker influence of cutoff frequency. The passband $\Delta\Omega$ sharply decreases in $0 \leq a(m) \leq 5 \cdot 10^{-4}$, that is connected with a decrease in ferrite filling of the waveguide. In rather narrow boundaries $5 \cdot 10^{-4} \leq a(m) \leq 1.5 \cdot 10^{-3}$ the passband is minimal and practically does not change, and when $a \geq (2\text{–}3) \cdot 10^{-3}\,\mathrm{m}$ an increase of $\Delta\Omega$ is observed, that is connected with the prevalence of dissipative factors in the ferrite film.

The role of ferromagnetic losses α and conductivity of metal coatings $\sigma_{1,3}$ in a beyond-cutoff waveguide is illustrated by Fig. 6.14, where $1 - \alpha = 1 \cdot 10^{-3}$, $2\text{–}5 \cdot 10^{-4}$, $3 - 2 \cdot 10^{-4}$, $4 - 1 \cdot 10^{-4}$, $5\text{–}5 \cdot 10^{-5}$, in a structure with parameters $\nu_\perp = 31.77 \cdot 10^9\,\mathrm{Hz}$, $h_1 = h_3, h_2 = 20 \cdot 10^{-6}\,\mathrm{m}, 4\pi M_S = 0.176\,\mathrm{T}, \sigma_{Cu}$ is the conductivity of copper; the conductivities of some metals are specified by points. It is obvious that the conductivity of coatings $\sigma_{1,3}$ most strongly influences the introduced signal depression in structures with rather a high level of ferromagnetic losses $\alpha_\| \geq 10^{-3} - 10^{-4}$ ($\Delta H_\| \approx 83.55\text{–}835.5\,\mathrm{A/m}$), and in structures with high-quality films at $\alpha_\| < (10^{-4} - 10^{-5})$ the influence of σ is weak.

In Fig. 6.15 dependences of K on the parameter of ferromagnetic losses α for SSMSW – 1 and SSSW – 2, 3 in a beyond-cutoff waveguide an with ideal ($\sigma = \infty$) and real conductivity of its metal coatings $1 - \sigma = \infty$, $2 - \sigma = \infty$, $3 - \sigma = 5.9 \cdot 10^7(\mathrm{Ohm} \cdot \mathrm{m})^{-1}$ for a structure on frequency $\nu_\perp = 31.72 \cdot 10^9\,\mathrm{Hz}, 4\pi M_S = 0.176\,\mathrm{T}$ are presented for comparison. For SSSW in a beyond-cutoff waveguide with $\sigma = \infty$ the losses on the center frequency ν_0 are essentially higher in comparison with SSSW, that is connected with an essential distinction of the group speeds of these waves near the resonant frequency and incorrectness of using the

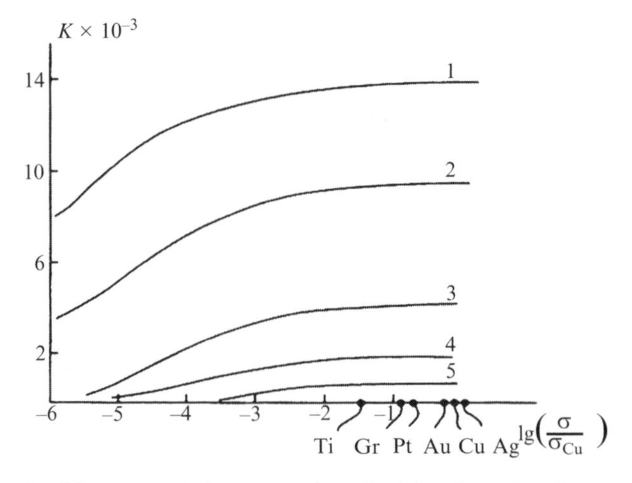

Fig. 6.14 The role of ferromagnetic losses α and conductivity of metal coatings $\sigma_{1,3}$ in a beyond-cutoff waveguide

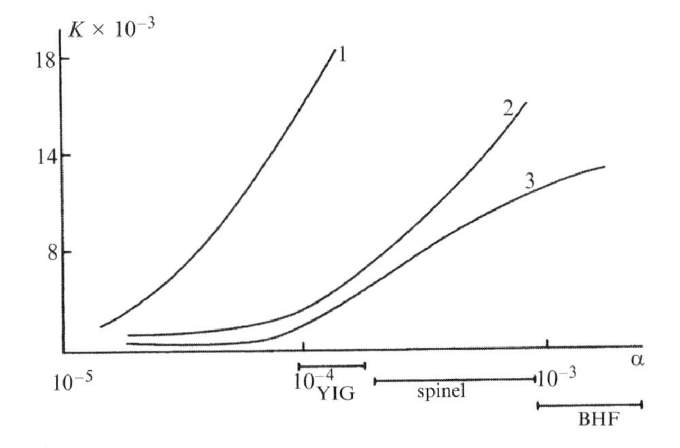

Fig. 6.15 Dependences of K on the parameter of ferromagnetic losses α for SSMSW – 1 and SSSW – 2, 3 in a beyond-cutoff waveguide an with ideal ($\sigma = \infty$) and real conductivity of its metal coatings: $1 - \sigma = \infty$, $2 - \sigma = \infty$, $3 - \sigma = 5.9 \cdot 10^7 (\text{Ohm} \cdot \text{m})^{-1}$ for a structure on frequency $\nu_{\perp} = 31.72 \cdot 10^9\,\text{Hz}, 4\pi M_S = 0.176\,\text{T}$

MSW (magnetostatic spin wave) approximation in the millimetric range near the resonant frequency of the structure (Chapter 4). Account of the real conductivity of metal coatings (curve 3) gives a more noticeable decrease of K for structures with the parameter $\alpha > 5 \cdot 10^{-4}$ (alloyed YIG, spinels, barium hexaferrite). The decrease in losses for SSSW in comparison with an idealized case ($\sigma = \infty$, curve 2) can be explained by an increase of the effective width of a beyond-cutoff waveguide at the expense of penetration of HF fields into the metal shields.

Dielectric losses in ferrite films and their dielectric bases at $t\delta_{1-3} \approx 10^{-4}$ practically did not affect the parameters of AFC in beyond-cutoff waveguides.

In Figs. 6.16 and 6.17 dependences of losses K and passbands $\Delta\nu_{3dB}$ on the internal magnetic field H_{0i} for various parameters $\Delta H_{\|}$ and $\alpha_{\|}$: $1-\Delta H_{\|}=417.7\,\mathrm{A/m}$, $2-2-\Delta H_{\|}=596.8\,\mathrm{A/m}$, $3-\Delta H_{\|}=815.6\,\mathrm{A/m}$, $4-\alpha_{\|}=10^{-3}$, $5-\alpha_{\|}=5\cdot10^{-4}$ in a structure with parameters $a_{bcs}=5\cdot10^{-4}\,\mathrm{m}$, $h_1=h_3$, $\frac{h_2}{a_{bcs}}=2\cdot10^{-2}$, $4\pi M_s=0.176\,\mathrm{T}$ are shown. With growing H_{0i} at $\Delta H_{\|}=const$ a decrease in K and $\Delta\nu_{3dB}$ is observed, and the steepness of these dependences with growing $\Delta H_{\|}$ at $\Delta H_{0i}=const$ practically does not vary. At $\alpha=const$ these dependences of K and $\Delta\nu_{3dB}$ have practically a linear character in the MF and EHF ranges.

The influence of saturation magnetization M_S of a ferrite film on the values of $K-1$ and $\Delta\nu_{3dB}-2$ is shown by Fig. 6.18 at $\alpha=5\cdot10^4$, $a=5\cdot10^{-4}\,\mathrm{m}$, $h_1=h_3$, $\frac{h_2}{a}=2\cdot10^{-2}$, $H_0=835.48\,\mathrm{kA/m}$. With growing M_S the attenuation on the central frequency ν_0 decreases, that is connected with an increase in the group speed in the structure, and the passband in the saturated regime $(H_{0i}\gg4\pi M_S)$ linearly increases.

In Fig. 6.19 theoretical (1, 2) and experimental (3, 4) dependences of the passband $\Delta\nu_{3dB}$ in two ranges of frequencies $\nu_H=30$–$40\,\mathrm{GHz}-1.3$ at $\Delta H_{\|}=397\,\mathrm{A/m}$ and $\nu_H=8$–$10\,\mathrm{GHz}-2.4$ at $\Delta H_{\|}=95.48\,\mathrm{A/m}$ on the thickness of the ferrite film h_2 at $\alpha_{\|}=5\cdot10^{-4}$, $4\pi M_S=0.176\,\mathrm{T}$ are presented. Experimental data reflect our results obtained on tmany YIG films. At advance into the millimeter range the steepness of the dependences $\Delta\nu_{3dB}(h_2)$ increases, that is connected with widening of FMR lines.

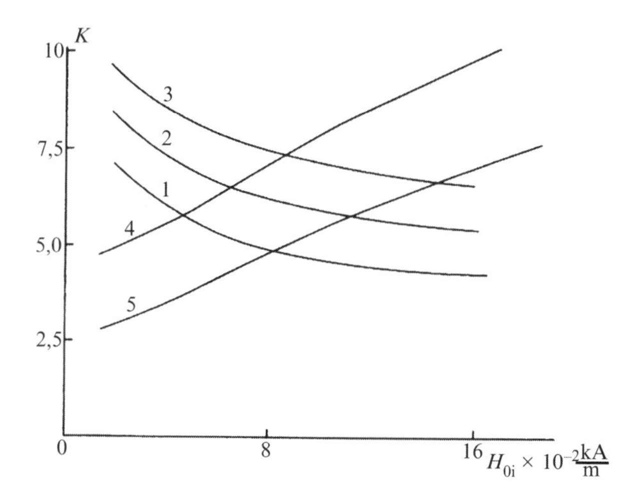

Fig. 6.16 Dependences of losses K on the internal magnetic field H_{0i} for various parameters $\Delta H_{\|}$ and $\alpha_{\|}$: $1-\Delta H_{\|}=417.7\mathrm{A/m}$, $2-\Delta H_{\|}=596.8\,\mathrm{A/m}$, $3-\Delta H_{\|}=815.6\,\mathrm{A/m}$, $4-\alpha_{\|}=10^{-3}$, $5-\alpha_{\|}=5\cdot10^{-4}$ in a structure with parameters $a_{bcs}=5\cdot10^{-4}\,\mathrm{m}$, $h_1=h_3$, $\frac{h_2}{a_{bcs}}=2\cdot10^{-2}$, $4\pi M_s=0.176\,\mathrm{T}$

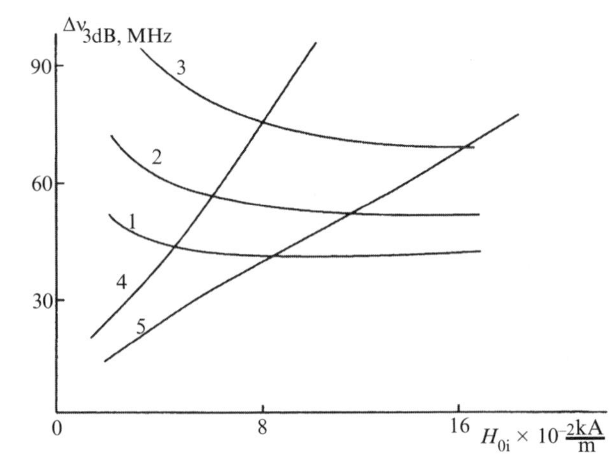

Fig. 6.17 Passbands $\Delta\nu_{3dB}$ on the internal magnetic field H_{0i} for various parameters $\Delta H_{\parallel}$ and $\alpha_{\parallel}$:$1-\Delta H_{\parallel}=417.7\,\mathrm{A/m}$, 2–$2-\Delta H_{\parallel}=596.8\,\mathrm{A/m}$, $3-\Delta H_{\parallel}=815.6\,\mathrm{A/m}$, $4-\alpha_{\parallel}=10^{-3}$, $5-\alpha_{\parallel}=5\cdot10^{-4}$ in a structure with parameters $a_{bcs}=5\cdot10^{-4}\,\mathrm{m}$, $h_1=h_3$, $\frac{h_2}{a_{bcs}}=2\cdot10^{-2}$, $4\pi M_s=0.176\,\mathrm{T}$

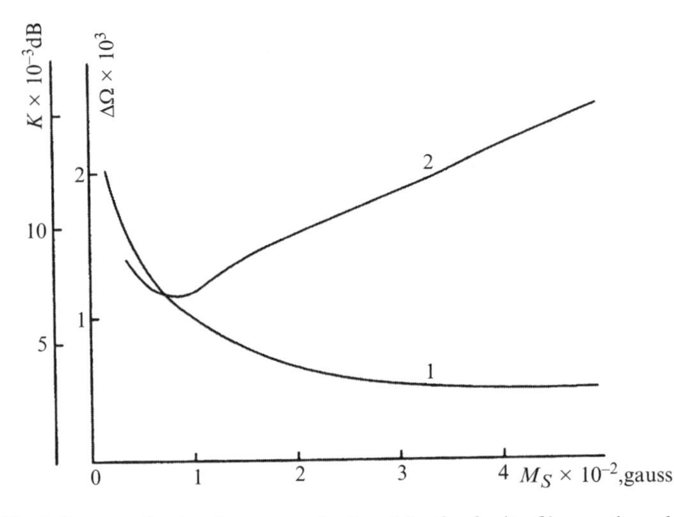

Fig. 6.18 The influence of saturation magnetization M_S of a ferrite film on the values of $K-1$ and $\Delta\nu_{3dB}-2$ is shown by Fig. 6.18 at $\alpha=5\cdot10^4, a=5\cdot10^{-4}\,\mathrm{m}$, $h_1=h_3$, $\frac{h_2}{a}=2\cdot10^{-2}$, $H_0=835.48\,\mathrm{kA/m}$

These results is in good agreement with the data in [324] obtained for YIG spheres in the MF and EHF ranges.

In Fig. 6.20 experimental dependences of $\Delta\nu_{3dB}$ for ferrite films of a varying thickness in a range of frequencies of 6–60 GHz, and $1-h_2=10.2\cdot10^{-6}\,\mathrm{m}$, $2-h_2=20\cdot10^{-6}\,\mathrm{m}$ are presented. The dependences $\Delta\nu_{3dB}(\nu)$ are linear, with a growing thickness of the film their steepness increases.

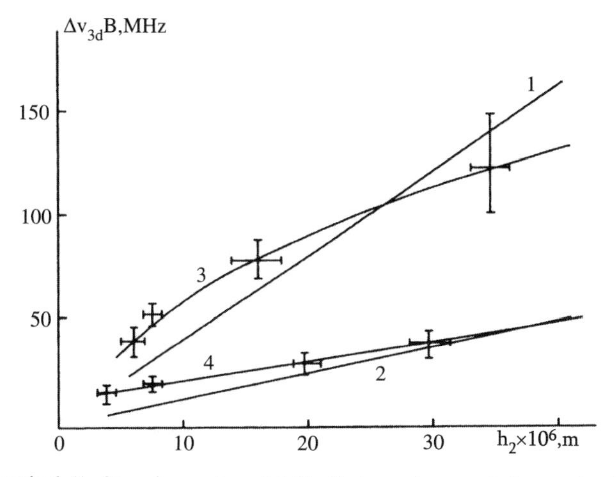

Fig. 6.19 Theoretical (1, 2) and experimental (3, 4) dependences of the passband $\Delta\nu_{3dB}$ in two ranges of frequencies $\nu_H = 30$–$40\,\text{GHz} - 1$, 3 at $\Delta H_{\|} = 397\,\text{A/m}$ and $\nu_H = 8$–$10\,\text{GHz} - 2$, 4 at $\Delta H_{\|} = 95.48\,\text{A/m}$ on the thickness of the ferrite film h_2 at $\alpha_{\|} = 5 \cdot 10^{-4}$, $4\pi M_S = 0.176\,\text{T}$

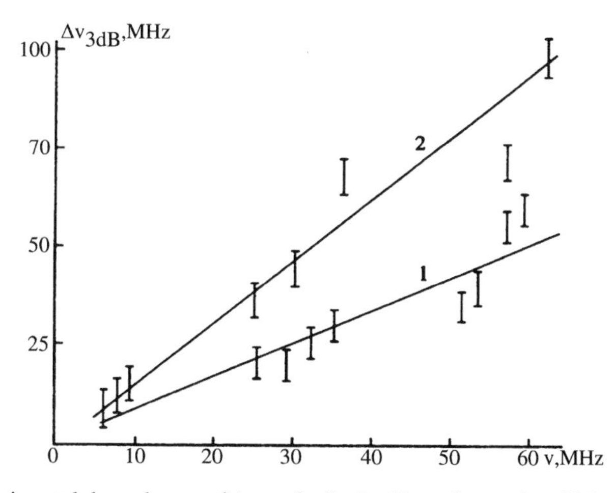

Fig. 6.20 Experimental dependences of $\Delta\nu_{3dB}$ for ferrite films of a varying thickness in a range of frequencies of 6–60 GHz, and $1 - h_2 = 10.2 \cdot 10^{-6}\,\text{m}$, $2 - h_2 = 20 \cdot 10^{-6}\,\text{m}$

The full transfer factor subject to transformation losses in the input and output elements of FDT and distribution losses in the transmission line oin a beyond-cutoff waveguide subject to major factors for the millimetric range – (a finite conductivity of screens, losses in layers, transverse gradient of losses, saturation magnetization and a field of crystallographic anisotropy) can be calculated within the developed approach.

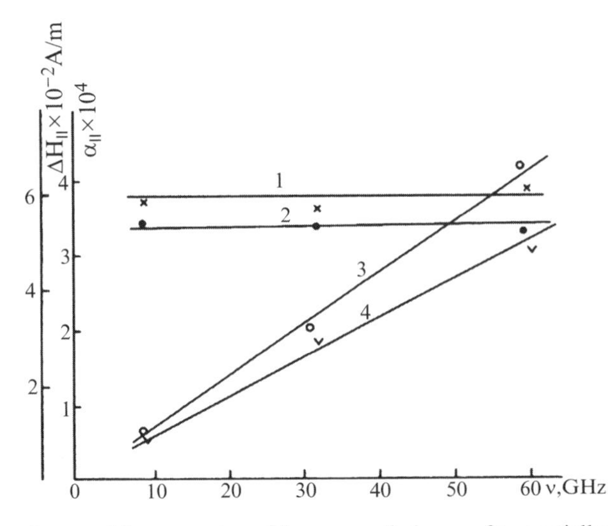

Fig. 6.21 Dependences of the parameters of ferromagnetic losses of tangentially magnetized YIG films $\alpha_{\|} - 1$, 2 and the line width of ferromagnetic resonance $\Delta H_{\|} - 3$, 4 in a range of frequencies of 6–60 GHz, and dependences 1, 3 – for $h_2 = 10.2 \cdot 10^{-6}$ m, 2, 4 – for $h_2 = 10.2 \cdot 10^{-6}$ m, $h_1 = 5 \cdot 10^{-4}$ m, $\varepsilon_{1,2} = 14$, $tg\delta_{1,2} = 10^{-4}$

This provides adequate processing of experimental AFC on the output of a measuring cell and determines a major parameter of the film, i.e., the parameter of ferromagnetic losses α and the line width of ferromagnetic resonance ΔH. Naturally, within the passband the amplitude κ'' and phase κ' constants of wave propagation, group speed, an internal magnetic field H_{0i} and related parameters can be determined as well.

In Fig. 6.21 dependences of the parameters of ferromagnetic losses of tangentially magnetized YIG films $\alpha_{\|} - 1$, 2 and the line width of ferromagnetic resonance $\Delta H_{\|} - 3$, 4 in a range of frequencies of 6–60 GHz, and dependences 1, 3 – for $h_2 = 10.2 \cdot 10^{-6}$ m, 2, 4 – for $h_2 = 10.2 \cdot 10^{-6}$ m, $h_1 = 5 \cdot 10^{-4}$ m, $\varepsilon_{1,2} = 14$, $tg\delta_{1,2} = 10^{-4}$ are presented. It is obvious that the parameter $\alpha_{\|}$ – *const* for the given structure in a wide range of frequencies.

6.3 Transparency Control in Layered Structures with an Absorbing Jacket

A waveguide with a wave H_{n0} whose E plane is filled with a layered structure with a lossy jacket on a part of the ferrite film, provides the regime of preferential signal passage near the resonant frequency of the structure.

In Fig. 6.22 a measuring device is shown, it contains: 1 – a waveguide, 2 – FDT on input and output, 3 – a lossy jacket or an embedding made of a lossy material, 4 – a layered structure with a ferrite film, 5 – a slot for a scanning structure.

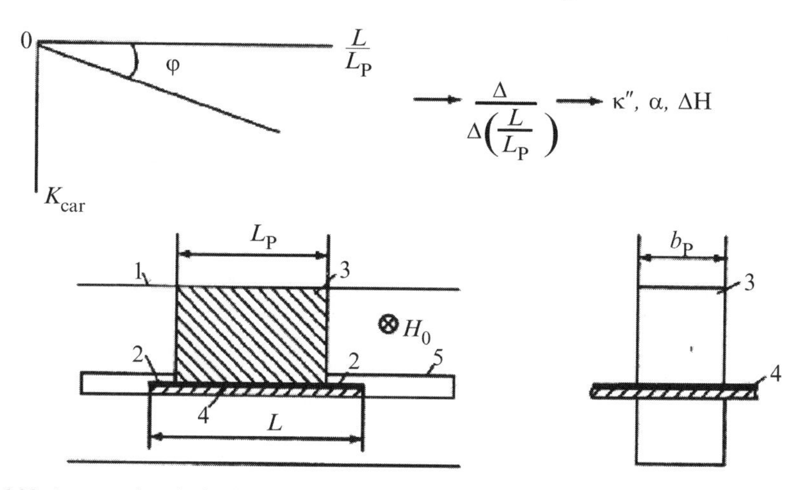

Fig. 6.22 A measuring device is shown, it contains: 1 – a waveguide, 2 – FDT on input and output, 3 – a lossy jacket or an embedding made of a lossy material, 4 – a layered structure with a ferrite film, 5 – a slot for a scanning structure

The parameters of FDT were investigated in Section 5.1. The e-transfer factor of the device $K_{car} = 2K_{con} + K_{TL}$. The amplitude constant κ'', parameter of losses in ferrite α and ΔH the steepness of the dependence $K_{car}(L/L_p)$ in the passband can be determined. The area under analysis of the structure is $S_a = b_p \times L_p$ (Fig. 6.22).

The amplitude constant of a propagating wave in the structure is determined from the relation

$$\kappa'' = \left[\frac{K_{TL}(dB) + 2K_{con}(dB)}{8.68} - 2\kappa''_{FDT}S\right]\frac{1}{L}, \tag{6.10}$$

where κ''_{FDT} is the amplitude constant of the wave in FDT, S the length of the input and output converter, κ''_{FDT} and S are determined by the technique from Section 5.1.

The relative error is

$$\left|\frac{\Delta\kappa''}{\kappa''}\right| = \left|\frac{\Delta L_{TL}}{L_{TL}} + \frac{\Delta K_{TL} + 2\Delta K_{con} + 17.36(\Delta\kappa''_{con}S + \kappa'_{con}\Delta S)}{K_{TL} + 2K_{con} - 17.36\kappa''_{con}S}\right|. \tag{6.11}$$

For high-quality ferrites $\alpha < 10^{-4}$, $\nabla_x\alpha < 10^{-1}$, $\nabla_x M_S < 10^{-1}$ with $L_p \gg S$ the value $K_{TL} \gg K_{con}$ and from Eq. (6.10) it follows that $\kappa'' \approx \frac{K_{TL}}{8.68L_p}$, and the error is $\frac{\Delta\kappa''}{\kappa''} = \left|\frac{\Delta K_{TL}}{K_{TL}} + \frac{\Delta L_p}{L_p}\right|$.

In Table 6.3 the results of determination of the amplitude constants of advancing waves κ'' in various structures on the basis of YIG films and spinels are given.

Table 6.3 The results of determination of the amplitude constants of advancing waves κ'' in various structures on the basis of YIG films and spines

Structure	Thickness of film, $h_2 \times 10^3$, m	Frequency, GHz	K_{TL}, dB	$\kappa'' \cdot 10^{-2}$, rad/m	Note
YIG	21	30–36.5	0.9	15	Carat®
YIG	35.2	–	1.7	28	
YIG	24.4	–	3.0	49	
YIG	24.5	–	4.7	77	
YIG	19.8	–	5.2	85	
spinel	13	30–36.5	3.5	58	Domain®
spinel	20	–	4.2	69	–
spinel	7	–	12.3	202	All-Russian scientific research institute of materials and electronic engineering

The dissipative parameters α and ΔH are determined from comparison of the steepness for both experimental and theoretical dependences $K_{TL}(L_p)$, and the angle of inclination is

$$\xi = arctg\left(m\frac{\Delta K - 2\Delta K_{con}}{L_p + m^2\frac{2\Delta K_{con}\Delta K}{L_p}}\right), \tag{6.12}$$

where m is the coefficient of proportionality with a dimension $\left[\frac{\text{rad}\cdot\text{m}}{\text{dB}}\right]$.

In practice usual дн $L_p \gg S$ and $K_{TL} \gg K_{con}$, therefore, from Eq. (6.12) it follows that

$$\xi \approx arctg\left(m\frac{\Delta K}{L_p}\right). \tag{6.13}$$

The relative errors from Eqs. (6.12) and (6.13), respectively, are

$$\left|\frac{\Delta\xi}{\xi}\right| = \left|\frac{[\Delta(\Delta K) + 2\Delta(\Delta K_{con})]L_p + m2\Delta K_{con}\Delta K L_p^{-1}}{\frac{\Delta K - 2\Delta K_{con}}{L + m^2 2\Delta K_{con}\Delta K L_p^{-1}}}\times\right.$$

$$\left.\times\frac{2m^2\left\{[\Delta(\Delta K_{con})\Delta K + \Delta(\Delta K)]L_p + \Delta L_p\Delta K_{con}\Delta K\right\}L_p^{-2}}{1-\left(\frac{\Delta K - 2\Delta K_{con}}{1+m^2 2\Delta K_{con}\Delta K L_p^{-1}}\right)^2}\right|, \tag{6.14}$$

$$\left|\frac{\Delta\xi}{\xi}\right| \approx \left|\frac{\Delta(\Delta K)}{\Delta K}\right| + \frac{\Delta L_p}{L_p}. \tag{6.15}$$

6.4 Control on Phase Inversion in Ferrite in Antiphase-Balanced Bridges

At advance of various ferrite inspection methods into the millimeter range one of the main tasks is searching for most sensitive methods and devices for control of registered signals. Natural widening of a FMR line leads to an increase of conversion (excitation) and propagation losses of electromagnetic waves in layered structures. There are many of structures with a FMR line than YIG. These are ferrites with the spinel structure, hexaferrites, etc. [27]. In polycrystalline structures this parameter is still higher [8]. In addition, a measuring cell and a usable method should provide measurement of the widest class of materials and their parameters. To this end, methods of parameter measurement of various ferrites on the basis of waveguides and strip bridges were developed. Bridge circuits and methods of their design are known for a long time.

This method is underlied by the effect of preferential signal passage in an antiphasly balanced bridge at phase inversion in the ferrite located in one of its arms, in the regime of preferential signal depression at $\nu \gg \nu_{cr}$ close to the resonant_frequency of the structure. Phase inversion occurs when the length of the structure reaches $L = \frac{\Delta\varphi}{\kappa'} = \frac{1+2n}{\kappa'}\pi$, $n = 0, 1, 2, \ldots$, where κ' is the phase constant of one or two waves (see Chapters 3, 4).

In Fig. 6.23 balancing circuits are presented: *a* – with the structures of waveguides $+\frac{\pi}{2} - \frac{\pi}{2}$, *b* – with a twisted joint by π, *c* – with a controllable phaser and attenuator.

In the trivil case in an antiphased balanced waveguide circuit (Fig. 6.23*b*) the following conditions are satisfied:

- For phases $(\kappa_1' L_1 + \kappa' L + \kappa_2' L_2) - \kappa_W' L_W = \pi$ (W – wave guide)
- For amplitudes $(\kappa_1'' L_1 + \kappa'' L + \kappa_2'' L_2) = \kappa_W'' L$

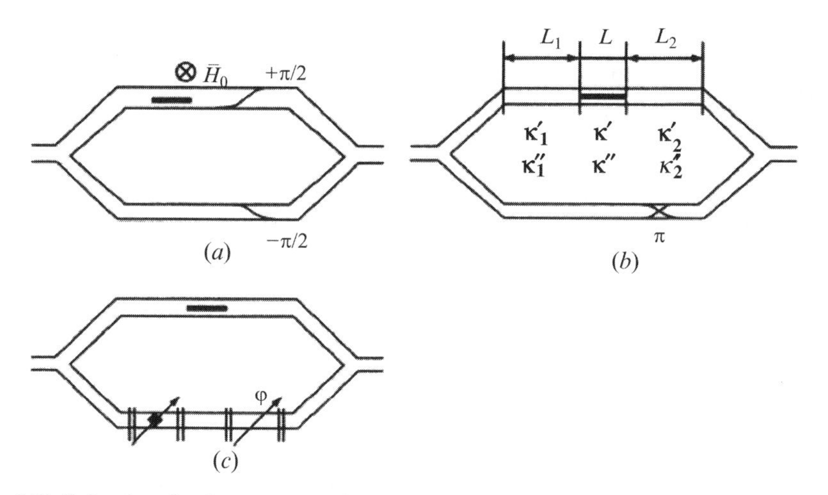

Fig. 6.23 Balancing circuits are rresrnted: (*a*) with the structures of waveguides $+\frac{\pi}{2} - \frac{\pi}{2}$, (*b*) with a twisted joint by π, (*c*) with a controllable phaser and attenuator

where $\kappa'_{1,2}$ and $\kappa''_{1,2}$ are the phase and amplitude constants of waves in the corresponding parts of the circuit, κ' and κ'' the wave constants in the ferrite-dielectric structure in the waveguide.

The amplitude constants $\kappa''_{1,2}$ are usually essentially lesser than those in ferrite, and phase balance in a bridge circuit is convenient to realize an arrangement of identical ferrite samples of in its two arms. Subject to these remarks the transfer factors are:

- On frequency ν_0

$$K_{car}(\nu_0) \approx 3\ \text{dB} + 8.68\kappa''(\nu_0)L + 20\lg\left|\cos\frac{\Delta\kappa'(\nu_0)L}{2}\right| \tag{6.16a}$$

- On frequency $\nu_1 > \nu_0$

$$K_{car}(\nu_1) \approx 8.68\kappa''(\nu_1)L + 20\lg\left|\cos\frac{\Delta\kappa'(\nu_1)L}{2}\right| \tag{6.16b}$$

- On frequency $\nu_2 < \nu_0$

$$K_{car}(\nu_2) \approx 8.68\kappa''(\nu_2)L + 20\lg\left|\cos\frac{\Delta\kappa'(\nu_2)L}{2}\right| \tag{6.16c}$$

In Fig. 6.24 the AFC of signals in weakly- $(\alpha \ll 1)$ and heavily dissipative $(\alpha < 1)$ ferrites are shown.

Let's exemplify some parameters to be determined in weakly dissipative ferrite structures.

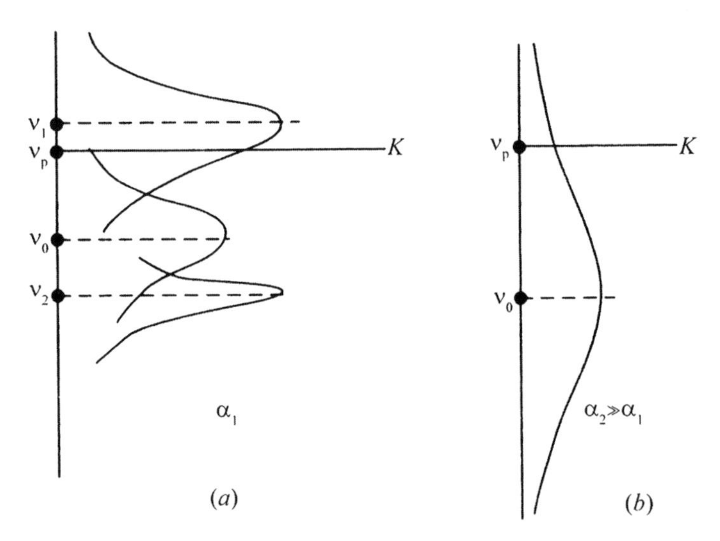

Fig. 6.24 The AFC of signals in weakly – $(\alpha \ll 1)$ and heavily dissipative $(\alpha < 1)$ ferrites

Amplitude constants are calculated from Eq. (6.16):

- On frequency ν_0

$$\kappa''(\nu_0) \approx \frac{\kappa(\nu_0) - 3\ \mathrm{dB}}{8.68L} \tag{6.17a}$$

- On frequency ν_1

$$\kappa''(\nu_1) \approx \frac{\kappa(\nu_1)}{8.68L} \tag{6.17b}$$

- On frequency ν_2

$$\kappa''(\nu_2) \approx \frac{\kappa(\nu_2)}{8.68L} \tag{6.17c}$$

The relative error for these values is

$$\left|\frac{\Delta\kappa''}{\kappa''}\right| \approx \left|\frac{\Delta\kappa}{\kappa} + \frac{\Delta L}{L}\right|. \tag{6.18}$$

Phase constants for the central frequencies K_{car} of a signal are determined.
For straight fast and slow waves

$$\kappa'(\nu_1) = \frac{1+2n}{L}\pi, \quad \kappa'(\nu_2) = \frac{1+2n}{L}\pi, \quad n = 0,\ 1,\ 2,\ \ldots \tag{6.19}$$

Mistiming of the phase constants of fast and slow waves at interference interaction is

$$\kappa'(\nu_{01}) = \frac{1+2n}{L}\pi, \quad n = 0,\ 1,\ 2,\ \ldots \tag{6.20}$$

The relative error of these values is

$$\left|\frac{\Delta\kappa''}{\kappa''}\right| \approx \left|\frac{\Delta\kappa}{\kappa} + \frac{\Delta L}{L}\right|. \tag{6.21}$$

Dissipative parameters of FDLS on the basis of films and plates are determined from comparison of the steepness of the theoretical and experimental dependences of the transfer factor $K_{car}(\nu_0, \nu_1, \nu_2)$ on the length of structure L.

For fast and slow waves on the frequency of interference interactions ν_0 the angle of inclination of the dependence $K_{car}(L)$ is

$$\xi \approx arctg\frac{\Delta K_{car}(\nu_0) - 3\ \mathrm{dB}}{8.68\Delta L}, \tag{6.22}$$

It determines α or $\Delta H(\nu_0)$.

For fast waves on frequency ν_1

$$\eta \approx arctg\frac{\Delta K_{car}(\nu_1)}{8.68\Delta L}, \tag{6.23}$$

for slow waves on frequency ν_2

$$\chi \approx arctg \frac{\Delta K_{car}(\nu_2)}{8.68 \Delta L}, \tag{6.24}$$

where η and χ determine the parameters of ferromagnetic losses α, $\alpha(\nu_2)$ and the line width of FMR $\Delta H(\nu_1)$ and $\Delta H(\nu_2)$ on the corresponding frequencies.

The relative errors for Eqs. (6.22–6.24) are

$$\left|\frac{\Delta \xi}{\xi}\right| = \left|\frac{\Delta(\Delta K_{car}(\nu_0)L + \Delta(\Delta L) \cdot \Delta K_{car}(\nu_0)}{[\Delta K_{car}(\nu_0) - 3\text{ dB}]\left[1 + \frac{\Delta K_{car}(\nu_0) - 3\text{ dB}}{8.68 \Delta L}\right]}\right|, \tag{6.25}$$

$$\left|\frac{\Delta \eta}{\eta}\right| = \left|\frac{\Delta(\Delta K_{car}(\nu_1))}{(\Delta K_{car}(\nu_1)} + \frac{\Delta(\Delta L)}{\Delta L}\right|, \tag{6.26}$$

$$\left|\frac{\Delta \chi}{\chi}\right| = \left|\frac{\Delta(\Delta K_{car}(\nu_2)}{\Delta K_{car}(\nu_2)} + \frac{\Delta(\Delta L)}{\Delta L}\right|. \tag{6.27}$$

6.5 Resonant Frequency and Magnetic Field Sensors

Consider the urgency and principal importance of the problem of resonant frequency measurement in FDLS in the centimeter and millimeter ranges for correct interpretation of experimental results at physical investigations of wave processes and, especially, at the design of magnetoelectronic devices on the basis of ferrite films.

Theoretical and experimental investigations have shown that in high-quality ferrite films with losses $\alpha < 10^{-4}$ and cross-section gradients $\nabla_x \alpha < 10^{-1}$, $\nabla_x H_{0i} < 10^{-1}$, effective processes of preferential excitation and propagation of signals by electromagnetic waves providing a loaded merit factor $Q_l \approx 10^3$–10^4(1 – loaded) for YIG and $Q_l \leq 100$–200 for $L_i - Z_n$ (L – lithium, Z – zinc) – spinels can be realized near the resonant frequency. It can be one – or two-wave regimes including fast and slow waves. As it was said in Introduction, within the millimeter range the problem of accuracy scissors between frequencies measured by usual MAW with an accuracy $\frac{\Delta \nu}{\nu} \approx 10^{-2}\%$ or by heterodyne wavemeters with an accuracy $\frac{\Delta \nu}{\nu} \approx 10^{-5}\%$) and the values of magnetic fields measured by Hall sensors for increased fields $H_0 > 0.8\,\text{MA/m}(H_0 > 10\text{kOe})$ with an error $\frac{\Delta H_0}{H_0} \approx 1$–$2\%$. This difference in measurement accuracies s of frequency and field by more than three orders of magnitude makes an essential error in the treatment of experimental dependences. NMR sensors provide a necessary accuracy of $\frac{\Delta H_0}{H_0} \approx 10^{-5}\%$ but have their cross-section dimensions $(10 \times 15) \cdot 10^{-6}\,\text{m}^2$ and require a raised uniformity of field $\frac{\delta H_0}{H_0} \leq 10^{-4}$, that can be obtained with special electromagnets for spectrometer measurements only. In millimeter-range magnetoelectronics fields of $H_0 > 10\,\text{kOe}$ are created by magnetic systems in a gap $h \leq 1 \cdot 10^{-3}\,\text{m}$ at nonuniformity reaching tens and hundreds of oersted.

Practically any of the considered (Chapter 5) waveguide or strip transformers, and devices of through-passage type (Chapter 6) can serve as a resonant frequency sensor RFS or fields RFDS. The basic correction at these measurements is the correction to the resonant frequency of the structure, providing its reduction to the resonant frequency of an idealized lost-free FDLS for which analytical expressions can be deribved. Measurements on reference magnetic boxes can be made, if necessary. However, at registration of a frequency in the AFC spectrum by means of MAW, direct measurement of the resonant frequency should be provided with the same accuracy $\frac{\Delta \nu_0}{\nu_0} \approx 10^{-2}$–$10^{-3}$, that will allow determination of an internal magnetic field from $\nu_r = \frac{\gamma}{2\pi} H_{0i}$ with the same accuracy $\frac{\Delta H_{0i}}{H_{0i}} \approx 10^{-2}$–$10^{-3}$ by a direct method. As $H_{0i}\left(\overset{\leftrightarrow}{N} H_0,\ \overset{\leftrightarrow}{N}_A H_A, \ldots\right)$, one of the values of interest can be expressed through ν_r, γ and other parameters, or as was done in [28], uncertain parameters can be excludeed from several equations. The last operation becomes simpler, when some independent measurements for various cases made.

The basic requirement in the millimeter range is an increase of the sensitivity of instruments. Therefore, it is most expedient to use waveguide FDT as RFS and RFDS in the regime of maximum interference interactions of fast and slow waves, at which the frequency resolution is maximal.

To decrease the deviation between the central frequency of the sensor and the resonant frequency of the investigated FDLS it is necessary to provide:

- Identical sizes and layout of the structure, namely, the arrangement of metal screens, dielectric spacers, etc., magnetic parameters of the field of crystallographic anisotropy H_{mon}, H_{bas}, (mon – monoaxial, bas – basic), orientation with respect to crystallographic axes and planes, dissipation, etc., external parameters – temperature, pressure, etc., and

$$\left(\overset{\leftrightarrow}{N} M_S\right)_{RFS} \approx \left(\overset{\leftrightarrow}{N} M_S\right)_{TL}, \quad \left(\overset{\leftrightarrow}{N}_{mon} H_{mon}\right)_{RFS} \approx \left(\overset{\leftrightarrow}{N}_{mon} H_{mon}\right)_{TL}$$

  $\alpha_{RFS} \approx \alpha_{TL}$, etc.
- A higher homogeneity of the outside exciting HF fields $\tilde{h}$ and currents $\tilde{j}$ in RFS, in comparison with FDLS (TL)

$$\left(\frac{\delta \tilde{h}}{\tilde{h}}\right)_{RFS} \ll \left(\frac{\delta \tilde{h}}{\tilde{h}}\right)_{TL}, \quad \left(\frac{\delta \tilde{j}}{\tilde{j}}\right)_{RFS} \ll \left(\frac{\delta \tilde{j}}{\tilde{j}}\right)_{TL}$$

- The same kind of magnetization and type of excited waves
- For narrowband operating modes of investigated spectras and AFC close laws of loss factors for transformation of the reactive power of radiation

$$(K_{los.B})_{RFS} \approx (K_{los.B})_{TL}$$

(B – reactive)

For some applications, especially at investigation of obliquely magnetized FDLS, at investigation of the distributions of parameters by the thickness of the ferrite film $M_S(x)$, $\alpha(x)$, $\Delta H(x)$, $H_{0i}(x)$ clamping to the resonant frequencies of FDLS is necessary:

- To frequency $\nu_\mathrm{H} = \frac{\gamma}{2\pi}(H_{0i})_\perp$ at normal magnetization
- To frequency $\nu_\perp = \frac{\gamma}{2\pi}(H_{0i})_\| = [\nu_\mathrm{H}(\nu + \nu_\mathrm{M})]^{1/2}$ at tangential magnetization

The sensor area S_S is subject to the characteristic sizes of the film in an external field H_0. For structures with tangential magnetization it is reached $S_S = dL_S$, d is the thickness of the film, L_S the length of the sensor, that determines spatial resolution at a level of 10–20 μm. At using a multi-layered film structure the spatial resolution is higher and the instantaneous distribution of resonant frequencies and the corresponding values of external fields H_0 is registered simultaneously.

In Fig. 6.25 various designs of sensors on the basis of ferrite films are shown: a – a sensor on a planar waveguide for measurements of the resonant frequency of obliquely magnetized structures $\nu_r(\varphi)$, b – a sensor for d_H and $\nu_\perp$ on a paired waveguide, c – a sensor for ν_H and $\nu_\perp$ on a rectangular waveguide, d – a sensor on a slot line. By means of such sensors the resonant frequencies of FDLS in the microwave and EHF ranges were determined, magnetic fields, including nonuniform ones in small working gaps of diminutive magnetic systems, were measured with a raised accuracy and spatial resolution.

Let's consider some applications of these sensors.

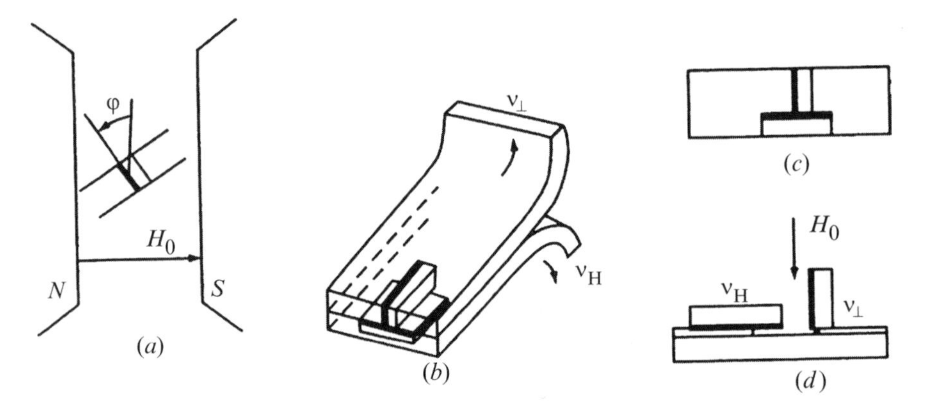

Fig. 6.25 Various designs of sensors on the basis of ferrite films are shown: (*a*) a sensor on a planar waveguide for measurements of the resonant frequency of obliquely magnetized structures $\nu_r(\varphi)$, (*b*) a sensor for ν_H and $\nu_\perp$ on a paired waveguide, (*c*) a sensor for ν_H and $\nu_\perp$ on a rectangular waveguide, (*d*) a sensor on a slot line

6.5.1 Gyromagnetic Ratio

By means of the calculated values of frequency corrections $\pm\delta\nu$ (signs "+" or "−" are taken depending on the used sensor of through-pass bor reflective type in the prelimiting or beyond-cutoff regimes) it is convenient to pass from the values of resonant frequencies ν_0 of the structure to the resonant frequency $\nu_r = \nu_0 \pm \delta\nu$ of an idealized structure ($\alpha = 0$) which is related to the field by a known relationship $\nu_r = \frac{\gamma}{2\pi} H_{ef}$ (ef – efficient), where γ is the gyromagnetic ratio,

$$H_{ef} = -\frac{\partial F}{\partial M} + \sum_{i=1}^{3} \frac{\partial}{\partial x_i} \cdot \frac{\partial F}{\partial(\partial M/\partial x_i)},$$

F the density of free energy at $T > 0$, M– magnetization [8].

Then, having theoretical and experimental data of both high-frequency and static parameters of the structure (losses, saturation magnetization, fields of anisotropy of the sample, etc.), allowing to get H_{ef}, and experimentally measured value of ν_o subject to calculated corrections $\pm\delta\nu$ or the value of ν_r we can put them in conformity and determine the value of γ, and, hence, that of g factor of the structure. The latter values will therefore be considered subject to the contribution of orbital moments and the influence of adjacent particles in the substance ($\gamma = \frac{ge}{2m_e s}$, e is the electron charge, m_e the rest mass of electron, s the light speed).

Thus, the relative error is

$$\left|\frac{\Delta\gamma}{\gamma}\right| = \left|\frac{\Delta H_{ef}}{H_{ef}} + \frac{\Delta\nu_0 + \Delta\delta(\nu)}{\nu_0 \pm \delta\nu}\right|, \tag{6.28}$$

and

$$\left|\frac{\Delta g}{g}\right| = \left|\frac{\Delta\gamma}{\gamma} + \frac{\Delta m_e}{m_e} + \frac{\Delta e}{e} + \frac{\Delta s}{s}\right|. \tag{6.29}$$

For example, consider weakly anisotropic YIG films in the millimeter range, for which $H_0 \gg 4\pi M_S$, $H_0 \gg H_A$ ($H_A = 3.342\,\mathrm{kA/m}, 4\pi M_S = 0.176\,\mathrm{T}$, $H_0 > 0.8\,\mathrm{MA/m}$). The films have their sizes $\frac{W}{h_2} \gg 1$ and $\frac{L}{h_2} \gg 1$, and their resonant frequencies are well described by limiting cases for flattened ellipsoids of revolution:

– At normal magnetization (n – normal)

$$\nu_n \pm \delta\nu_n = \frac{\gamma}{2\pi}(H_0 - 4\pi M_S) \tag{6.30}$$

– At tangential magnetization

$$\nu_t \pm \delta\nu_t = \frac{\gamma}{2\pi}\sqrt{H_0(H_0 + 4\pi M_S)} \tag{6.31}$$

From Eqs. (6.30) and (6.31) it follows that on derivation of $\nu_n \pm \delta\nu_n$, $\nu_t \pm \delta\nu_t$ and exception of H_0 or $4\pi M_S$ from these formulas, we get the value of γ. The values of $4\pi M_S$ and H_0 can be measured with an extra accuracy by a series of precision methods.

On elimination of $4\pi M_S$ from Eqs. (6.30) and (6.31) we have

$$\gamma = \frac{\nu_n \pm \delta\nu_n}{4H_0} \pm \left(\frac{\nu_n \pm \delta\nu_n}{4H_0}\right)^2 + \left(\frac{\nu_t \pm \delta\nu_t}{2H_0^2}\right)^2. \tag{6.32}$$

The relative error is

$$\left|\frac{\Delta\gamma}{\gamma}\right| = \Big| \{4H_0\,(\Delta\nu_n + \Delta\delta(\nu_n)) + \Delta H_0\nu_n A - \nu_n\,[\Delta\nu_n + \Delta\delta(\nu_0)H_0 + \Delta H_0\nu_n] + \\ + 8\nu_t\,[(\Delta\nu_t + \Delta\delta(\nu_t))H_0 + \Delta H_0\nu_t]\} \left[8H_0^3\left(\frac{\nu_n \pm \Delta\nu_n}{4H_0} + A) \times A\right)\right]^{-1}\Big|, \tag{6.33}$$

where

$$A = \left[\left(\frac{\nu_n \pm \Delta\nu_n}{4H_0}\right)^2 + \left(\frac{\nu_t \pm \Delta\nu_t}{2H_0^2}\right)^2\right]^{1/2}.$$

On elimination of field H_0 from Eqs. (6.30) and (6.31) we get

$$\gamma = -\frac{3}{16}\frac{\nu_n \pm \delta\nu_n}{4\pi M_S} + \left[\left(\frac{3}{16}\frac{\nu_n \pm \delta\nu_n}{\pi M_S}\right)^2 \frac{(\nu_t \pm \delta\nu_t)^2 - (\nu_n \pm \delta\nu_n)^2}{32\pi^2 M^2}\right]. \tag{6.34}$$

The relative error is

$$\left|\frac{\Delta\gamma}{\gamma}\right| = \Bigg| \Big\{96\pi M_S^2\,[(\Delta\nu_n + \Delta\delta(\nu_n))\,M_S + \Delta M_S]\cdot B + 18\pi M_S\nu_n \\ \times \left[(\Delta\nu_n + \Delta\delta(\nu_n))^2 M_S + \Delta M_S\,(\nu_n + \delta(\nu_n))\right] + 16\,[(\Delta\nu_t \pm \Delta\delta(\nu_t)) \\ + (\Delta\nu_n \pm \Delta\delta(\nu_n))]\,M_S^2 + 16M_S\Delta M_S\left[(\nu_t \pm \delta\nu_t)^2 + (\nu_n \pm \delta\nu_n)^2\right]\Big\} \\ \times \left\{512\pi^2 M_S^4\left(-\frac{3}{16}\frac{(\nu_n \pm \delta\nu_n)\,M_S}{\pi M_S}B\right)B\right\}\Bigg|, \tag{6.35}$$

where

$$B = \left[\left(\frac{3}{16}\frac{\nu_n \pm \delta\nu_n}{\pi M_S}\right)^2 + \frac{(\nu_t \pm \delta\nu_t)^2 - (\nu_n \pm \delta\nu_n)^2}{32\pi^2 M_S^2}\right]^{1/2}.$$

From Eqs. (6.28), (6.33), and (6.35) one can see that the higher the values of ν_n, ν_t and H_0, the lower the error of γ determination.

Neglecting the field of anisotropy H_A in Eqs. (6.30) and (6.31) leads to the relative error $\frac{\Delta\gamma}{\gamma} > 10^{-3}$ by Eq. (6.33), and $\frac{\Delta\gamma}{\gamma} > 10^{-2}$ by Eq. (6.35).

Saturation magnetization of a ferrite film can be determined from Eqs. (6.30) and (6.31) on elimination of H_0

$$M_S = \frac{3}{16}\cdot\frac{\nu_n \pm \delta\nu_n}{\gamma} + \left\{\left(\frac{3}{16}\cdot\frac{\nu_n \pm \delta\nu_n}{\gamma}\right)^2 - \right.$$
$$\left. -\frac{1}{32\pi\gamma^2}\left[(\nu_n \pm \delta\nu_n)^2 - (\nu_t \pm \delta\nu_t)^2\right]\right\}^{1/2}. \tag{6.36}$$

The relative error is

$$\left|\frac{\Delta M_S}{M_s}\right| = \left|3C\left[(\Delta\nu_n \pm \delta(\nu_n))\,\gamma + \Delta\gamma(\nu_n \pm \delta\nu_n)\right] + \frac{9}{16}\frac{\nu_n \pm \delta\nu_n}{\gamma}\right.$$
$$\times\left[(\Delta\nu_n \pm \Delta\delta(\nu_n))\,\gamma + \Delta\gamma(\nu_n \pm \delta\nu_n)\right]$$
$$+\frac{1}{16\pi}\frac{1}{\gamma^2}\{[(\nu_t \pm \delta\nu_t)\,(\delta\nu_t \pm \Delta\delta(\nu_t)) \; - (\nu_n \pm \delta\nu_n)(\Delta\nu_n \pm \Delta\delta\nu_n)]\,\gamma^2$$
$$\left.-\gamma\Delta\gamma\left[(\nu_t \pm \delta\nu_t)^2 - (\nu_n \pm \delta\nu_n)^2\right]\right\} \times \left(-\frac{3}{16}\frac{(\nu_n \pm \delta\nu_n)}{\gamma} + C\right)^{-1}\right|, \tag{6.37}$$

where

$$C = \left(\frac{3}{16}\cdot\frac{\nu_n \pm \delta\nu_n}{\gamma}\right)^2 + \left\{\frac{1}{32\pi\gamma^2}\left[(\nu_t \pm \delta\nu_t)^2 - (\nu_n \pm \delta\nu_n)^2\right]\right\}^{1/2}.$$

Neglecting the field of anisotropy H_A gives an error

$$\left|\frac{\Delta M_S}{M_S}\right| > 10^{-2}.$$

External magnetic field H_0

Let's exclude $4\pi M_S$ from Eqs. (6.30) and (6.31), then

$$H_0 = \frac{\nu_n \pm \delta\nu_n}{4\gamma} + \left[\left(\frac{\nu_n \pm \delta\nu_n}{4\gamma}\right)^2 + \left(\frac{\nu_t \pm \delta\nu_t}{2\gamma^2}\right)^2\right]^{1/2}. \tag{6.38}$$

The relative error is

$$\left|\frac{\Delta H_0}{H_0}\right| = \left|\left\{\left[4\gamma^2(\Delta\nu_n + \Delta\delta(\nu_n))\gamma + \Delta\gamma\nu_n\right]\cdot D + \frac{1}{8}\gamma(\nu_n \pm \delta\nu_n)\right.\right.$$
$$\left.\times\left[(\Delta\nu_n \pm \Delta\delta(\nu_n))\gamma + \Delta\gamma\nu_n\right] + 16\left[\nu_t(\Delta\nu_t + \Delta\delta(\nu_t))\gamma^2 - \gamma\Delta\gamma\nu_t^2\right]\right\}$$
$$\left.\times\, 32\gamma^4 D\cdot\left[\frac{\nu_n + \delta\nu_n}{4\gamma} + D\right]^{-1}\right|, \tag{6.39}$$

where

$$D = \left[\left(\frac{\nu_n \pm \delta \nu_n}{4\gamma} \right)^2 + \left(\frac{\nu_t \pm \delta \nu_t}{2\gamma^2} \right) \right]^{1/2}.$$

If the value of M_S for a ferrite film is known, ν_t has been measured and $\pm\delta\nu_t$ have been determined, then

$$H_0 = -2\pi M_S + \left[(2\pi M_S)^2 + \left(\frac{\nu_t \pm \delta \nu_t}{\gamma} \right)^2 \right]^{1/2},$$

and if the frequency ν_n has been measured and $\pm\delta\nu_n$ have been determined, then

$$H_0 = 4\pi M_S + \frac{\nu_n \pm \delta \nu_n}{\gamma}.$$

The relative errors are

$$\left| \frac{\Delta H_0}{H_0} \right| = \left| \{ 4\pi\gamma^3 \Delta M_S E + 8\pi^2 \gamma^3 M_S \Delta M_S + 2(\nu_t \pm \delta \nu_t) \right. \\ \left. \times \left[(\Delta\nu_t \pm \Delta\delta\nu_t)\gamma + \Delta\gamma(\nu_t \pm \Delta\nu_t) \right] \} 2E^{-1} \right|, \tag{6.40}$$

$$\left| \frac{\Delta H_0}{H_0} \right| = \left| \frac{\Delta\gamma}{\gamma} + [4\pi(M_S \Delta\gamma + \gamma M_S) + \Delta\nu_n \pm \delta\nu_n] (4\pi M_S + \nu_n \delta\nu_n) \right|, \tag{6.41}$$

where

$$E = \left[(2\pi M_S)^2 + \left(\frac{\nu_t \pm \delta \nu_t}{\gamma} \right)^2 \right]^{1/2}.$$

If at tangential magnetization of a ferrite film the vector of the field of anisotropy $\overline{H}_A$ is in one plane with $\overline{H}_0$, which is perpendicular to the vector of wave number $\bar{\kappa}$, then $\left| \frac{\Delta H_0}{H_0} \right| < 10^{-3}$, which even in such a trivial case is more exact by two orders of magnitude than in the case of Hall sensors measuring raised fields $H_0 > 0.8$.

A small volume of a ferrite film provides a raised spatial resolution of such sensors, that can be used for measurements of nonuniform external fields ∇H_0. At mechanical movement of the sensor by the corresponding coordinates either three components $\frac{\partial H_{0x}}{\partial x}$, $\frac{\partial H_{0y}}{\partial y}$, and $\frac{\partial H_{0z}}{\partial z}$ or two of them are measured, the third one to be calculated from the condition $\nabla H_0 = 0$ (Fig. 6.26). At using of two- (Fig. 6.27*a*) or multilayered (Fig. 6.27*b*) films with $M_{S1}, M_{S2}, \ldots$, instant measurement of H_0 in the desired sections is provided.

In Fig. 6.28 the results of comparison of magnetic field measurements in a system with holes of various diameters in the polepieces and by means of the developed RFS – curve 1(*a*) and a Sh1- 8 tester on the basis of a diminutive Hall sensor – curve 2(*b*) are presented. Figure 6.28*c* shows these large-scale dependences in the vicinity of the polepieceholes. It is obvious that the Hall sensors practically does not

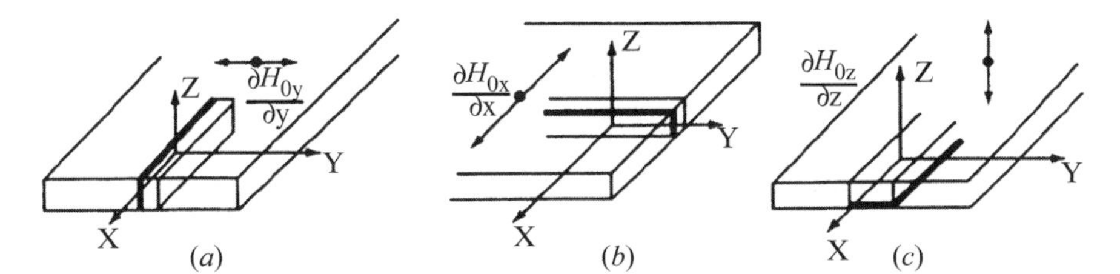

Fig. 6.26 At mechanical movement of the sensor by the corresponding coordinates either three components $\frac{\partial H_{0x}}{\partial x}$, $\frac{\partial H_{0y}}{\partial y}$, and $\frac{\partial H_{0z}}{\partial z}$ or two of them are measured, the third one to be calculated from the condition $\nabla H_0 = 0$

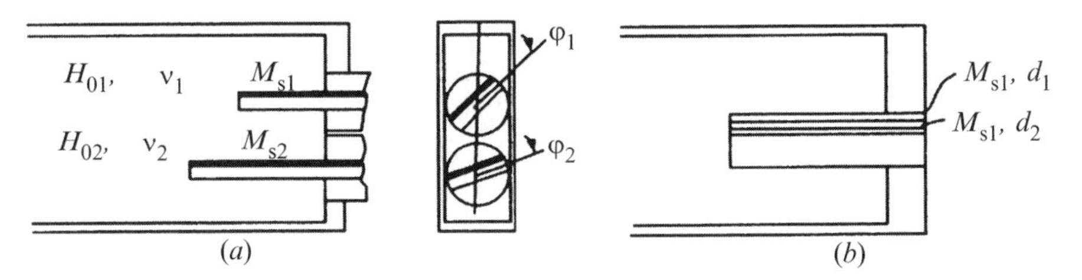

Fig. 6.27 Application at using of two- (*a*) or multilayered (*b*) films with $M_{S1}, M_{S2}, \ldots$, instant measurement of H_0 in the desired sections is provided

register a nonuniform field near the holes. In Fig. 6.29 dependences of a magnetic field $H_0(y) - 1$ and $H_0(x) - 2$, measured at a distance $1.5 \cdot 10^{-4}$ m from the surface of the magnetic system made of two *SmCo* plates (Sm – samarium, Co – cobalt) by means of the developed sensor are shown. The nonuniform magnetic field is well registered with a resolution not worse than 3%.

6.6 Magnetostatic Wave Control

Depending on the boundary conditions and, mainly, the direction of an external magnetic field with respect to the ferrite film plane, in layered structures it is possible to vary over a wide range the types of excited and ducted waves and the areas of their localization. Such a wave probe has a controlled spatial resolution in the range of several orders of magnitude and provides nondestructive control of the basic magnetic parameters of structures. The boundary frequencies of various types of MSW at $\kappa'_{\max} \cong 10^5$–$10^6\,\mathrm{m}^{-1}$ allow the value of internal magnetic field H_{0i} and related values (saturation magnetization M_S and field of anisotropy $\overline{H}_A$) to be determined in longitudinally homogeneous structures. These parameters can be averaged over the volume of a film – $\langle H_{0i}\rangle$, $\langle M_S\rangle$, $\langle H_A\rangle$, or their cross-section distributions $H_{0i}(x)$, $M_S(x)$, $H_A(x)$ can be determined, or the parameters in a narrow surface layer of the order of magnitude $(0.2$–$0.5) \times 10^{-6}$ m – $\langle H_{0i}(y,z)\rangle$, $\langle M_S(y,z)\rangle$, $\langle H_A(y,z)\rangle$. The time dispersion of signal delay τ in obliquely magnetized structures allow the dissipative parameters $\langle \Delta H_\varphi\rangle$, $\langle \alpha_\varphi\rangle$, $\langle \alpha_\varphi(y,z)\rangle$, and $\langle \Delta H_\varphi(x)\rangle$, $\langle \alpha_\varphi(x)\rangle$ to be

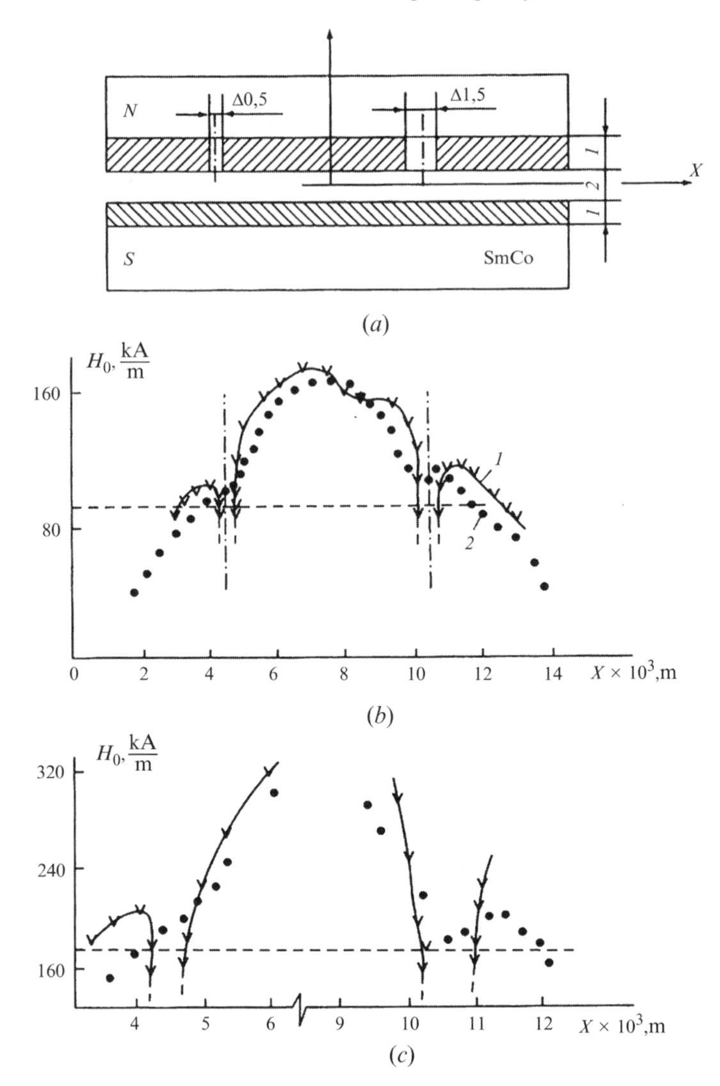

Fig. 6.28 The results of comparison of magnetic field measurements in a system with holes of various diameters in the polepieces and by means of the developed RFS – curve 1(*a*) and a Sh1-8 tester on the basis of a diminutive Hall sensor – curve 2(*b*) are presented, these large-scale dependences in the vicinity of the polepieceholes shows (*c*)

determined. In structures of finite sizes at reflection of MSW from the edge of the film and the points of turn (magnetic mirrors) the signal delay time and MSW dispersion influence the distribution of an internal end magnetic field $H_{0i}(x,y)$. Along with the orientation of the investigated FDLS with transformers in an external magnetic field H_0 at the angle φ the orientation of the structure at an angle Θ in the flat of transformers (Fig. 6.30) is used.

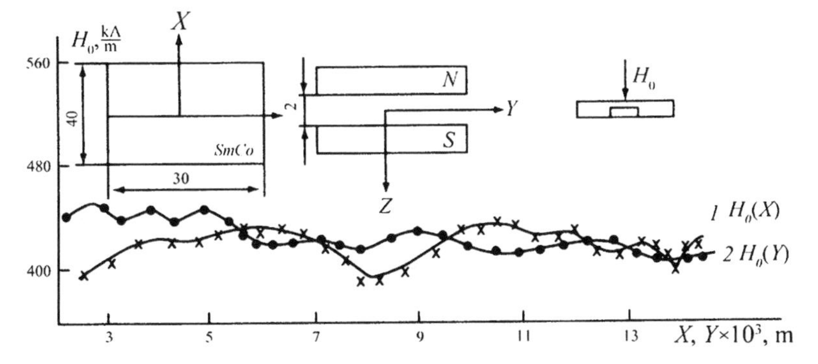

Fig. 6.29 Dependences of a magnetic field $H_0(y) - 1$ and $H_0(x) - 2$, measured at a distance $1.5 \cdot 10^{-4}$ m from the surface of the magnetic system made of two *SmCo* plates by means of the developed sensor are shown

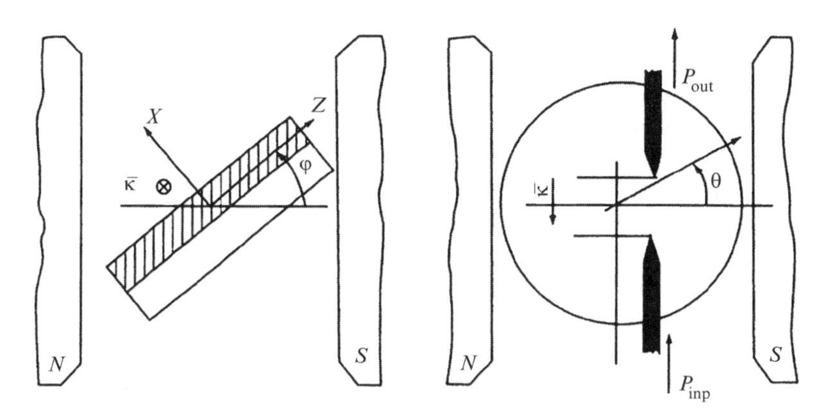

Fig. 6.30 Along with the orientation of the investigated FDLS with transformers in an external magnetic field H_0 at the angle φ and Θ in the flat of transformers

Orientational methods for measurement of ferrite parameters are known and have been used for diagnostics of volume ferrites based on resonant effects [27]. Wave methods have gained development in last years and were used for measurement of H_A in [311] and for measurement of $M_S(x)$ in the centimeter range in [311]. The latter method is not adequate, as the used way of account of the influence of distribution $M_S(x)$ on dispersive characteristics does not consider the influence of the distributions of fields of monoaxial_$\overline{H}_{mon}(x)$ and basic $\overline{H}_{bas}(x)$ anisotropy, metal loading of ferrite with a finite conductivity. Besides, the presence of cross-section dissimilarities $M_s(x)$, $H_{mon,bas}(x)$ is accompanied by cross-section dissimilarities of the dissipative factors $\Delta H(x)$ and $\alpha(x)$ in real structures.

Let's consider a symmetric FDLS (Fig. 6.31): a layer of ferrite (1) with its thickness d and penetrabilities $\overleftrightarrow{\mu}$ and ε; layers of dielectrics (2) and (3) of thicknesses t and l and penetrabilities $\varepsilon_{l,t}$; two metal screens (4) with a conductivity $(\sigma_{l,t})$.

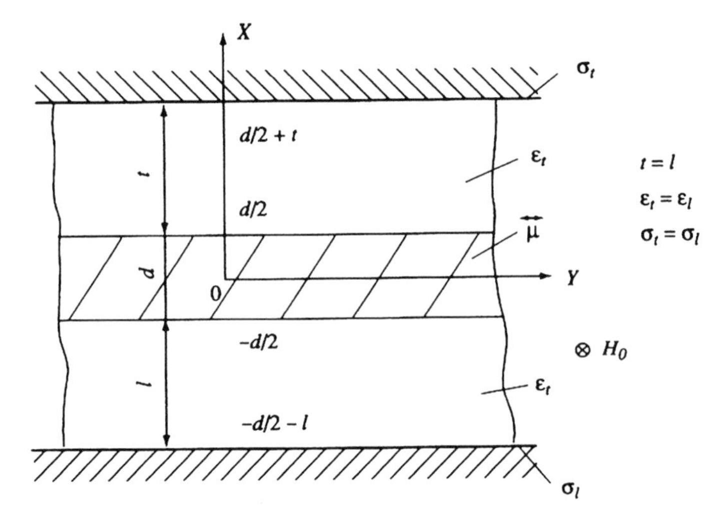

Fig. 6.31 A symmetric FDLS: a layer of ferrite (1) with its thickness d and penetrabilities $\overleftrightarrow{\mu}$ and ε; layers of dielectrics (2) and (3) of thicknesses t and l and penetrabilities $\varepsilon_{l,t}$; two metal screens (4) with a conductivity $(\sigma_{l,t})$

The high-frequency magnetic potential ψ for SMSW in a tangentially magnetized structure near the border $x_1 = d/2$ at $H_0 \uparrow\uparrow OZ$ is

$$\psi_1(x) = \begin{cases} \psi_1(0) \cdot e^{-\frac{\mu_{a1}}{\mu_1}\kappa'_{y1}x}, & -\frac{d}{2} \leq x \leq \frac{d}{2} \\ \psi_1(0) \cdot e^{-\kappa'_{y1}x}, & \frac{d}{2} < x \leq \frac{d}{2} + t \\ \psi_1(0) \cdot e^{\kappa'_{y1}x}, & -(\frac{d}{2}+l) \leq x < -\frac{d}{2}, \end{cases} \tag{6.42}$$

where

$$\mu_1 = 1 + 4\pi\chi_1, \ \chi_1 = \frac{\gamma M_1 \nu}{2\pi[\nu_n^2 - (1+\alpha_1^2)\nu^2 + 2j\alpha_1\nu\nu_n]},$$

$$\mu_{a1} = 4\pi\chi_{a1}, \ \chi_{a1} = \frac{\gamma M_1 \nu}{2\pi[\nu_n^2 - (1+\alpha_1^2)\nu^2 + 2j\alpha_1\nu\nu_n]},$$

$\alpha_1 = \frac{\Delta H_1}{H_{0i}}$ is the parameter of losses in the layer near the border $x_1 = \frac{d}{2}$, κ_{y1} the wave number of SMSW.

The high-frequency potential of SMSW is localized near the ferrite layer border $x_1 = \frac{d}{2}$ within the limits of the area

$$S_1(x,y) = \beta \int_0^{\frac{d}{2}+t} \psi_1(x)dx = \beta\psi_1\left(\frac{d}{2}\right)\frac{\mu_1}{\mu_{a1}\kappa'_{y1}}\left[1 - e^{-\frac{\mu_{a1}}{\mu_1}\kappa'_{y1}\frac{d}{2}} + e^{-\kappa'_{y1}\frac{d}{2}}(1-e^{-\kappa'_{y1}t})\right] \tag{6.43}$$

where $\psi_1\left(\frac{d}{2}\right)$ is the potential on the ferrite surface, β is the constant of proportionality.

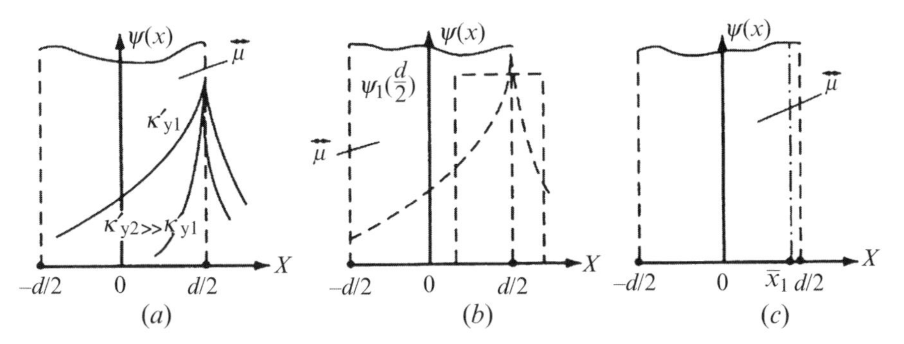

Fig. 6.32 Increases of localization of SMSW at the ferrite surface with $x = d/2$ at increase κ'_{y1} (*a*), $\overline{x_1} = \frac{\overline{x_{1t}} - \overline{x_{1d}}}{2}$ is the average coordinate (*b, c*)

From Eq. (6.43) it follows that at increase κ'_{y1} the localization of SMSW at the ferrite surface with $x = d/2$ (Fig. 6.32a) increases. The is an equal to Eq. (6.43) area of a rectangular with a side of $\psi_1\left(\frac{d}{2}\right)$

$$S(x,y) = \beta(d - \overline{x_1})\psi_1\left(\frac{d}{2}\right), \tag{6.44}$$

where $\overline{x_1} = \frac{\overline{x_{1t}} - \overline{x_{1d}}}{2}$ is the average coordinate (Fig. 6.32b, c).

Then, from Eqs. (6.43) and (6.44) we get

$$\overline{x_1} = d - \frac{\mu_1}{\mu_{a1}\kappa'_{y1}}\left[1 - e^{-\frac{\mu_{a1}}{\mu_1}\kappa'_{y1}\frac{d}{2}} + e^{-\kappa'_{y1}\frac{d}{2}}\left(1 - e^{-\kappa'_{y1}t}\right)\right]. \tag{6.45}$$

Similarly for $SMSW_3$ localized at the surface $x = -d/2$ at $H_0 \uparrow\downarrow 0Z$, we have an expression for the average coordinate

$$\overline{x_3} = \left\{d - \frac{\mu_1}{\mu_{a3}\kappa'_{y3}}\left[1 - e^{-\frac{\mu_{a3}}{\mu_1}\kappa'_{y3}\frac{d}{2}} + e^{-\kappa'_{y3}\frac{d}{2}}\left(1 - e^{\kappa'_{y3}l}\right)\right]\right\}. \tag{6.46}$$

For SSSLMSW of the lowest (basic) type the average coordinate is

$$\overline{x_2} = \frac{d}{2}. \tag{6.47}$$

For high-quality ($\alpha \ll 1$) weakly anisotropic ferrites ($H_A \ll 4\pi M_s \ll H_0$) in the millimeter range

$$\overline{x_1} \cong d - \frac{1}{2\kappa'_{y1}\left(1 + \frac{2H_0}{4\pi\langle M_{s1}\rangle}\right)}\left[1 - e^{-\left(1+\frac{2H_0}{4\pi\langle M_{s1}\rangle}\right)\kappa'_{y1}\frac{d}{2}} + e^{-\kappa'_{y1}\frac{d}{2}}\left(1 - e^{-\kappa'_{y1}t}\right)\right], \tag{6.48}$$

$$\overline{x_3} \cong -\left\{d - \frac{1}{2\kappa'_{y3}\left(1 + \frac{2H_0}{4\pi\langle M_{s3}\rangle}\right)}\left[1 - e^{-\left(1+\frac{2H_0}{4\pi\langle M_{s3}\rangle}\right)\kappa'_{y3}\frac{d}{2}} + e^{-\kappa'_{y3}\frac{d}{2}}\left(1 - e^{\kappa'_{y3}l}\right)\right]\right\}. \tag{6.49}$$

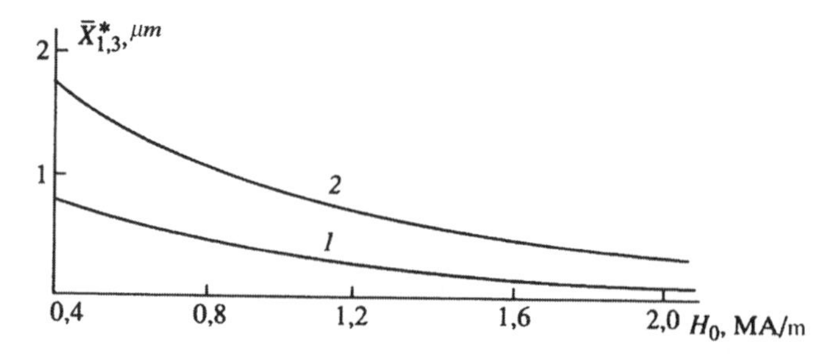

Fig. 6.33 The values of $\overline{x^*_{1,3}} \cong [2\kappa'_{y1,3}(1+\frac{2H_0}{4\pi\langle M_{s1}\rangle})]^{-1}$, determining the spatial resolution in Eqs. (6.48) and (6.49), on the external field H_0, are presented, where $1 - 4\pi\langle M_s\rangle = 0.176\,\mathrm{T}$; $2 - 4\pi\langle M_s\rangle = 0.5\,\mathrm{T}$ for $\kappa'_y = 10^5\,\mathrm{rad/m}$

From Eqs. (6.48) and (6.49) it is seen that the spatial resolution for SMSW determining averaging of the magnetic parameters of the structure is high at high values of $\kappa'_{y1,3}$ and ratio $\frac{H_0}{2\pi\langle M_{s1,3}\rangle}$.

In Fig. 6.33 the values of $\overline{x^*_{1,3}} \cong \left[2\kappa'_{y1,3}\left(1+\frac{2H_0}{4\pi\langle M_{s1}\rangle}\right)\right]^{-1}$, determining the spatial resolution in Eqs. (6.48) and (6.49), on the external field H_0, are presented, where $1-4\pi\langle M_s\rangle = 0.176\,\mathrm{T}$; $2–4\pi\langle M_s\rangle = 0.5\,\mathrm{T}$ for $\kappa'_y = 10^5\,\mathrm{rad/m}$. It is obvious that in the millimetric range for $H_0 > 0.8\,\mathrm{MA/m}$ the resolution several times higher than that in the centimeter range. So, at $H_0 \cong 0.8\,\mathrm{MA/m}$ $\overline{x^*_{1,3}} < (1-0.4)\times 10^6\,\mathrm{m}$, and at $H_0 \cong 1.6\,\mathrm{MA/m}$ $\overline{x^*_{1,3}} < (0.5-0.2)\times 10^{-6}\,\mathrm{m}$. For the transformers providing excitation of SMSW with $\kappa'_y \cong 10^6\,\mathrm{rad/sm}$ the resolution raises by an order of magnitude and makes $\overline{x^*_{1,3}} < (0.1–0.2)\times 10^{-6}\,\mathrm{m}$.

When a ferrite layer with transformers is arranged at various angles φ to the external magnetic field (Fig. 6.30), the following waves are excited and propagate in the structure:

- At $\varphi = 0$, $H_0 \uparrow\uparrow 0Z$ – SSMSW for $\kappa'_{\max}$ localized close to the surface $x = d/2$
- At $0 < \varphi \leq \varphi_1 - \delta\varphi$ – semi-surface magnetostatic waves[3], – SMSMSW$_1$, at $\kappa'_{\max}$ localized mainly in half-subspace of the ferrite layer of a thickness (0, *d*/2)
- At $\varphi_1 + \delta\varphi < \varphi < \pi/2$ – semi-spatial magnetostatic waves (SSMSW*), localized in the volume of the layer and weakly prevailing in a half-subspace of the ferrite (0, *d*/2)
- At $\varphi = \pi/2$, $H_0 \uparrow\uparrow 0Z$ – SSLMSW distributed in the volume of the layer
- At $\varphi_2 - \delta\varphi \geq \varphi > \pi/2$ – SSMSW$^*_2$ localized in the volume and weakly prevailing in a half-subspace of the ferrite $(0, -d/2)$
- At $\pi > \varphi > \varphi_2 + \delta\varphi$ – SMSMSW$_3$ for $\kappa'_{\max}$ localized mainly in a half-subspace of a thickness $(0, -d/2)$
- At $\varphi = \pi$ – SSMSW$_3$ for $\kappa'_{\max}$ localized near the surface $x = -d/2$

[3] Waves include superpositions $Ae^{-\kappa'_y x} + Be^{\kappa'_y x}$ and $C\cos\kappa'_y x + D\sin\kappa'_y x$, in which can prevail or exponential (surface), or trigonometrical (spatial) distributions of functions

Table 6.4 The values of frequencies at $\kappa'_{\max}$ and $\kappa'_{\min}$ averaged over the surface and volume layers are presented for various types of MSW

Wave type	$\kappa'_{\max}$		$\kappa'_{\min}$	
	Frequency	Internal field	Frequency	Internal field
$SSMSW_1$	ν_1	$\langle H_{0i}\rangle_1$	ν_1	$\langle H_{0i}\rangle_1^0$
$SMSMSW_1$	$\nu_1 \div \nu_{1\varphi}$	$\langle H_{0i}\rangle_{1\varphi}$	$\nu_1 \div \nu_{1\varphi}$	$\langle H_{0i}\rangle_{1\varphi}^0$
$SSMSW_1^*$	$\nu_{1\varphi} \div \nu_2$	$\langle H_{0i}\rangle_2$	$\nu_{1\varphi} \div \nu_2$	$\langle H_{0i}\rangle_2^0$
SLMSW	ν_2	$\langle H_{0i}\rangle$	ν_2	$\langle H_{0i}\rangle^0$
$SSMSW_3^*$	$\nu_2 \div \nu_{2\varphi}$	$\langle H_{0i}\rangle_{2\varphi}$	$\nu_2 \div \nu_{2\varphi}$	$\langle H_{0i}\rangle_{2\varphi}^0$
$SMSMSW_3$	$\nu_{2\varphi} \div \nu_3$	$\langle H_{0i}\rangle_{3\varphi}$	$\nu_{2\varphi} \div \nu_3$	$\langle H_{0i}\rangle_{3\varphi}^0$
$SSMSW_3$	ν_3	$\langle H_{0i}\rangle_3$	ν_3	$\langle H_{0i}\rangle_3^0$

In Table 6.4 the values of frequencies at $\kappa'_{\max}$ and $\kappa'_{\min}$ averaged over the surface and volume layers are presented for various types of MSW.

The internal fields include the external magnetic field H_0, the field of demagnetizing factors $\overleftrightarrow{N}M_s$, the field of axial anisotropy $\overleftrightarrow{N}_{mon}H_{mon}$, the field of anisotropy of the basic plane $\overleftrightarrow{N}_{bas}H_{bas}$, etc., and the frequency limit is

$$\nu_0 = F(H_0, \overleftrightarrow{N}M_s, \overleftrightarrow{N}_{mon}H_{mon}, \ \ldots). \tag{6.50}$$

In monoaxial crystals (barium hexaferrite), in crystals with the cubic structure (YIG, spinels) the field of axial anisotropy prevails over the field of anisotropy in the basic plane $\overline{H}_{mon} > \overline{H}_{bas}$ [28]. For crystals with the cubic structure the field of anisotropy is low $\overline{H}_{mon} \ll 4\pi M_S$, and in the EHF range $\overline{H}_{mon} \gg 4\pi M_S$. For monoaxial crystals $\overline{H}_{mon} \ll 4\pi M_S$ and $\overline{H}_{mon} \geq H_0$.

To separate the internal magnetic parameters and $\overleftrightarrow{N}_{mon}H_{mon}$ in Eq. (6.50) orientational operations (Fig. 6.30) by angles φ and θ are used.

Let's consider a case of weakly anisotropic strongly saturated films (YIG, spinels) with $\overline{H}_{mon} \ll 4\pi M_S$ and $H_0 \gg 4\pi M_S$. The orientation by angle θ (Fig. 6.30) in the plane of a tangentially magnetized film $\varphi = 0,\ \pi$ and $H_0 > 0,\ H_0 < 0$ gives two characteristic frequencies:

– At $\theta_1 = \theta_{10}$

$$\nu_{1,3}\Big|_{\theta\,=\,\theta_1} = \nu_{\min}, \quad \nu'_{1,3}\Big|_{\theta\,=\,\theta_1} = \nu'_{\min} \tag{6.51}$$

– At $\theta_{21} = \theta_{10} \pm \pi$

$$\nu_{1,3}\Big|_{\theta\,=\,\theta_2} = \nu_{\max}, \quad \nu'_{1,3}\Big|_{\theta\,=\,\theta_2} = \nu'_{\max} \tag{6.52}$$

and frequencies $\nu'_{1,3}$ are registered on $\nu_\perp$ and ν_H with due account of $\overline{H}_{mon}$.

The input and output strip transformers are located along the field $\overline{H}_0$ and work in the broadband regime ($\kappa'_{max} > 10^5 - 10^5\,\mathrm{rad/m}$).

Strict observtion of the condition (6.52) means that the investigated structure has homogeneous (by layer thickness) field of anisotropy ($H_{mon} = const$). At $|\theta_2 - \theta_{10}| \neq \pi$ there is a cross-section heterogeneity of the field of anisotropy $H_{mon}(x)$ and $\frac{\Delta\langle H_{mon1}\rangle}{\Delta\overline{x_1}} \neq \frac{\Delta\langle H_{mon3}\rangle}{\Delta\overline{x_3}}$, which are most precisely registered on the upper boundary of frequency at κ'_{max} and have the average coordinates $\overline{x}_1$ for $\nu_1(\theta_1)$ and $\nu_1(\theta_2)$ and $\overline{x}_3$ for $\nu_3(\theta_1)$ and $\nu_3(\theta_2)$.

At κ'_{min} the frequencies $\nu'_{1,3}(\theta_1)$ and $\nu'_{1,3}(\theta_2)$ determine $\langle H_{mon1}\rangle'$ and $\langle H_{mon3}\rangle'$, and the mean value is

$$\left.\frac{\langle H_{mon1}\rangle'_0 + \langle H_{A3}\rangle'_0}{2}\right|_{\theta_1} = \left.\frac{\langle H_{A1}\rangle'_0 + \langle H_{A3}\rangle'_0}{2}\right|_{\theta_2} = \langle H_A\rangle' \tag{6.53}$$

on frequency

$$\left.\frac{\nu'_1 + \nu'_3}{2}\right|_{\theta_1} = \left.\frac{\nu'_1 + \nu'_3}{2}\right|_{\theta_2} = \overline{\nu'}. \tag{6.54}$$

In structures with monoaxial anisotropy[4] at tangential magnetization ($\varphi = 0, \pi$) for measurement of saturation magnetization it is convenient to use the angle of orientation

$$\theta_3 = \theta_{10} \pm \frac{\pi}{2}. \tag{6.55}$$

For structures with cross-section heterogeneity of their monoaxial anisotropy

$$\theta_2 - \theta_1 = \pi \pm \frac{\delta\theta_{mon}}{2} \text{ and } \theta_3 = \theta_2 - \pi \mp \frac{\delta\theta_{mon}}{2}, \tag{6.56}$$

where $\delta\theta_{mon}$ is an angle considering $\overline{H}_{mon}$.

Then, subject to Eqs. (6.55) and (6.56), the formulas for resonant frequencies ignoring $\overline{H}_{mon}$ can be used.

Let's proceed with consideration of testing methods for ferrite films in the millimeter range.

6.7 Control of Cross-Section Distribution of Saturation Magnetization

Methods of discrete and continuous probing of a ferrite film by various types of superficial and volume MSW have been developed.

[4] YIG, spinels, strongly anisotropic films of barium hexaferrite with axis of facile magnetization $\bar{c}$ in film plane.

6.7.1 Method of Discrete Probing of a Ferrite Layer by Surface and Volume MSW

At discrete changing the direction of an external magnetic field $\overline{H}_0$ or turning the structure in a field $\overline{H}_0$ (at $\varphi = 0,\ \varphi = \pi/2,\ \varphi = \pi$) the following waves are excited and propagate in the structure (Fig. 6.34):

- $SSMSW_1$, localized at one surface of the film
- $SSLMSW_2$, distributed over the thickness of the film
- $SSMSW_3$, localized at the other surface of the film

Using $SSMSW_1$, $SSLMSW_2$ and $SSMSW_3$ at κ'_{max}, it is possible to determine three values of magnetization in the layer by their boundary frequencies

$$\langle M_{s1} \rangle = \frac{1}{2}\left(\frac{\nu_1}{\gamma} - \frac{H_0}{\pi}\right) \text{ at } \bar{x}_1, \tag{6.57}$$

$$\langle M_{s2} \rangle = \frac{\pi}{\gamma^2 H_0}\left[(\nu_1^1)^2 + (\nu_3^1)^2\right] - 2H_0 \text{ at } \bar{x}_2, \tag{6.58}$$

$$\langle M_{s3} \rangle = \frac{1}{2}\left(\frac{\nu_3}{\gamma} - \frac{H_o}{\pi}\right) \text{ at } \bar{x}_1, \tag{6.59}$$

where $\nu_{1,3}$ is the upper boundary frequencies at κ'_{max}; $\nu'_{1,3}$ the boundary frequencies of homogeneous precession determined by RFS resonance frequency; $\bar{x}_{1,2,3}$ the average coordinates from Eqs. (6.46–6.48).

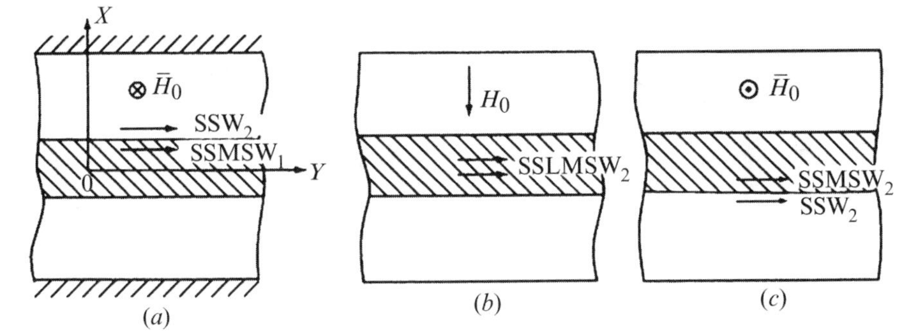

Fig. 6.34 At discrete changing the direction of an external magnetic field $\overline{H}_0$ or turning the structure in a field $\overline{H}_0$ (at $\varphi = 0,\ \varphi = \pi/2,\ \varphi = \pi$) the following waves are excited and propagate in the structure:
– $SSMSW_1$, localized at one surface of the film, *a*
– $SSLMSW_2$, distributed over the thickness of the film, *b*
– $SSMSW_3$, localized at the other surface of the film, *c*

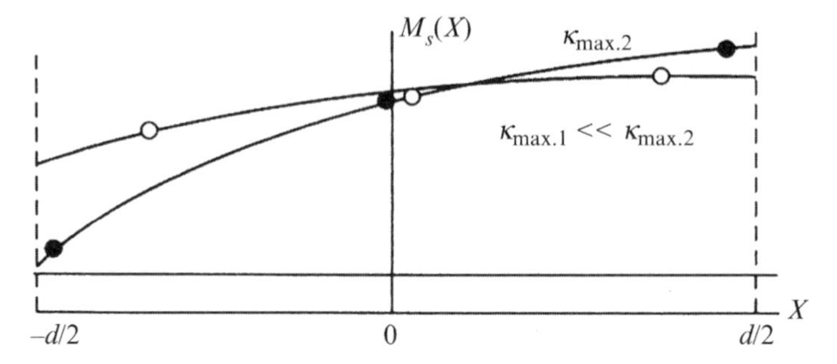

Fig. 6.35 Dependences $M_S(x)$ for measuring cells with various κ'_{max} are presented

The relative errors are

$$\left|\frac{\delta\langle M_{s1,3}\rangle}{\langle M_{s1,3}\rangle}\right| = \left|\frac{\delta v_{1,3}\gamma + \delta\gamma v_{1,3}}{\gamma^2} + \frac{1}{\pi}\delta H_0\left(\frac{v_1}{\gamma} - \frac{H_0}{\pi}\right)^{-1}\right|, \tag{6.60}$$

$$\left|\frac{\delta\langle M_{s2}\rangle}{\langle M_{s2}\rangle}\right| = 2\left|\frac{v_1'\delta v_1' + v_3'\delta v_3' + \delta H_0}{(v_1')^2 + (v_3')^2 - 2H_0} + \frac{2\gamma\delta\gamma H_0 + \gamma^2\delta H_0}{\gamma^2 H_0}\right|. \tag{6.61}$$

By the three values $\langle M_{s1}\rangle$, $\langle M_{s2}\rangle$, $\langle M_{s3}\rangle$ the extrapolated dependence $M_s(x)$ is plotted. In Fig. 6.35 such dependences for measuring cells with various κ'_{max} are presented.

The flowchart of an experimental installation includes a standard panoramic instrument of SWR$_e$ and attenuations or WB with an oscillographic indicator and a wavemeter of a required accuracy, an instrument H_0 or our developed sensors of resonant frequencies, a measuring cell with stripline antennas allowing turning in an external magnetic field at ab angle φ (angles $\varphi = 0, \pi/2, \pi$) and at an angle φ in the plane of the structure and the device of angle registration φ and θ.

Having obtained the dependence $M_s(x)$ by the given technique, using the orientations (6.51) and (6.52) and corresponding relationships for resonant frequencies [27] we can determine $\overline{H}_{mon}(x)$.

Let's emphasize that at discrete probing of SMSW it is possible to explore $M_s(x)$ in multilayered structures by determination of piecewise-linear functions from the values of frequencies at κ'_{max} on the borders (Fig. 6.36a–c).

6.7.2 Method of Continuous Probing of a Ferrite Layer by Various Types of MSW

If a continuous distribution of magnetic parameters of the structure, both near the borders of the film and inside the layer, should be obtained, the method of comparison with a reference for whom these parameters are constant by layer thickness or vary much more weaker than in the investigated samples is used.

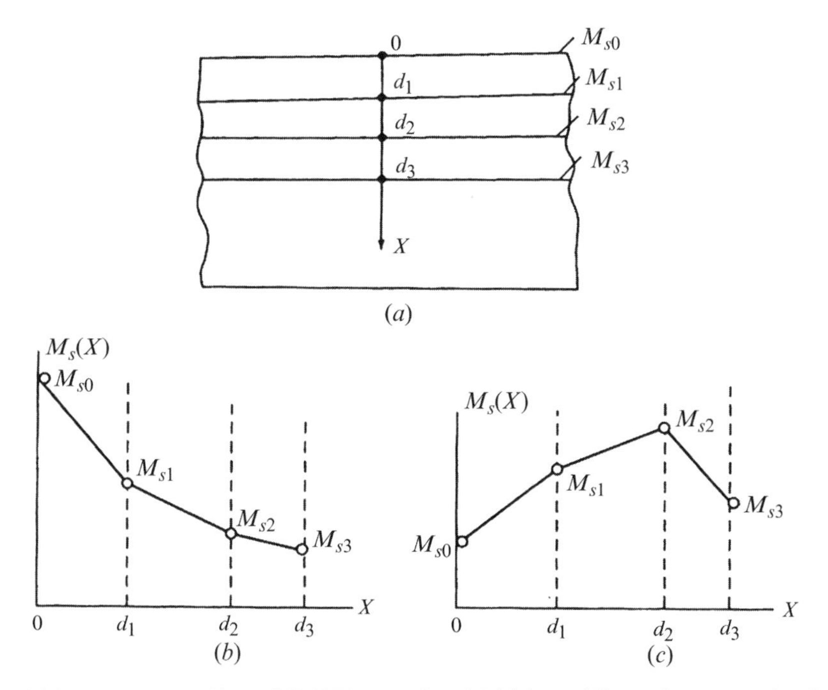

Fig. 6.36 At discrete probing of SMSW to explore $M_s(x)$ in multilayered structures by determination of piecewise-linear functions from the values of frequencies at κ'_{max} on the borders (*a, b, c*)

In this case, functional dependences of the upper boundary frequencies are compared at κ'_{max} for the investigated and reference samples (Fig. 6.37). Their difference (the sector of vertical hatching) determines the internal magnetic parameters. The reference sample with $M_S = const$, $H_{mon} = const$, $H_{bas} = const$ should have, in comparison with the investigated one, close parameters and sizes by all values, except that under study. For example, at research of $M_s(x)$ for a reference $M_S = const$, and fields H_{mon} and H_{bas} can be either comparable for them or $(H_{bas}(x))_{et} \ll H_{mon}(x)$ and $(H_{bas}(x))_{et} \ll H_{bas}(x)$ (et-etalonr – reference). Naturally, both for the reference and the investigated sample, performance of orientational operations by the angles φ and θ is necessary.

For weakly anisotropic ferrites ($H_{mon} \ll 4\pi M_S$) in the EHF range ($H_0 \gg 4\pi M_S$) at $\theta = \theta_3$ we have

$$\varphi = 0,\ \bar{x} = 0,\ \langle M_s \rangle = \frac{2\pi}{\gamma^2 H_0}\left[\nu_{et}^2(0) - \nu^2(0)\right] - \langle M_{set} \rangle, \tag{6.62}$$

$$0 < \varphi < \varphi_1 - \delta\varphi,$$

$$\frac{\pi}{2} > \varphi > \varphi_1 + \delta\varphi, 0 < \bar{x} < \frac{d}{2}, \tag{6.63}$$

$$\langle M_s \rangle = \frac{4\pi^2 \nu^2(\varphi) - (\gamma H_0)^2}{4\pi^2 \nu_{et}^2(\varphi) - (\gamma H_0)^2} \langle M_{s\,et} \rangle$$

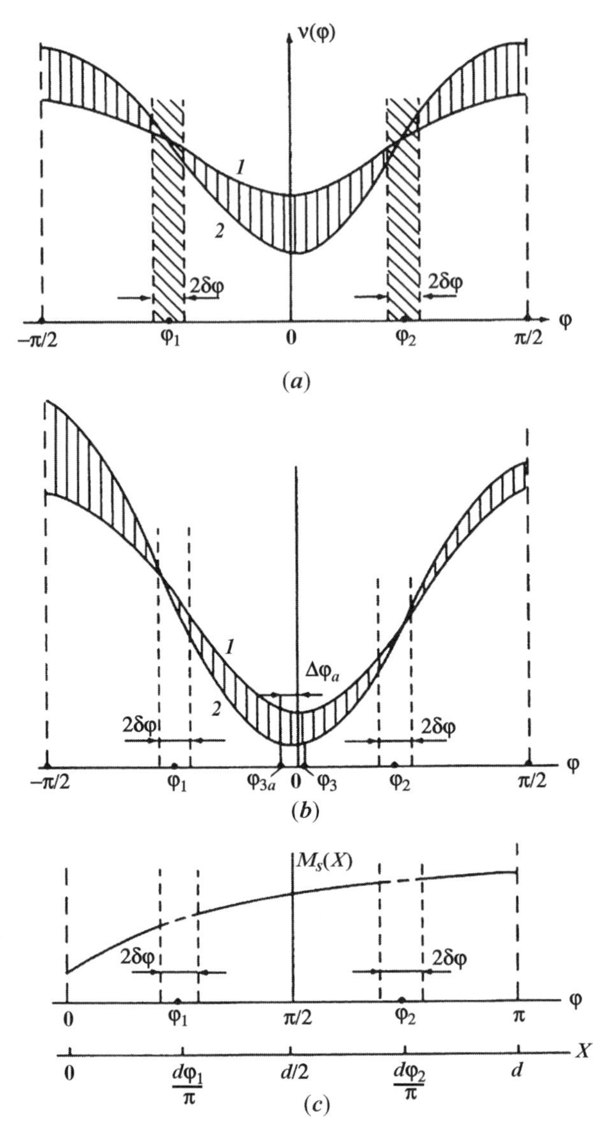

Fig. 6.37 Functional dependences of the upper boundary frequencies are compared at $\kappa'_{\max}$ for the investigated and reference samples

$$\varphi = \frac{\pi}{2},\ \bar{x} = \frac{d}{2}, \qquad \langle M_s \rangle = \tfrac{1}{2\gamma}\left[\nu_{et}(\tfrac{\pi}{2}) - \nu(\tfrac{\pi}{2})\right] + \langle M_{s\,et} \rangle, \tag{6.64}$$

$$\frac{\pi}{2} < \varphi < \varphi_2 - \delta\varphi, \qquad -\tfrac{d}{2} < \bar{x} < 0, \tag{6.65}$$

$$\pi > \varphi > \varphi_2 + \delta\varphi, \qquad \langle M_s \rangle = \frac{4\pi^2\nu^2(\varphi) - (\gamma H_0)^2}{4\pi^2\nu_{et}^2(\varphi) - (\gamma H_0)^2} \langle M_{s\,et} \rangle$$

$$\varphi = \pi,\ \bar{x} = d \qquad \langle M_s \rangle = et\left[\nu_{et}^2(\pi) - \nu^2(\pi)\right] - \langle M_{s\,et} \rangle, \tag{6.66}$$

where $2\delta\varphi$ is the range of angles within which the dependences $\nu_{et}(\varphi)$ and $\nu(\varphi)$ and tensors of demagnetized factors $\overleftrightarrow{N}(\varphi_1) = \overleftrightarrow{N}(\varphi_2) = 0$ are crossed, $\bar{x} = \frac{d}{\pi}\varphi$ the coordinate in the layer.[5]

In the intervals

$$\frac{d}{\pi}(\varphi_1 - \delta\varphi) < \bar{x} < \frac{d}{\pi}(\varphi_1 + \delta\varphi) \text{ and } \frac{d}{\pi}(\varphi_2 - \delta\varphi) < \bar{x} < \frac{d}{\pi}(\varphi_2 + \delta\varphi)$$

the dependence $M_S(x)$ is extrapolated by its extreme points (Fig. 6.37*c*).

In Fig. 6.37 dependences of the upper boundary frequencies $\nu(\kappa_{\max})$ (curve 1) and $\nu_{et}(\kappa_{\max})$ (curve 2) for weaky anisotropic structures in the EHF range are presented at $\theta = \theta_3$ (Fig. 6.37*a*) for $(H_{mon})_{et} \ll (4\pi M_S)_{et}$, $H_{mon} \ll 4\pi M_S$, $H_{bas} \gg (4\pi M_S)_{et}$ and $H_0 \gg 4\pi M_S$ at $\theta = \theta_2$ (Fig. 6.37*b*) when the influence of $(H_{mon})_{et}$ and H_{mon} near the borders $\bar{x} = 0$ and $\bar{x} = d$ is registered (at $\theta = 0$ and $= \pi$) and at the center of the layer $\bar{x} = \frac{d}{2}$, and, finally, the resulting dependence $M_S(x)$ (Fig. 6.37*c*).

The distinction of the angles θ_3 and θ_{3a} at normal magnetization of structures, for which the minimum values of frequencies are $(\nu_{\min}(\varphi_3))_{et}$ and $\nu_{\min}(\varphi_3)$ allows the direction of the vector $\overline{H}_{mon}$ to be determined

$$\varphi_a = (\varphi_3)_{et} - \varphi_3,$$

which is in the plane of angle θ_3.

Table 6.5 summarizes the results of our experimental investigations of the distribution $M_S(x)$ by the above method.

In Fig. 6.38 experimental dependences $\nu_{et}(\varphi)$ and $\nu(\varphi)$ (Fig. 6.38*a*), and the dependence of the distribution $M_S(x)$ are presented. It is obvious that the character of the distribution $M_S(x)$ is most complex in one of the hemi-layers of the structures. Registration of a small splash at a level of 1% of the value $M_S(x)$ at $x = 30 \times 10^{-6}$ m shows a high resolution of the method.

Table 6.5 The results of our experimental investigations of the distribution $M_S(x)$ by the above method

φ°	$\nu_{et}(\kappa'_{\max})$, *MHz*	$\nu_{et}(\kappa'_{\max})$, *MHz*	$x \times 10^6$, m	$\langle 4\pi M_s \rangle$, T	The note
+90	10,900	10,900	0	0.1709	Diameter of YIG
+62.5	10,100	10,100	5.87	0.1709	structure –
+44	8,150	81,500	12.01	0.1709	76×10^{-3} m, $d =$
+17	4,620	46,200	19.06	0.1709	47×10^{-6} m. Thickness
−5	3,850	39,200	24.80	0.1690	of reference film
−15	4,600	49,300	27.42	0.1590	$d_{et} = 10.2 \times$
−35	7,300	76,500	32.64	0.1382	10^{-6} m, $(4\pi M_S)_{et} =$
−70	10,320	9,980	41.78	0.1332	0.1709 T Transformer
−80	10,700	10,200	44.39	0.1282	$W_{MSL} = 30 \times 10^{-6}$ m
−90	10,900	10,280	47.0	0.1219	

[5] In comparison with previous case, the reference mark of layer thickness lays on one of its surfaces.

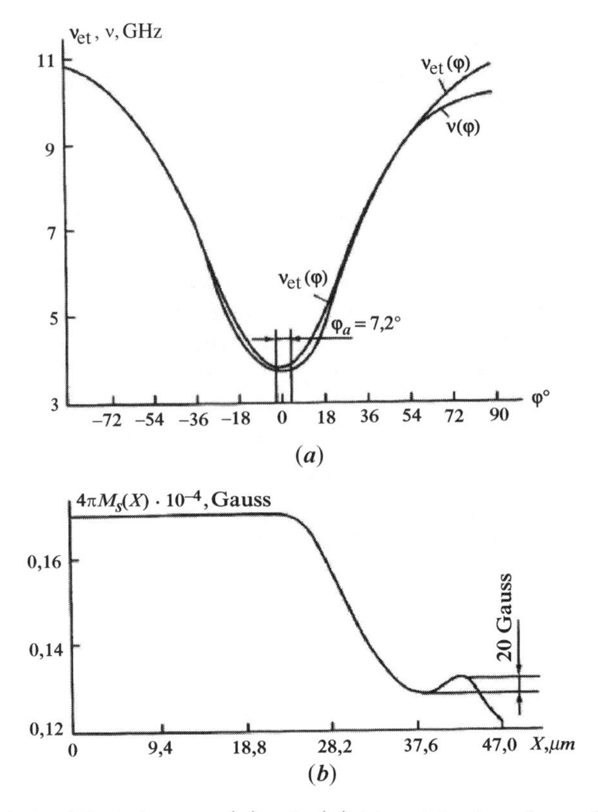

Fig. 6.38 Experimental dependences $\nu_{et}(\varphi)$ and $\nu(\varphi)$ (a), and the dependence of the distribution $M_S(x)$ are presented

6.8 Control of Internal Magnetic End Field Distribution

One of the problems essentially constrained practical application of magnetoelectronic devices in MMR is connected with the strong irregularity of AFC signals at a power difference up to 20–40 dB. Our investigations have shown that a major factor of such irregularity is wave reflection from the end faces of a ferrite film, entrance for the edges of the input and outputgoing strip transformers. One of ways to eliminate this irregularity is creation of conditions for a wave at its propagation towards the film side to effectively fade. Thus, the law of change of an external magnetic field $H_0(y)$ (the wave extends along the axis $0Y$) should be opposite to the law of change of internal magnetic field $H_{0i}(y)$ in the ferrite film. From the calculation [8] it follows that the demagnetizing factors $\overleftrightarrow{N} M_S$ effectively reduce thge internal magnetic field at a distance $\Delta y \cong (3–4)d$, d being the thickness of the film.

Therefore, for estimation of the internal magnetic field $H_{0i}(y)$ a special method was developed. It uses sounding the investigated film edge by pulses, and

$$\frac{\Delta y(H_{0i}(\nu_{1,2}))}{L} = \frac{\tau_2^2(\nu_2) + \tau_1^2(\nu_1)}{2\tau_1(\nu_1)\tau_2(\nu_2)}, \tag{6.67}$$

where Δy is the increment of TL along the direction of wave distributionat which a differential delay $\Delta\tau$ is provided; $\tau_1(\nu_1)$ and $\tau_2(\nu_2)$ the signal delays on frequencies. ν_1 and ν_2, respectively; L the length of the film edge hanging for the transformer.

In practical measurements it is convenient to choose frequencies corresponding to the maxima of resonances, for which $[\kappa'(\nu_2) - \kappa'(\nu_1)] \cdot L = n\pi,\ n = 1,\ 2, \ldots$. So, for example, from the data in Figs. 7.17a and 7.21 it follows that in a passband from $\nu_H = 35.6\,\text{GHz}$ up to the upper boundary of frequency $\nu = 36.8\,\text{GHz}$, subject to Eq. (7.23), the gradient value of the internal magnetic field is $\frac{\Delta H_{0i}}{\Delta y} \cong -0.11\,\text{kOe/mm}$ at $\Delta\tau \cong 70 \cdot 10^{-9}\,\text{s}$. For eliminating the irregularity of AFC of such a DL, it is necessary to apply an external magnetic field which falls down outside of the strip transformers with a gradient $\frac{\Delta H_{0i}}{\Delta y} \geq +0.11\,\text{kOe/mm}$.

6.9 Conclusions

1. A method and resonant devices for control, including nondestructive one, of key parameters of layered structures on the basis of mono- and polycrystalline ferrite plates and films in a wide range of frequencies (above three to five octaves) have been developed.
2. The method of preferential signal excitation in the near zone of a strip transformer with a ferrite-dielectric structure in the prelimiting $(\nu \gg \nu_{cr})$ regime allows dissipative factors –(the parameters of ferrite α and ΔH, amplitude and phase constants, wave distributions, an internal magnetic field) to be determined from data of theoretical analysis. The oeriodic character of the dependence of reflection factor on the length of the ferrite-dielectric structure and the difference between the maxima and minima determine the quality of ferrite on microwave frequencies.
3. The testing method for layered film ferrite-dielectric structures on the basis of the transparency effect is based on double filtration of signals – in the near zone of the input slot-guide transformer with orthogonal orientation of the ferrite film to the plane of the exciting slot due to increasing the homogeneity of outside exciting fields, and in its distant zone or in a beyond-cutoff waveguide due to increasing the steepness of the dispersion of amplitude wave constants at tuning out from the central frequency. The method provides multi-octave nondestructive diagnostics on microwave frequencies and EHF.
4. The testing method for ferrite films on the transparency effect of layered structures with an absorptive covering allows the parameters of extending waves and properties of structures to be controlled in a wide range of frequencies on a frequency near the resonant one.

5. The method of antiphased balanced bridge at phase inversion on a ferrite-dielectric structure provides control of mono- and hypocrystalline ferrite plates and films on microwave frequencies and EHF.
6. The existence of a two-wave regime in layered structures on the basis of weakly dissipative ferrite films was substantiated. The amplitude and phase constants of fast and slow waves with close dispersions in the field of resonant frequencies have been determined.
7. It is shown that the parameter of ferromagnetic losses α is independent of the range of working frequencies, and the line width of FMR for ferrite films increases linearly as frequency grows, and the steepness of the dependence $\frac{\Delta H}{\Delta \nu} \sim 10^{-2}\,\mathrm{Oe/GHz}$, that restricts the level of accessible losses for signal distribution in the millimeter range.
8. The basic high-frequency dissipative parameters of epitaxial films of various structures of YIG, ferrites, spinels, BHF – $\alpha, \Delta H$, their topology by the area and thickness of the layer, similar dependences in the microwave and EHF ranges have been investigated by non-destructive methods.
9. Methods and devices for non-destructive control of ferrite films have been developed with sounding an investigated site by surface MSW near one surface at tangential $H_0 > 0$, volume MSW in the layer at normal, superficial MSW near to other surface at tangential magnetization ($H_0 < 0$), and also semi-surface ($\mathrm{SMSMSW}_{1,2}$) and semi-spatial (SSMSW^*) waves in aslant magnetized layered structures, top boundary frequencies which bear data on an internal magnetic field and the parameters defining it in a layer of localization MSW, characterized in the average coordinate of the analysis $x^*_{1,2,3}$.
10. The direct express method of control of saturation magnetization distributions over the thickness of the layer d of an investigated ferrite film in MMR is underlied by sounding by:

    – Surface MSW localized at one film surface (SMSW_1 at $H_0 > 0$)
    – Spatial MSW distributed in the layer (SSLMSW, at H_0)
    – Surface MSW localized at the other surface of the film (SMSW_2 at $H_0 < 0$)

    for which the upper boundary frequencies $\nu_1(\kappa'_{SMSW_1})$, $\nu_2(\kappa'_{SSLMSW_2})$, $\nu_3(\kappa'_{SMSW_3})$ are registered and from which the values of $M_S(x_1^*)$, $M_S(x_2^*)$, $M_S(x_3^*)$ are determined at the corresponding average coordinates of the layer x_1^*, x_2^*, $x_3^* \in [0,d]$, and the dependence $M_S(x)$ is restored.
11. The testing method for $M_S(x)$ control is underlied by comparison of the upper boundary frequencies of the above MSW types with those in a film with close sizes (thickness, length, width) field of anisotropy, and saturation magnetization being constant or close to constant $M_S(x)$ which is obtained by the express method.
12. Our investigation of the distributions $M_S(x)$ in film ferrites in MMR has revealed a spatialre solution not worse than 1%.
13. Sounding of the film edge by MSW pulses on provides control of the gradient of an internal magnetic field along the direction of wave distribution.

Chapter 7
Controlled Magnetoelectronic Devices

Our theoretical and experimental research has allowed the following requirements to be formulated:

- To ferrite materials and film structures and methods of their control
- To magnetic systems on the basis of high-power-consuming rare-earth (RE) alloys
- To converters on the basis of waveguide and strip lines for excitation and reception of waves in multilayered dissipative structures
- To transmission line on the basis of ferrite films

Parameters of the designed operated devices of low and high power levels in the millimeter range are discussed.

7.1 Radiophysical Aspects of Millimeter-Range Magnetoelectronics

Let's consider the basic scientific problems which development has been directed on the design of magnetoelectronics devices of the millimeter range of radiowaves.

Promotion of fundamental and applied research into the field of MMR magnetoelectronics requires solution of the following questions:

- Choice of a working material
- Induction of the highest achievable external magnetic fields within the limits of small dimensions and weights
- Excitation of waves by waveguide and planar converters
- Development of an electrodynamic theory of wave excitation by extraneous sources
- Selection of specific parameters and features most typical for MMR
- Diagnostics of film parameters in MMR
- Control over external and internal magnetic fields with an increased accuracy and spatial resolution

A.A. Ignatiev, *Magnetoelectronics of Microwaves and Extremely High Frequencies in Ferrite Films*.
DOI: 10.1007/978-0-387-85457-1_8,

- Control of wave dispersions in MMR structures
- Electrodynamic analysis of characteristic wave dispersions in MMR structures
- Determination of amplitude, phase, and frequency distortions, ways of their reduction or acceleration elimination
- Technological aspects

Below we consider most prominent aspects of MMR magnetoelectronics.

7.1.1 Magnetic Materials and Film Structures for MMR

Key parameters of some ferrites used now in the millimeter range are resulted in Table 7.1. Extremely achievable passbands $\Delta\nu_{\max}$ and working frequencies $\nu_{\max}$ are specified in brackets in columns.

Choice of a working material and magnetic film structure is subject to:

1. The necessary magnitude of internal magnetic field – $H_{0i}(H_0,\ M_S,\ H_A,\ H_s,\ldots)$ related to the working range of frequencies $\omega \geq \gamma H_{0i}$
2. The required size of saturation magnetization M_S determining the band properties and average value of group speed in a strip
3. The field magnitude of the axial $\overline{H}_A$ and basic $\overline{H}_c$ anisotropy, having the most essential importance for high-anisotropy materials $(H_A,\ H_s \gg 4\pi M_S)$
4. The admissible value of dissipation parameter $\Delta H_{||,\perp}$ or $\alpha_{||,\perp}$, determining both the band properties near the resonant frequencies and restriction on achievable losses for excitation and propagation of a signal in structures
5. The thermal stability of the internal field $\frac{\Delta H_{0i}(T)}{H_{0i}} = const$ and the frequency $\frac{\Delta\nu(T)}{\nu} = const$ related to magnetic parameters $M_S(T),\ H_A(T),\ H_s(T)$ as well as to the field of a magnetic system $H_0(T)$
6. Requirements to the mechanical strength and degradation influences (ionizing radiation, temperature and pressure differences, humidity, etc.)

From the data in Table 7.1 it follows that the range of working frequencies for film structures of garnet and spinels reaches nearly identical values of 70–75 GHz, and for selective devices Li–Zn-spinels have a passband approx. by four to seven times wider but compare unfavourably with YIG in terms of ΔH. Now these materials allow a range of frequencies up to 60–70 GHz to be utilized. By mechanical durability YIG films essentially surpass spinel materials. Magnetoelectronic devices based on YIG-films will effectively work in reflection modes and on signal passage in both narrow-band modes (by selective effects near the resonant frequencies of FMDLS) and broadband ones, including frequency ranges of the existence of magnetostatic spin waves.

For advance into the middle (frequencies up to 100–120 GHz) and short-wave (up to 240 GHz at $H_A \geq 5.2\,\mathrm{MA/m}$ and $H_0 \approx 1.6\,\mathrm{MA/m}$) parts of the millimeter range strongly anisotropic ferrite films are promising. However, rather a small thickness of the existing BHF films (down to $15 \cdot 10^{-6}$ m) and a wide line of FMR require searching for special modes of effective signal propagation. Antiferromagnetic films are still under laboratory research.

Table 7.1 The range of working frequencies for film structures of garnet and spinels reaches nearly identical values of 70-75 GHz, and for selective devices Li-Zn-spinels have a passband approx.

Structure type	$4\pi M_S \times 10, \mathrm{T}$; $(\Delta\nu_{\max}, \mathrm{GHz})$	H_A, H_D, H_E, $H^2\Delta$, A/m ($\Delta\nu_{\max}$, GHz at $H_0 = 1.9\,\mathrm{MA/m}$)	$2\Delta H$, A/m	The note
Garnets	0.05–2.0 (0.07–2.8)	$H_A \leq 0.8–8(65–70)$	≤80–140	$Y_3Fe_5O_{12}$ Films parameters to volumetric crystals are approach. High mechanical durability, endurance to ionization. Structures diameter is 45 and 76 mm. Films thickness up 100 μm
Spinels	3 (2.8–7)	$H_A \leq 8–24(70–75)$	≤400	Li–Zn, $MgAl_2O_3$ Films quality is worse in comparison with volumetric crystals. Low mechanical durability. Model size is $10 \times 20\,\mathrm{mm}^2$. Films thickness up 15–17 μm
Hexaferrite	2–5 (2.8–7)	$H_A \leq 2 \cdot 10^6(85–126)$	$\leq(8–30) \times 10^2$	Hexaferrite of barium. Model size is 25 μm. Films thickness up 13–17 μm
Antiferromagnetics	0.1 0.14	$H_A \cong 0.8–8$	–	$\alpha \cdot Fe_2O_3$ – hematite $H_E = 730\,\mathrm{MA/m}$, $H_D = 1.8\,\mathrm{MA/m}$, $H^{\Delta}{}_2 = 1.1\,\mathrm{MA/m}^2$ Monocrystals
		$H_E \cong 8(10^3–10^5)$		$FeBO_3$–iron carbonate $H_E = 2.5\,\mathrm{MA/m}$, $H_D = 72\,\mathrm{MA/m}$, $H^{\Delta}{}_2 = 80\,\mathrm{kA/m}^2$ Films models $0.5 \times 0.5\,\mathrm{mm}^2$
		$H_D \cong 8(10^2–10^3)$		$NiCO_3$–nickel carbonate $H_E = 19\,\mathrm{MA/m}$, $H_D = 72\,\mathrm{MA/m}$ Monocrystals

7.1.2 Magnetic Systems for Magnetoelectronic Millimeter-Range Devices: Heatset Field Reorganization

In the design MS for MED of the millimeter range it was required:

- To provide fields not lower than 1 MA/m at small weight and dimensions ($m \leq$ 0.1–0.3 kg, $V < 10^{-3}\,\mathrm{m}^3$)
- To control the temperature of the resultant field of a working frequency in MED not worse than $(10^{-5}–10^{-6})\frac{\%}{^\circ\mathrm{C}}$ in a temperatures range wider $(-30,\ +60)^\circ$C, and $(-60,\ +30)^\circ$C
- To realize mechanical and electric (including high-speed one with a speed of reorganization $160\frac{\mathrm{kA/m}}{\mathrm{mc}}$) field and frequency reorganization

Figure 7.1 presents photos of the designs of MS made of materials KS-17 and KS-25 (the smallest MS provided a field $H_0 \cong 1\,\mathrm{MA/m}$ in a gap $0.9 \cdot 10^{-3}\,\mathrm{m}$ on the area of active zone $(2 \cdot 5) \cdot 10^{-6}\,\mathrm{m}^2$ at a weight $40 \cdot 10^{-3}\,\mathrm{kg}$ and dimensions $(12 \cdot 20 \cdot 8) \cdot 10^{-9}\,\mathrm{m}^3$), including:

1 – Polepiece
2 – Plates made of samarium-cobalt alloys
3 – Magnetic circuit
4 – Adjustable magnetic shunt for mechanical field reorganization over a wide range
5 – Coils for electric high-speed field reorganization
6 – Temperature-compensated magnetic shunt

Such a design (Fig. 7.2) provides:

- Effective concentration of dispersion fields
- A combination of a high speed and required range of field reorganization

Fig. 7.1 Photos of the designs of MS made of materials KS-17 and KS-25

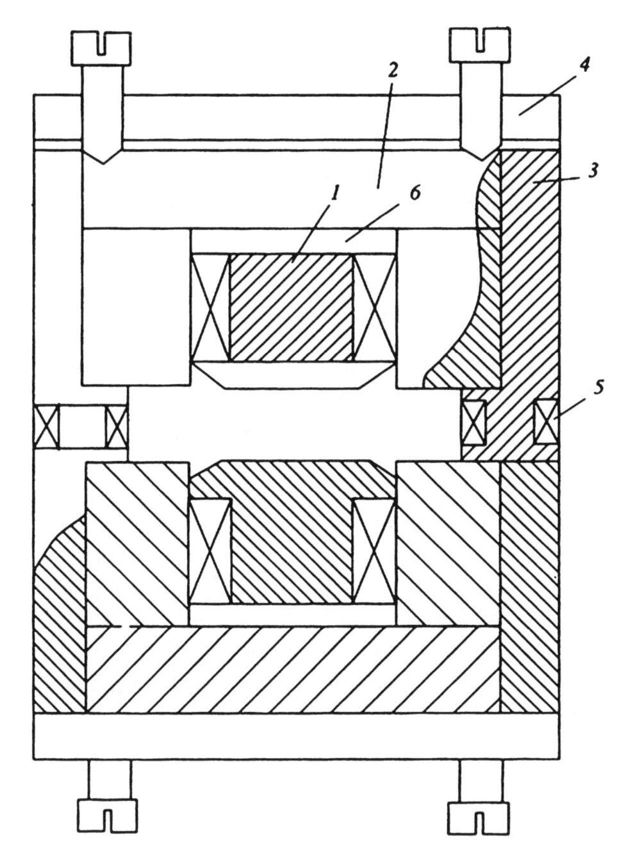

Fig. 7.2 Construction of miniature magnetic system

- Temperature control within the limits of the preset weight and dimensions of the system
- Assembly of the system on a magnetoelectronic device
- A high mechanical durability

For technical applications of MED the (TFC) or a related temperature field coefficient (TFDC) are important parameters, $TFC = \frac{\Delta v/v_0}{\Delta T}$ and $TFDC = \frac{\Delta H_{0i}/H_{0i}}{\Delta T}$. The internal magnetic field H_{0i} contains a number of factors depending on temperature:

1. Own magnetic parameters of the ferrite layer

$$F_1(T) = f_i(\overleftrightarrow{N} M_s, \overleftrightarrow{N}_A H_A, \overleftrightarrow{N}_s H_s, \ldots)$$

2. An external magnetic field

$$F_2(T) = f_l(H_0(T)).$$

Generally, in the field of temperatures $210 \leq T(K) \leq 370$ for the basic magneto-electronic materials (YIG and spinels) the specified functional dependences do not coincide and, on the average, have the same sign of their slope $\frac{dF_1}{dT} < 0$ and $\frac{dF_2}{dT} < 0$. For monoaxial and cubic structures such orientations in the external field can be chosen, that the slope signs F_1 and F_2, will be opposite.

Let's consider ways to control the temperature of fields in the EHF range:

- Due to proper orientation of the structure in the external field
- Introduction of mechanical and electric tuning elements
- Application of thermal noises with preset magnetic properties

In flattened ellipsoids and ferrite films at switching from tangential to normal magnetization there is such an angle of structure magnetization $0 < \varphi_T < \pi/2$ at which the most essential factor (e.g., related to saturation magnetization for weakly anisotropy ferrites) can be compensated. Selection of such an angle φ_T is possible, for which $\frac{dF_1}{dT} = -\frac{dF_2}{dT}$ in some interval of temperatures ΔT.

In Fig. 7.3 the dependences of temperature frequency drifts in a YIG film ($d = 25 \cdot 10^{-6}$ m, $4\pi M_S = 0.1760$ T, $H_A = 3.34$ kA/m) on the angle of inclination φ in an external magnetic field H_0 are presented: $1 - \varphi = 0°$; $2 - \varphi = 60°$; $3 - \varphi = 36°$. At tangential magnetization $\varphi = 0(1)$ the value of $\frac{\Delta\nu}{\Delta T} < 0$ and the steepness of the dependence correspond to the known ones [304]. In the field of negative temperatures the dependence $\nu(T)$ has a nonlinear character, that is an undesirable

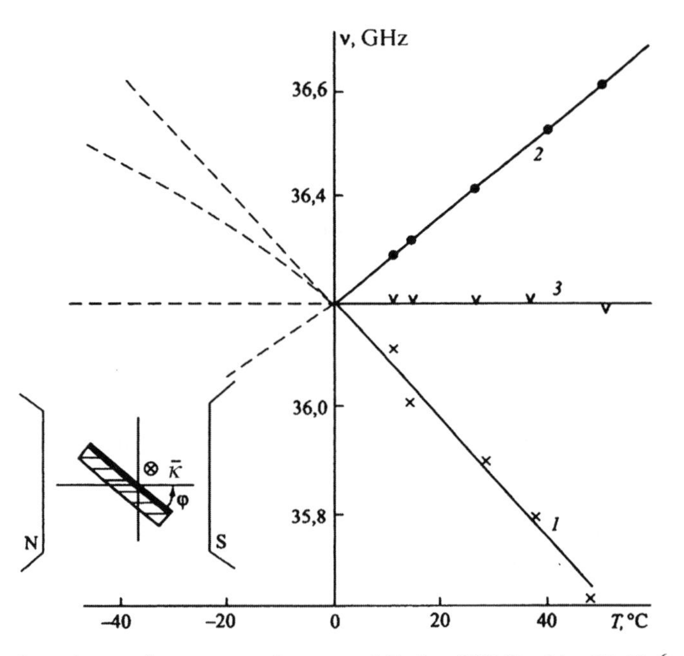

Fig. 7.3 The dependences of temperature frequency drifts in a YIG film ($d = 25 \cdot 10^{-6}$ m, $4\pi M_S = 0.1760$ T, $= 3.34$ kA/m) on the angle of inclination φ in an external magnetic field H_0 are presented: $1 - \varphi = 0°$; $2 - \varphi = 60°$; $3 - \varphi = 36°$

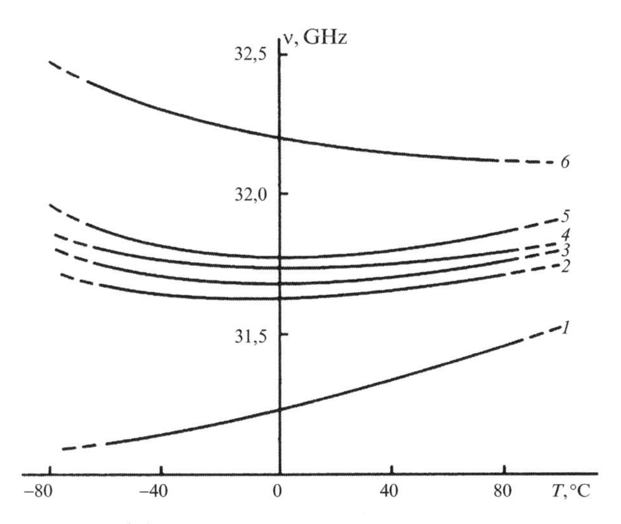

Fig. 7.4 The dependences $\nu(T)$ are presented for angles close to that of temperature compensation $\varphi \cong \varphi_{T_\kappa}$, where: $1-\varphi=40°$; $2-\varphi=36°$; $3-\varphi=36.5°$; $4-\varphi=35°$; $5-\varphi=34°$; $6-\varphi=30.5°$

factor and is related to the dependence $M_S(T)$. Such orientations of the structure are possible, at which $\nu = const$ ($\varphi \cong 36°$, curve 3), and $\frac{\Delta\nu}{\Delta T} > 0$ ($\varphi \cong 60°$, curve 2).

In the latter case the temperature field drift of the magnetic system $H_0\ (T)$ can be compensated. The vector of a monoaxial anisotropic field $\overline{H}_A \perp \overline{H}_0$ lies in one plane with $\bar{\kappa}$. The layer of ferrite was thermally insulated from the poles of the permanent magnet. Changing the temperature of the ferrite film was carried out by means of a heating element.

In Fig. 7.4 the dependences $\nu(T)$ are presented for angles close to that of temperature compensation $\varphi \cong \varphi_{T_\kappa}$, where: $1-\varphi=40°$; $2-\varphi=36°$; $3-\varphi=36.5°$; $4-\varphi=35°$; $5-\varphi=34°$; $6-\varphi=30.5°$.

There is an angle $\varphi_T = 36.5°$ for which the frequency drift $\pm\Delta\nu$ from ν_0 is minimal. In an interval of temperatures $T = (-70 \text{ to } 100)°\text{C}$ the frequency drift $\Delta\nu < 100\,\text{MHz}$, $\nu_0 = 31.78\,\text{GHz}$, and $TFC < 1.8 \cdot 10^{-5}\ 1/°\text{C}$.

Another way of thermal control is connected with adjustment of the value of external magnetic field H_0 by means of changing the current in the tuning coil of MS by condition H_{0i} = const and ν_0 = const in the set interval of temperatures. In Fig. 7.5 the dependences of the own temperature frequency drifts $\nu(T)$ (Fig. 7.5*a*) in a tangentially magnetized FDLS in a magnetic system made on materials KS-17, and the values of current in the tuning coil (Fig. 7.5*b*) at $\nu_0 = 21.852\,\text{GHz}$ with an accuracy of frequency registration with a panoramic SWR_e and attenuation measuring instrument of type R2-65 not worse than $10^{-2}\%$ are presented. Measurements were made in a chamber of controllable temperature MS-71 (JAPAN TABAI MFG.CO. LTD). The average steepness of the dependence was $\Delta\nu/\Delta T \cong -7\,\text{MHz/K}$. In an interval of temperatures $\Delta T = \pm 50\,°\text{C}$ the frequency drift was no more $\pm 1\,\text{MHz}$, that corresponded to $TFC < 9 \cdot 10^{-7}\ 1/\text{K}$.

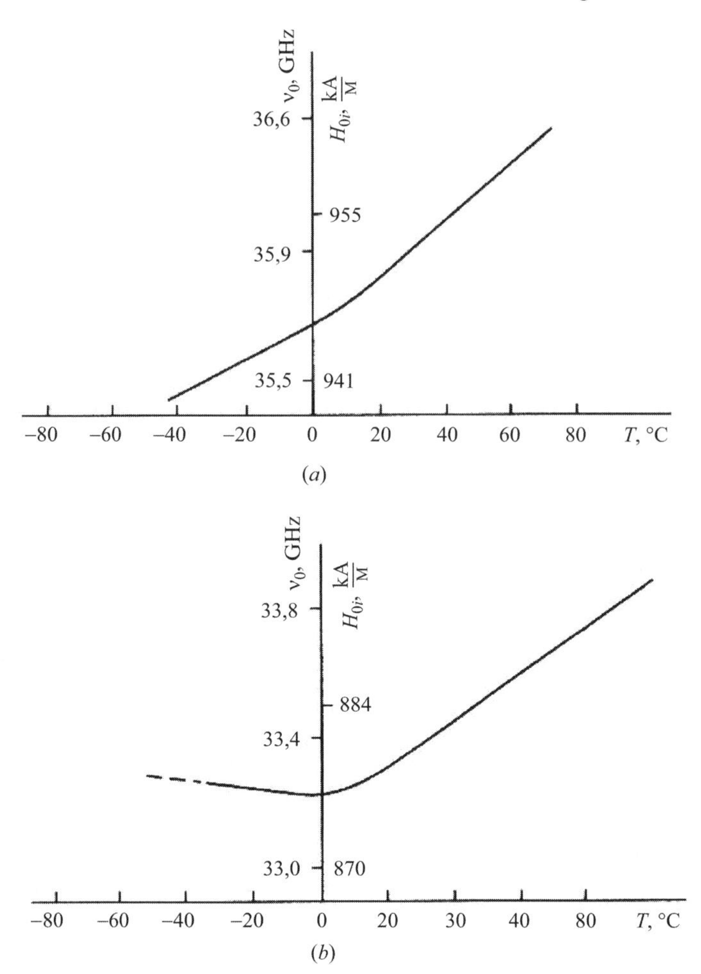

Fig. 7.5 The dependences of the own temperature frequency drifts $\nu(T)$ (*a*) in a tangentially magnetized FDLS in a magnetic system made on materials KS-17, and the values of current in the tuning coil (*b*) at $\nu_0 = 21.852\,\text{GHz}$

A third way is connected with application of thermal shunts to provide a change of the law $F_2(T)$ to be opposite to $F_2(T)$ by sign. It is provided due to materials having a Curie temperature $T_K = (110–170)^\circ\text{C}$ and a certain kind of nonlinear magnetization.[6]

The steepness of the dependence $H_0/\Delta T$ and its kind of nonlinearity are determined by the volume of the material of thermo-shunts and their arrangement over a magnetic system.

In Fig. 7.6 the dependences of changes of the resonant frequency $\nu_0(T)$, measured by means of our designed sensors on the basis of YIG films (sensors were

[6] Thermal shunts were developed by Central Research Institute of steels and alligations.

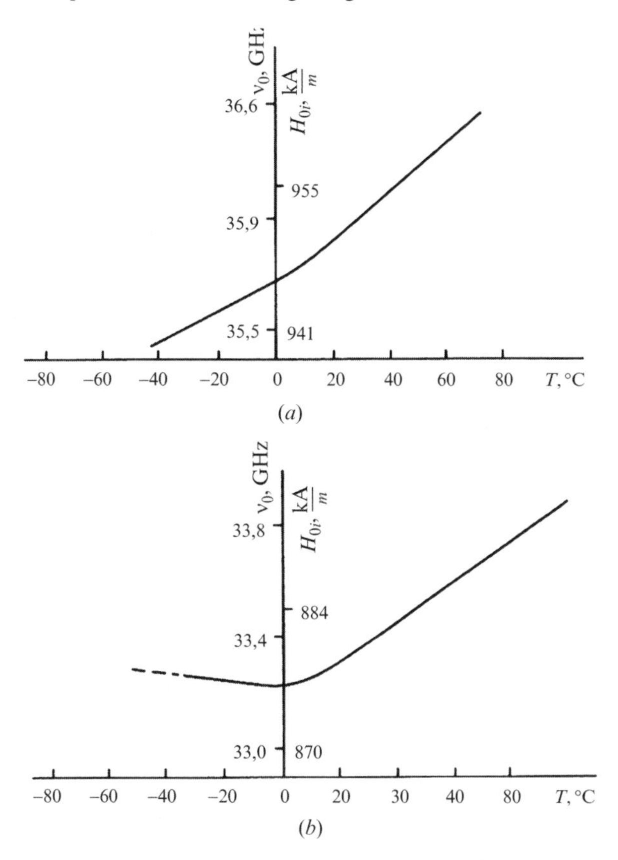

Fig. 7.6 The dependences of changes of the resonant frequency $\nu_0(T)$, measured by means of our designed sensors on the basis of YIG films

heat-insulated from MS), in MS with a varying number of shunts (the volume of a material) (Fig. 7.6*a* – three shunts, Fig. 7.6*b* – two shunts) are presented. With three shunts (the curve in Fig. 7.6*a*) the average steepness of changes of external field is $\Delta H_0/\Delta T \cong +195\frac{\text{A/m}}{^\circ\text{C}}$. With two shunts (the curve on Fig. 7.6*b*) in the field of $T \cong 0^\circ$ a sign reversal of the dependence $\nu(T)$ is observed, and

$$\left(\frac{\Delta H}{\Delta T}\right)_{T<0} \cong -147\frac{\text{A/m}}{^\circ\text{C}}.$$

The point of inflection close to $T \cong 0^\circ$ is related to the magnetic properties of thermal shunts. Thus, due to proper choice of a material, the volume of thermal shunts, and their arrangements in MS it is possible to change the course of the dependence $H_0(\Delta T)$ and to realize a temperature-compensated mode in a preset interval of temperatures.

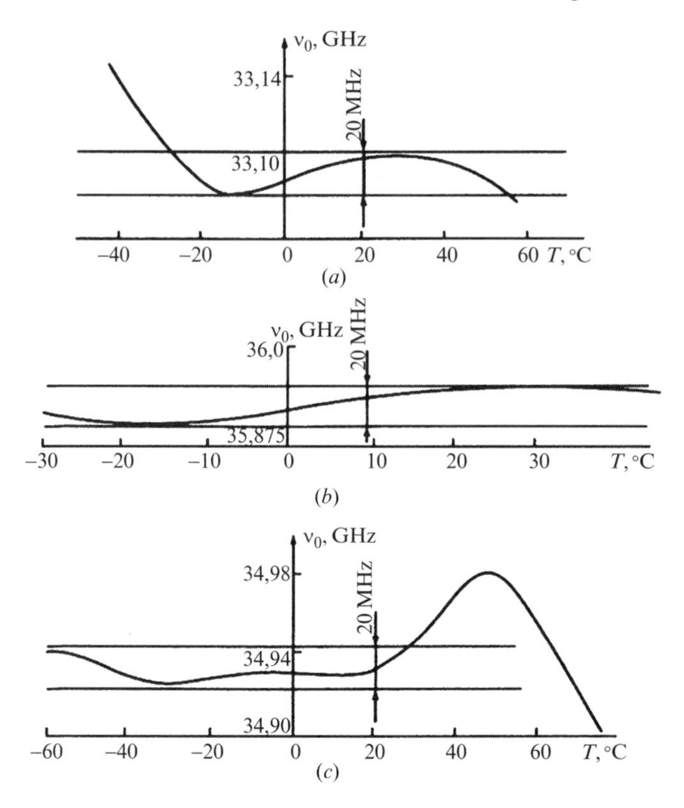

Fig. 7.7 Dependences illustrating selection of a suitable material and the volume thermal shunts for thermal control of working frequency ν_0 in a preset interval of temperatures are presented: ((*a*) 2 31 N4H shunts, (*b*) 4 31 N6H shunts, (*c*) 2 31 N6H shunts)

In Fig. 7.7 some dependences illustrating selection of a suitable material and the volume thermal shunts for thermal control of working frequency ν_0 in a preset interval of temperatures are presented: (Fig. 7.7: *a* –2 31 N4H shunts, *b* – 4 31 N6H shunts, *c* – 2 31 N6H shunts). In an interval of temperatures $\Delta T = (-30 \div +50)$°C a drift $\Delta\nu_0 < \pm 25\,\text{MHz}$ and a $TFC < 4.7 \cdot 10^{-6}\ 1/°\text{C}$ were reaches, while for an interval $\Delta T = (-60 \div +30)$°C these values were $\Delta\nu_0 < \pm 10\,\text{MHz}$ and $TFC < 1.6 \cdot 10^{-6}\ 1/°\text{C}$.

A basic advantage of MED is electric reorganization of frequency and other parameters due by adjustment of an external field H_0.

In Fig. 7.8 the dependences of frequency deviation $\Delta\nu$ on current I in the operating coils in MED in a range of frequencies of 20–25 Hz are presented.

In Fig. 7.9 the dependences determining the speed of reorganization of the central frequency of MED on the frequency of modulating current $\Delta\nu_{\text{mod}}$ are given, $I_{\text{mod}} = const$. The speed of reorganization in a range of frequencies of 35–40 GHz in FDLS with tangential magnetization on frequency is $\beta_{\nu_0} \cong (3.5\text{–}5.0)$ GHz/ms, that by field is $\beta_{H_{0i}} \cong (100\text{–}140)$ kA/ms.

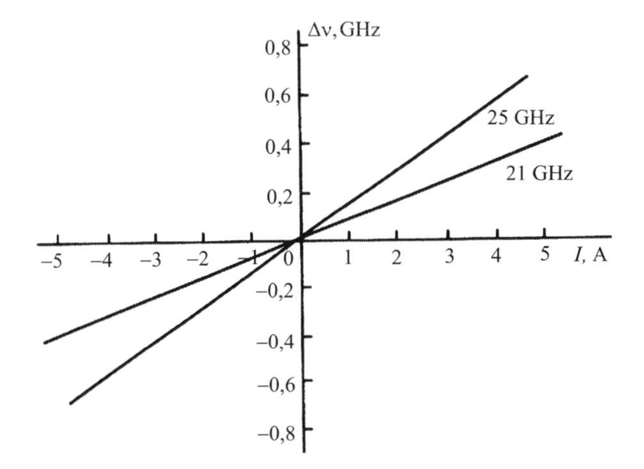

Fig. 7.8 The dependences of frequency deviation $\Delta\nu$ on current I in the operating coils in MED in a range of frequencies of 20–25 GHz are presented

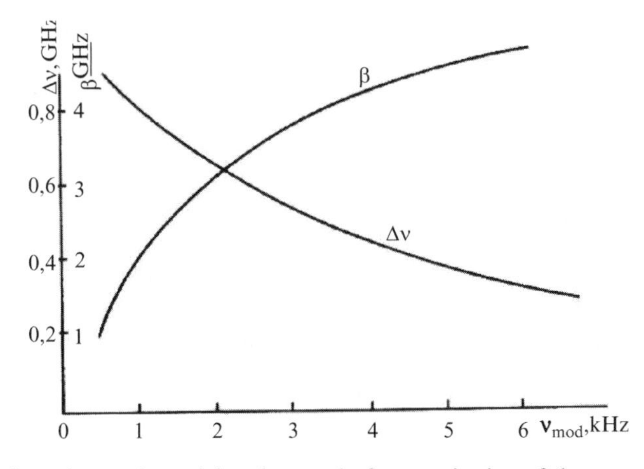

Fig. 7.9 The dependences determining the speed of reorganization of the central frequency of MED on the frequency of modulating current $\Delta\nu_{\mathrm{mod}}$ are given

Let's note that the extremely achievable values of external magnetic fields H_0 created by permanent magnets made of alloys with a high coercitive force and specific magnetic energy (platinum-cobalt alloys, intermetallic compounds [56]) are limited by the value of saturation induction B_S of the materials used in polepieces and magnetic circuit. So, for example, for iron-cobalt alloys $B_S \leq 2.4\,\mathrm{T}$ (50% Fe and 50% Co), that is a limiting factor at the design of tiny magnetic systems in the millimeter range.

7.1.3 Film Diagnostics in MMR

At utilization of new frequency ranges of magnetoelectronic devices, and at physical research of wave processes a basic place belongs to new means and methods of diagnostics and control. For adequate processing of experimental data (AFC parameters of signals passed through an investigated structure or locally excited in a required area of analysis and those reflected from the structure) an understanding of the physics of processes of excitation, absorption of a signal and its distribution in various types of converters and transfer lines is necessary. This requires development of the most strict (at present) theoretical model and application of the electrodynamic method of analysis. However, development of correct physical models of processes requires passage through various approximations in theory and experiment.

Figure 7.10 presents a table illustrating most essential (for MMR) control parameters of ferrite films and new developed methods and devices of control, providing nondestructive and interoperational diagnostics in the UHF and EHF ranges in a band of frequencies wider than three to five octaves. The most essential conclusion from our research was formulation of criteria determining the quality of ferrite films in MMR. These are:

- A low level of ferromagnetic losses ($\alpha < 10^{-4}$) or line width of ferromagnetic resonance ($\Delta H \leq 10^{-4} H_{0i}$)
- A low level of the cross-section gradient of losses in a film ($\nabla\alpha < 10^{-1}, \nabla(\Delta H) < 10^{-1}$)
- A low cross-section gradient of an internal magnetic field ($H_{0i} < 10^{-1}$)
- A low cross-section gradient of magnetization ($\nabla M_S < 10^1$) and fields of anisotropy ($\nabla H_A < 10^1$)

These circumstances are related to the presence of a natural though thin transition layer in a film on the border with the dielectric base and caused by penetration of ions of the substrate into the ferrite layer, by mechanical pressure. The developed quality monitoring techniques allow the distribution of the specified parameters by thickness and areas of film structures to be investigated. Our researches has shown that the line width of FMR. As frequency grows, linearly increases with a steepness

$$\frac{\Delta(\Delta H)}{\Delta\nu} \cong 10^{-2} \quad \frac{\text{Oe}}{\text{GHz}} \left(0.8\frac{\text{A/m}}{\text{GHz}}\right),$$

that is related to the increase in the angle between the vectors of magnetic induction and field in the ferrite, and does not depend on power levels in the expanded dynamic range (up to several watt in the near MMR). The later feature is caused by an increased magnetic rigidity of the ferrite film in MMR, for which characteristic are:

- For HF fields $\tilde{h} \ll H_0$
- For magnetizations $\tilde{m} \ll 4\pi M_S$

Development of various methods and devices for wide-range control provided not only "sewing" of our results together with the data by other techniques in the UHF-range but also getting independent results of measurements in the EHF range.

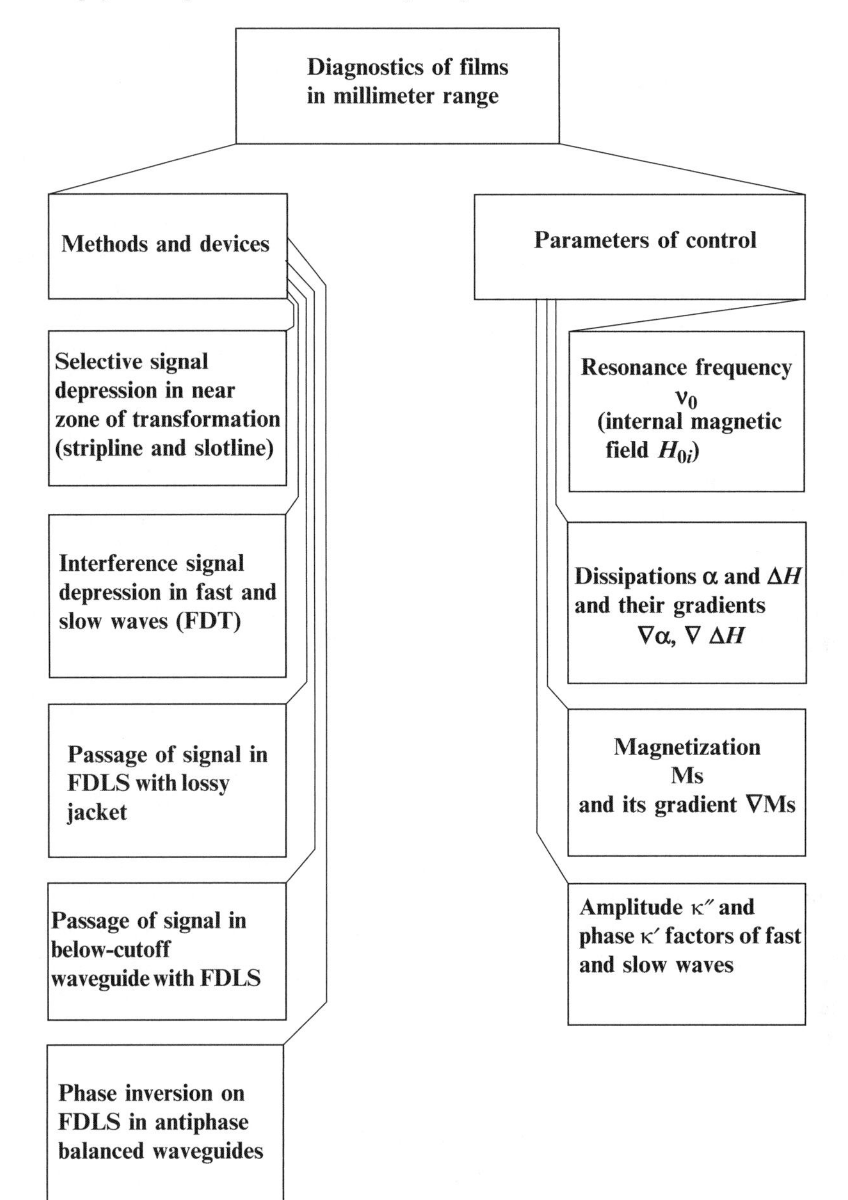

Fig. 7.10 A table illustrating most essential (for MMR) control parameters of ferrite films and new developed methods and devices of control, providing nondestructive and interoperational diagnostics in the UHF and EHF ranges in a band of frequencies wider than three to five octaves

7.1.4 Electrodynamics of Excitation and Propagation of Electromagnetic Waves in Ferrite Films and Multilayered Bigyrotropic Structures in the Millimeter Range

In the recent years of the development of MMR magnetoelectronics an essential restriction of the MSW approximation widely applied in the UHF range for calculations of the parameters of converters and transmission lines in the millimeter range of radiowaves was obvious. First, this is problems of correct description of wave processes near the resonant frequencies of FDLS connected with the necessity to develop highly effective selective devices of low and high power levels (LPL and HPL). The necessity to explain the experimentally observed essential increase in the yield of the wave spectrum excited and propagating in high-quality $(\alpha < 10^{-4})$ ferrite films, at promotion into the short-wave centimeter range and into the millimeter one. The design of resonant frequency sensors for layered ferrite-dielectric structures with arbitrary magnetization by an external field H_0 is necessary. Besides, our research has revealed several specific factors, namely, an increase of the magnetic rigidity and the cross-section gradient of losses in FDLS, which are naturally accompanied (or vice versa) by a cross-section gradient of the internal magnetic field and the related parameters. Another essential factor which constrained promotion of research is connected with a high (1–2%), in comparison with the error of frequency measurements, error of measurement of increased magnetic fields by means of Hall sensors. Besides, successes in the development of new materials (film magnetic semiconductors, dielectrics, opportunities of using semiconductors) determined the urgency of studying wave processes in heterostructures magnetized by an external field.

These circumstances have determined the necessity of development of an electrodynamic theory of wave excitation by extraneous currents–converters with parallel and orthogonal orientations of multilayered bigyrotropic structures screened with impedance surfaces, in view of specific MMR factors (losses in layers, their cross-section gradients, and cross-section gradients of electric and magnetic parameters) (Figs. 7.11, 7.12).

7.1.5 Waves in Layered Structures on the Basis of Magnetoarranged Films in MMR

Our research made for *LE* and *LM* waves representing the full basis of characteristic waves in multilayered bigyrotropic structures have shown that, along with forward and return surface and volume slow waves whose boundary frequencies tend to a constant at $\kappa' \to \infty$ or, for some cases, pass into the boundary frequencies of magnetostatic spin waves, there are fast waves which frequencies increase and decrease beyond all bounds at $\kappa' \to \infty$ and for which the phase constants $\kappa' < \kappa'_0$, where κ'_0 is the constant in an empty waveguide. In the field of resonant frequencies ν_φ, ν_{H}, and $\nu_\perp$, depending on the distance a to the metal screens, the prelimit $(\lambda \ll 2a)$

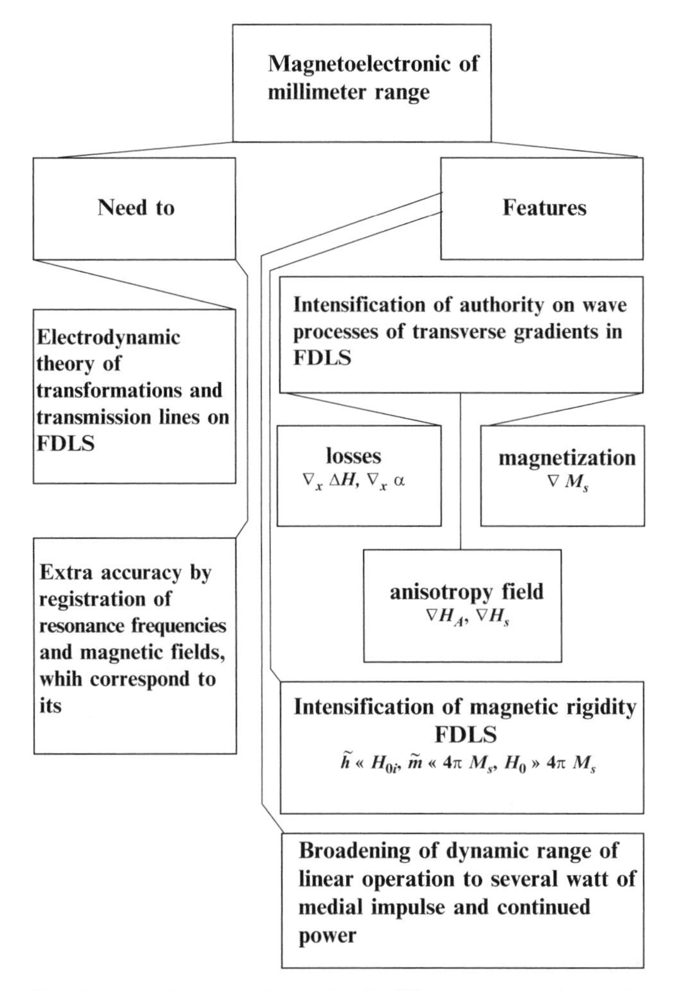

Fig. 7.11 Specific character of magnetoelectronic of millimeter ranges of wave-length

or beyond-cutoff ($\lambda \gg 2a$) modes of a screened ferrite-dielectric or bigyrotropic layered structure are realized, at which either selective attenuation of absorption or selective reduction of attenuation of passage of those or other waves occurs. Besides. In the prelimit mode between fast and slow waves at low ferromagnetic losses ($\alpha < 10^{-4}$) interference interaction exists, namely, wave hybridization, which leads to periodic attenuation of waves in the structure on its certain length. Hybridization of waves was also observed for modes of the dielectric type in FDLS. The basic conclusion of our research shows that wave dispersions essentially fall outside the bounds determined by the MSW approximation. The limits of applicability of the MSW approximation at promotion from the centimeter into millimeter range of radiowaves were established. In Fig. 7.13 a summary table of the basic waves in FDLS in MMR is given.

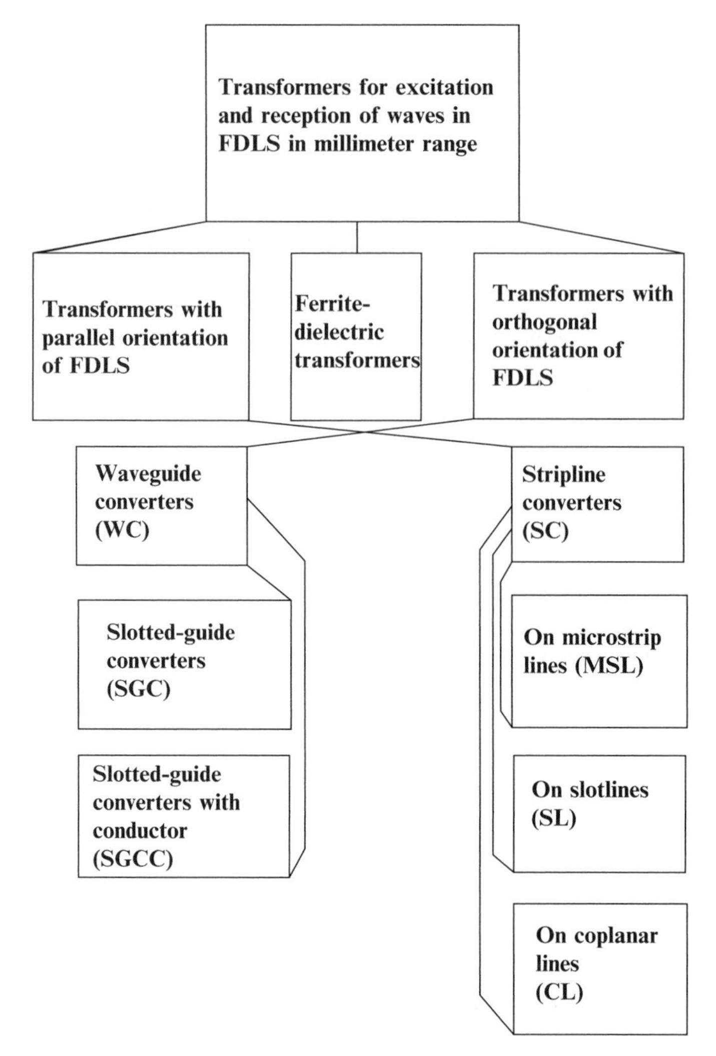

Fig. 7.12 Demand of transformers for excitation and reception of waves in FDLS in millimeter range

7.1.6 Magnetoelectronic MMR Devices

Magnetoelectronic MMR devices represent a new class of solid-state operated wave devices for information processing in real time. These devices realize:

- Reorganization of output signal parameters, including high-speed electric one, by frequency, amplitude and signal phase, control of delay time and passband
- An expanded dynamic range of the linear mode up to levels of some watt of continuous and average pulse power

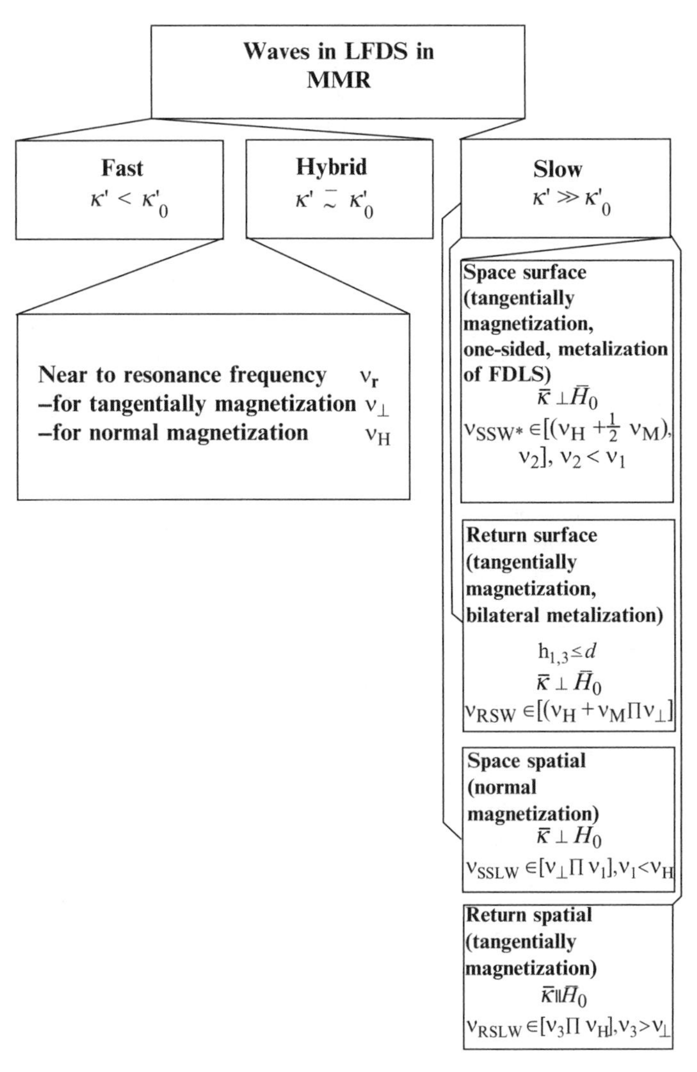

Fig. 7.13 A summary table of the basic waves in FDLS in MMR

- Formation of required AFC and wave dispersion laws in converters and transfer lines
- A low level of introduced losses (no higher than 1 dB)
- A high density of accommodation of working channels

a simple interface to the existing element base.

In Fig. 7.14 most important operating modes are listed, in the table of Fig. 7.15 are our developed magnetoelectronic devices of the millimeter range.

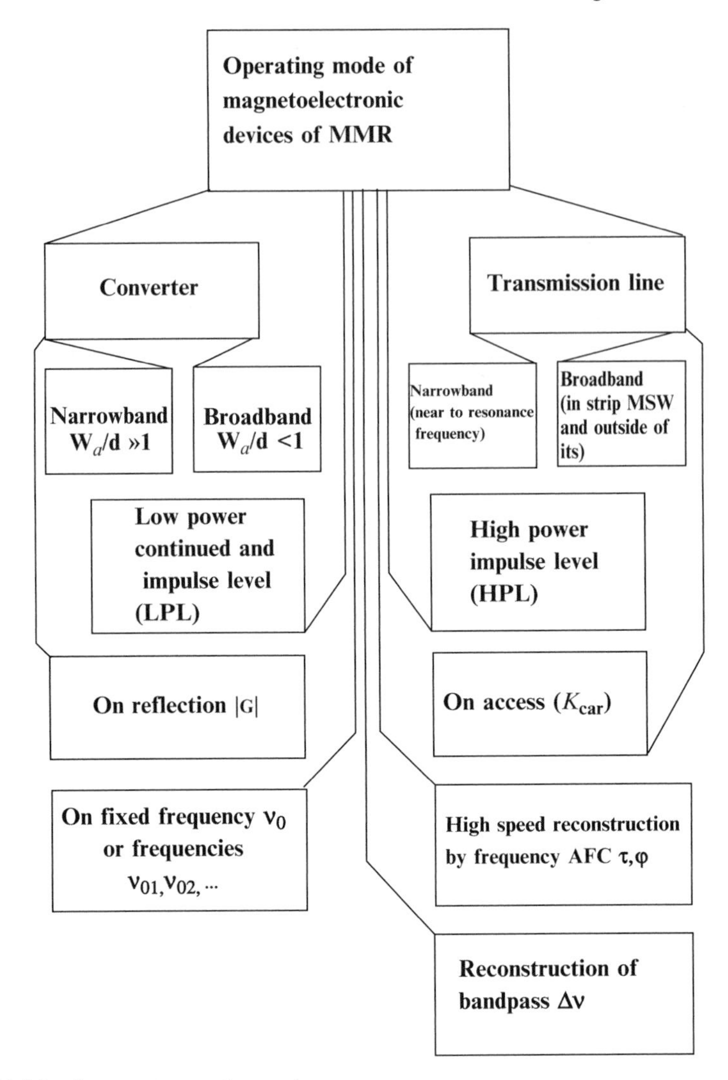

Fig. 7.14 Most important operating modes

7.2 Magnetoelectronic Devices of Low and Average Power Levels

Key parameters of our devices of signal filtration and delay, an onboard multichannel receiver in the millimeter range are presented.

For MED of low and average power levels (up to several watt in the continuous mode and several kW in the pulse one, at $Q \approx 10^3$) admissible is application of:

- Small cross-section sections of metal and dielectric waveguides, strip lines
- Absorbing coverings
- Short-circuited designs

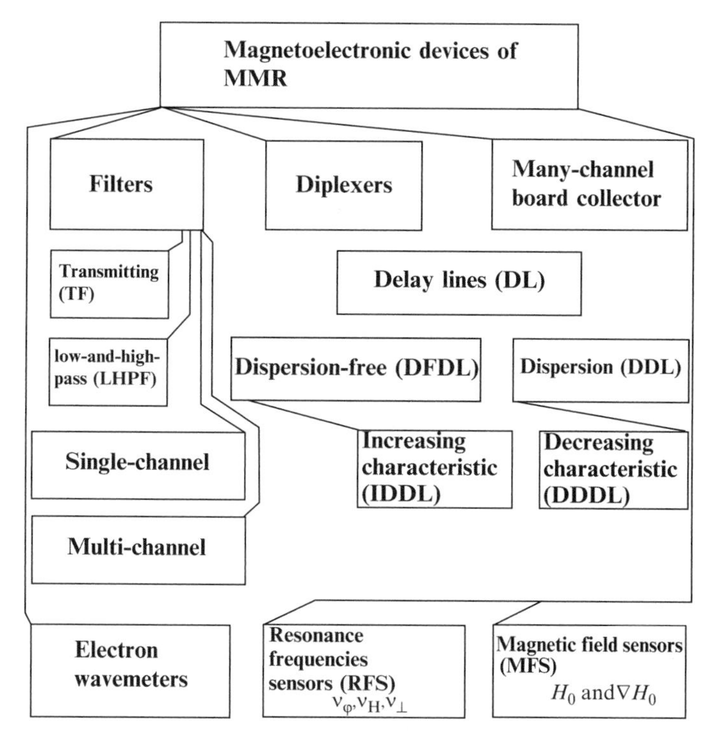

Fig. 7.15 Developed magnetoelectronic devices of the millimeter range

The following operated devices have been developed:

- On slow waves, including magnetostatic ($\kappa \leq 10^5$–10^6 rad/m)
- On slow and fast waves ($\kappa \leq 10^5$ rad/m)
- On fast waves ($\kappa \leq 10^3$ rad/m)

The most broadband modes are used at the design of delay lines of signal and multichannel filters. Narrow-band modes are used in the development of various single-channel filters.

7.2.1 *Magnetoelectronic MMR Delay Lines*

The major objective at the design of a DL consists in an increase of the delay time of a signal, expansion of the working band of frequencies, and a decrease of introduced losses. Commercial epitaxial YIG films are most suitable. Spinel and BHF structures are still under research development. Their basic demerit for today is rather a wide FMR line and small sizes of structures

$$\Delta H_{sp} \approx 3-5 \text{ Oe}(238-398 \text{ A/m}),$$
$$\Delta H \approx 7-15 \text{ Oe}(0.56-1.20 \text{ kA/m}).$$

To achieve the maximum time of a delay it is necessary to provide excitation and propagation of waves in FDLS in a wide range of frequencies and wave numbers $\kappa' \leq (10^5\text{–}10^6)\,\text{rad/m}$.

From this point of view, most comprehensible are MSL, SL, CL with a width of the strip line $W \approx \frac{2\pi}{\kappa'_{\max}}$. At $W \approx (5\text{–}10)\cdot 10^{-6}\,\text{m}$ we have for DL $\kappa'_{\max} \approx (1.2 \div 0.6)10^6\,\text{rad/m}$, that corresponds to the broadband mode of a converter.

SSSLW in normally magnetized structures have the minimum level of introduced losses. The minimum weight of MS and dimensions of devices are reached automatically.

From calculation and optimization of the parameters of a MSA and the dispersive properties of SSSLW under a preset list of DL parameters, basic sizes of FDLS follow. As edge absorbers a non-uniform magnetic field on the external sites of MSA is used. The arrangement topology of both input and output aerials and applied devices of signal discrimination between input and output provide a required block.

In Fig. 7.16*a* the design of DL on SSSLW of the millimeter range[7] is shown, it includes: 1 – an input rectangular waveguide; 2 – a dielectric waveguide; 3 – an input transition from a dielectric waveguide onto a microstrip line; 4 – an input

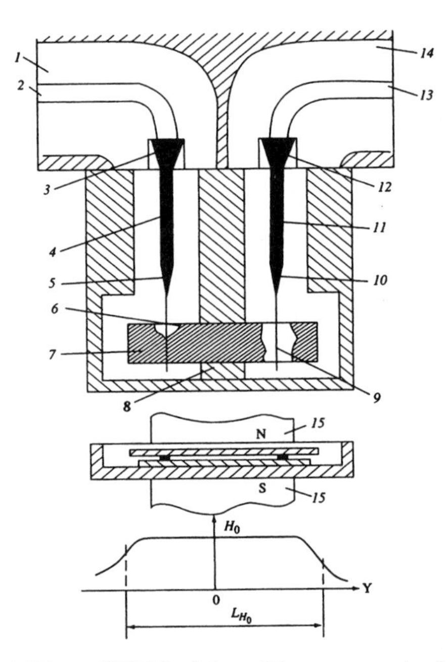

Fig. 7.16*a* The design of DL on SSSLW of the millimeter range is shown, it includes: 1 – an input rectangular waveguide; 2 – a dielectric waveguide; 3 – an input transition from a dielectric waveguide onto a microstrip line; 4 – an input MSL; 5 – a matching transition of MSL; 6 – an input MSA; 7 – FDLS; 8 – a discrimination device; 9 – an output MSA; 10 – a matching adapter of MSL; 11 – an output MSL; 12 – an output adapter from a microstrip line to a dielectric waveguide; 13 – a dielectric waveguide; 14 – an output rectangular waveguide; 15 – a magnetic system

[7] In cooperation with scientific research institute "Domain".

Fig. 7.16*b* DL in permanent magnet

MSL; 5 – a matching transition of MSL; 6 – an input MSA; 7 – FDLS 8 – a discrimination device; 9 – an output MSA; 10 – a matching adapter of MSL; 11 – an output MSL; 12 – an output adapter from a microstrip line to a dielectric waveguide; 13 – a dielectric waveguide; 14 – an output rectangular waveguide; 15 – a magnetic system. The magnetic system of DL (Fig. 7.16*a*,*b*) is designed so that to provide a decreasing field $H_0(y)$ compensating the internal regional magnetic field $H_{0i}(y)$ in FDLS that provides wave absorption on these parts of DL in the area of edges of FDLS and reduces the irregularity of AFC. Figure 7.17 shows the AFC of a signal in DL without an absorber – a and with an absorber made of ceramics CT-30 in the field of DL discrimination and at the edges of FDLS – *b*.

In Fig. 7.18*a* photo of the circuit providing detection of resonant frequency ν_H in the given DL by means of a designed RFS on the basis of a ferrite film, completely identical to that used in DL is shown. From Fig. 7.17*a* one can see that SSLMSW belong to a band of frequencies in the field of $\kappa'_{\max}$ comprising 50% from the total passband of DL.

In Fig. 7.19*a* flowchart of our installation made in a superheat version with reduction of an intermediate frequency with a sensitivity not worse than 50–80 dB/mW[8] is shown. The duration of a probing impulse is 0.1 μs. The measuring installation in Fig. 7.19 contains: 1 – a klystron generator G4–91; 2 – a generator of rectangular impulses G5–48; 3 – an oscillograph C1–65; 4 – a DL on FDLS; 5 – a mixing diode A123; 6 – an oscillator on the basis of klystron generators G4–91 and GC-115; 7 – a power supply 12 W; 8 – a broadband amplifier of 1–2 GHz M42136; 9 – an amplifier U-2; 10 – a detector.

Figure 7.20 presents oscillograms of both probing and delayed impulses in DL-1.[9]

In Fig. 7.21 the dependence of introduced losses K on delay time τ is given, from which it is possible to estimate a key parameter – the line width of FMR used in FDLS

[8] Installation developed in scientific research institute "Domain".

[9] Registered in scientific production association "Flight" by eng. A.V. Krupin.

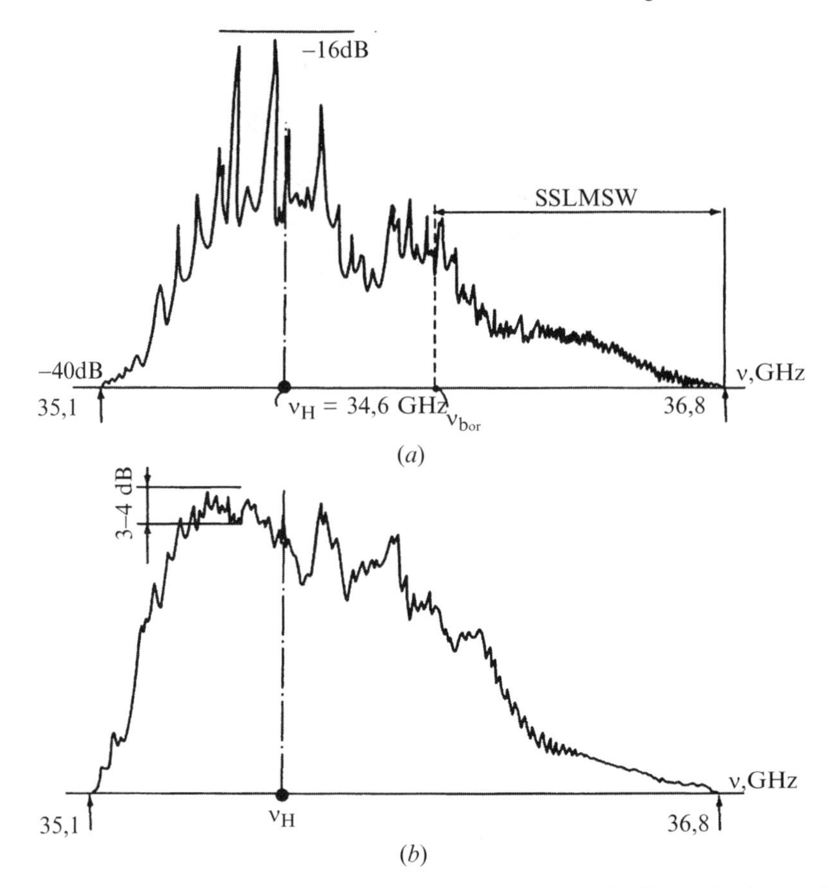

Fig. 7.17 The AFC of a signal in DL without an absorber (*a*) and with an absorber made of ceramics CT-30 in the field of DL discrimination and at the edges of FDLS (*b*)

$$\Delta H(\mathrm{Oe}) = \frac{\Delta K(\mathrm{dB}) - [2K_{los.con}(\mathrm{dB}) + 6\ \mathrm{dB}]}{76.4 \cdot \Delta\tau(\mathrm{ns})} 10^{-3}, \tag{7.1}$$

where $K_{los.con}$ is the losses on the converter.

Extrapolation of the dependence $k(\tau)$ to $\tau = 0$(dotted line in Fig. 7.21) determine the value of $2k_{los.con} + 6\,\mathrm{dB}$ from which it follows that the losses on transformation in the used MSL are $K_{los.con} \approx 9\,\mathrm{dB}$. From experimental data (Fig. 7.21) and Eq. (7.1) it follows that $\Delta H_{\perp} \approx 1.7\,\mathrm{Oe}\,(135.3\,\mathrm{A/m})$, that agrees well with the data by other methods developed in our work.

An advantage of magnetoelectronic DL consists in the opportunity to form the preset shape of dispersion $\tau(\nu)$. In Fig. 7.22 are presented decreasing – *a*, dispersion-free – *b* and increasing – *c* characteristics of DL on SSSLW in a range of frequencies 22–26 GHz. In Table 7.2 key parameters of the developed DL are collected.

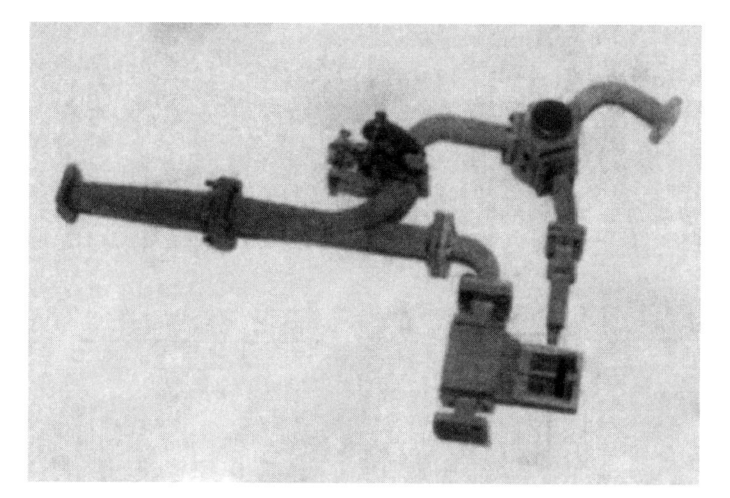

Fig. 7.18 A photo of the circuit providing detection of resonant frequency ν_H in DL by means of a designed sensor RFS on the basis of a ferrite film, completely identical to that used in DL

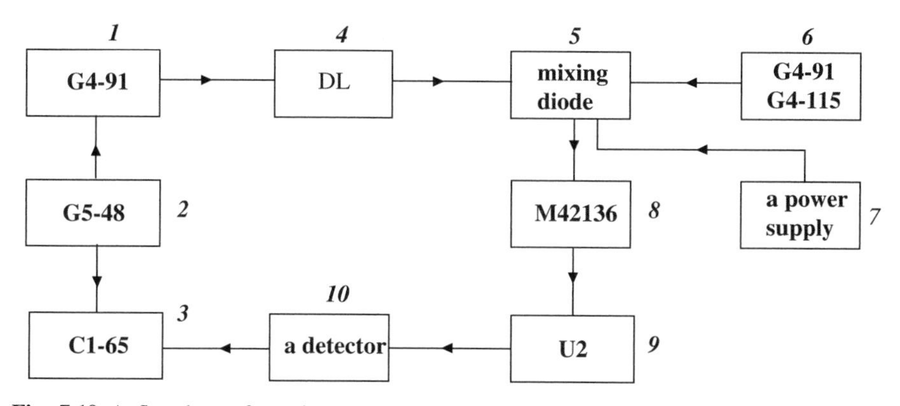

Fig. 7.19 A flowchart of our installation made in a superheat version with reduction of an intermediate frequency with a sensitivity not worse than 50–80 dB/mW

7.2.2 Magnetoelectronic MMR Filters

Magnetoelectronic filters were developed on the basis of selective excitation and propagation of electromagnetic waves near the resonant frequencies of layered structures based on ferrite films at using various types of strip and waveguide converters. Waveguide-slot converters and FDT in the mode of interference interactions between fast and slow waves in weakly dissipative FDLS are most simply realized in MMR and unidirectional. In such devices there are no additional losses on input and output of 6 dB, that is important for utilizing MMR.

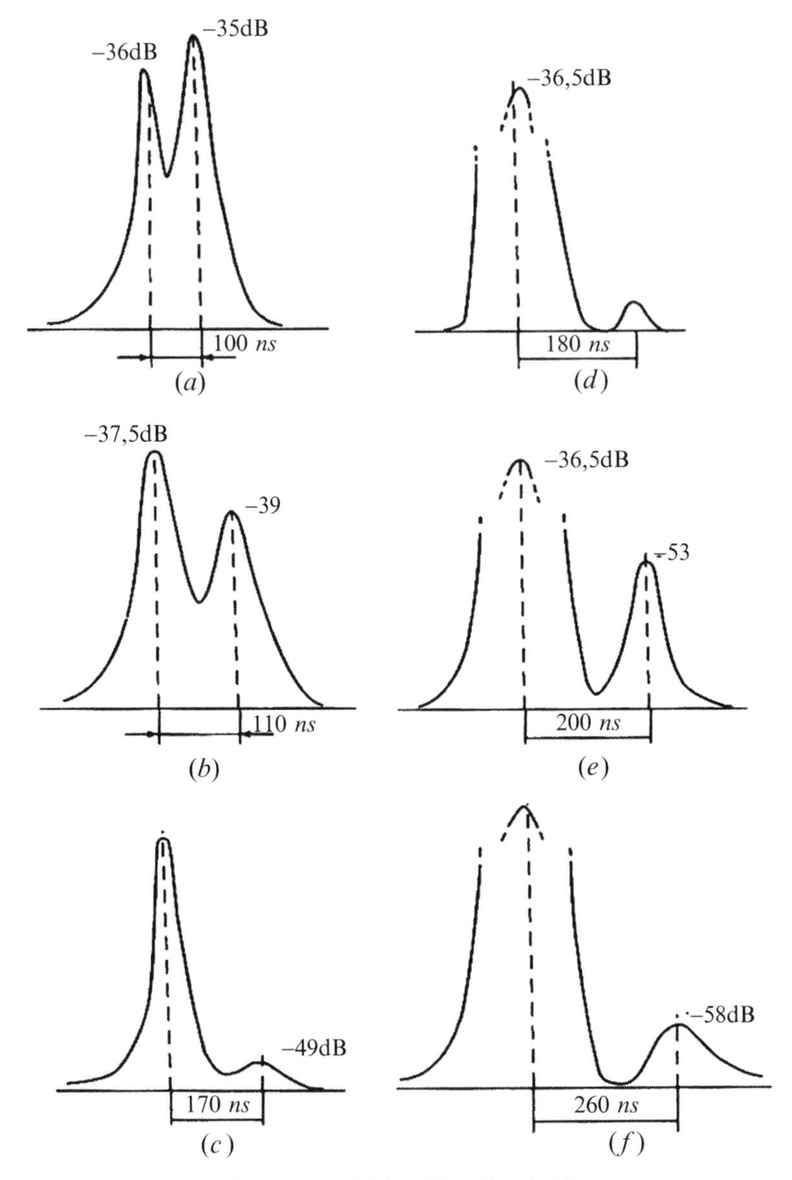

Fig. 7.20 Oscillograms of both probing and delayed impulses in DL

7.2.3 A Filter on a Beyond-Cutoff Waveguide

In such a device the following effects are realized:

- Filtration of a signal in the near zone of an input waveguide-slot, slot or strip converter with orthogonal orientation of FDLS

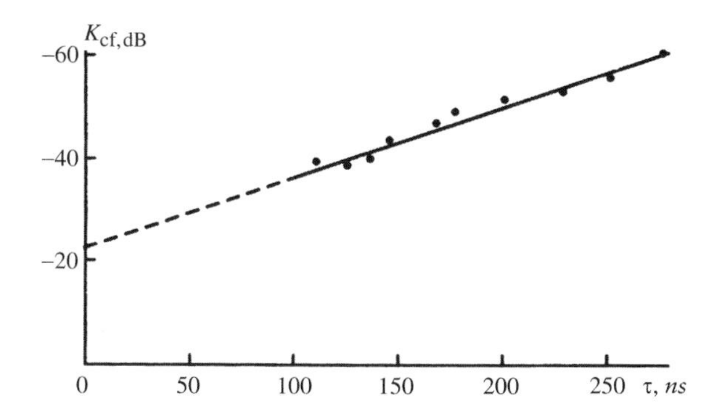

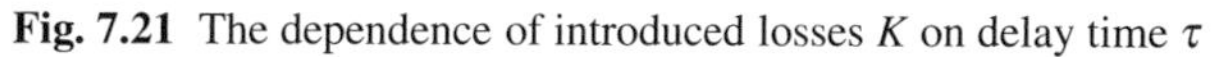
Fig. 7.21 The dependence of introduced losses K on delay time τ

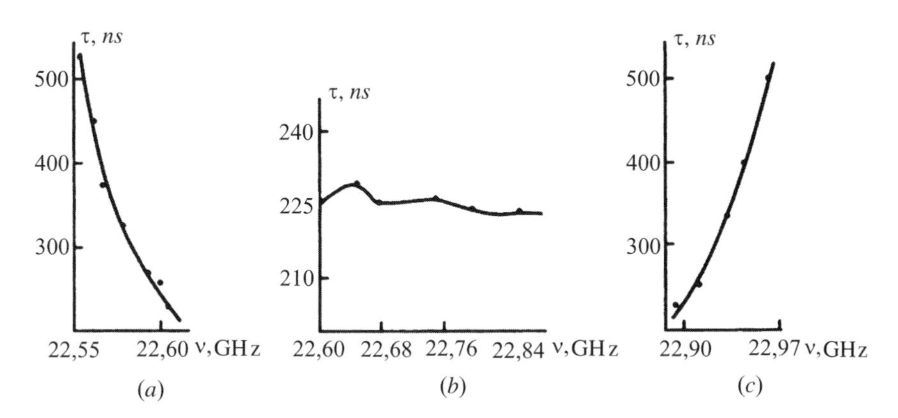

Fig. 7.22 Decreasing (*a*), dispersion-free (*b*) and increasing (*c*) characteristics of DL on SSSLW in a range of frequencies 22–26 GHz

Table 7.2 Key parameters of the developed DL

Delay line type	$\tau \cdot 10^9$, s	Diapason, GHz	$\Delta\tau \cdot 10^9$, s	$\tau(\nu)$, MHz	$\tau(\nu) \cdot \Delta\nu$
IDDL	225–540	25.55–22.62	315	70	22
DDDL	225–500	22.90–22.97	275	70	19
DFDL	225	22.65–22.87	–	220	–
DDL	65–160	24.80–25.72	95	920	87
DDL	100–300	34.6–36.5	200	1,900	380

- Filtration of a signal in a transfer line on a beyond-cutoff waveguide due to an increase in the steepness of the amplitude constant of a slow wave at tuning-out from the central frequency
- Enhanced out-of-band attenuation of a signal due to a jump of the sections of both bringing and allocating waveguides and the beyond-cutoff section, and due to an increase in the length of the beyond-cutoff section as well
- Enhanced heat removal, that requires no additional cooling of FDLS and allows such devices to be uses in an expanded dynamic power range

Various variants of filters on a beyond-cutoff waveguide have been realized in a range of frequencies 2–80 GHz.

In the long-wave part of the centimeter range on input and output of the beyond-cutoff section MSL with an increased uniformity of exciting HF fields $\frac{W_{MSA}}{d} = 80$, $d = 30 \cdot 10^{-6}$ m, $4\pi M_S = 0.176$ T were used at longitudinal-orthogonal arrangement of FDLS under the conducting strip and with an active length of the converter $L_a = \lambda/4$, λ being the wavelength in MSL. The length of the beyond-cutoff section of $L_{bcs} \approx 3 \cdot 10^{-3}$ m provided out-of-band blocking of a signal at a level of 55 dB. The passband was $\Delta\nu_{3dB} = (4\text{–}6)$ MHz, the introduced losses on the central frequency are $K(\nu_0) \approx 4$ dB.

In the short-wave part of centimeter frequencies ($\nu > 20$–35 GHz) and in the millimeter range, filters were made in a waveguide variant. As input and output devices matching FDT (protruding parts of FDLS) and waveguide-slot converters with orthogonal orientation of FDLS were used.

Let's report the results of our experimental examination of the target parameters of such filters in a range of frequencies (30–40) GHz.

$$1 - h_2 = 5.9 \cdot 10^{-6}\,\text{m}, 2 - h_2 = 16.9 \cdot 10^{-6}\,\text{m}, 3 - h_2 = 21.2 \cdot 10^{-6}\,\text{m},$$
$$4 - h_2 = 37 \cdot 10^{-6}\,\text{m}, 4\pi M_S = 0.176\,\text{T}$$

In Fig. 7.23 the dependences of FDT-introduced losses on the length of the protruding part of FDLS S for films of various thickness are presented:

$$1 - h_2 = 5.9 \cdot 10^{-6}\,\text{m}, 2 - h_2 = 16.9 \cdot 10^{-6}\,\text{m}, 3 - h_2 = 21.2 \cdot 10^{-6}\,\text{m},$$
$$4 - h_2 = 37 \cdot 10^{-6}\,\text{m}, 4\pi M_S = 0.176\,\text{T}.$$

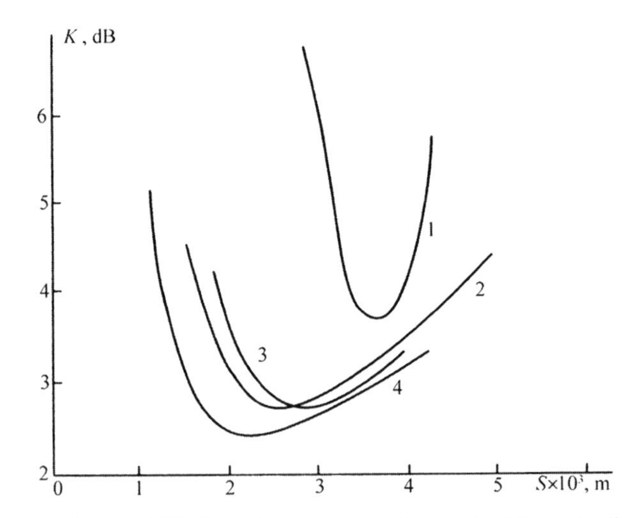

Fig. 7.23 The dependences of FDT-introduced losses on the length of the protruding part of FDLS S for films of various thickness are presented

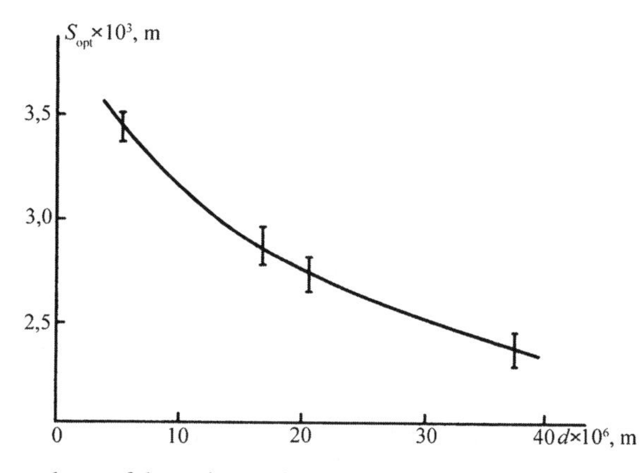

Fig. 7.24 The dependence of the optimum size S_{opt}, at which FDT-introduced losses are minimal, on the thickness of the ferrite film h_2

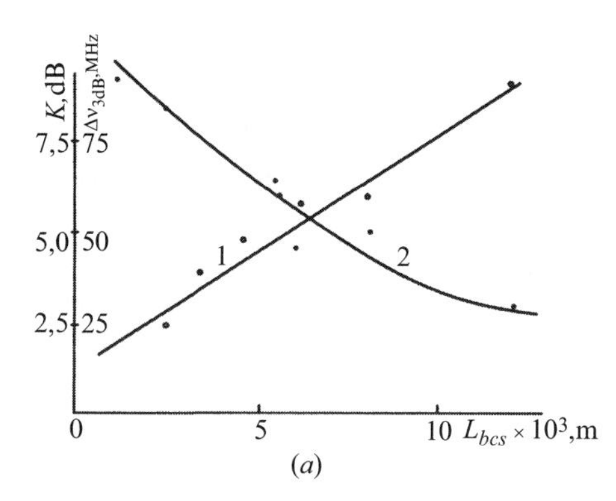

Fig. 7.25*a* Experimental dependences of the introduced losses $K - 1$ and the passbands $\Delta\nu_{3dB} - 2$ on the length of the beyond-cutoff section L_{bcs}

Figure 7.24 shows the dependence of the optimum size S_{opt}, at which FDT-introduced losses are minimal, on the thickness of the ferrite film h_2.

In Fig. 7.25*a*, *b* experimental dependences of the introduced losses $K - 1$ and the passbands $\Delta\nu_{3dB} - 2$ on the length of the beyond-cutoff section L_{bcs} (Fig. 7.25*a*) and on the thickness of the ferrite film h_2 (Fig. 7.25*b*) are presented. As follows from theoretical analysis, the passband of DL on a beyond-cutoff waveguide decreases at increasing its length L_{bcs} ($\Delta\nu_{3dB} \sim L_{bcs}^{-1}$), and the introduced attenuation is $\sim L_{bcs}$.

Table 7.3 contains the parameters of waveguide-beyond-cutoff filters in MMR for three values of the central frequency ν_0 in an electromagnet, in a permanent magnet. In the note the basic geometrical sizes of the filter elements, target parameters, weight and dimensions are given.

In Fig. 7.26 typical dependences of AFC – 1 and SWR$_e$ an entrance – 2 of a waveguide-beyond-cutoff filter in a range of frequencies 30–37 GHz are shown. It is obvious well enough that the central frequency of a signal reflected from the input FDT with a waveguide-slot converter ν_{ref} does not coincide with the central

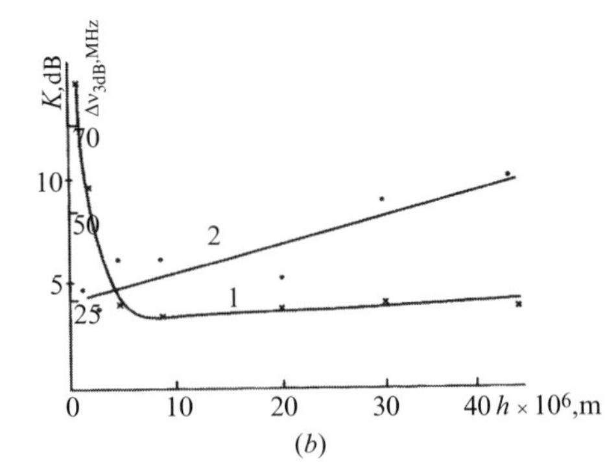

Fig. 7.25*b* Experimental dependences of the introduced losses $K - 1$ and the passbands $\Delta\nu_{3dB} - 2$ on the thickness of the ferrite film h_2

Table 7.3 The parameters of waveguide-beyond-cutoff filters in MMR for three values of the central frequency ν_0 in an electromagnet, in a permanent magnet

Parameters	Central frequency, GHz	In electromagnet	In permanent magnet	The note
Losses on ν_0, dB	30	3.5	4.5	
	35	3.0	4.5	$L_n = 5.5 \cdot 10^{-3}\,\text{m}$
	37	3.5	5.0	$h_2 = 35.5 \cdot 10^{-6}\,\text{m}$
Pass band on level 3 dB, MHz	30	80	110	$L_{bcs} = 0.4 \cdot 10^{-3}\,\text{m}$
	35	80	110	$S = (2 \div 2.5) \cdot 10^{-3}\,\text{m}$
	37	80	115	$a = (0.5\text{–}1) \cdot 10^{-3}\,\text{m}$
				$h_1 = (3\text{–}5) \cdot h_2$
Level of parasitic constituents, dB	30			$\Delta H_{\parallel} < 40\,\text{A/m}$
	35	30	25	$h_2 \leq (30 \div 40) \cdot 10^{-6}$
	37			
Out-of-band barrage, dB at tuning out from ν_0 on $\pm(3\text{–}5)\Delta\nu_{3dB}$	$27 \div 37$	40–50	$40 \div 50$	$SWR_e(\nu_0) < 1.1 \div 1.2$ $SWR_e(\nu) < 10 \div 20$ $P_{\text{imp}} = 3\,\text{kW}$, $Q = 10^3$ $V = (7\text{–}20) \cdot 10^{-6}\,\text{m}^3$ $M = 0.03\text{–}0.2\,\text{kg}$

frequency of a signal ν_0 on output of the device, and their tuning out is $\nu_0 - \nu_{ref} \approx 138\,\text{MHz}$. As well as theoretical analysis predicts, the band of a signal $(\Delta\nu_{ref})_{3\,\text{dB}}$ reflected from the input of the device essentially is wider than the passband on the output $(\Delta\nu_0)_{3\,\text{dB}}$. The level of out-of-band blocking $K_{bloc} < -40\,\text{dB}$ is also below the level of sensitivity of the used panoramic measuring R2–65 instrument.

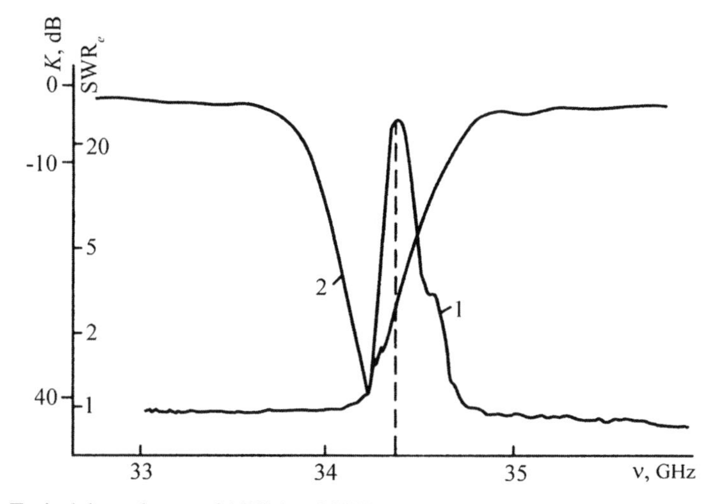

Fig. 7.26 Typical dependences of AFC-1 and SWR$_e$ an entrance – 2 of a waveguide-beyond-cutoff filter in a range of frequencies 30–37 GHz

7.2.4 A Waveguide Filter on a Ferrite Film with an Absorbing Covering

Let's consider a waveguide filter in which a layered structure with a ferrite film with an absorbing covering on its central part is located in the *E* plane. Such a device contains both input and output FDTs working in a mode of interference attenuation of fast and slow waves on a certain length *S*, and a transfer line on FDLS providing selective passage of a signal on a frequency close to the resonant one given a certain type of absorbing covering is deposited.

For such a type of filters characteristic are:

- Detuning of the central frequencies of selective attenuation of a signal in the input FDT and passage of a signal in TL on a ferrite film with an absorbing covering
- Effective broadband attenuation of a signal outside of the working passband, determined by the material of the absorbing covering and its extent
- A mode of running wave in all the band of frequencies at $SWR_{inp} \to 1$
- A simple design

Waveguide filters with a partial absorbing covering on a ferrite film were tested in a range of frequencies 6–80 GHz.

Figure 7.27 shows the dependences of AFC (curve 1) of a signal and SWR$_e$ (curve 2) of a filter with an absorbing covering, and Fig. 7.28 gives its photo. Unlike waveguide-beyond-cutoff designs, such filters are more broadband and possess an improved selectivity.

In Table 7.4 parameters of packaged filters-preselector on the effect of transparency of their ferrite-dielectric structure with an absorbing covering for YIG films and Li–Zn-spinels in MMR are given for comparison.

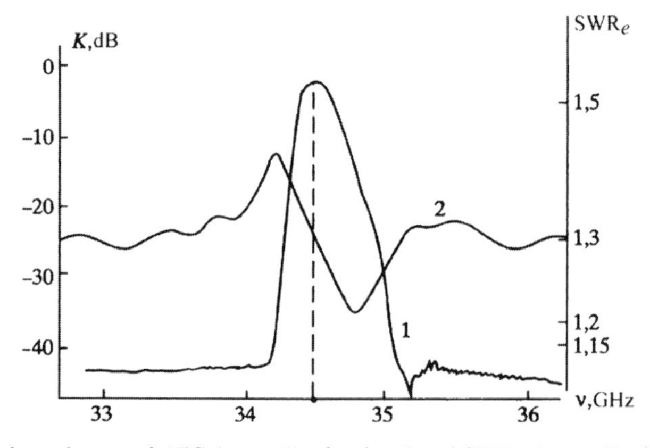

Fig. 7.27 The dependences of AFC (curve 1) of a signal and SWR$_e$ (curve 2) of a filter with an absorbing covering

Fig. 7.28 Photo of filter with lossy jacket

7.2.5 Filters on FDLS-Connected Waveguides in MMR

The opportunity of using of selective connection of two waveguides by means of a ferrite ellipsoid is considered in [8], with self-consistent analysis allowing to determine the transfer factors of such devices in view of the sizes of the sample, its magnetization, and ferromagnetic losses. In [367] an attempt to consider connection of two waveguides by means of a ferrite-dielectric structure is made.

Various variants of realization of waveguide filters connected with a layered structure on the basis of ferrite films by the wide wall of waveguides were investigated:

– Monoaxial waveguides
– Crossed waveguides

Table 7.4 Parameters of packaged filters-preselector on the effect of transparency of their ferrite-dielectric structure with an absorbing covering for YIG films and Li-Zn-spinels in MMR

Parameters	$h_2 = 10 \cdot 10^{-6} m$ $4\pi M_S = 0.176\text{T}$	$h_2 = 25 \cdot 10^{-6} m$ $4\pi M_S = 0.176\text{T}$	$h_2 = 50 \cdot 10^{-6} m$ $4\pi M_S = 0.176\text{T}$	$h_2 = 18 \cdot 10^{-6} m$ $4\pi M_S = 0.3\text{T}$
Losses on ν_0, dB	−6	−4	−2	−3
Pass band on level 3 dB, MHz	40	80	120	700
Level of parasitic constituents, dB	−50	−45	−40	−45
Out-of-band barrage at tuning out from ν_0 on $\pm(3\text{–}5)\Delta\nu_{3dB}$	−50	−50	−50	−50

Table 7.5 Parameters of filters on FDLS-connected waveguides in MMR

Constitutive execution	YIG film thickness 10^{-6}m	Band at level 3 dB, MHz	Introduced losses, dB	Attenuation outside pass band, dB	Excursion of tuning out, GHz	SWR_e
Coaxial waveguides, communication through FFDS at crack in center of wide wall	50	$(60 \div 80)$ 100	3[10] 5.3	55	$27 \div 37$[10] $31 \div 35$	1.4–1.5
Lossy jacket on ferrite film	20[11]	90[10]	3.4[10]	50	$27 \div 37$	1.7
Crossed waveguides, communication through FFDS at wide wall	55	50[11]	10	50	$27 \div 37$	1.5

The sizes of communication window and parameters of FDLS were varied. The filter works on the effect of homogeneous precession of magnetization in an longitudinally-magnetized ferrite film.

In Table 7.5 parameters of these filters are resulted.

7.2.6 Low-and-High-Passage MMR Filters

The effects of selective excitation of a signal with its subsequent attenuation in FDLS can be obtained on various types of strip converters (concentrated elements

[10] Data are received in an electromagnet.

[11] Spinel

Table 7.6 Key parameters and characteristics of several breadboard models of reconstructed LHPF MMR on various types of planar converters (SLC, CLC, MSCL), and waveguides with FDLS

Transformer type	Orientation of ferrite layer to flatness of transformer	MSW type	Thickness of ferrite layer $h_2 \times 10^{-6}$ m	Width of transformer, 10^{-3} m
SL	Orthogonal, in the line of crack	SSMSW	15	0.35
SL	Orthogonal, across crack	SSMSW	15	0.35
CL	Orthogonal, across crack	SSMSW	15	0.7
MSL	Parallel	SSLMSW	17	0.75
Waveguide with FDLS	E – flatness of waveguide			

Transformer type	Introduced losses K, dB	Level of barrage, K_{bar}, dB	Band at level 3 dB, MHz	Figure of merit
SL	3	−40	139	3,400, ..., 3,600
SL	3	−40	135	2,800, ..., 3,000
CL	3	−35	143	2,900, ..., 3,600
MSL	2	−40	196	1,900, ..., 2,000
Waveguide with FDLS				

with extraneous sources) and on pieces of waveguides with FDLS in the prelimit mode $(\nu \gg \nu_{cr})$ – ferrite-dielectric transformers. Filters of a protecting type on strip converters work for reflection, and waveguide filters with FDLS do for passage. As well as for the above considered magnetoelectronic devices, high-speed reorganization by frequency and an expanded dynamic range of input power in a linear mode will be characteristic of LHPF on the basis of ferrite films.

In Table 7.6 key parameters and characteristics of several breadboard models of reconstructed LHPF MMR on various types of planar converters (SLC, CLC, MSCL), and waveguides with FDLS are given.

7.2.7 A Multichannel Onboard Receiver of Direct Amplification with a Magnetoelectronic Filter

As an example illustrating the advantages of MED in MMR in comparison with other solid-state and waveguide devices, a multichannel onboard receiver of direct amplification with a waveguide-beyond-cutoff filter switched by frequency (by channels) on the basis of a YIG film has been designed. The receiver (Fig. 7.29) contains: (1) a reception aerial of the cone type; (2) a waveguide polarizer providing transformation of an accepted wave into wave H_{10} of a rectangular waveguide; (3) a matching adapter to a waveguide channel $5.7 \cdot 0.9\,\text{mm}^2$; (4) a multichannel filter on a

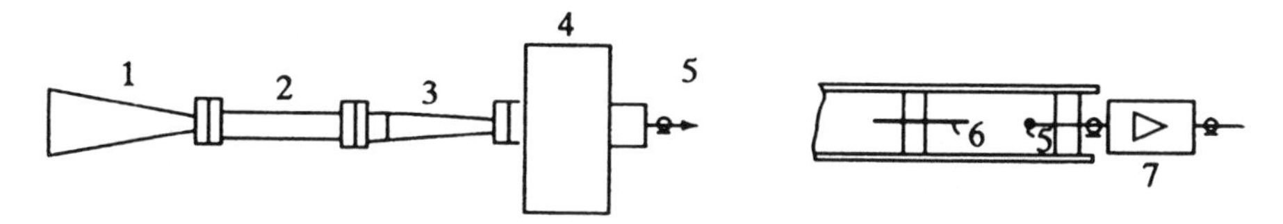

Fig. 7.29 The receiver of direct intensification contains: (1) a reception aerial of the cone type; (2) a waveguide polarizer providing transformation of an accepted wave into wave H_{10} of a rectangular waveguide; (3) a matching adapter to a waveguide channel $5.7 \cdot 0.9\,\text{mm}^2$; (4) a multichannel filter on a beyond-cutoff waveguide of FDLS; (5) an EHF detector; (6) FDLS; (7) an amplifier of videofrequency

Fig. 7.30 The receiver of direct intensification

beyond-cutoff waveguide of FDLS; (5) an EHF detector; (6) FDLS; (7) an amplifier of videofrequency.

The receiver (Fig. 7.30) has the following parameters:

A range of frequencies	35–37 GHz
Number of channels	5
Passband in the channel	100 MHz
Frequency tuning out of channels	150–200 MHz
Losses on the central frequency	5–8 dB
Blocking outside of the passband	more than 40 dB
Nonreciprocal "entrance-exit" of filter	20 dB
Input sensitivity	55 dB/W
Passive heatset with magnetic shunts not worse than $2 \cdot 10^{-6}\%/\text{K}$ in an interval of temperatures	$(60 \div +30)^\circ$C, $(30 \div +60)^\circ$C
A displacement current of the detector	100 mA
Weight	100 g
Dimensions	$15 \cdot 20 \cdot 30\,\text{mm}^3$

7.3 Magnetoelectronic Devices of a High Power Level

The increased magnetic rigidity, increased level of threshold powers, low ferromagnetic losses provide the opportunity to use FDLS in modes of excitation and transfer of high power levels. Fast and slow waves with increased phase and group speeds provide the achievement of low introduced losses near the resonant frequencies of FDLS.

For MED of an increased power level (above several watt in the continuous mode and several tens kilowatt in the pulse mode) required are:

- An increased or standard cross section of the waveguide
- An air gap between FDLS and waveguide walls
- An excess pressure
- Accommodation of FDLS on a wall of the waveguide for better heat removal
- No absorbing coverings with a low conductivity
- A mode of running wave ($SWR_{inp} < 1.5$–2.0)
- Physical principles providing the minimum losses in FDLS

At first experimental research made on YIG plates in waveguide-beyond-cutoff devices, it has been shown that the level of output power P_{out} linearly depends on the input power P_{inp} for volume and surface slow waves. In Fig. 7.31 experimental dependences $P_{\text{out}} = P_{\text{inp}} \cdot K_{\text{cf}}$ on the level of input power P_{inp} for SSSLW (curve 1, $\nu = 23.5\,\text{GHz}$) and for RSSW (curve 2, $\nu = 24.5\,\text{GHz}$) in the continuous mode are presented. Note that the total transfer losses are $K_{\text{cf}} = -10\,\text{dB}$, the transformation losses in FDT are $K_{los.con} = -4\,\text{dB}$.

In Fig. 7.32 dependence of the level of output power $P_{out}^{pm} = P_{pmx}^{pm} \cdot K_{cf}$ on the input one P_{inp}^{pm} in the pulse mode is given at a relative pulse duration $Q = 10^3$ on a waveguide-beyond-cutoff filter with a YIG film of a thickness $h_2 = 28 \cdot 10^{-6}\,\text{m}$, $K_{\text{cf}} = -4\,\text{dB}$. As well as for continuous power, the dependence $P_{out}^{pm}(P_{inp}^{pm})$ is

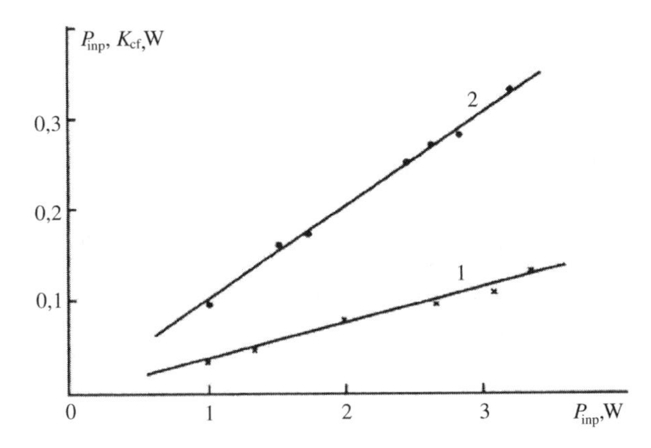

Fig. 7.31 Experimental dependences $P_{out} = P_{inp} \cdot K_{cf}$ on the level of input power P_{inp} for SSSLW (curve 1, $\nu = 23.5\,\text{GHz}$) and for RSSW (curve 2, $\nu = 24.5\,\text{GHz}$) in the continuous mode

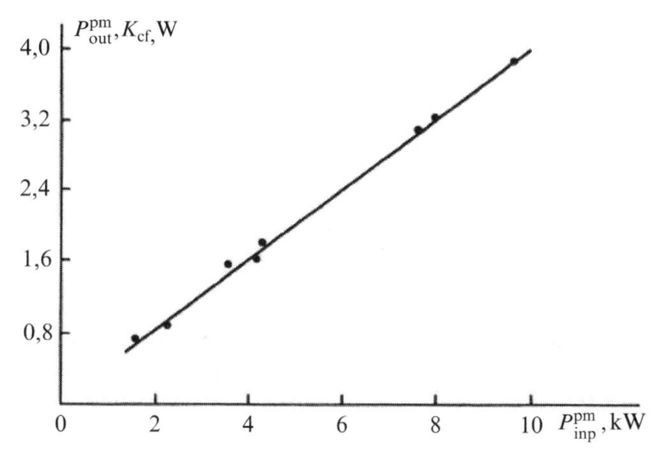

Fig. 7.32 Dependence of the level of output power $P^{pm}_{out} = P^{pm}_{inp} \cdot K_{cf}$ on the input one P^{pm}_{inp} in the pulse mode is given at a relative pulse duration $Q = 10^3$ on a waveguide-beyond-cutoff filter with a YIG film of a thickness $h_2 = 28 \cdot 10^{-6}$ m, $K_{cf} = -4$ dB

linear. The linear character of the dependence P^{pm}_{out} on P^{pm}_{inp} in FDLS can be violated due to two causes, namely:

- Direct thermal losses caused by a finite size of the structure
- Excitation of spin waves at HF power levels exceeding $h_{thresholdpower}$ and their dissipation into heat

The presence of thermal processes, including utmost permissible ones, in FDLS and the device as a whole is simply registered by the drift of the central frequency of the device. Another, more exact way is associated with registration of the spectral structure of a signal on the output of FDLS.

The pulse operating mode of FDLS may appear most interesting. In book [367] thermodynamic processes in ferrite films under the action of HF magnetic field $\tilde{h}$ are analyzed. The heat-removing properties of FDLS to the metal wall of the waveguide and heat exchange with the environment are considered. The power absorbed by ferrite during the action of an impulse is

$$P_f = \frac{1}{2}\mu_0\omega^2 V_f \omega_M(T)\alpha(T)$$
$$\times \frac{[2\omega_0 + \omega_M(T)]\,[\omega_0 + \omega_M(T)] + \omega^2 + \omega_0^2 + \omega^2\alpha^2(T) - \omega_0\omega_M(T)}{[\omega^2 - \omega_0^2 + \omega^2\alpha(T) - \omega_0\omega_M(T)]^2 + \{\omega\alpha(T)[2\omega_0 + \omega_M(T)]\}^2},$$

where μ_0 is the permeability of vacuum, ω the frequency of UHF, V_f the volume of ferrite, h the amplitude of a HF magnetic field, $\omega_M = 4\pi\gamma M_s$ the characteristic frequency, γ the gyromagnetic ratio, $\alpha(T)$ the parameter of ferromagnetic losses, $\tau = H_{he}/C_f$, C_f the specific thermal capacity of ferrite.

The temperature of the film at the end of an impulse is

$$T_1 = \frac{P_f}{H_{he}}(1 - e^{-\frac{\tau_{pm}}{\tau}}),$$

and at the beginning of the next one is

$$T_2 = \frac{P_f}{H_{he}}(e^{-\frac{\tau_{pm}}{\tau}} - 1) \cdot e^{-\frac{T_{pm}}{\tau}},$$

where T_{pm} is the period of impulses, τ_{pm} their duration, H_{he} the factor of heat exchange, P_f the power absorbed by ferrite during the action of an impulse.

In Table. 7.7 and 7.8 data on temperature changes of ferrite films with various parameters of ΔH for 1 min for various levels of pulse power P_{pm} at a relative pulse duration $Q = 10^3$ and for various relative pulse durations at $P_{pm} = 3\,\text{kW}$ are given. It is obvious that for high-quality YIG films with $\Delta H = 80\,\text{A/m}$ the power pf several tens kW dissipates without any essential change of the temperature of the ferrite film.

The threshold value of HF field of a low mode of spin wave is

$$h_{thr} = \frac{2\omega}{\omega_M}\Delta H.$$

Our estimations and preliminary experiments have shown the opportunity to design magnetoelectronic filters of increased pulse power levels.

Table 7.7 Temperature changes of ferrite films with various parameters of ΔH for various levels of pulse power P_{pm} at a relative pulse duration Q = 10^3

Parameter ΔH, oersted (A/m)	Input power P_{pm}, kW, T/$\tau_{pm} = 10^3$, ΔT °C for 1 min			
	3	10	40	100
1(80)	0	0.5	2	11
5(400)	0.1	2.5	10	25
$200(16 \cdot 10^3)$	3	93	373	Destroyed

Table 7.8 Temperature changes of ferrite films with various parameters of ΔH for various levels of pulse power P_{pm} for various relative pulse durations at P_{pm} = 3 kW

Parameter ΔH, oersted (A/m)	Porosity T/τ_{pm}, $P_{pm} = 3\,kW$, ΔT °C for 1 min			
	10^3	10^2	10	2
1(80)	0	0.2	1.8	9
5(400)	0.1	0.8	7.8	41
$200(16 \cdot 10^3)$	3	30	300	Destroyed

7.3.1 A Waveguide HLP Filter with Phase Inversion in a Ferrite Film Structure in an Antiphased Balanced Waveguide Bridge

Ring bridges can be constructed on coaxial, strip, and waveguide lines, they find wide application in radiolocation and signal processing circuits in the UHF and EHF ranges. Strip hybrid bridge devices have been intensely developed in last years for TDC balancing dividers, power adders, ring bridges, "magic" connections, etc.

As is specified in [20], "hybrid bridge devices are a directed coupler, using the principle of circuit arm interface for reception of mutually untied signals with equal amplitudes on two output circuit arms". Variants of design of narrow-band bridges (sometimes they are called "$\frac{3}{2}\lambda$ bridges") and broadband ones, when the phase in one of the circuit arms is varied by means of a concentrated phase shifter from 0 up to π, or due to spatial turning of the line by 180° are known. In [20] a "technology" of making such strip lines is described. Each ring bridge contains a phase-shifting element.

The primary task consisted in determination of such wave modes near the resonant frequencies of FDLS at which a turn of phase of this or that wave by π would take place on the length of the structure L, phase inversion would be carried out with $\Delta\varphi = \kappa' L = \pi/(1+n)$, $n = 0, 1, 2, \ldots$. Thus, from the point of view of decreasing introduced losses, it is expedient to use extremely short FDLS, for which $n = 0$.

Waveguide ring bridges in both narrow-band (with a length difference of circuit arms $\pm n \cdot \lambda/4$) and broadband (with such a difference of $\pm n \cdot \lambda$) modes were used. The broadband mode was realized on waveguides having a spatial turn by $+\pi/2$ and $-\pi/2$ each. Original HPL broadband waveguide adders and dividers constructed by the principle of power summation and division in waveguides with smoothly changing connection were used. These devices had an electric durability $P_{pm} > 40$ kW, provided detuning of channels in the balancing circuit up to 20 dB at the own losses not higher than 1 dB. The T-bridges led to a strong irregularity of AFC in the balancing circuit, that was due to detuning between the channels at a level 3 dB. At the usage of 3 dB waveguide directed couplers it was reached in a strip of frequencies of 10 GHz, a level of detuning is $-(40 \div 50)$ dB.

The photo (Fig. 7.33*b*) demonstrates some variants of PIWF: on T-bridges (Fig. 7.33*a*); on waveguides with spatial change of wave phase by $\pm\pi/2$ (Fig. 7.33*b*); on commercial directed couplers (Fig. 7.33*c*).

In Fig. 7.34 the dependences of AFC of the output signal (Fig. 7.34*a*) and SWR on input (Fig. 7.34*b*) for PIWF made on adders of an original design under the scheme with changing spatial orientation of waveguides in the circuit arms by $\pm\pi/2$ are presented. It is obvious that $(K_{cf})_{\max}$ corresponds to SWR_{inp} on the same frequency. Figure 7.35 shows the AFC of output signal in PIWF with commercial 3 dB directed couplers. One can see that the level of out-of-band attenuation is essential lower than that in the previous case.

In Table 7.9 key parameters of PIWF for non-standard and commercial power adders are listed. It is obvious that the own losses in the filter make not higher than 1.0–1.5 dB, that is due to the absence of matching sites in the form of dielectric wedges of FDLS.

Fig. 7.33*a* Some variants of PIWF: on T-bridges

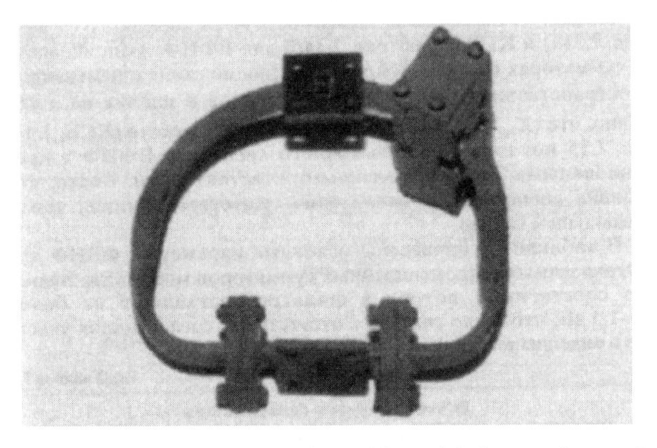

Fig. 7.33*b* Some variants of PIWF: on waveguides with spatial change of wave phase by $\pm\pi/2$

Fig. 7.33*c* Some variants of PIWF: on commercial directed couplers

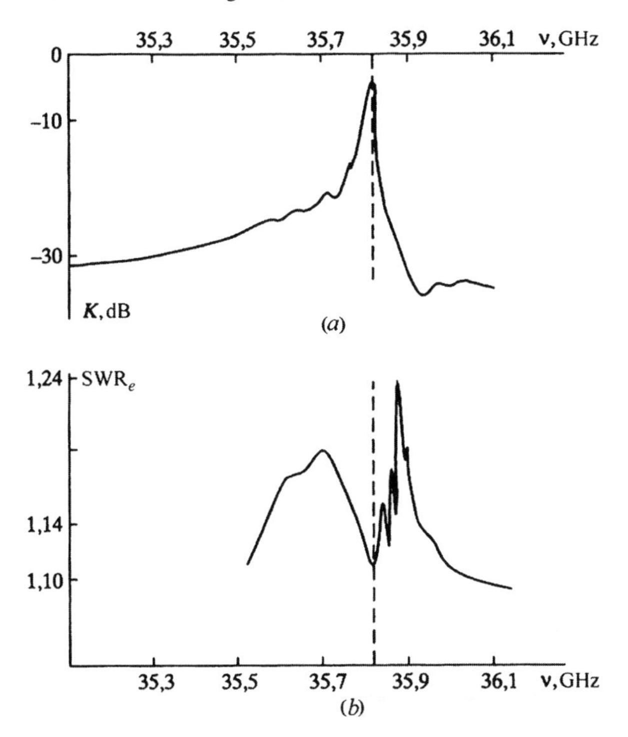

Fig. 7.34 The dependences of AFC of the output signal (a) and SWR_e on input (b) for PIWF made on adders of an original design under the scheme with changing spatial orientation of waveguides in the circuit arms by $\pm\pi/2$

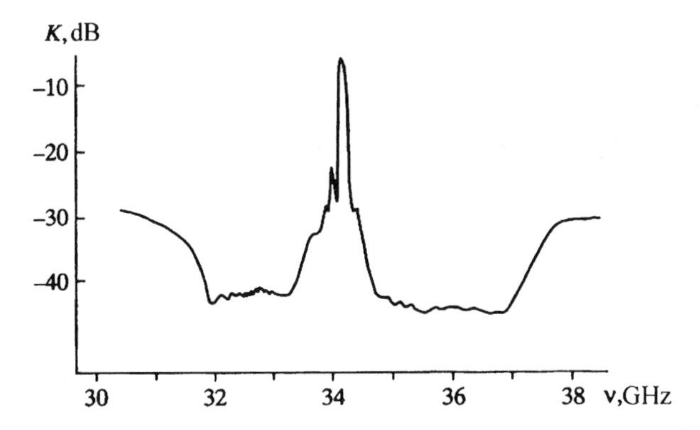

Fig. 7.35 The AFC of output signal in PIWF with commercial 3 dB directed couplers

Table 7.9 Key parameters of PIWF for non-standard and commercial power adders

Parameters	Waveguide filter with phase inversion		The note
	Non-standard integrator	Industrial integrator	
Frequency band of tuning out, GHz	27–37	27–37	$h_2 = 25 \cdot 10^{-6}\,\text{m}$
Losses on ν_0, dB	$-(4.0\text{–}4.5)$	-2.3	$4\pi M_S = 0.176\,\text{T}$
Pass band $\Delta\nu_{3dB}$, MHz	40	40	$\Delta H_{\|} = 1.1$ Oe
Level of barrage, dB	-20	-40	$L = 8 \cdot 10^{-3}\,\text{m}$
			$c = 1.5 \cdot 10^{-3}\,\text{m}$
Losses in denominators, dB	-1	$-(0.5\text{–}0.6)$	

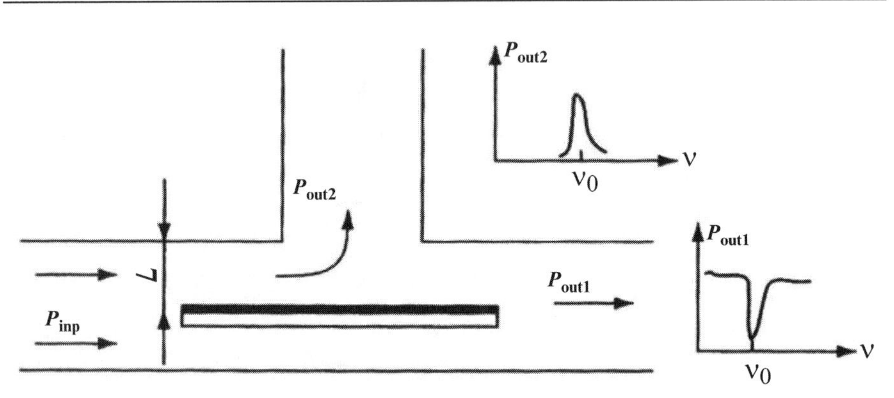

Fig. 7.36 A waveguide bridge of the H-type containing FDLS in the E plane in the field of communication window provides selective signal branching from the basic to the side channel on frequencies close to the resonant one

7.3.2 A Waveguide Filter with a Selective Directed Power Coupler FDLS in MMR

A waveguide bridge of the H-type containing FDLS in the E plane in the field of communication window provides selective signal branching from the basic to the side channel on frequencies close to the resonant one (Fig. 7.36). In the case of the absence of an external magnetic field ($H_0 = 0$), FDLS acts as a piece of a layered dielectric waveguide, in which HF fields concentrate. It provides improvement of discrimination between the basic and side channels. In Fig. 7.37 experimental dependences of the level of discrimination P_{out}/P_{inp} in a such device for various distances of FDLS from the communication window plane are presented: $1 - 0.5 \cdot 10^{-3}\,\text{m}$; $2 - 1 \cdot 10^{-3}\,\text{m}$; $3 - 0.5 \cdot 10^{-3}\,\text{m}$. There exists such a distance t, at which $(P_{out}/P_{inp})_{\min}$. The discrimination level at t_{opt} also depends on the angle of inclination of the side channel waveguide P_{out} to the basic waveguide axis (Fig. 7.38). It is experimentally shown that $\varphi_{opt} \cong (72\text{–}75)^{\circ}$.

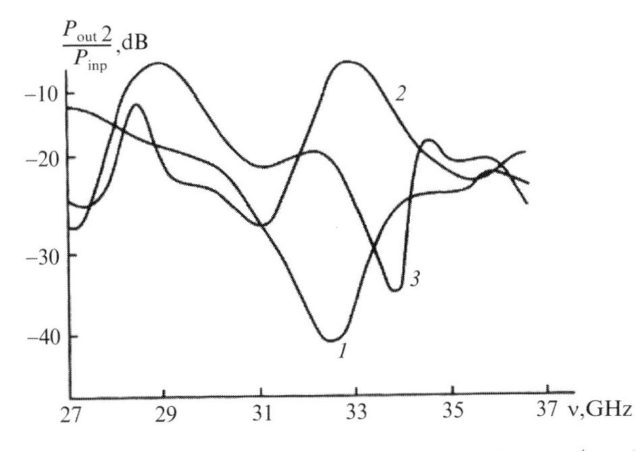

Fig. 7.37 Experimental dependences of the level of discrimination P_{out}/P_{inp} in a such device for various distances of FDLS from the communication window plane are presented: $1 - 0.5 \cdot 10^{-3}$ m; $2 - 1 \cdot 10^{-3}$ m; $3 - 0.5 \cdot 10^{-3}$ m

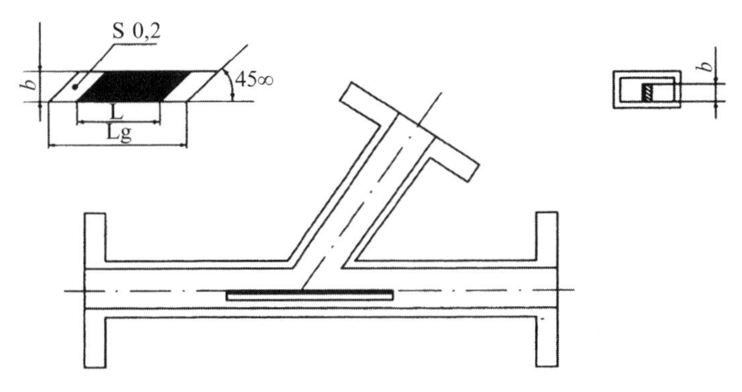

Fig. 7.38 Waveguide filter

In Fig. 7.39 are shown the AFC of a signal (Fig. 7.39*a*) on the output of such a filter, the introduced losses in the basic channel after FDT (Fig. 7.39*b*), and SWR on the input (Fig. 7.39*c*). It is obvious that the filter on the effect of selective power branching provides low losses in the basic channel $K_1 = P_{out}/P_{inp}$ outside of the working passband $K_1(\nu) \leq (0.5–1.3)$ dB, low losses on the central frequency $K_1(\nu_0) < 2$ dB, blocking at a level $K_3 \cong 20 \div 25$ dB and $SWR_{inp} < 1.87$ on the central frequency ν_0.

Such filters allow cascading in the basic, side and both channels simultaneously. This allows effective filters with practically any level of blocking to be designed, for example, with $K_3 > 70–100$ dB, selectivity $K_{sel} > 0.7–1.0$ dB/MHz, and $K_{sq70/3} \leq (4.5 - 2)$ in the EHF range.

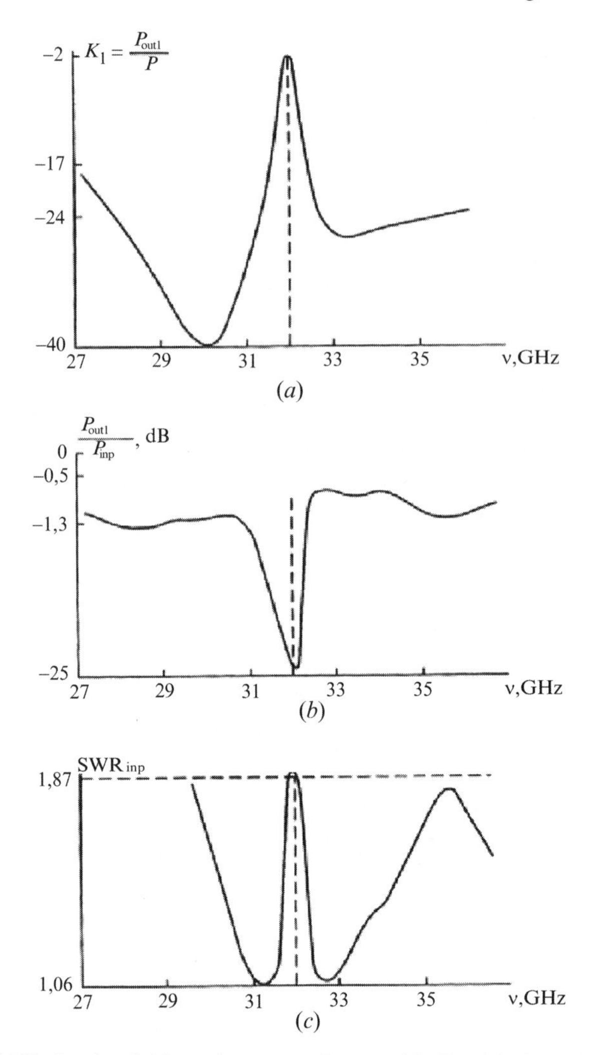

Fig. 7.39 The AFC of a signal (*a*) on the output of waveguide filter, the introduced losses in the basic channel after FDT (*b*), and SWR on the input (*c*)

In Fig. 7.40 possible variants of compact module design are shown:
a – a four-cascade filter with a blocking level $K_{bloc} \geq 100\,\text{dB}, K_{sq} = 2, \beta = (1.4\text{–}2.0)\text{dB/MHz}$; *b* – a five-channel filter; *c* – a five-channel filter with double cascading in the side channels.

It has been shown experimentally that the shape of FDLS and the sizes of the ferrite layer influence the selective properties of the filter. Structures with slope angles of the dielectric base under 45° were used. The ferrite layer which length L was less than that of the dielectric base L_g had similar cants. In Fig. 7.41 the dependence of introduced losses K_{los} on the parameter $\frac{L}{L_g}$ for an YIG film is presented with

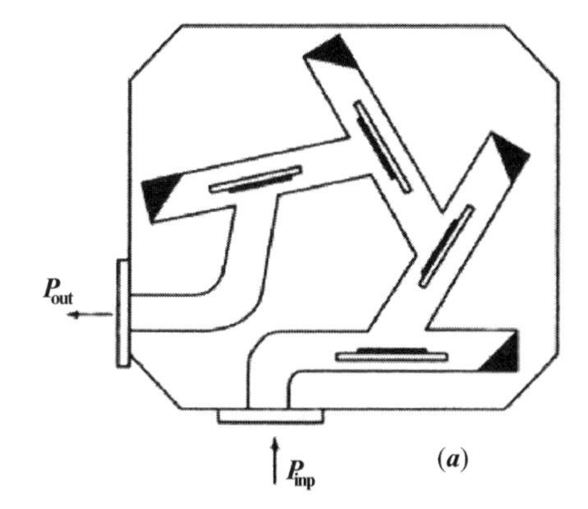

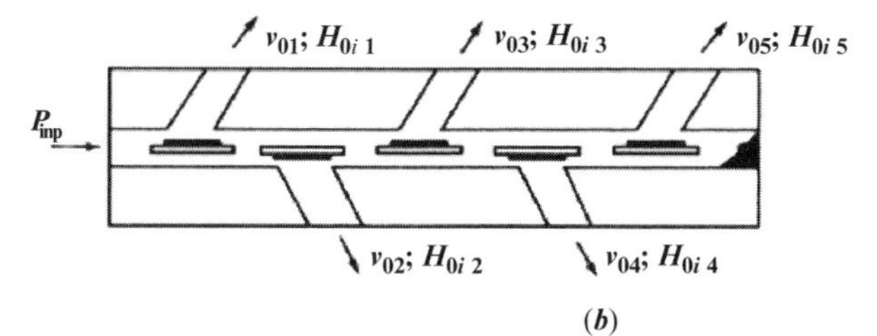

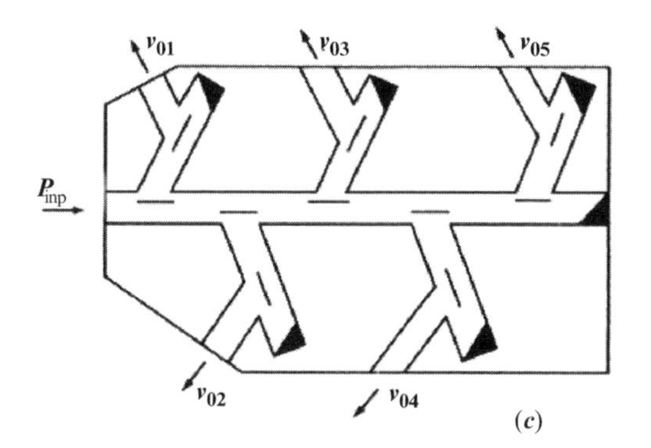

Fig. 7.40 Variants of compact module design are shown:
a – a four-cascade filter with a blocking level $K_{bloc} \geq 100\,\text{dB}$, $K_{sq} = 2, \beta = (1.4\text{–}2.0)\text{dB/MHz}$;
b – a five-channel filter; c – a five-channel filter with double cascading in the side channels

a thickness $d = 21 \cdot 10^{-6}\,\text{m}$, $4\pi M_S = 0.176\,\text{T}$. As the parameter $\frac{L}{L_g}$ increases, the losses K_{los} decrease.

In Fig. 7.42 are shown the AFC of a one-cascade filter with a ferrite-dielectric structure of a special shape to allow return slow waves (on lower frequencies) and space surface magnetic waves (on upper frequencies) to be suppressed. The following parameters are reached in a range of frequencies 33.6–35.7 GHz: $K_{sq} = 1.9$, $K_{los} = -2\,\text{dB}$, $\Delta\nu_{3dB} = 105\,\text{MHz}$, $SWR_e \leq 2$, $\beta = 0.36\,\text{dB/MHz}$.

At double cascading by the side channel (Fig. 7.40*c*):

$$K_{sq} = 2,\ \beta = 1\ \text{dB/MHz}.$$

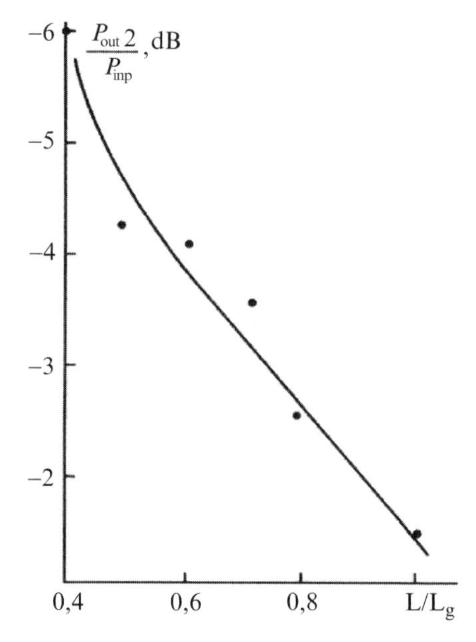

Fig. 7.41 The dependence of introduced losses K_{los} on the parameter $\frac{L}{L_g}$ for an YIG film is presented with a thickness $d = 21 \cdot 10^{-6}\,\mathrm{m}$, $4\pi M_S = 0.176\,\mathrm{T}$

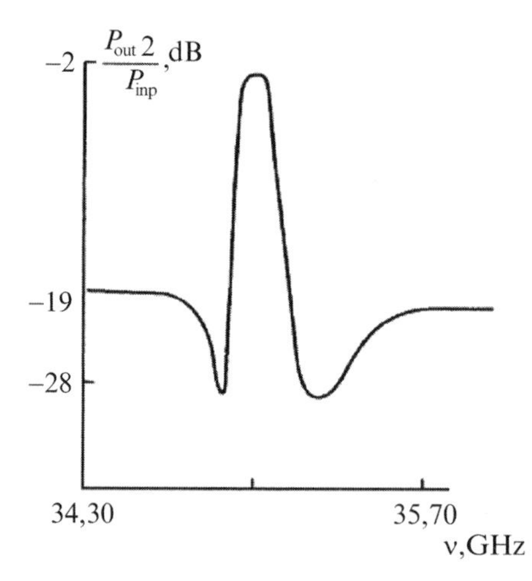

Fig. 7.42 The AFC of a one-cascade filter with a ferrite-dielectric structure of a special shape

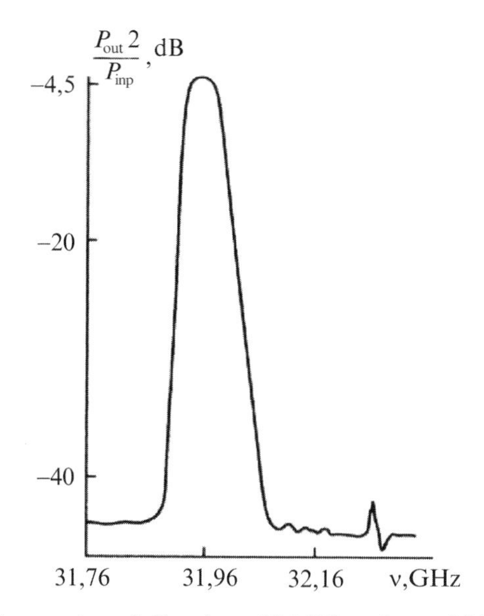

Fig. 7.43 The AFC of a two-channel filter ($\nu_1 = 30.5\,\text{GHz}$ and $\nu_2 = 32.06\,\text{GHz}$), in which each central frequency can be modulated with a speed 3.5 GHz/ms

Figure 7.43 shows the AFC of a two-channel filter ($\nu_1 = 30.5\,\text{GHz}$ and $\nu_2 = 32.06\,\text{GHz}$), in which each central frequency can be modulated with a speed 3.5 GHz/ms.

The number of working channels of the filter is $N = \frac{K_0}{\kappa_{los/cas}}$, where K_0 is the admissible level of signal attenuation of the filter in the whole, $K_{los/cas}$ the level of losses in one cascade.

At $K_0 = 10\,\text{dB}$ and $\text{K}_{\text{los}/cas/} = 1.3\,\text{dB/channel}$ it is possible to realize a HPL filter-preselector with $N = 7$, at a level $K_{bloc} > (20 \div 25)$ dB and $SWR_e \leq 1.87$ on the central frequency.

Filters on the specified principle of action enable cascading in the basic, side, and both channels simultaneously. This allows effective filters with practically any level of blocking (for example, $K_{bloc} > 70$–$100\,\text{dB}$, selectivity $\beta \approx (0.7$–$40)$ dB/MHz and $K_{sq70/3} \leq (4.5 - 2)$ in the EHF range (Fig. 7.44) to be designed.

7.4 Conclusions

1. Magnetoelectronic LPL and HPL devices on the basis of high-quality epitaxial ferrite films have been developed for the millimeter range of radiowaves.
2. Base designs of broadband magnetoelectronic devices covering the spectra of fast and slow magnetostatic waves have been developed, their band of existence considerably falls outside the bounds determined by the MSW approximation.

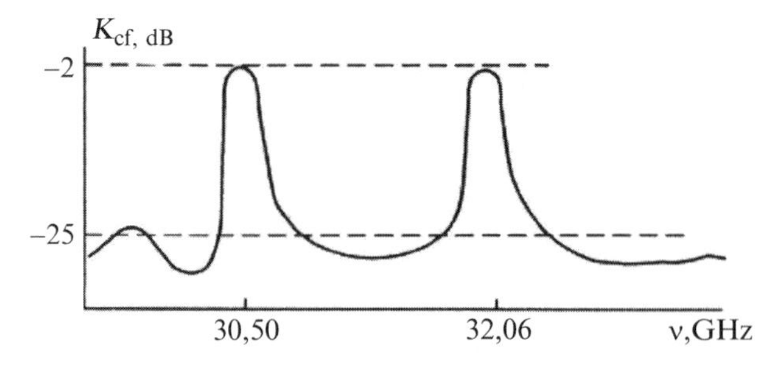

Fig. 7.44 Effective filters with practically any level of blocking (for example, $K_{bloc} > 70$–$100\,\mathrm{dB}$, selectivity $\beta \approx (0.7$–$40)\,\mathrm{dB/MHz}$ and $K_{sq70/3} \leq (4.5-2)$ in the EHF range

They can serve a basis for the design of delay lines with increasing, decreasing, and dispersion-free characteristics, tranversailles filters, multichannel filters-preselectors in MMR.

3. A design of DL on SSSLW, providing achievement of a delay time up to 300–500 ns in a band of frequencies 1.5–1.8 GHz has been developed, the losses being 12–60 dB.
4. Base designs of selective one- and multichannel devices of LPL have been developed on:

   - A beyond-cutoff waveguide
   - A rectangular waveguide with an absorbing covering on its ferrite film
   - Waveguides connected with a ferrite film

5. Low-and-high-pass filters of MMR on the basis of strip and waveguide converters have been designed. It is shown that strip converters with orthogonal orientation of the ferrite film and the exciting plane provide most sharp selectivity in a band of frequencies of reorganization with an external magnetic field above 50%.
6. An onboard multichannel receiver with application of a magnetoelectronic filter to switch working channels by an external magnetic field, favourably differing from its commercial prototype has been developed.
7. Designs of HPL filters-preselectors on the effects of phase inversion in structures with ferromagnetic films in antiphased balanced waveguide bridges and on the effect of selective signal branching on frequencies close to the resonant ones have been developed. The level of introduced losses is below (1–1.5) dB, the out-of-band blocking above $-70\,\mathrm{dB}$ were reached at pulse power levels above 10 kWt and a relative pulse duration $Q \geq 10^3$.

Conclusion

A complex of fundamental and applied, theoretical and experimental studies of wave processes in magneto arranged film structures in the millimeter range of radio waves has been carried out. The following basic scientific problems have been solved:

- Electromagnetic wave excitation and reception in multilayered bigyrotpropic structures screened with impedance surfaces, without any restriction by losses, their cross-section distribution, the distribution of electric and magnetic parameters
- Electromagnetic wave propagation in multilayered bigyrotropic structures in flat waveguides and at unilateral metallization with impedance surfaces
- Nondestructive measurements of the basic dissipative and magnetic parameters of film ferrites within the UHF and EHF ranges
- Development of the radiophysical basis for the design of millimeter-range magnetoelectronic devices

The following results have been obtained:

1. A generalized theory of electromagnetic wave excitation in multilayered structures screened with impedance surfaces with parallel and orthogonal orientations to the exciting plane has been developed to analyze the properties of magnetoelectronic converters and transmission lines made of layered bigyrotropic structures containing magnetized layers of ferrites, semiconductors, magnetic semiconductors, dielectrics, ferroelectrics and their various combinations in the UHF and EHF ranges.
2. The properties of microstrip, slot and waveguide converters with various orientations and magnetization of their structures made of ferrite films in the UHF and EHF ranges have been investigated theoretically and experimentally. It is shown that:

   – For selective excitation and reception of fast and slow electromagnetic waves near the resonant frequencies an enhanced uniformity of the exciting fields is required, which is most simply implemented at the orthogonal orientation of the structures to the exciting plane.

DOI: 10.1007/978-0-387-85457-1_9,

- Microstrip converters are the most universal ones and allow both the broadband and narrow-band modes to be realized at excitation of fast and slow electromagnetic, and magnetostatic waves.
- Slot converters provide excitation of the full spectrum of waves in the broadband modes and selective excitation of fast and slow electromagnetic waves near the resonant frequencies at the orthogonal orientation of the structure.

3. Most essential factors determining the principal features at excitation and propagation of waves near the resonant frequencies in the millimeter range have been revealed, namely:

   - Reduction of the maximum values of the HF magnetic susceptibilities of the imaginary parts of both the diagonal and off-diagonal components $\chi_r'' = \chi_{ar}'' = \frac{M_S}{2\Delta H(\nu)}$ and the real part of the diagonal component $\chi_r' = \frac{M_S}{2\Delta H_{0i}}$
   - Losses in the layers and a rise of their cross gradients at higher frequencies
   - Cross gradients of the static electric and magnetic parameters
   - Back influence of the excited waves in the structures on the source fields
   - A more important role of metal screens

4. Our experiments have shown waveguide and strip converters, and transmission lines made of ferrite films to provide an expanded dynamic range of the linear mode up to several tens watt of continuous and average powers in the pulse modes.
5. A significant excess of the band of waves of various types excited by strip and waveguide converters in both the broadband and narrow-band modes and ducted in ferrite-dielectric layered structures with a low level of ferromagnetic losses ($\alpha < 10^{-4}$) over the band determined from the MSW approximation has been experimentally found and theoretically confirmed. The same applies to an essential linear shift of the lower frequency border of the MSW approximation from which the model of dipole–dipole interactions is valid towards higher wave numbers ($\kappa' > 50 - 100\,\mathrm{rad/cm}$) at advance into the millimeter range.
6. Selective processes at excitation and propagation of various wave types in arbitrary magnetized structures based on weakly dissipative ferrite films ($\alpha < 10^{-3} \div 10^{-4}$) in ranges near the resonant frequencies have been shown, theoretically and experimentally, to be described by a two-wave model to determine:

   - The effects of selective signal attenuation at interference of fast and slow electromagnetic waves, signal transmission by slow waves in structures with an absorbing covering, selective directed power branching by the structure, selective phase inversion on the fast, slow, or both waves in an antiphase balanced circuit (in the pre-limit mode)
   - The effects of transmission under concurrence of fast and slow waves depending on the frequency range (in the post-limiting/beyond-cutoff mode)

7. On the basis of the selective effects of fast and slow electromagnetic wave excitation in layered structures with ferrite films (including multilayered ones) the following types of sensors have been designed:

   - Sensors of the resonant frequencies for structures with arbitrary magnetization
   - Sensors of external and internal magnetic fields to provide an accuracy and spatial resolution two or three orders of magnitude higher in comparison with semiconductor Hall sensors

8. New methods and devices of nondestructive estimation of the following key parameters of ferrite film structures have been developed: FMR line width, cross gradients of saturation magnetization, anisotropy field, and internal field. These devices are based on the effects of selective absorption and propagation of fast and slow waves, discrete and continuous sounding of the structures with various types of magnetostatic and weakly-delayed electromagnetic waves with solid and surface distributions of fields, the delay time dispersion of signals in a boundary field.
9. Physical principles of the design of magnetoelectronic devices of low and high power levels have been developed, they include:

   - Design and optimization of converters
   - Design of transmission lines with required dispersions
   - Design of miniature magnetic systems with a heatset, high-speed electric and discrete mechanical field reorganization
   - Devices for coordination with waveguides
   - Ways to reduce the irregularity and shape of AFC
   - Increasing the electric strength of devices

10. A new class of controllable magnetoelectronic devices on the basis of ferrite films in the millimeter range of low and high power levels has been developed, namely:

    - Single-channel and multichannel band-transmitting and band-blocking filters, including preselectors with high-speed reorganization and small introduced losses of LPL and HPL
    - Controllable lines for signal delay with decreasing, increasing, and weakly dispersive dependencies
    - A miniature multichannel receiver for direct amplification

11. There are some promising leads in the field of EHF magnetoelectronics, namely:

    - EHF magnetooptics
    - Non-linear processes at increased levels of the continuous and pulse power in the EHF range
    - Design of active devices of LPL and HPL on the basis of EHF magnetoelectronic elements

- Processes in structures with HTSC
- Electromagnetic radiation scattering on finite-sized ferrite-dielectric structures
- Design of controllable solid-state PA.

Appendix 1

Tensors $\overleftrightarrow{\mu}_{Tn}$ and $\overleftrightarrow{\varepsilon}_n$ of the magnetic field $\overline{H}_0$:

- In the *OZ* direction

$$\overleftrightarrow{\mu}_{nz} = \begin{Vmatrix} \mu_{Tn} & j\mu_{Nn} & 0 \\ -j\mu_{Nn} & \mu_{Tn} & 0 \\ 0 & 0 & \mu_{Ln} \end{Vmatrix}, \quad \overleftrightarrow{\varepsilon}_{nz} = \begin{Vmatrix} \varepsilon_{Tn} & j\varepsilon_{Nn} & 0 \\ -j\varepsilon_{Nn} & \varepsilon_{Tn} & 0 \\ 0 & 0 & \varepsilon_{Ln} \end{Vmatrix} \tag{Ap. 1a}$$

- In the *OX* direction

$$\overleftrightarrow{\mu}_{nx} = \begin{Vmatrix} \mu_{Ln} & 0 & 0 \\ 0 & \mu_{Tn} & j\mu_{Nn} \\ 0 & -j\mu_{Nn} & \mu_{Tn} \end{Vmatrix}, \quad \overleftrightarrow{\varepsilon}_{nx} = \begin{Vmatrix} \varepsilon_{Ln} & 0 & 0 \\ 0 & \varepsilon_{Tn} & j\varepsilon_{Nn} \\ 0 & -j\varepsilon_{Nn} & \varepsilon_{Tn} \end{Vmatrix} \tag{Ap. 1b}$$

- In the *OY* direction

$$\overleftrightarrow{\mu}_{ny} = \begin{Vmatrix} \mu_{Tn} & 0 & j\mu_{Nn} \\ 0 & \mu_{Ln} & 0 \\ -j\mu_{Nn} & 0 & \mu_{Tn} \end{Vmatrix}, \quad \overleftrightarrow{\varepsilon}_{ny} = \begin{Vmatrix} \varepsilon_{Tn} & 0 & j\varepsilon_{Nn} \\ 0 & \varepsilon_{Ln} & 0 \\ -j\varepsilon_{Nn} & 0 & \varepsilon_{Tn} \end{Vmatrix} \tag{Ap. 1c}$$

Given losses, the components of the tensors $\overleftrightarrow{\mu}_n$ and $\overleftrightarrow{\varepsilon}_n$ are complex with:

- Their diagonal components

$$\mu_{Tn} = \mu'_{Tn} - j\mu''_{Tn}, \; \varepsilon_{Tn} = \varepsilon'_{Tn} - j\varepsilon''_{Tn}$$

- Their off-diagonal components

$$\mu_{Nn} = \mu'_{Nn} - j\mu''_{Nn}, \; \varepsilon_{Nn} = \varepsilon'_{Nn} - j\varepsilon''_{Nn}$$

Given low losses in the ferrite ($\alpha_n << 1$), the components of $\overleftrightarrow{\mu}_n$ are

$$\mu'_{Tn} = \left\{ \left[\left(1+\alpha_n^2\right) \omega_{\mathrm{H}n}^2 - \omega^2 \right] \cdot (\mu_0 - 1) + 4\alpha^2 \mu_0 \omega^2 \omega_{\mathrm{H}n} \right\} D_{\mu n}^{-1},$$
$$\mu''_{Tn} = \left[\left(1-\alpha_n^2\right) \omega_{\mathrm{H}n}^2 - \omega^2 \right] \cdot \gamma M_{sn} \omega D_{\mu n}^{-1},$$
$$\mu'_{Nn} = 2\alpha_n \omega\, \omega_{\mathrm{H}n} D_{\mu n}^{-1},$$
$$\mu''_{Nn} = \left[\left(1+\alpha_n^2\right) \omega_{\mathrm{H}n}^2 + \omega^2 \right] \cdot \alpha_n \omega_{\mathrm{H}n} D_{\mu n}^{-1},$$
$$D = \mu_0 \left[\left(1+\alpha_n^2\right) \omega_{\mathrm{H}n}^2 - \omega^2 \left(4\alpha_n^2 - \omega_{\mathrm{H}n}^2 - 1\right) \right], \tag{Ap. 2}$$
$$\omega_n = \gamma (H_{0i})_n,$$

γ – gyromagnetic ratio ($\gamma < 0$),
H_{0i} – internal field intensity,
$\alpha_n = \left(\frac{\Delta H}{H_{0i}}\right)_n$ – phenomenological parameter of ferromagnetic losses,
ΔH – line width of ferromagnetic resonance,
ω – signal frequency,
M_{sn} – saturation magnetization,
μ_0 – magnetic constant ($\mu_0 = 4\pi \cdot 10^{-7}\,\mathrm{H/m}$).

For polar semiconductors with free charge carriers in a magnetic field $H_0||OZ$ such a model is used in which a dielectric lattice constant ε_{cn} independent of frequency and magnetic field is introduced, and

$$\overleftrightarrow{\varepsilon}_n = \begin{Vmatrix} \varepsilon_{Tn} + \varepsilon_{cn} & j\varepsilon_{Nn} & 0 \\ -j\varepsilon_{Nn} + \varepsilon_{cn} & \varepsilon_{Tn} + \varepsilon_{cn} & 0 \\ 0 & 0 & \varepsilon_{Ln} + \varepsilon_{cn} \end{Vmatrix}, \tag{Ap. 3}$$

where

$$\varepsilon'_{Tn} = \varepsilon_{cn} + \omega_{pn} \left(\omega_{cn}^{*2} - \omega^2 - \omega_{\tau n}^2 \right) F_{\varepsilon n}^{-1},$$
$$\varepsilon''_{Tn} = \frac{\omega_{pn}}{\omega} \left(2\omega_{cn}^{*2} - 3\omega^2 \right) \omega_{\tau n}^2 F_{\varepsilon n}^{-1},$$
$$\varepsilon'_{Nn} = -2\omega_{pn}^2 \omega_{\tau n} \omega_{cn}^2 F_{\varepsilon n}^{-1},$$
$$\varepsilon''_{Nn} = \frac{\omega_{pn}^2}{\omega^2} \left(\omega_{cn}^{*2} \omega_{\tau n}^2 - \omega^2 \omega_{cn}^{*2} + \omega_{cn}^{*4} \right) F_{\varepsilon n}^{-1},$$
$$\varepsilon'_{Ln} = \varepsilon_{cn} - \frac{\omega_{pn}^2}{\omega_{\tau n}^2 + \omega^2},$$
$$\varepsilon''_{Ln} = \frac{\omega_{pn}^2}{\omega} \cdot \frac{\omega_{\tau n}}{\omega_{\tau n}^2 + \omega^2},$$
$$F_{\varepsilon n} = \omega_{\tau n}^2 \left(1 + 4\omega^2\right) - \omega^2 + \omega_{cn}^{*2},$$
$$\omega_{pn} = \frac{e_{0n}^2 n_n}{\varepsilon_0 m_n^*},$$

$\omega_{\tau n} = \frac{1}{\tau_n}$ – plasma frequency,
$\omega_c^* = \frac{qB}{\varepsilon_0 m_n^*}$ – collision rate,
n_n – electron concentration,
B – magnetic induction,
e_{0n} – electron charge,
τ_n – free path time,
m_n^* – effective mass.

Appendix 2

Estimation of Gelder's parameters for the difference $|f(\xi) - f(\xi^T)|$.

$$\begin{aligned}
&\text{For} \quad f(\xi) = e^{-j\frac{\kappa_0 w}{2}\xi\sqrt{1-\xi^2}},\ f(\xi^T) = e^{-j\frac{\kappa_0 w}{2}\xi^T t\sqrt{1-(\xi^T)^2}} \\
&\left|f(\xi) - f(\xi^T)\right| = \left|e^{-j\frac{\kappa_0 w}{2}\xi\sqrt{1-\xi^2}} - e^{-j\frac{\kappa_0 w}{2}\xi\sqrt{1-\xi^2-2\xi\varepsilon-\varepsilon}}\right| \\
&\qquad \leq \left|1 - e^{-j\frac{\kappa_0 w}{2}\varepsilon\sqrt{1-\varepsilon^2}}\right| \leq \frac{1}{2}\varepsilon^2.
\end{aligned}$$

So, Gelder's parameters are:

$$A = \frac{1}{2},\ \lambda = 2. \tag{Ap. 4}$$

Appendix 3

Solution of the non-uniform integral second-type Fredholm equation with an expressed kernel.

After multiplication of Eq. (1.130) by $e^{j\frac{\kappa_0 w}{2}\sin t}$ and integration over the range $(-\pi/2,\ \pi/2)$, we have $x(t^T) = B + E \cdot x(t^T) \cdot D$, where

$$x(t^T) = \int_{-\pi/2}^{\pi/2} J(t^T)\, e^{j\frac{\kappa_0 w}{2}\sin t^T} \mathrm{d}t^T,$$

$$B = \int_{-\pi/2}^{\pi/2} e^{j\frac{\kappa_0 w}{2}\sin t} \mathrm{d}t, \qquad \text{(Ap. 5)}$$

$$D = \int_{-\pi/2}^{\pi/2} e^{j\frac{\kappa_0 w}{2}\sin t} \left[e^{-j\frac{\kappa_0 w}{2}\sin t} \cos t \cdot \ln \frac{1-\sin t}{1+\sin t} - \sin t \right] \mathrm{d}t.$$

Then, from Eq. (Ap. 5) we get

$$x(t^T) = \frac{B}{1-ED}. \qquad \text{(Ap. 6)}$$

Substitution of Eq. (Ap. 6) into Eq. (1.130) produces

$$J(t) = 1 + E \left[e^{-j\frac{\kappa_0 w}{2}\sin t} \cos t \cdot \ln \frac{1-\sin t}{1+\sin t} - \sin t \right] \cdot \frac{B}{1-ED}. \qquad \text{(Ap. 7)}$$

References

1. Gurevich, A.G., Ferrites on ultrahigh frequencies, Moscow, 1960.
2. Ferrites in nonlinear superhigh-frequency devices: Collected translations, Ed. A.G. Gurevich, Moscow, 1961.
3. Ferromagnetic resonance, Ed. S.V. Vonsovsky, Moscow, 1961.
4. Nonlinear properties of ferrites in UHF fields: Collected translations, Ed. A.L. Mikaelyan, Moscow, 1963.
5. Mikaelyan A.L., Theory and application of ferrites on ultrahigh frequencies, Moscow, 1963.
6. Laks B., Batton K., Superhigh-frequency ferrite and ferrimagnetics, Translation from English into Russian under the editorship of A.G. Gurevich, Moskow, 1965.
7. Monosov A.L., Nonlinear ferromagnetic resonance, Moscow, 1971.
8. Gurevich A.G., Magnetic resonance in ferrites and antiferromagnetics, Moscow, 1973.
9. Bochkarev A.I., Modern ferrite devices of millimetric range, Radio Eng. (Russ.), 1984, Vol. (12), p. 16.
10. Watkins, Jonson Company Catalogue, YIG Dev., 1985, No. (5), p. 24.
11. Nicol'skiy V.V., Nikol'skaya T.I., Electrodynamics and distribution of radiowaves, Moscow, 1989.
12. Gigoyan S.S., Meriakri V.V., Murmuzhev B.A., Functional elements of millimetric range on mirror dielectric waveguides, Moscow, 1988.
13. Gigoyan S.S., Meriakri V.V., Murmuzhev B.A., Scientific instrument making in millimeter and submillimeter ranges: Collected scientific papers, Moscow, 1988, p. 5.
14. Faucets and circulators of Midisco®, MSM, 1989, Vol. 19(9), p. 81.
15. Built-in faucets of Local Microwave-Narda West®, MSM, 1989, Vol. 19(9), p. 77.
16. Miniature faucets and circulators, Microw. RF, 1989, Vol. 28(12), p. 142.
17. Faucets and circulators of Sievra Microwave Technology®, Microw. RF, Vol. 28(7), p. 98.
18. Faucets and circulators of Loral Microwave-Narda West®, Microw. RF, Vol. 28(12), p. 14.
19. Builtiin faucets of Sierra Microwave Technology®, Microw. RF, Vol. 28(11), p. 160.
20. Gvozdev V.I., Nefedov E.I., Volume integrated UHF circuits, Moscow, 1985.
21. Meiss J.A. et al., IEEE MTT-S', Int. Microw. Symp. Dig., 1989, Vol. 1, p. 145.
22. Owens J.M. et al., IEEE MTT-S', Int. Microw. Symp. Dig., 1989, Vol. 1, p. 141.
23. Akhumyan A.A. et al., Ferrite faucets of millimetric range on a mirror dielectric polybark waveguide, Radio Eng. (Russ.), 1990, Vol. (2), p. 41.
24. Neklepaev I.G., Sloushch A.V., Research and calculation of broadband Y-switches with a ferrite loose leaf as a triangular prism, Quest. Radio Electron. Ser. OVR (Russ.), 1990, fascicle (5), p. 89.
25. Kazantsev B.I., Choice of type of ferrite untying devices for UHF-power amplifier chains, Electron. Tech. Ser. 1 UHF Electron. (Russ.), 1990, Vol. 5(429), p. 18.
26. Waveguide faucets of Manry Microwave Corp.®, Microw. J, 1990, Vol. 33(6), p. 226.
27. Yakovlev Yu.M., Gendelev S.Sh., Ferrite monocrystals in radio electronics, Moscow, 1975.

28. White R.L., Solt I.H., Multiple ferromagnetic resonance in ferrite spheres, Phys. Rev., 1956, Vol. 104(1), p. 56.
29. Walner L.R., Magnetostatic modes in ferromagnetic resonans, Phys. Rev., 1957, Vol. 105(2), p. 390.
30. Dillon J.E., Magnetostatic modes in ferrimagnetic spheres, J. Appl. Phys., 1958, Vol. 112(1), p. 59.
31. Dillon J.F., Magnetostatic modes in disks and rods, J. Appl. Phys., 1965, Vol. 31(9), p. 1605.
32. Damon R.V., Van de Vaart H., Propagation of magnetostatic spin waves at microwave frequencies in a normally-magnetizes disc, J. Appl. Phys., 1965, Vol. 36(11), p. 3453.
33. Olson F.A., Yaeger J.R., Microwave delay techiques using YIG, IEEE Trans. MTT, 1965, Vol. 13(1), p. 63.
34. Damon R.W., Van de Vaart H., Propagation of magnetostatic spin waves at microwave frequencies, J. Appl. Phys., 1966, Vol. 37(6), p. 2445.
35. Le Graw R.C., Comstock R.L., Physical acoustics, oskow, 1968, Vol. 3, Part B, Chapter 4.
36. Lebed B.M., Lopatin V.P., Experimental research of spectrum of magnetostatic fluctuations of an yttrium ferrogarnet, FTT, 1971, Vol. 13(5), p. 1397.
37. Vashkovskiy A.V., Zubkov V.I., Kil'dishev V.N., Murmuzhev B.A., Interaction of surface magnetostatic waves with charge carriers on ferrite-semiconductor interface, Letters to ZETF (Russ.), 1972, Vol. 4(16), p. 4.
38. Lukomskiy V.V., Tsvirko Yu.A., Amplification of magnetostatic waves in ferrite plates by drift stream of carriers, FTT, 1973, Vol. 15(3), p. 700.
39. Bespyatykh J.I., Zubkov V.I., Convective instability of surface waves in a multilayered structure made of ferromagnetic, semiconductor, and dielectric components, ZTF (Russ.), 1975, Vol. 11, p. 2386.
40. Zilberman P.E., Landau mechanism of electronic absorption and amplification of spin waves, FTT (Phys. Solids), 1977, Vol. 19(10), p. 2986.
41. Vashkovskiy A.V., Nonlinearity and existence conditions of convective instability in ferrite-semiconductor structures, FTT (Phys. Solids), Vol. 19(11), p. 3475.
42. Hittal C., Excitation of spin waves in a ferromagnet by a uniform field, Phys. Rev., 1958, Vol. 110(6), p. 1295.
43. Stavey M.H., Tannenwald P.E., Direct observation on spin-wave resonance, Phys. Rev. Lett., 1958, Vol. 1(5), p. 168.
44. Tannenwald P.E., Weber R., Exchange integral in cobalt from spin-wave resonance, Phys. Rev., 1961, Vol. 121(3), p. 715.
45. Akhiezer A.I., Barjahtar V.G., Peletminskiy S.V., Spin waves, Moscow, 1967.
46. Renaldi S., Digest Intermag' 84: International Magnetic Conference, Hamburg, April 10–13, New York, 1984, p. 403.
47. Hexeferrite monocrystal films for UHF techniques: Information to leading experts, Comp. Shentsova S.V., Inf. Ref. LAOI, 1984, No. (22), p. 6.
48. Thick monocrystal films of ferrogarnets for UHF devices on MSW: An abstract review of foreign materials, Comp. Shentsova S.V., Inf. Ref. LAOI, 1985, No. (3), p. 10.
49. Yakovlev J.M., Garnet epitaxial structures for spin-wave electronics, Rev. Electron. Tech. Ser. 6 Mater., 1986, Vol. 7(1227), p. 56.
50. Ferrospinel films for UHF devices of millimeter range, Comp. Shentsova S.V., Inf. Ref. LAOI, 1987, No. (4), p. 17.
51. Glass H.L., Ferrite films for microwave and MM-wave applications, Microw. J., 1987, Vol. 30(6), p. 19.
52. Glass H.L., Ferrite films for microwave and millimeterwave devices, Proc. IEEE, 1988, (2), p. 151.
53. Glass H.L., Ferrite films for UHF devices, TIIER, 1988, Vol. 76(2), p. 64.
54. Cheparin V.P., Prokhorenko V.I. et al., Hexaferrite films on weakly magnetic substrates, ZTF, 1989, Vol. 59(12), p. 143.
55. Mishin D.D., Magnetic materials, Moscow, 1981, p. 335.
56. Kurushin E.P., Nefedov E.I., Application of thin monocrystal ferrite films in UHF microelectronic devices, Microelectronics (Russian), 1977, Vol. 6(6), p. 549.

57. Collins J.H., Owens J.M., Magnetostatic wave and SAN devices – similarities, differences and trade-off, IEEE Int. Circuits Syst. Proc., New York, 1978, p. 536.
58. Lebed B.M., Lopatin V.P., Magnetostatic oscillations in ferrites and their use in UHF technics, Electron. Eng. (Russ.) Ser. 1 UHF Electron., 1978, Vol. 12(561), p. 60.
59. Vashkovskij A.V., Lebed B.M., Zubtsov B.I. et al., Properties of layered ferrite-semiconductor structures. Application on ultrahigh frequencies, Electron. Eng. Ser. 1 UHF Electron., 1979, Vol. 6(620), p. 56.
60. Shekhtman F.I., Outlook of use of devices on magnetostatic waves for analog processing of UHF signals: A review, Radio Electron. Abroad (Russ.), 1979, Vol. 25(893), p. 9.
61. Owens J.M., The index translation. A series 2, 1982, No. (10), p. 78; IEEE International Symposium on Circuits and Systems Proceedings, 1979, p. 568.
62. Vashkovskij A.V. et al., Properties of layered structures "ferrite-semiconductor", application on ultrahigh frequencies, Electron. Eng. Ser. 1 UHF-Electron., 1979, Vol. 6 (620), p. 48.
63. Adem D., Daniel M.R., Shreder D.K., Application of devices on magnetostatic waves as a way of microminiaturization of UHF devices, Electronics, 1980, No. (11), p. 36.
64. Owens J.M., Caster R.L., Smith C.V., Magnetostatic waves, microwave SAW? I Ultrasonics Symposium Proceedings, Boston, MA, 1980, New York, 1981, Vol. I P, p. 506.
65. Adam J.D., Daniel J.R., The status of magnetostatic devices, Digest Intermag' 81: International Magnetic Conference on Grenoble, New York, 1981.
66. Nikitov V.A., Nikitov S.A., Research and development of devices on magnetostatic waves, Foreign Radio Electron. (Russ.), 1981, No. (12), p. 41.
67. Manes C., Owens J.M., Microwave signal processing using magnetostatic wave devices, Alta Freg., 1982, Vol. 51(2), p. 53.
68. Kats L.I., Safonov A.A., Interaction of UHF electromagnetic oscillations with charge carrier plasma in a semiconductor, Saratov, Saratov University Press, 1979.
69. Adam J.D., Daniel M.R., O'Kefle T.W., Magnetostatic wave devices, Microw. J, 1982, Vol. 25(2), p. 95.
70. Carter R.L., Capacitance topology for frequency modeling of bipolar transistors, Microw. Syst., 1983, Vol. 13(3), p. 103.
71. Shekhtman F.I., Experimental devices for information processing on magnetostatic waves, Radio Electron. Abroad (Russ.), 1983, Vol. 2(974), p. 5.
72. Yakovlev J.M., Merkulov A.I., Magnetic semiconductors for devices of functional electronics, Electron. Eng. Ser. 6 Mater., 1989, Vol. 9(985), p. 99.
73. Khartemyann P., Planar YIG iron–yttrium garnet (?) devices on MSW, Digest Intermag'84: International Magnetic Conference, Hamburg, 1984, p. 410.
74. Ishak W.S., Microwave signal processing using magnetostatic wave devices, Ultrasonics Symposium Proceedings, Dallas, TX, November 14–16, New York, 1984, Vol. 1, p. 152.
75. Takigo K., Tamagava daiganu coga cube kie, Met. Fac. Eng. Tamagawa Univ., 1985, No. (20), p. 201.
76. Ishak W.S., Chang H.M.-W., Magnetostatic wave devices for microwave signal processing, Hewlett-Packard J., 1985, Vol. 36(2), p. 10.
77. Gulyaev Yu.V., Zilberman P.E., Magnetoacoustic waves in laminas and films of ferromagnetics, Proc. Inst. High. Educ. Phys., Tomsk University Press, 1988. Vol. 31(11), p. 6.
78. Owens J.M., Colhins J.H., Carter P.A., System application of magnetostatic wave devices, Circuits Syst. Signal Process., 1985, Vol. 4(1/2), pp. 316–333.
79. Szilard M., Microhulamn ferrite es ferrites eszhozoh kutatasa, fejleszfese, Hiradas-technica, 1986, Vol. 37(6), p. 262.
80. Yakovlev J.M., Salyganov V.I., FMR in planar structures and problems of creation of planar ferrite filters, Electron. Eng. Ser. 1 Microw. Freq. Electron., 1987, Vol. 17(1309), p. 64.
81. Magnetostatic waves in electronics: A Review, Results of science and technics. Series "Electronics", 1987, Vol. 19.
82. Ishak W.S., Magnetostatic wave technology: A review, Proc. IEEE, 1988, Vol. 76(2), p. 171.
83. Adam J.D., Analog signal processing by means of UHF ferrites, TIIER, 1988, Vol. 76(2), p. 73.

84. Ishak V.S., Application of magnetostatic waves, TTIER, 1988, Vol. 76(2), p. 86.
85. Vugalter G.A., Gilinskiy I.A., Magnetoastaic waves, Proc. Inst. High. Educ. Radiophys., 1989, Vol. 32(10), p. 1187.
86. Kuzyakov V.G., Compound ferrite-dielectric components for radio-electronic devices, Electron. Eng. Ser. 1 UHF-Electron., 1990, Vol. 2(1520), p. 25.
87. Magnetoelectronic materials: Features of their technology and outlook of application/ Bichurin M.I., Venevtsev J.N., Didkovskaya O.S. et al., Seignetto-magnetic materials, Moscow, 1990, p. 118.
88. Adam J.D., Magnetostatic-wave and surface-acoustic-wave signal processing, Digest Intermag'89: International Magnetic Conference, Washington, DC, March 28–31, 1989, New York, 1989, p. AC5.
89. Shurupova N.G., Ferrite UHF filters: A review of patents and copyright certificates for 1984–1985, Moscow, 1987, p. 22.
90. Shurupova N.G., Devices on magnetostatic spin waves: A review of patents published in 1982–1988. Moscow, 1989, p. 27.
91. Shurupova N.G., Ferrite UHF filters: A review of patents and copyright certificates for 1986–1987, Moscow, 1990, p. 18.
92. Ferrite UHF technology abroad: Information reference: From materials of foreign conferences in 1986–1987. Moscow, 1989, No. (3), p. 45.
93. Yakovleva T.V., YIG devices: Reference table; Data on foreign development of YIG devices for 1989, Moscow, 1990, p. 8.
94. Military UHF Electronics Conference, Moscow, 1986.
95. Livshits I.I., Rozhkov V.M., Ryabov B.A., Use of artificial Earth satellites for communication in millimeter range, Foreign Radio Electron., 1987, No. (5), p. 41.
96. Shcherbak V.I., Vodyanin I.I., Reception devices of radio-electronic struggle systems, Foreign Radio Electron., No. (5), p. 50.
97. Ferrite UHF filters: A review of patents and copyright certificates for 1984–1985, Laboratory of patent research, Comp. Shurupova N.G., Moscow, 1987, p. 22.
98. UHF technology on the threshold of 1990ies, IEEE Trans. MTT, 1989, Vol. 37(6), p. 1040.
99. Devices on magnetostatic spin waves: A review of patents in 1982–1988, Research of directions of active patenting, Comp. Shurupova N.G., Moscow, 1989, p. 27.
100. Ferrite technology abroad: (On materials of foreign conferences in 1986–1987), Information reference of LAOI, 1989, No. (3), p. 45.
101. Kats L.I., Research of electromagnetic wave distribution features in millimeter and submillimeter ranges in magnetoactive plasma of charge carriers in a semiconductor, Doctoral thesis, Saratov, 1980.
102. Demchenko N.P., Kats L.I., Nefedov I.S., Feature of wave distribution in magnetic semiconductors, Proc. Inst. High. Educ. Radiophys., 1989, Vol. 32(7), p. 891.
103. Vapne G.M., Forecast of military expenses for UHF equipment in the USA, Inf. Ref. LAOI, 1990, No. (4), p. 3.
104. UHF facilities in priority technology plans of the Ministry of Defence of the USA, Comp. Vapne G.M., Inf. Ref. LAOI, No. (8), p. 16.
105. Chuvilin V.P., Some leads of development of foreign radio-electronic systems of military purpose: An Abstract review (On materials of the open press), Inf. Ref. LAOI, 1990, No. (3), p. 23.
106. MSW engineering abroad: (On materials of 1988–1989), Comp. Vapne G.M., Inf. Ref. LAOI, No. (5), p. 25.
107. Nikolsky V.V., Variational methods for internal tasks of electrodynamics, oscow, 1967.
108. Nikolsky V.V., Gyrotropic perturbation of a waveguide, Radio Eng. Electron., 1957, No. (2), p. 157.
109. Gandulu A.K., Webb D.C., Microstrip excitation of magnetostatic surface wave microstrip transducers: Theory and experiment, IEEE Trans. MTT, 1975, Vol. 23(12), p. 998.
110. Wu H.J., Smith C.V., Collins J.H., Owens J.M., Bandpass filtering with multibar magnetostatic surface wave microstrip transducers, Electron. Lett., 1977, Vol. 3(20), p. 610.

111. Gandulu A.K., Webb D.C., Banks C., Complete radiation impedance of microstrip encited magnetostatic surface waves, IEEE Trans. MTT, 1978, Vol. 26(6), p. 444.
112. Adam J.D., Patterson R.W., O'Keefe T.W., Magnetostatic wave interdigital transducer, J. Appl. Phys., 1978, Vol. 49(3), p. 1797.
113. Emtage P.E., Interaction of magnetostatic waves with a current, J. Appl. Phys., Vol. 49(8), p. 4475.
114. Parekh J.P., Theory for magnetostatic forward volume wave excitation, J. Appl. Phys., 1979, Vol. 50(3), p. 2452.
115. Parech J.P., Tian H.S., Meander line excitation of magnetostatic surface waves, Proc. IEEE, 1979, Vol. 67(1), p. 182.
116. Sethares J.C., Weinberg I.J., Apodization of variable coupling MSSW transducers, J. Appl. Phys., 1979, Vol. 50(3), p. 2458.
117. Sethares J.C., Magnetostatic surface wave transducers, IEEE Trans. MTT, 1979, Vol. 27(11), p. 902.
118. Robbins W.P., Approximate theory of magnetostatic surface wave transducers on JIG, IEEE Trans. Son. Ultrason., 1979, Vol. 26(3), p. 230.
119. Wu H.J., Smith C.V., Owens J.M., Bandpass filtering and input impedans characterization for driven multielement transducer pair delay line magnetostatic wave devices, J. Appl. Phys., 1979, Vol. 50(3), Part 2, p. 2455.
120. Vashkovskiy A.V., Gerus S.V., Dikshteyn I.E., Tarasenko V.V., Excitation of surface magnetostatic waves in ferromagnetic plates, ZTF Russ. J. Tech. Phys., 1979, Vol. 49(3), p. 628.
121. Seminozhenko V.P., Sobolev V.P., Theory of spin wave excitation by a variable magnetic field in antiferromagnetics with magnetic anisotropy of "an easy plane" type, FTT Russ. Phys. Solids, 1980, Vol. 22(3), p. 829.
122. USA Patent 4199737, MKI 333/154 (HO3 H9/26). Magnetostatic wave device, Patterson R.W., O'Keffe T.W., Adam, J.P., Appl. 18.10.78, No. 952432. Published 22.04.80.
123. Parekh J.P., Tuan H.S., Excitation of magnetostatic backward volume waves, IEEE Trans. Mag., 1980, Vol. 16(5), p. 1165.
124. Kalinikos B.A., Excitation of propagating spin waves in ferromagnetic films, IEEE Proc. H., 1980, Vol. 127(1), p. 4.
125. Kalinikos B.A., Spectrum and linear excitation of spin waves in ferromagnetic films, Proc. Inst. High. Educ. Phys., 1981, Vol. 24(8), p. 42.
126. Shurilenko B.E., Moskalev V.M., Oboznenko Yu.L., Excitation of magnetostatic oscillations by hypersound waves, XV All-Union Conference on the Physics of Magnetic Phenomena: Abstract, Perm, 1981, p. 141.
127. Makhmudov Z.Z., Sultanov N.M., To the theory of spin wave excitation in a magnetic semiconductor by a high-frequency electric field: (At any direction of its distribution), Proc. Acad. Sci. Ser. Physicotech. Math. Sci., 1981, Vol. 11(1), p. 72.
128. Emtage P.R., Generation of magnetostatic surface waves by a microstrip, J. Appl. Phys., 1982, Vol. 53(7), p. 5122.
129. Vashkovskiy A.V., Zubkov V.I., Kildishev V.N., Research of efficiency of magnetostatic wave excitation in films by converters of various types, Methods of functioning in realization of radio engineering devices, Kiev, 1982, p. 48.
130. Vashkovskiy A.V., Zubkov V.I., Kildishev V.N., Feature of magnetostatic wave excitation in iron–yttrium garnet films caused by the type of a converter, Proceedings of 6th International Conference on Gyromagnetic Electronics and Electrodynamics, Varna, October 3–10, 1982, Sofia, 1982, p. 83.
131. Vashkovsky A.V., Gerus S.V., Influence of capacitance on magnetostatic wave transducer effectivity, Proc. 7th Colloq. Microw. Commun., Budapest, 1982, Vol. 2, p. 768.
132. Zyuzin A.M., Kudelkin N.N. et al., A new mechanism of spin-wave resonance excitation by a homogeneous field in two-layer magnetic films, Letters to ZTF Russ. J. Tech. Phys., 1983, Vol. 9(3), p. 177.

133. Ilchenko M.E., Mogilnyi S.V., Excitation of magnetostatic oscillations of a gyromagnetic resonator by a non-uniform field of vitkovogo a coiled element, 29th International Wisconsin Colloquium Ilminan, October 29–November 2, 1984, Ilminan, 1984, H. 3: Vortragsr. A4, A5, A6, pp. 69–72.
134. Kalinikos B.A., Excitation of dipole-exchange spin waves in ferromagnetic films: Green's spin-wave functions, ZTF Russ. J. Tech. Phys., 1984, Vol. 54(9), p. 1846.
135. Vugalter G.A., Makhalin V.N., Reflection and excitation of forward volume magnetostatic waves by a metal strip, Radio Eng. Electron., 1984, Vol. 29(7), p. 1252.
136. Bogun A.V., Kandyba P.E., Lavrenev A.A., Sorokin V.G., Experimental research of excitation of surface magnetostatic waves by a microstrip line, Physical phenomena in instrument making of electronic and laser engineering: Int. Coll. Book, Moscow, 1985, p. 114.
137. Vugalter G.A., Makhalin V.N., Reflection and excitation of surface magnetostatic waves by a metal strip, ZTF Russ. J. Tech. Phys., 1985, Vol. 55(3), p. 497.
138. Gilinskiy I.A., Shcheglov I.M., Theory of surface magnetostatic wave excitation by a microstrip line: Numerical results, Novosibirsk, 1985, pp. 6–85.
139. Gilinskiyj I.A., Shcheglov I.M., Theory of surface magnetostatic wave excitation, Russ. J. Tech. Phys., 1985, Vol. 55(12), p. 2323.
140. Vugalter G.A., Gilinskiy I.A., Excitation and reception of surface magnetostatic waves by a microstrip converter, Novosibirsk, 1985, p. 8.
141. Dmitriev V.F., Kalinikos B.A., Kovshikov N.G., Observation of exchange oscillations of radiation resistance at excitation of spin waves in normally magnetized films of iron–yttrium garnet, Russ. J. Tech. Phys., 1985, Vol. 55(10), p. 2051.
142. Dmitriev V.F., Kalinikos B.A., Kovshikov N.G., Experimental research of radiation resistance of microstrip aerials to spin waves, Russ. J. Tech. Phys., 1986, Vol. 56(11), p. 2169.
143. Goldberg L.B., Excitation of surface magnetostatic waves by a point current element, Russ. J. Tech. Phys., 1986, Vol. 56(10), p. 1893.
144. Sorokin V.G., Bogun P.V., Kandyba P.E., Radiation resistance of a microstrip line at excitation of magnetostatic waves, Russ. J. Tech. Phys., 1986, Vol. 56(12), p. 2377.
145. Gusev B.N., Gurevich A.G., Vugalter G.A., Krasnov E.S., Excitation and reception of surface spin waves by asymmetrical coplanar converters, Letters to Russ. J. Tech. Phys., 1986, Vol. 12(9), p. 537.
146. Vugalter G.A., Gusev B.N., Gurevich A.G., Chivileva O.A., Excitation of a surface magnetostatic wave by a coplanar converter, Russ. J. Tech. Phys., 1986, Vol. 56(1), p. 149.
147. Gondurov S.A., Dmitriev V.F., Kalinikos B.A., Feature of frequency dependences of radiation resistance of microstrip aerials to spin waves in YIG films, Problems of integrated UHF electronics: Coll. of sci. papers, Leningrad, 1986, p. 29.
148. Lysenko V.A., Transfer function of UHF devices on magnetostatic waves, Radio Eng. Electron., 1986, No. (8), p. 1627.
149. Goldberg L.B., Penzyakov V.V., Excitation of space volume magnetostatic waves by a point current element, Russ. J. Tech. Phys., 1986, Vol. 56(6), p. 1049.
150. Vugalter G.A., Gusev B.N., Gurevich A.G., Excitation and reception of surface spin waves by arbitrarily loaded converters, Russ. J. Tech. Phys., 1987, Vol. 57(7), p. 1348.
151. Vugalter G.A., Gilinskiy I.A., Excitation and reception of surface magnetostatic waves by a segment of a microstrip line, Radio Eng. Electron., 1987, Vol. 32(3), p. 465.
152. Gilinskiy I.A., Shcheglov I.M., Excitation and reception of surface magnetostatic waves. A multielectrode problem, Novosibirsk, 1987, p. 43.
153. Dudenko A.V., Glushkov E.V., Vyzulin S.A., Power characteristics of magnetostatic waves excited by a microstrip converter, Krasnodar, 1987, p. 20.
154. Zolotavitskiyj A.B., Excitation of magnetostatic waves in a layered structure by ferrite-semiconductor structural current, Letters to Russ. J. Tech. Phys., 1987, Vol. 13(2), p. 98.
155. Bogdanov M.N., Gusev B.N., Krasnov E.S., Designing of MSW filters with slot converters, Prob. Radio Electron. Ser. Prod. Technol. Equip., 1988, No. (2), p. 23.
156. Dmitriev V.F., Kalinikos B.A., Excitation of extending magnetization waves, Proc. Inst. High. Educ. Phys., 1988, Vol. (11), p. 24.

157. Goldberg L.B., Spatial-frequency characteristic of radiators of return volume magnetostatic waves, Electron. Tech. Ser. 1 UHF Electron., 1988, Vol. 5(409), p. 29.
158. Vugalter G.A., Rogozhina M.B., Gusev B.N., Gurevich A.G., Specific impedance of a converter, Russ. J. Tech. Phys., 1988, No. (4), p. 839.
159. Lysenko V.A., Radtsen Yu.Yu., Input impedance of UHF devices on spin waves: Analysis and synthesis of radio engineering devices and signals, Novgorod, 1987, No. 8852-B-88.
160. Vashkovskiy A.V., Stalmakhov A.V., Excitation of magnetostatic waves by a source of finite sizes, 9th International Conference on Microwave Ferrites: ICMF'88, Esztergom, September 19–23, 1988, Proceedings, Budapest, 1988, p. 162.
161. Zubkov V.I., Kildishev V.N., Excitation of magnetostatic waves in iron–yttrium garnet films by an antenna of variable width, 9th International Conference on Microwave Ferrites: ICMF'88, Esztergom, September 19–23, 1988, Proceedings, Budapest, 1988, p. 181.
162. Martynov B.A., Tretiakov S.A., Specific parameters and input resistance of a strip aerial for surface magnetostatic waves, Leningrad polytechnical institute, 1988, No. 2.
163. Dudenko A.V., Glushkov E.V., Vyzumen S.A., Energy characteristic of waves excited by a microstrip converter, Leningrad polytechnical institute, 1988, No. 3.
164. De D.K., Coop transducer matching for low insertion loss magnetostatic surface wave resonator, J. Appl. Phys., 1988, Vol. 64(10), Part 1, p. 5210.
165. Barak J., Lachish U., Study of the excitation of magnetostatic modes in yttrium–iron-garnet films by a microstrip line, J. Appl. Phys., 1989, Vol. 65(4), p. 1652.
166. Dmitriev V.F., Kalinikos B.A., To self-consistent theory of spin wave excitation by multielement aerials, Russ. J. Tech. Phys., 1989, Vol. 59(1), p. 197.
167. Gilinskiy I.A., Shcheglov I.M., Excitation and reception of surface magnetostatic waves by multielement converters. Part I, Russ. J. Tech. Phys., 1989, Vol. 59(7), p. 66.
168. Gilinskiyj I.A., Shcheglov I.M., Excitation and reception of surface magnetostatic waves by multielement converters. Part 2, Russ. J. Tech. Phys., 1989, Vol. 59(7), p. 74.
169. Goldberg L.B., Exchange oscillations of radiation impedance of surface magnetostatic waves, Radio Eng. Electron., 1989, Vol. 34(4), p. 702.
170. Rogozin V.V., Saledinov S.U., Kharina T.G., Current density distribution over a strip activator of spin waves in a ferrite layer with unilateral metallization, Leningrad polytechnical institute, Leningrad, 1989, p. 15.
171. Bogun P.V., Kandyba P.E., Lavrenev A.A. et al., Research of microstrip converters of magnetostatic waves and their using in devices of functional electronics, Microelectronics and semiconductor devices, Moscow, 1989, No. (10), pp. 75–85.
172. Smekhov M.V., Experimental research of single-mode excitation of magnetostatic waves by a loopback converter, Radio Eng. Electron., 1989, Vol. 34(10), p. 2222.
173. Pankrats A.I., Smyk A.F., Selective excitation of magnetostatic oscillations in a iron–yttrium garnet film, Russ. J. Tech. Phys., 1989, Vol. 59(9), p. 150.
174. Kolodin P.A., Excitation dispersion of spin waves in ferromagnetic film structures with cross non-uniform magnetic parameters, Ph.D. thesis, Leningrad, 1990.
175. Dmitriev V.F., Selective properties of spin-wave devices on the basis of slot and coplanar lines, Radio Eng. Electron., 1990, Vol. 35(9), p. 1821.
176. Zilberman P.E., Shishkin V.G., Excitation efficiency of running exchange spin waves by UHF current in a ferrite film, Radio Eng. Electron., Vol. 35(1), p. 204.
177. Goldberg L.B., Calculation of spatial-frequency characteristics of microstrip radiators of magnetostatic waves, Radio Eng. Electron., Vol. 35(6), p. 1212.
178. Shcheglov N.M., Distribution and excitation of surface magnetostatic wave in layered structures, Ph.D. thesis, Institute of physics, COSiberial Branch of Academy of Sciences USSR. Series 19 "Electronics. Radio engineering. Communication", 1990, No. 3, p. 42.
179. Preobrazhenskiy V.L., Fetisov Yu.K., Magnetostatic waves in a time-varying medium, Proc. Inst. High. Educ. Phys., Tomsk, 1988, Vol. 31(11), p. 54.
180. Makeyeva G.S., An electrodynamic theory of integrated wave-leading structures with wave excitations in thin-film ferrite and semiconductor layers, Radio Eng. Electron., 1989, Vol. 34(6), p. 1184.

181. Beregov A.S., Kudinov E.V., Distribution of magnetostatic waves in screened layered structures, Analysis and engineering realization of functional electronics systems of UHF range, Kiev, 1980, pp. 59–67.
182. Glushchenko A.G., Magnetostatic wave in a layered gyrotropic structure at any orientation of bias field, Proc. Inst. High. Educ. Radio Eng., 1981, Vol. 24(3), p. 351.
183. Kiyohiro Kanasaki, Hirotaks Takagi, Masuyeshi Umeno, Passband control of surface magnetostatic waves by spacing a metal plate apart from the ferrite surface, IEEE Trans. MTT, 1979, p. 924.
184. Vasiliev I.V., Makeyeva G.S., Distribution of magnetostatic waves in a metallized ferrite structure of finite sizes, Radio Eng. Electron., 1984, Vol. 29(3), pp. 419–423.
185. Yesikov O.S., Toloknov N.A., Fetisov Yu.N., Surface magnetostatic waves in a ferrite-dielectric-metal structure of finite sizes, MIFI, Moscow Engineering-Physical Institute, Moscow, 1986, No. 483–81.
186. He Huahui, Shig Xingwa, Beny Zekun, Magnetostatic wave (MSV) dispersion in the multiple layer structures magnetized in an arbitrary direction, Digest Intermag'89: International Magnetic Conference, Hamburg, April 10–13, 1989, New York, 1989, p. 470
187. Beregov A.S., Magnetostatic waves in multilayered structures in view of ferrite film width, Proc. Inst. High. Educ., 1982, Vol. 25(8), p. 36.
188. Samaguti Ya. et al., Characteristics of distribution of surface magnetostatic waves in a goffered plate of yttrium garnet, 1977, Vol. 60(B11), pp. 897–899; WCP No. A-82378, The decree. translations. Series 1, 1980, No. 6, p. 5.
189. Gulyaev Yu.V., Nikitov S.A., Plesskiy V.P., Distribution of magnetostatic waves in a normally-magnetized ferrite plate with periodically rough planes, FTT Russ. Phys. Solids, 1980, Vol. 22(9), p. 2831.
190. Caj Minczi, Seshardi S.R., Factor of reflection and transmission of return waves extending in goffered YIG films, TIIER, 1980, Vol. 68(2), p. 95.
191. Gulyaev Yu.V., Nikitov S.A., Plesskiy V.P., Reflection of surface magnetostatic waves from a periodically rough surface area of ferrite, Russ. Phys. Solids, 1981, Vol. 23(4), p. 1231.
192. Gulyaev Yu.V., Nikitov S.A., Bragg's reflection of a surface magnetostatic wave from a periodic surface area of ferrite at oblique incidence, Russ. Phys. Solids, 1981, Vol. 23(12), p. 3678.
193. Beregov A.S., Diffraction of magnetostatic waves in a structure with a ferromagnetic layer/functional UHF electronics, Kiev, 1984.
194. Vugalter G.A., Makhalin V.N., Reflection and excitation of surface magnetostatic waves by a metal strip, Russ. J. Tech. Phys., 1985, Vol. 55(3), p. 497.
195. Tuan H.S., Parekh J.P., IEEE Trans. Magn., 1987, Vol. 23(5), Part 2, p. 3331.
196. Filippov V.V., Yan O.V., Reflection of a cross elastic and non-uniform magnetostatic wave on border with a ferromagnetic, Report to Academy of Sciences Byeloruss. SSR, 1987, Vol. 31(3), p. 213.
197. Vashkovskiy A.V., Stalmakhov A.V., Shakhnazaryan D.G., Formation, reflection, and refraction of wave beams of magnetostatic waves, Proc. Inst. High. Educ. Phys., 1988, Vol. 31(11), p. 67.
198. Zubkov V.I., Yepanechnikov V.A., Influence of metal surfaces on surface magnetostatic wave spectrum in two-layer ferromagnetic films, 9th Conference on Microwave Ferrites: ICMF'88, Esztergom, September 19–23, 1988, Proceedings, Budapest, 1988, p. 176.
199. Bespyatyhh Yu.I., Dikshteyn I.E., Simonov A.D., Scattering of surface magnetostatic waves by a metal half-plane, 9th Conference on Microwave Ferrites: ICMF'88, Esztergom, September 19–23, 1988, Proceedings, Budapest, 1988, p. 71.
200. Vugalter T.A., Khvosheva N.S., Reflection of magnetostatic waves by a microstrip line segment, Radio Eng. Electron., 1988, Vol. 33(10), p. 2055.
201. Bespyatykh Yu.I. et al., Reflection of surface magnetostatic waves from ferrite film surface, Radio Eng. Electron., 1989, Vol. 34(1), p. 41.
202. Zubkov V.I., Lokk E.G., Shcheglov V.I., Passage of surface magnetostatic waves under a metal strip located above ferrite film surface, Radio Eng. Electron., Vol. 34(7), p. 1381.

203. Bespyatykh Yu.I., Dikshteyn I.E., Simonov A.D., Scattering of magnetostatic waves by an ideally-conducting half-plane, Russ. J. Tech. Phys., 1989, Vol. 59(2), p. 10.
204. Bespyatykh Yu.I., Dikshteyn I.E., Simonov A.D., Distribution of surface magnetostatic waves in a ferrite plate with a metal half-plane, Radio Eng. Electron., 1989, Vol. 59(8), p. 1618.
205. Vugalter G.A., Korovin A.G., Full internal reflection of return volume magnetostatic waves from a metallized segment of a ferrite film, Letters to Russ. J. Tech. Phys., 1989, Vol. 15(21), p. 73.
206. Zubkov V.I., Yepanechnikov V.A., Influence of metal planes on surface magnetostatic wave spectrum in two-layer ferromagnetic films, Russ. J. Tech. Phys., 1989, Vol. 59(9), p. 53.
207. Zavislyak I.V., Pismennyi A.Yu., Perpendicular incidence of magnetostatic waves on an interface, Prog. Phys. Sci., 1990, Vol. 35(1), p. 98.
208. Bogatko A.V., Vlaskin S.V., Diffraction of spin waves in an aslantly-magnetized ferromagnetic layer, Electron. Eng. Ser. UHF Electron., 1990, Vol. 2(426), p. 27.
209. Vashkovskiy A.V., Gerus S.V., Kharitotnov V.D., Spectrum of surface magnetostatic waves extending in a spatially-periodic magnetic field, Proceedings of VI International Conference on Gyromagnetic Electronics and Electrodynamics, Varna, October 3–10, 1982, Sofia, 1982, p. 450.
210. Vasiliev I.V., Spectrum of characteristic waves of a magnetostastic waveguide with stepwise-inhomogeneous bias, Radio Eng. Electron., 1984, Vol. 24(5), p. 908.
211. Bespyatykh Yu.I., Kharitonov V.D., Scattering of surface magnetostastic waves in a ferromagnetic on a magnetization heterogeneity, Russ. Phys. Solids, 1985, Vol. 27(10), p. 3132.
212. Kamenetskiy E.O., Filtration of spin waves in ferrite films with a periodic internal magnetic field, Electron. Eng. Ser. UHF Electron., 1986, Vol. 3(387), p. 12.
213. Golovach G.P., Zavkelian I.V., Dispersion of magnetostastic waves in a non-uniform ferrite layer: Forward and return tasks, Abstract of Regional Conference on Spin Phenomena of UHF Electronics, Krasnodar, 1987, p. 13.
214. Goldberg L.B., Penziakov V.V., Dispersive characteristic of surface magnetostastic waves in periodically irregular ferromagnetic films, Radio Eng. Electron., 1988, Vol. 33(6), p. 1232.
215. Myasoyedov A.N., Fetisov Yu.K., Scattering of volume magnetostastic waves on a dynamic magnetic lattice, Russ. J. Tech. Phys., 1989, Vol. 59(6), p. 133.
216. Zilberman P.E., Umanskiy A.V., Magnetostatic waves in a periodically non-uniform ferrite plate, Radio Eng. Electron., 1990, Vol. 35(8), p. 1628.
217. Burlak G.N., Kotsarenko N.Ya. et al., Magnetostatic waves in films with cross non-uniform magnetization, Proc. Inst. High. Educ. Radio Eng., 1990, Vol. 33(2), p. 74.
218. Burlak G.N., Magnetostatic waves in ferromagnetic films at a non-uniform magnetic field, Letters to Russ. J. Tech. Phys., 1986, Vol. 12(24), p. 1476.
219. Burlak G.N., Grimalskiy V.V., Kotsarenko N.Ya., Magnetostatic waves in ferromagnetic films in a non-uniform field, Russ. J. Tech. Phys., 1989, Vol. 59(8), p. 32.
220. Korovkin V.Yu., Amplitude–frequence characteristics of an MSW path in a non-uniform magnetic field, Russ. J. Tech. Phys., 1989, Vol. 59(1), p. 166.
221. Vashkovskiy A.V., Zubkov V.I., Lokk E.G., Shcheglov V.I., Influence of heterogeneity of a constant magnetic field on surface magnetostatic wave trajectory, Letters to Russ. J. Tech. Phys., 1989, Vol. 15(4), p. 1.
222. Zubkov V.I., Lokk E.G. et al., Distribution of surface magnetostatic waves in a non-uniform constant magnetic field with a shaft-like structure, Radio Eng. Electron., 1990, Vol. 35(8), p. 1617.
223. Prokushkin V.N., Influence of specific conductivity of loading on dispersive characteristics of magnetostatic waves, Proc. Inst. High Educ. Radio Electron., 1989, Vol. 32(7), p. 73.
224. Lutsev L.V., Berezin I.L., Thermal stability of parameters of magnetostatic waves extending in films with an arbitrary bias direction, Electron. Eng. Ser. UHF Electron., 1989, Vol. 6(420), p. 3.
225. Lebed B.M., Novikov G.M., Popov S.N., Temperature stabilization of devices on spin waves, Electron. Eng. Ser. Quality Manag. Standard. Metrol. Test., 1989, Vol. 5(137), p. 19.

226. Vyzumin S.A., Rozenson A.E., Shekh S.A., Temperature factor of frequency of magnetostatic waves at oblique magnetization, Radio Eng. Electron., 1990, Vol. 35(1), p. 202.
227. Kalinikos B.A., Spectrum and linear excitation of spin waves in ferromagnetic spots, Proc. Inst. High. Educ. Phys., 1980, No. (8), p. 42.
228. Bugaev A.S., Gulyaev Yu.V., Zilberman P.E., Filimonov Ya.A., Magnetoelastic interaction in thin ferromagnetic layers, XV All-Union Conference on the Physics of Magnetic Phenomena: Abstract, Perm, 1981, No. (21), p. 65.
229. Gulyaev Yu.V., Zilberman P.E., Lugovskoj A.V., Mednikov A.I., Interaction of dipoles and exchange magnetostatic waves in thin ferromagnetic layers, XV All-Union Conference on the Physics of Magnetic Phenomena: Abstract, Perm, 1981, Part 4, p. 133.
230. Burtyka M.V., Yakovenko V.M., Bound magnetoplasma and sound waves on semiconductor-piezodielectric interface, Russ. Phys. Solids, 1981, Vol. 23(1), p. 318.
231. Mednikov A.M., Nikitov S.A., Popkov A.F., Scattering of volume magnetostatic waves on a surface acoustic wave, Russ. Phys. Solids, 1982, Vol. 24(10), p. 3008.
232. Mednikov A.M., Popkov A.F., Modulation of spin waves in a YIG film by a volume acoustic wave, Letters to Russ. J. Tech. Phys., 1983, Vol. 9(8), p. 485.
233. Andreyev A.S., Zilberman P.E., Kravchenko V.B. et al., Effects of interaction magnetostatic and elastic waves in structures with a tangentially magnetized film of submicronic iron–yttrium garnet, Letters to Russ. J. Tech. Phys., 1989, Vol. 10(2), p. 90.
234. Gulyaev Yu.V., Andreyev A.S., Zilberman P.E. et al., Magnetoelastic effects in tangentially-magnetized iron–yttrium garnet films, Radio Eng. Electron., 1985, No. (10), p. 1992.
235. Kalinikos B.A., Dipole-exchange spin waves in ferromagnetic films, Doctoral thesis, Leningrad, 1985.
236. Anfinogenov V.B., Verbitskaya T.N., Kazakov G.T. et al., Distribution of magnetostatic waves in a ferrite-ferroelectric structure, Letters in Russ. J. Tech. Phys., 1986, Vol. 12(8), p. 454.
237. Filippov V.V., Yan O.V., Resonant interaction of volume and surface acoustic waves with a bound magnetostatic wave extending along the gap between ferromagnetics, Russ. J. Tech. Phys., 1988, Vol. 58(8), p. 1617.
238. Kryshpan R.G., Medved L.V., Osipenko V.A., Transformation of modes of magnetostatic waves at their scattering on a surface acoustic wave in a YIG film, Russ. J. Tech. Phys., Vol. 58(12), p. 2315.
239. Mitsay J.N., Freedman Ya.A., Magnetoelastic waves in strongly anisotropic ferromagnetics, Prog. Phys. Sci., 1990, Vol. 35(3), p. 459.
240. Anfinogenov V.B. et al., Hybrid electromagnetic-spin waves in contacting layers of a ferroelectric and ferrite. II. Experiment, Radio Eng. Electron., 1990, Vol. 35(2), p. 320.
241. Beregov A.S., Magnetostatic waves in a tangentially magnetized layer of a cubic ferromagnetic in view of monoaxial anisotropy, Kiev, 1984, pp. 30–38.
242. Belyakov S.V., Kalinikos B.A., Kozhus N.V., Dispersion of dipole-exchange spin waves in anisotropic monocrystal magnetic films. Part I. Spectrum of dipole-exchange waves, Electron. Eng. Ser. UHF Electron., 1989, Vol. 1(415), p. 22.
243. Zavislyak I.V., Talalaevskiy V.M., Chevnyuk L.V., Feature of spectra of magnetostatic waves due to anisotropy, Russ. Phys. Solids, 1989, Vol. 31(5), p. 319.
244. Surin V.V., Shevchenko Yu.A., Nonlinear phenomena in ferrite UHF delay lines, Radio Eng. Electron., 1972, Vol. 17(11), p. 2065.
245. Mednikov A.M., Nonlinear effects at distribution of surface spin waves in YIG films, Russ. Phys. Solids, 1981, Vol. 23(1), p. 242.
246. Zvezdin A.K., Popkov A.F., To the nonlinear theory of magnetostatic spin waves, Russ. J. Exp. Theor. Phys., 1983, Vol. 84(2), p. 606.
247. Chubukov A.V., A nonlinear theory of spin waves for ferromagnetic chains with monoaxial easy-plane anisotropy, Cryogenics, 1989, Vol. 10(11), p. 1166.
248. Melkov G.A., Nonlinear interactions of spin waves in ferrites, Phys. Multifreq. Syst., 1984, No. (6), p. 62.

249. Gulyaev Yu.V., Nikitov S.A., Three-magnon interaction of magnetostatic waves, 7th International Conference on Microwave Ferrites, Smolenice, September 17–22, 1984, Conference on Transactions, Bratislava, 1984. S. A. 46.
250. Zvezdin A.K., Kostyuchenko V.V., Popkov A.F., Nonlinear spin waves extending along domain border, Russ. Phys. Solids, 1985, Vol. 27(10), p. 2936.
251. Zilberman P.E., Nikitov S.A., Temiriazev A.G., Four-magnon disintegration and kinetic instability of running magnetostatic waves in iron–yttrium garnet films, Letters to Russ. J. Exp. Theor. Phys., 1985, Vol. 42(3), p. 92.
252. Genkin G.M., Golubeva N.G., Nonlinear properties of surface magnetostatic waves, Proc. Inst. High. Educ. Radiophys., 1985, Vol. 28(3), p. 387.
253. Gulyaev Yu.V., Zilberman P.E. et al., Nonlinear effects at distribution of magnetostatic waves in normally-magnetized thin iron–yttrium garnet films, Russ. Phys. Solids, 1986, Vol. 28(9), p. 2774.
254. Gurzo V.V., Prokushkin V.N., Reykhel V.V., Sharaevskiy Yu.P., Characteristics of an attenuator with dynamic nonlinearity on surface magnetostatic waves, Proc. Inst. High. Educ. Radio Electron., 1986, No. (4), p. 7.
255. Boardman A.D., Gulyaev Yu.V., Nikitov S.A., Jpn. J. Phys., 1988, Vol. 27(12), Part 2, pp. 12438–12441.
256. Pylaev E.S., Distributed excitation of a running magnetostatic wave at nonlinear scattering of electromagnetic radiation, Radio Eng. Electron., 1988, Vol. 33(12), p. 2492.
257. Chivileva O.A., Anisimov A.N., Gurevich A.G., Excitation and reception of surface magnetostatic waves behind nonlinearity threshold, Russ. J. Tech. Phys., 1988, Vol. 58(6), p. 1204.
258. Izyumov Yu.A., Solitons in quasi one-dimensional magnetics and their research by neutron scattering, Prog. Phys. Sci., 1988, Vol. 155(4), p. 553.
259. Bordman A.D., Gulyaev Yu.V., Nikitov S.A., Nonlinear surface magnetostatic waves, Russ. J. Exp. Theor. Phys., 1989, Vol. 91(6), p. 2140.
260. Dudko G.M., Filimonov Ju.A., Development of modulation instability of magnetostatic waves in ferrite films, Letters to Russ. J. Exp. Theor. Phys., 1989, Vol. 15(2), p. 55.
261. Bordman A.D., Nikitov S.A., Love's nonlinear magnetoelastic waves, Russ. Phys. Solids, 1989, Vol. 31(4), p. 143.
262. Pat. Appl. 6451801, MKI HOI P/23. 24.08.87. Published 28.02.89.
263. Wenzhong Hu., Effects of various parameters on magnetostatic surface waves, Proceedings of International Conference on Electronic Components and Materials (ICECM), Beining, China, November 7–10, 1989, pp. 432–435.
264. Vashchenko V.I., Zavisliak I.V., Three-wave interaction of magnetostatic waves, Proc. Inst. High. Educ. Radiophys., 1989, Vol. 32(1), p. 41.
265. Adamashvili T.T., Nonlinear waves in ferromagnetics, Theor. Math. Phys., 1989, Vol. 80(3), p. 461.
266. Anisimov A.N., Chivileva O.A., Gurevich A.G., Influence of a high-amplitude wave on attenuation of a weak surface magnetostatic wave, Russ. Phys. Solids, 1990, Vol. 32(6), p. 1622.
267. Taranenko A.Yu., Kinetic instability of spin waves in ferrites, Ph.D. thesis, Kiev, 1990.
268. Krutsenko I.V., Melkov T.A., Parametrical excitation of a second group of spin waves in ferrites, Russ. Phys. Solids, 1979, Vol. 21(1), p. 271.
269. Zhibnyuk V.S., Processes of relaxation at parametrical excitation of spin waves in ferrites, Bulletin of Kiev university. Physics, Kiev, 1980, No. 21, p. 83.
270. Zilberman P.E. et al., Parametrical excitation of short exchange spin waves in tangentially magnetized films of iron–yttrium garnet in a non-uniform UHF field, Letters in Russ. J. Tech. Phys., 1988, Vol. 14(7), p. 585.
271. Melkov G.A., Parametrical excitation of spin waves by a surface magnetostatic wave, Russ. Phys. Solids, 1988, Vol. 30(8), p. 2533.
272. Vugalter G.A., Threshold of parametrical instability at excitation of surface magnetostatic waves in a ferrite film, Russ. J. Exp. Theor. Phys., 1990, Vol. 97(6), p. 1901.

273. Zilberman P.E., Golubev N.S., Temiriazev A.G., Parametrical excitation of spin waves in tangentially magnetized films of iron–yttrium garnet by spatially-localized pumping, Russ. J. Exp. Theor. Phys., Vol. 97(2), p. 634.
274. Radmanesh M., Chiao-Min Chu, Haddad G., Magnetostatic wave propagation in a finite YIG-loaded rectangular waveguide, IEEE Trans. MTT, 1986, Vol. 34(12), p. 1377.
275. Kosiba Ji Long, Kosiba Masanor, IEEE Trans. MTT, 1989, Vol. 37(4), p. 680.
276. Ivanov V.N., Shchuchinskiy A.G., Electrodynamic analysis of waves in a tangentially magnetized ferrite layer, All-Union Scientific and Technical Conference on the Problems of Integrated UHF Electronics: Abstract, Leningrad, 1984, p. 126.
277. Shchuchinskiy A.G., Surface waves in a tangentially magnetized ferrite layer, Radio Eng. Electron., 1984, Vol. 29(9), p. 1700.
278. Ivanov V.N., Demchenko N.P., Nefyodov I.S. et al., Wave in a tangentially magnetized ferrite layer: (Electrodynamic calculation and uniform asymptotic forms), Proc. Inst. High. Educ. Radiophys, 1989, Vol. 32(6), p. 764.
279. Gerson T., Nadan J., Surface electromagnetic modes of a ferrite plate, IEEE Trans. MTT, 1974, Vol. 22, p. 757.
280. Deil F., Surface waves in normally-magnetized ferrite, IEEE Trans. MTT, 1974, Vol. 22, p. 743.
281. Bardati F., Samparielo P., The modal spectrum of a sossy ferrimagnetic slab, IEEE Trans. MTT, 1979, Vol. 7(7), p. 679.
282. Albuquerque M.R.M.L., D'Assuncao A.G., Maia Marcio R.G., Criarola Attilio, J. IEEE Trans. Magn., 1989, Vol. 25(4), p. 2944.
283. Rother D., Beyer A., IEEE MTT, International Microwave Symposium Digest, New York, 1988, Vol. 2, p. 761.
284. Golovko A.D., Zavislyak I.V. et al., Analysis of energy fluxes of magnetostatic waves in a ferromagnetic layer in view of delay effects, Russ. J. Tech. Phys., 1990, Vol. 60(5), p. 150.
285. Morgentaler F.R., Electromagnetic and spin waves in ferrite media, TIIER, 1988, Vol. 76(2), p. 50.
286. Gromova L.I., Kurushin E.P., Diffraction of waves on a metal half-plane in a flat waveguide with cross-magnetized ferrite plate, 16 All-Union Seminars on Gyromagnetic Electronics and Electrodynamics, Kuybyshev, 1990, pp. 29–30.
287. Nikolsky V.V., Measurement of ferrite parameters on ultrahigh frequencies, Radio Eng. Electron., 1956, Vol. 1(4), p. 447; Vol. 1(5), p. 638.
288. Nikolsky V.V., Calculation of phase shifts of gyrotropic imperfections in a waveguide by perturbation method, Radio Eng. Electron., 1957, Vol. 7(2), p. 833.
289. Gurevich A.G., Bogomaz N.A., Nonmutual phase shifts and attenuation factor in a waveguide with a ferrite plate, Radio Eng. Electron., 1958, Vol. 7(9), p. 1133.
290. Gusev B.N., Chivileva O.A., Gurevich A.G. et al., Attenuation of a surface magnetostatic wave, Letters to Russ. J. Tech. Phys., 1983, Vol. 9(3), p. 159.
291. Petrov V.V., Galaktionova G.M., Bushueva T.N., Ferromagnetic resonance in real crystals of ferrite-garnets, Electron. Eng. Ser. 6, 1984, Vol. 4(189), p. 23.
292. Chivileva O.A., Emirian L.M., Gusev N.B. et al., Temperature dependences of magnetostatic wave attenuation, Russ. Phys. Solids, 1985, Vol. 27(2), p. 534.
293. Bogun P.V., Lavrenov A.A., Sorokin V.G., Measurement of distribution losses of magnetostatic surface waves, Electron. Eng. Ser. 6 Mater., 1985, Vol. 2(201), p. 43.
294. Belyakov S.V., Gorodaykina O.A., Temperature dependence of ferromagnetic resonance frequency of a ferrite ellipsoid with cubic crystallographic anisotropy, Electron. Eng. Ser. UHF Electron., 1986, Vol. 7(391), p. 28.
295. Nikitov S.A., Attenuation of magnetostatic waves in a normally-magnetized film of ferromagnetic caused by spin–phonon interaction, Russ. J. Tech. Phys., 1988, Vol. 58(8), p. 1576.
296. Nikitov S.A., Gavrilin S.N., Dispersion of surface magnetostatic waves at distribution on a rough surface, Radio Eng. Electron., 1979, Vol. 30(11), p. 2415.

297. Shichang Zhou, Shaoping Li, Yibing Li, Huahui He, A study on ferromagnetic resonance (FRM) for (la, Ga): YIG single crystal films, 9th International Conference on Microwave Ferrites: ICMF'88, Esztergom, September 19–23, 1988, Proceedings. Budapest, 1988, p. 50.
298. Afsar M.N., Button K.J., Multimeter wave ferromagnetic resonance in cuble and hexagonal ferrites, IEEE MTT Int. Microw. Symp. Dig. New York, May 25–27, 1988/New York, 1988, Vol. 1, p. 1211.
299. Lutsev L.V., Berezin I.L., Yakovlev Yu.M., Spin-wave resonance in films with a linear magnetization profile, Electron. Eng. Ser. UHF Electron., 1988, Vol. 5(419), p. 5.
300. Kudryashkin I.G., Krutogin D.G., Ladygin E.A. et al., Ionic implantation of iron–yttrium garnet films and its influence on distribution of surface magnetostatic waves, Russ. J. Tech. Phys., 1989, Vol. 59(3), p. 70.
301. Balinskiy M.G., Characteristics of attenuation of magnetostatic waves in small saturating fields, Electron. Eng. Ser. 1 UHF Electron., 1989, Vol. 10(424), p. 11.
302. Pomelov A.V., Gorskiy V.B., Sorokin V.G., Measurement of attenuation parameter of MSW by wave and resonant methods, Electron. Eng. Ser. 1 UHF Electron., 1989, Vol. 7(421), p. 43.
303. Slavin A.M., Zinovik M.A., Influence of substrate material on properties of ferrite samples; Exchange of technological experience, Moscow, 1989, No. 7, p. 77.
304. Belyakov S.V., UHF parameters of a perpendicularly-magnetized anisotropic ferrite disk, Electron. Eng. Ser. 1 UHF Electron., 1989, Vol. 3(417), p. 24.
305. Zotov N.I., Kozhukhar A.J., Poloneychik I.I., Temperature dependence of parameters of spectra of spin waves of iron–yttrium garnet films, Electron. Eng. Ser. 6 Mater., 1988, Vol. 2(237), p. 69.
306. Silantiev N.N., Measurement technique of parameters of monocrystal ferrite-garnets, Moscow, 1989.
307. Author's certificate 1508179 USSR, MKI G 0IR 33/04. A way of determination of parameters of ferrite-garnet films, Bury Yu.A., Pronina N.V., Simferopol state university, Filed 11.02.1987. Published 15.09.1989, Discovery. Invention, 1989, No. 34.
308. Gurevich A.G., Chivileva O.A., To determination of dissipation parameter of running spin waves, Letters to Russ. J. Tech. Phys., 1989, Vol. 15(11), p. 7.
309. Nikitov S.A., To the theory of relaxation of magnetostatic waves in ferromagnetic films, Letters to Russ. J. Tech. Phys., 1990, Vol. 16(5), p. 30.
310. Author's certificate 1539698 USSR, MKI G0IR 33/05. A way of local measurement of saturation magnetization of a ferrite film, Gorskiy V.B., Pomyalov A.V., Institute of radio engineering and electronics, AS of USSR, Filed 23.01.1988, Published 30.01.1990, Discovery. Invention, 1990, No. 4.
311. Beregov A.S. et al., Determination of magnetization and oscillations of anisotropy of epitaxial YIG films with orientation (III), Proc. Inst. High Educ. Radio Electron., 1990, Vol. 33(3), p. 66.
312. Yakovlev Yu.M., Noskov Yu.N., Measurement of ferromagnetic resonance parameters of ferrite monocrystals, Electron Eng. Ser. Ferrite Eng., 1970, No. (3), p. 87.
313. Lebed B.M., Lavrovich V.A., Khokhlyshev I.O., Ferrite filters and their application in devices with magnetic reorganization of frequency, Electron. Eng. Ser. 1 UHF Electron., 1982, Vol. 10(914), p. 87.
314. Research of an opportunity to create an installation for studying parameters of high anisotropic ferrite materials by methods of magnetic resonance, AS of USSR, Vladivostok, 1989, p. 138.
315. Kornilov A.K., Soloukhin N.G., Shelukhin I.V., An installation for automated local measurement of distribution characteristics of surface magnetostatic waves in epitaxial ferrite-garnet structures, Electron. Eng. Ser. Quality Manag. Standard. Metrol. Tests, 1989, Vol. 5(137), p. 47.
316. Kozhuhar A.Yu., Radiospectroscopy of epitaxial ferrite-garnet films. Part 1. Research methods, Electron. Eng. Ser. 6 Mater., 1988, Vol. 5(234), p. 3.
317. Isvenko L.A., Method and equipment for determination of parameters of ferrite-garnet films, Electron. Eng. Ser. 6 Mater., 1988, Vol. 1(246), p. 67.

318. Vashkovskiy A.V. et al., A delay line on a photosensitive planar ferrite-semiconductor structure, Letters to Russ. J. Tech. Phys., 1978, Vol. 4(20), p. 1231.
319. Courtois L., Mahicu J.P., Magnetostatic surface wave group delay egnalizer, 8th European Microwave Conference 78, Paris, 1978, Conference on Proceedings, Sevenoaks, 1978, p. 210.
320. Vashkovskiy A.V., Zubkov V.I., Tsurukian T.M., Input resistance of ferrite delay lines in conditions of nonlinear ferromagnetic resonance, Radio Eng. Electron., 1979, No. (11), p. 2290.
321. Daniel A.B., Adam J.D., O'Keefe T.M., Linearly dispersive delay lines at microwave frequencies using magnetostatic waves, Ultrasonics Symposium Proceedings, New Orleans, LA, 1979, New York, 1979, p. 806.
322. Kudinov E.V., Pulse characteristics of a delay line on the basis of magnetostatic waves, Methods of functional electronics in design of radio engineering devices, Kiev, 1982, pp. 19–20.
323. Makoto T., Yoshiniko M., Takoshi O., Nobuaki K., A new technique for magnetostatic wave delay lines, Ferrites Proceedings ICF3, Kysto, September 29–October 2, Tokyo, 1980, Dordrecht, 1982, p. 847.
324. Masaoka Yoshiniko, Tsutsumi Makoto, Ohira Iakaski, Magnetostatic wave delay lines using an inhemogeneously magnetized YIG slab, Trans. Inst. Electron. Commun. Eng. Jpn., 1980, Vol. 65(11), p. 1117.
325. Ursulyak N.D., Rudy Yu.B., Volobuev N.M. et al., Research of a delay line on magnetostatic waves, Electron. Eng. Ser. UHF Electron., 1982, No. (2), p. 12.
326. Belitskiy A.V., Miasnikov A.V., Nadeyev M.M. et al., Experimental research of delay lines and a phase shifter on magnetostatic waves, Electron. Eng. Ser. UHF Electron., 1984, Vol. 3(363), p. 19.
327. Volluet G., Unidirectional magnetostatic forward volume wave transducers, Digest Intermagnetic Conference, Boston, MA, 1980, New York, 1980, p. 36/5.
328. Reed H.W., Owens Y.M., Smith C.V., Carter R.L., Simple magnetostatic delay lines in microwave: Pulse compression loops, III MTT-S International Microwave Symposium Digest, New York, 1980, p. 40.
329. Tsutsumi Nakoto, Masaoka Yoshiniko, Chira Takashi, Humagai Nobuaki, A new technique for magnetostatic wave delay lines, IEEE Trans. Microw. Theor. Technol., 1981, Vol. 29(6), Part 1, p. 588.
330. Adam Y.D., Daniel Michael R., Hothwick C.E., MCW dispersive delay lines in a compressive receiver, Ultrasonics Symposium Proceedings, San Diego, CA, October 27–29, 1982, Vol. 1, New York, 1982, p. 533.
331. Lysenko V.A., Coherent processing of UHF radio signals by devices on magnetostatic waves, Problems of synthesis and processing of signals in information systems, Novgorod, 1983, p. 118–128.
332. Chang Kok Wai, Owens J.M., Carter R.L., Linearly dispersive time-delay control of magnetostatic surface wave variable ground-spacing, Electron. Lett., 1983, No. (14), p. 546.
333. Ogrin Yu.F., Lugovskoy A.V., Exchange oscillations of delay time of magnetostatic wave (MSW) pulses in thin films of iron–yttrium garnet (YIG), Letters to Russ. J. Tech. Phys., 1983, Vol. 9(7), p. 421.
334. Kalinikos B.A., Kovshikov N.G., Kozhus N.V., An approximate method of calculation of delay lines on spin waves, Radio Eng., 1983, Vol. 26(10), p. 77.
335. Baraulin I.A., Bezmaternykh A.N., Temerov V.L., Magnetoelastic and magnetostatic delay lines of decimeter range, Krasnoyarsk, 1984, p. 16.
336. Smelov M.V., Kirsanov J.A., Denisov D.S., Makarova N.Ya. et al., Experimental research of delay lines on magnetostatic waves, Radiophysics and research of substance properties, Omsk, 1984, p. 26.
337. Smelov M.V., Kirsanov J.A., Denisov D.S., Makarova N.Ja., Experimental research of delay lines on magnetostatic waves, Radio Eng., 1985, No. (77), p. 121.
338. Adkins L.R., Giass H.L., Jin K.K. et al., Electronically varyable time delays using magnetostatic wave technology, Microw. J., 1986, Vol. 29(3), p. 109.

339. De Gasperis P., Miccoly G., Non-dispersive delay line at microwave frequency by magnetostatic forward volume waves, Electron. Lett., 1986, No. (20), p. 1065.
340. Shcheglov V.I., Research of spectral and time characteristics of a ferrite delay line at increased temperatures, Radio Eng. Electron., 1986, Vol. 31(4), p. 84.
341. Willems D.A., Owens I.M., A Hu band MSW delay line, IEEE-MTT: 5th International Microwave Symposium, Baltimore, MD, June 2–4, 1986, p. 477.
342. Korovin V.Yu., Levin M.D., Marekhin A.V., Obtaining a long delay of magnetostatic waves in thin ferromagnetic films, Radio Eng., 1987, No. (7), p. 65.
343. Balinskiy M.T., Yereshchenko I.N., Kozyr I.N., Mongolov B.D., Minimization of losses in MSW lines with a set phase delay, Devices of integrated and functional UHF electrons, Kiev, 1988, p. 799.
344. Bajapai S., Carter R.L., Owens J.M., Insertion loss of magnetostatic surface wave delay lines, IEEE Trans. Microw. Theor. Technol., 1988, Vol. 36(1), pp. 132–136.
345. Karmanov M.V., Kravtsov I.A., Surguchev I.A., A spin resonator as a delay line on the basis of a symmetric microstrip course, Prob. Radio Electron. Ser. Prod. Eng. Process. (PEP), 1988, No. (1), p. 3.
346. Feng Zekun, Su Yun, Zhou Shichang, He Huahui. Study of magnetostatic forward volume wave (MSFVW) delay lines in the multiple layers, 9th International Conference on Microwave Ferrites: ICMF'88. Esztergom, September 19–23, 1988, Proceedings, Budapest, 1988, p. 286.
347. Popina S.M., Dubovitskiy S.A., Simanchuk B.P., Tutchenko A.A., Development of warranty dispersive delay lines on magnetostatic waves in a range 8 ... 12 GHz, Ferrite UHF devices and materials: XV All-Union Scientific and Technical Conference on UHG Ferrite Engineering "Spin (Magnetostatic) Waves, Nonlinear Processes. UHF Devices on Magnetostatic Oscillations and Waves", Leningrad, 1990, p. 118.
348. USA patent 3913039, MKI HOIP 1/20, 3/08, 7/00. A ferrite filter of a high power level on the basis of iron–yttrium garnet. Filed 21.08.1974, Published 14.10.1975. The United States of America as represented by the Secretary of the Army.
349. Koine Takur, Tuneable bandpass filters utilizing the magnetostatic wave propagation, Ultrasonics Symposium Proceedings, Cherry Hill, New York, 1978, p. 689.
350. Koike Takuro, Tunable low-loss microwave filters up to X-band utilizing magnetostatic surface wave propagation, Ultrasonics Symposium Proceedings, New Orleans, LA, 1970, New York, 1979, p. 810.
351. Sasaki H., Mikoshiba N., Tunable magnetostatic surface wave demiltiplexing, filter/switch, Electron. Lett., 1980, No. (16/18), p. 700.
352. Carter R.L., Owens J.M., Smith C.V., Reed K.W., Ion-implanted magnetostatic wave reflective array filters, J. Appl. Phys., 1982, Vol. 53(3), Part 11, p. 2655.
353. Morgenthaler F.R., Novel devices based upon filed gradient control of magnetostatic modes and waves, Ferrites: Proceedings of ICF3, Kyoto, September–October, 1980, Tokyo, Dordrecht, 1982, p. 839.
354. Chang N.S., Microwave tunable filter using ultra-thin magnetic film, IEEE Trans. Magn., 1982, Vol. 18(6), p. 1604.
355. Castera J.P., Hartemann P., A multipole magnetostatic volume wave resonator filter, IEEE Trans. Magn., 1982, Vol. 18(6), p. 1601.
356. Williams D.F., Schwarz S.E., Design and performance of coplonar waveguide bandpass filters, IEEE Trans. Microw. Theor. Technol., 1983, Vol. 31(7), p. 558.
357. Reed K.W., Owens J.M., Carter R.L., Smith C.V., An oblique incidence ion implanted MSFVM RAF with linear group delay, IEEE MTT-S International Microwave Symposium Digest, Boston, MA, May 31–June 3, 1983, New-York, 1983, p. 256.
358. Schloemann E., Blight R.E., YIG-filter recovery after exposure to high power and X-band frequency-stepped YIG-filter, IEEE MTT-S International Microwave Symposium Digest, Boston, MA, May 31–June 3, 1983, New-York, 1983, p. 329.
359. Babenko V.E., Shabunin V.M., Calculation of a reconstructed UHF filter on the basis of MSW, Physical phenomena in devices of electronic and laser engineering: Collected Book, Moscow, 1985, p. 124.

360. Iliechenko M.E., Designing of UHF solid-state filters, Electron. Eng. Ser. 1 UHF Electron., 1986, No. (1), p. 13.
361. Gulyaev Yu.V., Ignatiev I.A., Babenko B.E. et al., Electric management of amplitude–frequency characteristics of a band-elimination filter on MSW, Letters to Russ. J. Tech. Phys., 1986, Vol. 12(8), p. 473.
362. Nathin I.I., Babichev R.K., Ivanov V.N. et al., Research of tunable low-and-high-pass filters on magnetostatic waves, Electron. Eng. Ser. 1 UHF Electron., 1986, Vol. 9(393), p. 13.
363. Osipov N.V., Shvedov O.C., Ezhov A.V. et al., Features of designing and experimental research of a 4-resonator ferrite filter, Electron. Eng. Ser. 1 UHF Electron., 1987, Vol. 8(402), p. 54.
364. Ohgihara T., Murakemi Y., Okamoto T., A 0.5–2.0 GHz tunable bandpass filter YIG film grown by LPE, IEEE Trans. Magn., 1987, Vol. 23(5), p. 3745.
365. Murakami Y., Takahiro Cl., Okamoto T., A 0.5–4.0, 0-GHz tunable bandpass filter using YIG film grown by LPE, IEEE Trans. Microw. Theor. Technol., 1987, Vol. 35(12), p. 1192.
366. Chang N.S., Erkin S., Characteristics of a high q filter composed of a magnetic thin film layered structure with periodic corrigation, IEEE Trans. Magn., 1987, Vol. 23(5), p. 3337.
367. Bogdanov G.B., Basics of the theory and applications of ferrites in measurement and control, Moscow, 1967.
368. Author's certificate 1418830 USSR, MKI HOIP 1/215. The frequency-selective device on spin waves, Kobychenkov A.F., Shavrov V.G., Filed 04.06.1986, Published 23.08.1988, Discovery. Invention, 1988, No. 31.
369. Development of a low-and-high-transmitting reconstructed ferrite filter in a microstrip implementation for reception modules of onboard equipment: A report on development, "Electronics Institute", Moscow, 1988, 33 p.
370. USA patent 4746884, MKI HOIP 7/00. A YIG thin film microwave apparatus, Murakami, Yoshikazu, Tanaka Hideo, Sony Corp. № 883603, Filed 09.07.1986. Published 24.05.1988.
371. Stitzer S.N.A microwave circuit model for a magnetostatic wave filter, IEEE MTT International Microwave Symposium Digest, New York, May 25, 1988, Vol. 2, p. 875.
372. Hanna S.M., IEEE Ultrasonics Symposium, Chicago, IL, October 2–5, 1988, Proceedings, Pittsburgh, PA, 1988, Vol. 1/2, p. 241.
373. Author's certificate 1885167 USSR, MKI H03H. A filter on magnetostatic waves, Vlaskin S.V., Dubovitsky S.A., Novikov G.M. et al., Application No. 4115062/24-09 (22), Filed 04.06.1986, Published 08.03.1988, Discovery. Invention, 1988, No. 12.
374. Andrusevich L.K., Ruchkan L.N., Berger M.N. et al., A quick-tunable UHF filter, Electron. Eng. Ser. UHF Electron., 1989, Vol. 8(422), p. 26.
375. Nishikawa Toshio, Wakino Kikuo, Tanaka Hiroaki et al., A low-loss magnetostatic wave filter using parallel strip transducer, IEEE MTT-S International Microwave Symposium Digest, Long Beach, CA, June 13–15, 1989, New York, 1989, Vol. 1, p. 153.
376. Vlasenko S.V. et al., Reconstructed low-and-high-transmitting filters on spin waves, Electron. Eng. Ser. 1 UHF Electron., 1990, Vol. 1(425), p. 8.
377. Author's certificate 1571752 USSR, MKI HO3H 9/30, HOIP 1/218. An operated UHF filter, Balinskiy M.T., Beregov A.S., Yereshchenko I.N. et al., Kiev polytechnical institute, Filed 20.07.1987, Published 15.06.1990, Discovery. Invention, 1990, No. 22.
378. Afanasiev A.I., Rudy Yu.B., Selection of acoustic modes by MSW devices, Electron. Eng. Ser. 1 UHF Electron., 1990, Vol. 5(429), p. 9.
379. High-speed filters of Om niyig Inc., Microw. RF, 1990, Vol. 29(65), p. 60.
380. Author's certificate 1538198 USSR, MKI HOIP 1/218. A filter on magnetostatic waves, Aliev T.D., Zubkov V.I., Lokk E.G. et al., Physicotechnical institute of Academy of Sciences of TSSR Tatarstan, Application No. 4400955/24-09, Filed 31.03.1988, Published 23.01.1990, Discovery. Invention, 1990, No. 3.
381. Adam J.D., Patterson R.W., Interdigital structures for magnetostatic wave transversal filter, IEEE International Symposium on Circuits and Systems Proceedings, New York, 1978, p. 569.
382. Daniel M.R., Adam Y.D., Compact magnetostatic wave channelizer, IEEE-MTT'S International Microwave Symposium, Baltimore, MD, June 2–4, 1986, New York, 1986, p. 481.

383. Author's certificate 1278997 USSR, MKI HOIP 3/218. A multichannel UHE device on magnetostatic waves, Vashkovskiy A.V., Zubkov V.I. et al., Application No. 3396487/24-09, Filed 12.02.1982, Published 04.06.1986, Discovery. Invention, 1986, No. 47.
384. Ataligan Y.J., Owens J.M., Reed K.W. et al. MSSW transversal filters based on current weighting in narrow (10 um) transducers, IEEE-MTT'S International Microwave Symposium, Baltimore, MD, June 2–4, 1986, New York, 1986, p. 575.
385. Adam J.D., Daniel Michael R., Talisa S.H., A 13-channel magnetostatic wave filterbank, IEEE MTT International Microwave Symposium Digest, New York, May 25–27, 1988, Vol. 2, pp. 879–882.
386. Adam Y.D., Daniel M., MSW filter banks, Microw. J., 1988, Vol. 31(2), p. 107.
387. Author's certificate 1563544 USSR, MKI HOIP 1.215. A multichannel UHF device on MSW, Novikov G.M., Petrunkin E.V., Bogatko A.V., Filed 01.04.1988, Published 17.08.1990, Discovery. Invention, 1990, No. 44.
388. Caster J.P., Harterman P., Magnetostatic surface wave oscillators and resonators, 8th European Microwave Conference – 78, Paris, 1978.
389. Castera J.P., Magnetostatic volume wave resonators, IEEE MTT-S International Microwave Symposium Digest, New York, 1979, p. 157.
390. Sveshnikov Yu.A.., Launets V.L., Pokrovskiy V.I. et al., Research of spectrum of magnetostatic oscillations of a monocrystal hexaferrite resonator; Problems of quality improvement and effective manufacture of radio-electronic equipment, Moscow, 1979, p. 112.
391. Vugalter G.A., A resonator on surface spin waves, Radio Eng. Electron., 1980, No. (1), p. 1376.
392. Rothe L., Benedix A., Unersuching der Ancopplung magnetostatischer Resonatoren an Microstripstrunturen, Nachrichten-techn. Electron., 1980, Vol. 30(6), S. 244.
393. Rothe L. Experimentelle Untersuchung der Vercopplung magnetostatischer Resonatoren mit dem microwellen feld einer Schlitzleitung., Nachrichten-techn. Electron., 1980, Vol. 30(6), S.238.
394. Poston T.D., Stancil L.D., A new microwave ring resonator using guided magnetostatic surface waves, J. Appl. Phys., 1984, Vol. 55(6), Part 2B, p. 2521.
395. Chang Kok Wai, Ishak Waguin, The effect of width modes on the performance of MSSW resonators, Ultrasonics Symposium on Proceedings, Dallas, TX, November 14–16, New York, 1984, Vol. 1, p. 164.
396. USA patent 4528529, MKI HOIP 7/08, HO3H 9/15; NKI 333/219. Magnetostatic wave resonator, Huijer E., Hewlett-Packard Co, 12.12.1983. Published 09.07.1985, Application No. 560625.
397. Ishak W., Chang K., Tunable microwave resonators using magnetostatic wave in YIG films, IEEE Trans. MTT, 1986, No. (12), p. 1383.
398. Vendik O.G., Kalinikos B.A., Wave processes in film ferrite layered structures, physical basics of spin-wave electronics, Proc. Inst. High. Educ. Phys., Published by Tomsk university, 1988, Vol. 31(11), p. 3.
399. USA patent 4782312, MKI HOIP 7/06. Mode selective magnetostatic wave resonators, Chang Kok Way, Guiseppe, M., Ishak, W., Hewlett-Packard Co, Filed 22.10.1987. Published 01.11.1988, Bull. No. 50.
400. Ohwi K., Okada F., A three-port ferrite resonator circuit using coplanar waveguide, IEEE Trans. Magn., 1987, Vol. 23(5), Part 2, p. 3748.
401. Japan patent 883605, NKI 333/239. A thin-film ferrite resonator, Murakami Yoshikazu, Tanaka Hideo. Song Corp, Filed 09.07.1986. Published 17.05.1988.
402. De D.K., Coordination of loopback converters for loss decrease of a resonator of magnetostatic surface waves, J. Appl. Phys., 1988, Vol. 64(10), Part 1, p. 5210.
403. USA patent 4774483, MKI HOIP 7/00. A diagonally coupled magnetostatic wave resonator, M. Cinseppe, Chang Kok Wai, Hewlett-Packard Co, Filed 09.09.1987. Published 27.09.1988.
404. Sanders A.F., Stancil D.D., Magnetostatic wave ring resonator with a rotating thin film, IEEE Trans. Magn., 1988, Vol. 24(6), p. 2805.

405. Hanna S.M., Microwave filters based on coupled MSW resonators and their applications, IEEE Ultrasonics on Symposium, Chicago IL, October 2–5, 1988, Proceedings, Pittsburgh, PA, Vol. 1/2, 1988, p. 241.
406. Chewnpu J., Parench J.P., Tuan H.S., MsFVW dimensional resonances of a YIG-film cavity, IEEE Ultrasonics on Symposium, Chicago IL, October 2–5, 1988, Proceedings, Pittsburgh, PA, Vol. 1/2, 1988, p. 233.
407. Karmanov M.V., Kravtsov I.A., Surguchev I.A., Determination of parameters of a spin-wave resonator with thin ferromagnetic films, Prob. Radio Electron. Ser. Prod. Eng. Equip., 1988, No. (2), p. 18.
408. France patent 8708742, MKI HOIP 1/218, HO3H 9/22. Filtre passe-bande a bille de granat d'accord, Barratt C., Christopher, Marcous J., Enertec Co, Filed 22.06.1987. Published 23.12.1988.
409. Hanna S.M., Zerong S., Single and coupled MSW resonators for microwave channelizers, IEEE Trans. Magn., 1988, Vol. 24(6), p. 2808.
410. De D.H., High-magnetostatic surface wave planar yttrium iron garnet resonator, J. Appl. Phys., 1988, Vol. 64(4), p. 2144.
411. Author's certificate 1510027 of USSR, MKI HOIP 1/218. A resonator on magnetostatic waves, Grechushkin K.V., Kulikov M.N., Prokushkin V.N., Institute of mechanics and physics of Saratov state university, Application No. 4274909/24-09, Filed 01.07.1987, Published 23.09.1989, Discovery. Invention, 1989, No. 35.
412. USA patent 94963, MKI HO3H 2/06. A MSW resonator with mode selection, Chang K.W., Miccoly G., Ishak W.B., Hewlett-Packard Co, Filed 19.10.1988, Published 26.04.1989.
413. Balinskiy M.G., Berezov A.S., Yeremenko I.N., A two-input resonator with coaxial converters on direct volume magnetostatic waves, Electron. Eng. Ser. 1 UHF Electron, 1990, Vol. 1(425), p. 17.
414. Niccoli G., Chang K.W., Mode-selective magnetostatic wave resonator, Electron. Lett., 1989, Vol. 23(6), p. 420.
415. Tsutsumi M., Takeda T., Magnetostatic wave resonators of microstrip type, IEEE MTT-S International Microwave Symposium Digest, Long Beach, CA, June 13–15, New York, 1989, Vol. 1, p. 149.
416. Japan patent 64-64410, MKI HO3H 9/25. A resonator of magnetostatic waves, Kokaj tekke koho. Series 701, 1989, Vol. 62, p. 43.
417. Tsutsumi M., Takeda T., IEEE MTT-S: International Microwave Symposium Digest, Long Beach, CA, June 13–15, 1989, Vol. 1, p. 149.
418. Prudkin V.P., Vasiliev K.J., Dobrovolskaya L.F., Ring resonators with non-uniform gyrotropic filling, Methods and means of designing of radioelectronic elements REA, Dnepropetrovsk, 1989, p. 20.
419. Gastora J.P., Harterman P., Adjustible magnetostatic surface-wave multistrip directional coupler, Electron. Lett., 1980, Vol. 16(5), p. 195.
420. Gastera J.P., Harterman P., Adjustible magnetostatic surface wave directional coupler, IEEE MTT'S: International Microwave Symposium Digest, New York, 1980, p. 37.
421. Schlomann E., Blight R.E., YIG-filter recovery after exposure to high power and X-band frequency-stepped YIG-filter, IEEE MTT'S: International Microwave Symposium Digest, Boston, MA, May 31–June 3, 1983, New York, 1983, p. 329.
422. Gastera J.P., Hastermann P., Le Y. et al., A tunable magnetostatic volume wave oscillator, IEEE MTT'S: International Microwave Symposium Digest, Boston, MA, May 31–June 3, 1983, New York, 1983, p. 318.
423. Lander V., Parekh J.P., MSSW delay line based oscillators, Proceedings of the 37th Annual Frequency Control Symposium, Philadelphia, PA, June 1–3, New York, 1983, p. 473.
424. Ishak W., Digest Intermag'84: International Magnetic Conference, Hamburg, April 10–13, New York, 1984, No. 4, p. 472.
425. Tsutsulin M., A microwave generator with a delay line on MPSW, Trans. Inst. Electron., 1986, Vol. 169-C(8), p. 1061.

426. Osbrink N., Preliminary tunable amplifiers and oscillators on iron–yttrium garnet for radio-prospecting ELINT system, Radio engineering of ultrahigh frequencies, Moscow, 1987, p. 19.
427. Duko G.M., Kazakov G.T., Kozhevnikov A.V., Filimonov Yu.A., Self-modulation and chaos at tetra-magnon disintegration of running MSW in YIG films, 9th International Conference on Microwave Ferrites: ICMF'88, Estergom, September 19–23, 1988: Proceedings. Budapest, 1988, p. 12.
428. Ishak W.S., Chang Kok-Wai, Kunz W.E., Moscoli G., Tunable microwave resonators and oscillators using magnetostatic waves, IEEE Trans. Ultrason. Ferroelec. Freq. Contr., 1988, Vol. 35(3), p. 396.
429. Tanbakuchi H., Nicholson D., Kunz B., Ishak W., Magnetically tunable oscillators and filters, Digest Intermag'89: International Magnetic Conference, Washington, DC, March 28–31, New York, 1989, p. AC3.
430. Tupikin V.D., Vandyshev K.G., Novikov G.M., A programmable UHF generator on magnetostatic waves, Electron. Eng. Ser. 8 Quality Manag. Standard. Metrol. Tests, 1989, Vol. 5(137), p. 13.
431. Japan patent 1109904, MKI HO3B 7/06, HOIP 7/00. A generator on magnetostatic waves,Mazyra Hirouki, Yokogava denki K.K. No. 62-267719, Filed 23.10.1987. Published 26.04.1989, Kokei tokha koho. Series 7(3), 1989, Vol. 108, p. 15.
432. Asao Hideki, Ohashi Hideynki, Ishida Osami, Hashimoto Tsutonnu. Broad-band tunable oscillator with two MSW resonators, 19th European Microwave Conference, London, September 4–7, 1989, Microwave'89: Conference on Proceedings of Tunbridge Wells, 1989, p. 999.
433. Maliniak D., Tiny generators on reconstructed YIG and overlapping ranges from 3 up to 6 and from 2 up to 8 GHz, Electronics, 1989, No. (23), p. 108.
434. Low-noise generators on YIG Microsource Inc., Microw. RF, 1989, Vol. 27(12), p. 37.
435. Kuzinov E.V., Yereshchenko I.N., Modelling of tunable UHF generators on magnetostatic waves, Electron. Eng. Ser. UHF Electron., 1989, Vol. 10(424), p. 35.
436. Belyaev A.I., Skabovskiy M.S., Titoykina N.V., Noise of UHF generators on bipolar transistors with a ferrite resonator, Electron. Eng. Ser. UHF Electron., 1989, Vol. 8(422), p. 71.
437. Chen C., Chu A., Oscillators using magnetostatic-wave active taped delay lines, IEEE Trans. MTT, 1989, Vol. 37(1), p. 239.
438. A thin-film YIG resonator of Omniying Inc.®, MSN, 1989, Vol. 19(8), p. 68.
439. Chen Chang-Lee, Chu A., Mahoney L.J. et al., Oscillators using magnetostatic-wave active tapped delay lines, IEEE Trans. Microw. Theor. Technol., 1989, Vol. 37(1), p. 239.
440. Author's certificate USSR 1469537, MKI HO3B 7/14. An UHF generator on magnetostatic waves, Novikov G.M, Application No. 4272409/24-09, Filed 30.06.1987, Published 30.03.1989, Discovery. Invention, 1989, No. 12.
441. Generators on YIG with filters of harmonics, Microw. RF, 1990, Vol. 29(1), p. 141.
442. An YIG generator of Microsource Inc.®, Microw. J., 1990, Vol. 33(6), p. 219.
443. Sokolskiy A.S., Afanasiev A.I., Rudy Yu.B., Research of noise characteristics of a generator on magnetostatic waves, Electron. Eng. Ser. 1 UHF Electron., 1990, Vol. 1(425), p. 67.
444. Yamada S., Chang N.S., Matsuo Y., Experimental investigation of magnetostatic surface wave amplification in Ga As ittrium iron garnet layered structure, J. Appl. Phys., 1982, Vol. 53(8), p. 5979.
445. Barkekhtar V.G., Grinev B.V., Sapogov S.A., Seminozhenko V.P., Distribution and amplification of spin waves in ferrite semiconductors in a strong constant electric field, Kiev, 1984, 27 p.
446. Kotsarenko N.J., Rapoport Yu.G., On the opportunity of superheat amplification of magnetostatic waves in ferromagnetics, Letters to Russ. J. Tech. Phys., 1984, Vol. 10(14), p. 843.
447. Krutsenko I.V., Melkov G.A., Ukhanov S.A., Amplification of return exchange magnetostatic waves in YIG films, Kiev, 1988, p. 68.

448. Kazakov G.T., Sukharev A.G., Filimonov Yu.A., Magnetoelastic "amplification" of MSW in YIG films, Digest Intermag'89: International Magnetic Conference, Washington, DC, March 28–31, New York, 1989, p. BP-7.
449. Melkov G.A., Sholom S.V., Amplification of surface magnetostatic waves by parametrical excitation, Russ. J. Tech. Phys., 1990, Vol. 60(8), p. 118.
450. Vashkovskiy A.V., Zubkov V.I., Tsarukhian T.M., Research of convolution on volume magnetostatic waves in a ferrite-semiconductor structure, Radio Eng. Electron., 1979, No. (1), p. 146.
451. Vlasov A.B., Zhukovskiy V.V., Chirina N.M., Convolution of radio signals in a delay line on the basis of an YIG monocrystal, Proc. Inst. High. Educ. Radio Electron., 1984, Vol. 27(5), p. 71.
452. Parekh Jayant P., Tuan Hang-Sheng, Chang K.W., Convolution with magnetostatic waves in YIG films, IEEE Trans. Son. Ultrason., 1985, Vol. 82(5), p. 690.
453. Pershin V.T., Account of dispersion in convolution devices on the basis of magnetoelastic waves, Radio Eng. Electron., 1986, No. (15), p. 87.
454. Miyazaki Y., Bhandari R., Optical intergrated switch and modulator using mode conversion induced by magnetostatic surface waves in three-dimensional waveguides, Ultrasonics Symposium Proceedings, Dallas, TX, November 14–16, New York, 1984, Vol. 1, p. 171.
455. Markov G.T., Chaplin A.F., Excitation of electromagnetic waves, Moscow, 1983.
456. Bronshteyn I.N., Semendiaev K.A., Handbook on mathematics for engineers and students, Moscow, 1986.
457. Kurushin E.P., Nefedov E.I., Electrodynamics of anisotropic waveguarding structures, Moscow, 1983.
458. Korn G., Korn T., Mathematical handbook for scientists and engineers, McGraw-Hill, 1968.
459. Felsen L., Marcuviz N., Radiation and scattering of waves, Moscow, 1978, Vol. 1.
460. Felsen L., Marcuviz N., Radiation and scattering of waves, Moscow, 1978, Vol. 2.
461. Muskhemishvili N.I., Singular integral equations, Moscow, 1962.
462. Wilkinson J.H., Reinsch C., Handbook for automatic computation: Linear algebra. Heidelberg/New York/Berlin, 1974.
463. Grinchishin J.T., Yefimov V.I., Lomakovich A.N., Algorithms and programs on BASIC, Moscow, 1988.
464. Grosse P., Free electrons in solids, Moscow, 1982.
465. Smirnov V.I., A course in higher mathematics, Moscow, 1953, Vol. 4.
466. Numerical methods in plasma physics, Translation from English into Russian edited by Dnestrovskiy Ju.N. and Kostomarov D.P., Moscow, 1974.
467. Stalmakhov V.S., Ignatiev A.A., Kulikov M.N., Research of excitation of magnetostatic waves in millimeter range, V International Congress on Gyrotropic Electronics, Vilnius, 1980, Vol. 1, p. 190.
468. Ignatiev A.A., Lepestkin A.N., Stalmakhov V.S., Synthesis of a transfer line on surface magnetostatic waves, XXXVI All-Union scientific session devoted to the Day of Radio: Abstract, Moscow, 1981, Part 3, p. 22.
469. Stalmakhov V.S., Ignatiev A.A., Kulikov M.N., Research of excitation of magnetostatic waves on frequencies 30–40 GHz, Radio Eng. Electron., 1981, Vol. 26(26), p. 2381.
470. Ignatiev A.A., Lepestkin A.N., Stalmakhov V.S., Research of excitation and management features of magnetostatic wave characteristics in YIG films in millimeter range, IV School-Seminar on Surface Waves in Solids, Novosibirsk, 1982, p. 89.
471. Stalmakhov V.S., Ignatiev A.A., Lepestkin A.N., Features of distribution of magnetostatic waves (MSW) in multilayered structures on frequencies above 30 GHz, I Scientific and Technical Conference on Integrated UHF Electronics, Novgorod, 1982, p. 58.
472. Mostovoy A.A., Ignatiev A.A., On the opportunity of research of MSW characteristics in millimeter range, XX All-Union Student's Conference on Student and the Scientific and Technological Revolution, Novosibirsk, 1982, p. 74.
473. Stalmakhov V.S., Ignatiev A.A., Outlook of promotion of devices on magnetostatic spin waves in millimeter range, Functional UHF electronics. Abstract, Kiev, 1983, p. 35.

474. Stalmakhov V.S., Ignatiev A.A., Lectures on spin waves, Saratov, 1983.
475. Ignatiev A.A., Lepestkin A.N., Stalmakhov V.S., Research of magnetostatic waves in YIG films on frequencies 30–40 GHz, I All-Union Meeting-Seminar on Spin-Wave Electronics and Electrodynamics, Saratov, 1982, pp. 19–21.
476. Ignatiev A.A., Grechushkin K.V., Lepestkin A.N., Properties of a beyond-cutoff waveguide partially filled with ferrite, XII All-Union Conference on Acoustoelectronics and Quantum Acoustics: Abstract, Saratov, 1983, Part 2, p. 66.
477. Ignatiev A.A., Mostovoy A.A., Specificity of optical methods for research of magnetostatic waves in ferrite films in short-wave range of radiowaves, All-Union Scientific and Technical Conference on Design and Application of Radio-Electronic Devices on Dielectric Waveguides and Resonators: Abstract, Saratov, 1983, p. 252.
478. Ignatiev A.A., Lepestkin A.N., Mostovoy A.A., Features of excitation of MSW in ferrite films in short-wave range, All-Union Scientific and Technical Conference on Design and Application of Radio-Electronic Devices on Dielectric Waveguides and Resonators: Abstract, Saratov, 1983, p. 275.
479. Ignatiev A.A., Lepestkin A.N., Mostovoy A.A., Matched elements in devices on MSW in short-wave range, All-Union Scientific and Technical Conference on Design and Application of Radio-Electronic Devices on Dielectric Waveguides and Resonators: Abstract, Saratov, 1983, p. 276.
480. Stalmakhov V.S., Ignatiev A.A., Magnetostatic spin waves on high frequencies, 6th Winter School-Seminar for Engineers on Lectures on UHF Electronics and Radiophysics, Saratov, 1983, p. 134.
481. Ignatiev A.A., Mostovoy A.A., Research of characteristics of a magnetostatic wave converter on a slot line, All-Union Scientific and Technical Conference on Problems of Integrated UHF Electronics: Abstract, Leningrad, 1984, p. 128.
482. Ignatiev A.A., Lepestkin A.N., A planar transfer line on magnetostatic waves of short-wave part of cm range, All-Union Scientific and Technical Conference on Problems of Integrated UHF Electronics: Abstract, Leningrad, 1984, p. 132.
483. Stalmakhov V.S., Ignatiev A.A., Grechushkin K.V., Lepestkin A.N., Volume magnetostatic waves on high frequencies, VI International congress on gyromagnetic electronics and electrodynamics, Bratislava, 1984, p. 121.
484. Ignatiev A.A., Lepestkin A.N., A nondestructive quality control plant for epitaxial ferrite films, Prospectus of the World's fair of achievements of young inventors "Bulgaria-85", Plovdiv, 1985, p. 47.
485. Stalmakhov V.S., Ignatiev A.A., Lepestkin A.N., Outlook of magnetoelectronics of millimeter range, Lecture at All-Union School-Seminar on Back-Wave UHF Electronics, Ashkhabad, 1985, p. 24.
486. Ignatiev A.A., Lepestkin A.N., Mostovoy A.A., Stalmakhov V.S., A microstrip converter of magnetostatic spin waves in short-wave range, Electron. Eng. Ser. UHF Electron., 1985, Vol. 4(376), p. 24.
487. Ignatiev A.A., Lepestkin A.N., Beginin V.I., Filtration of signals in devices on epitaxial ferrite films in short-wave part of UHF range, II All-Union School-Seminar on Spin-Wave UHF Electronics, Ashkhabad, 1985, p. 127.
488. Ignatiev A.A., Mostovoy A.A., Properties of magnetostatic wave converters on the basis of a slot line in short-wave part of UHF range, II All-Union School-Seminar on Spin-Wave UHF Electronics, Ashkhabad, 1985, pp. 115–116.
489. Ignatiev A.A., Lepestkin A.N., Stalmakhov V.S., Kiselyova T.V., Synthesis of a transfer line on surface MSW, II All-Union School-Seminar on Spin-Wave UHF Electronics, Ashkhabad, 1985, p. 34.
490. Ignatiev A.A., Lepestkin A.N., Mostovoy A.A., Research of interference of magnetostatic waves in limited-size media, II Seminar on Functional Magnetoelectronics: Abstract, Krasnoyarsk, 1986, p. 214.
491. Ignatiev A.A., Lepestkin A.N., Nondestructive control of parameters of ferromagnetic films, II Seminar on Functional Magnetoelectronics: Abstract, Krasnoyarsk, 1986, p. 141.

492. Ignatiev A.A., Lepestkin A.N., A plant for nondestructive quality control of epitaxial ferrite films, Saratov, 1986, p. 2.
493. Ignatiev A.A., Lepestkin A.N., Research of excitation of magnetostatic waves in barium hexaferrite on frequencies 20, 80 GHz, Russ. J. Tech. Phys., 1986, Vol. 56(9), p. 1829.
494. Ignatiev A.A., Lepestkin A.N., Nondestructive UHF diagnostics of parameters of magnetic films, Regional Conference of Backs-Wave Phenomenon of UHF Electronics: Abstract, Krasnodar, 1987, p. 142.
495. Ignatiev A.A., Lepestkin A.N., Nondestructive UHF control of saturation magnetization of film ferrites, XV All-Union Seminar on Gyromagnetic Electronics and Electrodynamics: Abstract, Kuibyshev, 1987, p. 73.
496. Ignatiev A.A., Lepestkin A.N., Nondestructive UHF diagnostics of ferrites, Brochure of international Leipzig autumn exhibition-fair (September, 1988), Leipzig, 1988, p. 4.
497. Lepestkin A.N., Ignatiev A.A., Features of wave processes in a waveguide with a ferrite film near cross resonance frequency, II All-Union School-Seminar on "Interaction of Electromagnetic Radiation with Semiconductor and Semi-dielectric Structures", Saratov, 1988, Part 3, p. 64.
498. Stalmakhov V., Ignatiev A., Lepestkin A., The nondestruactive diagnostik of the ferrite films at high frequency, International Conference on Magnetics: ICM'88, Paris, 1988, p. 252.
499. Ignatiev A.A., Lepestkin A.N., Research of parameters of film ferrites by electrodynamic sounding, 9th International Conference on Microwave Ferrites (September 1988): ICMF'88, Esztergom, 1988, p. 118.
500. Ignatiev A.A., Mostovoy A.A., Converters of magnetostatic waves on the basis of slot and cross strip lines in millimeter range, Electron. Eng. Ser. UHF Electron., 1990, Vol. 5(429), p. 11.
501. Ignatiev A.A., Lepestkin A.N., A sensor of magnetic induction, Conference on Microelectronic Sensors: Abstract, Ulyanovsk, 1988, p. 189.
502. Beginin V.I., Ignatiev A.A., Mostovoy A.A., Selective properties of planar and waveguide converters of magnetostatic waves, Regional Conference on Spin-Wave Phenomena in UHF Electronics: Abstract, Krasnodar, 1987, p. 160.
503. Mostovoy A.A., Ignatiev A.A., Excitation of magnetostatic waves by equivalent magnetic current, XV All-Union Seminar on Gyromagnetic Electronics and Electrodynamics, Kuibyshev, 1988, p. 13.
504. Mostovoy A.A., Ignatiev A.A., Electrodynamic properties of one-side metallized converters, XV All-Union Seminar on Gyromagnetic Electronics and Electrodynamics, Kuibyshev, 1988, p. 417.
505. Mostovoy A.A., Ignatiev A.A., The transducer of the magnetostatic volume wave on the basis of the slot line in millimetre range, 6th International School on Microwave Physics and Technology, Varna, 1989, p. 143.
506. Mostovoy A.A., Ignatiev A.A., Transfer lines on magnetostatic waves with set parameters on the basis of slot converters, 4 All-Union Schools-Seminars on Backs-Wave UHF Electronics: Abstract, L'vov, 1989, p. 18.
507. Mostovoy A.A., Ignatiev A.A., Excitation of forward volume magnetostatic waves by converters on the basis of a slot line, 2 All-Union School-Seminar on Interaction of Electromagnetic Waves with Semiconductor and Semiconductor-Dielectric Structures: Abstract, Saratov, 1988, Part. 3, p. 66.
508. Stalmahov V., Ignatiev A., Lepestkin A., Nondestructive diagnostics of ferrite film at high frequencies, Phys. Lett., 1988, Vol. 133(7–8), p. 430.
509. Ignatiev A.A., Stalmakhov V.S., Experimental research of magnetostatic waves in millimeter range, Proc. Inst. High. Educ. Phys., 1988, Vol. 31(11), p. 86.
510. Ignatiev A.A., Problems of UHF magnetoelectronic. Lecture, IV All-Union School-Seminar on Spin-Wave UHF Electronics, L'vov, 1989, p. 20.
511. Ignatiev A.A., Lepestkin A.N., Diagnostics of film ferrites on microwave and UHF frequencies: Lecture, IV All-Union School-Seminar on Spin-Wave UHF Electronics, L'vov, 1989, p. 19.

512. Ignatiev A.A., Magnetoelectronic devices of UHF range, IV Seminar on Functional Magnetoelectronics: Abstract, Krasnoyarsk, 1990, pp. 11–12.
513. Ignatiev A.A., Fast and slow waves in structures on the basis of film ferrites in UHF and EHF ranges, Abstract, Krasnoyarsk, 1990, p. 190.
514. Ignatiev A.A., Fast and slow waves in structures on the basis of film ferrite in UHF and EHF ranges: Lecture, Abstract, Krasnoyarsk, 1990, p. 30.
515. Mostovoy A.A., Ignatiev A.A., Excitation of magnetostatic waves in ferrite films with orthogonal orientation to slot converter plane in EHF range, Abstract, Krasnoyarsk, 1990, p. 300.
516. Lepestkin A.N., Ignatiev A.A., High-speed management of distribution of high-frequency fields in a waveguide with a ferrite film, Abstract, Krasnoyarsk, 1990, p. 302.
517. Beginin V.I., Ignatiev A.A., Waveguide UHF filters on the basis of film ferrites, Abstract, Krasnoyarsk, 1990, p. 304.
518. Ignatiev A.A., Problems of UHF magnetoelectronics, 10th International Conference on Microwave Ferrites: X ICMF, Poland, Szczyrk, 1990, p. 18.
519. Mostovoy A.A., Ignatiev A.A., Excitation of electromagnetic waves in layered bigyrotropic structures by extraneous surface currents, III All-Union School-Seminar on Interaction of Electromagnetic Waves with a Solid, Saratov: Publishing house of Saratov State University, 1991, p. 102.
520. Ignatiev A.A., Waves in layered ferrite-dielectric structures in the UHF range, III All-Union School-Seminar on Interaction of Electromagnetic Waves with a Solid, Saratov: Publishing house of Saratov State University, 1991, p. 107.
521. Ignatiev A.A., Mostovoy A.A., Excitation of electromagnetic waves in layered bigyrotpopic structures with nonlinear orientation, II All-Union Scientific and Technical Conference on Devices and Methods of Applied Electrodynamics, Moscow, 1990, p. 18.
522. Mostovoy A.A., Ignatiev A.A., Excitation of electromagnetic waves in layered bigyrotpopic structures by sources with orthogonal orientation, II All-Union Scientific and Technical Conference on Devices and Methods of Applied Electrodynamics, Moscow, 1990, p. 17.
523. Ignatiev A.A., Lepestkin A.N., High-power filters with selective phase inversion on the basis of ferrite films, V All-union school on spin-wave UHF electronics, Moscow, 1991, p. 153.
524. Mostovoy A.A., Ignatiev A.A., Excitation of electromagnetic waves in layered structures on the basis of ferrite, V All-union school on spin-wave UHF electronics, Moscow, 1991, p. 155.
525. Ignatiev A.A., Features of excitation and distributions of MSW in millimeter range: Lecture, V All-union school on spin-wave UHF electronics, Moscow, 1991, p. 30.
526. Ignatiev A.A., Magnetoelectronic devices of low and high power levels in millimeter range, V All-union school on spin-wave UHF electronics, Moscow, 1991, p. 35.
527. Ignatev A.A., Mostovoy A.A., Theory of bigyrotropic multilayered converters, International Conference on Gyromagnetic Electronics and Electrodynamics ICMF'92: Abstract, Alushta, 1992, p. 4.
528. Ignatiev A.A., Lepestkin A.N., Mostovoy A.A., Waves in layered ferrite-dielectric structures in millimeter range, International Conference on Gyromagnetic Electronics and Electrodynamics ICMF'92: Abstract, Alushta, 1992, p. 8.
529. Ignatiev A.A., Full spectrum of electromagnetic waves in structures on the basis of weakly dissipative ferrite films in millimeter range, International Conference on Magnetic Electronics ICMF'92, Krasnoyarsk, 1991, p. 41.
530. Ignatiev A.A., Mostovoy A.A., Beginin V.I., Bigyrotropic multilayered magnetoelectronic converters for millimeter range, International Conference on Magnetic Electronics ICMF'92, Krasnoyarsk, 1991, p. 48.
531. Ignatiev A.A., Lepestkin A.N., Beginin V.I., Controllable magnetoelectronic delay lines for millimeter range, International Conference on Magnetic Electronics ICMF'92, Krasnoyarsk, 1991, p. 51.
532. Ignatiev A.A., Magnetoelectronic devices and transfer lines of UHF range, All-Union Seminar on Mathematical Modelling of Physical Processes in Antenna-Feeding Pathes, Saratov, October 3–5, 1990, Saratov: Publishing house of Saratov State University, 1990, p. 36.

533. Ignatiev A.A., Features of excitation and distributions of fast and slow waves in flat ferrite-dielectric structures in UHF and EHF ranges, All-Union seminar on mathematical modelling of physical processes in antenna-feeding pathes, Saratov, October 3–5, 1990, Saratov: Publishing house of Saratov State University, 1990, p. 37.
534. Author's certificate USSR 1467615, MKI HOIP 1/215. An UHF divisor of channels, Ignatiev A.A., Mostovoy A.A., Institute of mechanics and physics of SGU, Application No. 4284590/24-09, 14.07.1987. Published 23.03.1989, Discovery. Invention, 1989, No. 11.
535. Author's certificate USSR 1356053, MKI HOIP 1/215. A superhigh-frequency filter, Ignatiev A.A., Mostovoy A.A., Institute of mechanics and physics of SGU, Application No. 3871559/24-09, 08.04.1985. Published 30.11.1987, Discovery. Invention, 1987, No. 14.
536. Author's certificate USSR 1223317, MKI HOIP 1/215. A device on magnetostatic waves, Ignatiev A.A., Mostovoy A.A., Bakhtin A.I., Institute of mechanics and physics, Application No. 3727397/24-09, 09.01.1984. Published 07.04.1986, Discovery. Invention, 1986, No. 13.
537. Author's certificate USSR 1241316, MKI HOIP 1/24. A waveguide-coplanar transition, Ignatiev A.A., Mostovoy A.A., Institute of mechanics and physics of SGU, Application No. 3734159/24-09, 27.04.1984, Published 01.03.1986, Discovery. Invention, 1986, No. 24.
538. Author's certificate USSR 1149196, MKI G OIR 33/16. A device for measurement of linewidth of ferromagnetic UHF resonance of ferrites, Ignatiev A.A., Lepestkin A.N., Stalmakhov V.S., Institute of mechanics and physics of SGU, Application No. 3668033/24-09, 29.11.1983. Published 07.04.1985, Discovery. Invention, 1985, No. 13.
539. Author's certificate USSR 1394163, MKI GOIP 27/36. A way of determination of saturation magnetization of a ferrite, Ignatiev A.A., Lepestkin A.N., Institute of mechanics and physics of SGU, Application No. 4076828, 10.06.1986. Published 08.01.1988, Discovery. Invention, 1988, No. 17.
540. Author's certificate USSR 1226559, MKI HOIP 1/23. An absorber of magnetostatic waves, Ignatiev A.A., Lepestkin A.N., Stalmakhov V.S., Institute of mechanics and physics of SGU, Application No. 3776586/24-09, 27.07.1984. Published 23.04.1986, Discovery. Invention, 1986, No. 15.

Index

Printed in the United States of America